C0-AMZ-006

ENERGY SCIENCE, ENGINEERING AND TECHNOLOGY SERIES

AIR CONDITIONING SYSTEMS: PERFORMANCE, ENVIRONMENT AND ENERGY FACTORS

ENERGY SCIENCE, ENGINEERING AND TECHNOLOGY SERIES

Oil Shale Developments
Ike S. Bussell (Editor)
2009. ISBN: 978-1-60741-475-9

Power Systems Applications of Graph Theory
Jizhong Zhu
2009. ISBN: 978-1-60741-364-6

Bioethanol: Production, Benefits and Economics
Jason B. Erbaum (Editor)
2009. ISBN: 978-1-60741-697-5

Bioethanol: Production, Benefits and Economics
Jason B. Erbaum (Editor)
2009. ISBN: 978-1-61668-000-8
(Online Book)

Introduction to Power Generation Technologies
Andreas Poullikkas
2009. ISBN: 978-1-60876-472-3

Handbook of Exergy, Hydrogen Energy and Hydropower Research
Gaston Pélissier and Arthur Calvet (Editors)
2009. ISBN: 978-1-60741-715-6

Energy Costs, International Developments and New Directions
Leszek Kowalczyk and Jakub Piotrowski (Editors)
2009. ISBN: 978-1-60741-700-2

Radial-Bias-Combustion and Central-Fuel-Rich Swirl Pulverized Coal Burners for Wall-Fired Boilers
Zhengqi Li
2009. ISBN: 978-1-60876-455-6

Syngas Generation from Hydrocarbons and Oxygenates with Structured Catalysts
Vladislav Sadykov, L. Bobrova, S. Pavlova, V. Simagina, L. Makarshin, V. Julian, R. H. Ross, and Claude Mirodatos
2009 ISBN: 978-1-60876-323-8

Corn Straw and Biomass Blends: Combustion Characteristics and NO Formation
Zhengqi Li
2009. ISBN: 978-1-60876-578-2

Computational Techniques: The Multiphase CFD Approach to Fluidization and Green Energy Technologies (includes CD-ROM)
Dimitri Gidaspow and Veeraya Jiradilok
2009. ISBN: 978-1-60876-024-4

Cool Power: Natural Ventilation Systems in Historic Buildings
Carla Balocco and Giuseppe Grazzini
2010. ISBN: 978-1-60876-129-6

Air Conditioning Systems: Performance, Environment and Energy Factors
Tobias Hästesko and Otto Kiljunen (Editors)
2010. ISBN: 978-1-60741-555-8

ENERGY SCIENCE, ENGINEERING AND TECHNOLOGY SERIES

AIR CONDITIONING SYSTEMS: PERFORMANCE, ENVIRONMENT AND ENERGY FACTORS

TOBIAS HÄSTESKO
AND
OTTO KILJUNEN
EDITORS

Nova Science Publishers, Inc.
New York

For permission to use material from this book please contact us:
Telephone 631-231-7269; Fax 631-231-8175
Web Site: http://www.novapublishers.com

LIBRARY OF CONGRESS CATALOGING-IN-PUBLICATION DATA

Air conditioning systems : performance, environment, and energy factors / editors, Tobias Hästesko and Otto Kiljunen.
p. cm.
Includes index.
ISBN 978-1-60741-555-8 (hardcover)
1. Air conditioning. I. Hästesko, Tobias. II. Kiljunen, Otto.
TH7687.7.A375 2009
697.9'3--dc22

2009034521

Published by Nova Science Publishers, Inc. ✢ New York

CONTENTS

Preface vii

Chapter 1 Alternative Refrigerating, Heat-Pumping and Air-Conditioning Systems on the Basis of the Open Absorption Cycle and Solar Energy 1
Alexander V. Doroshenko, Leonid P. Kholpanov and Yury P. Kvurt

Chapter 2 Principle of Low Energy Building Design: Heating, Ventilation and Air Conditioning 149
Abdeen Mustafa Omer

Chapter 3 Design and Implementation of a Solar-Powered Air-Conditioning System 195
J. N. Lygouras and V. Kodogiannis

Chapter 4 Room Air Distribution, Indoor Environmental Quality and Energy Performance 225
Zhang Lin and Wai Sun Shum

Chapter 5 Solar Thermal Energy Technology: Experimental Study on the Design and Performance of an Indirect Solar Dryer for Agricultural Products 243
Sabah A. Abdul-Wahab

Chapter 6 Passive and Available House Air-Conditioning by Exploiting Renewable Energy Sources 267
Majdi Hazami, Sami Kooli, Meriem Lazaar, Abdel Hamid Farhat and Ali Belghith

Chapter 7 Multi-Purpose Operation Planning of the Micro-Energy Source for Air Conditioning with Renewable Energy 293
Shin'ya Obara

Chapter 8 Improving Design in Reversible Heat Pumps 315
C. J. Renedo, A. Ortiz and J. Carcedo

Chapter 9 Load Predictor for Air Conditioning Systems Using Artificial Neural Networks Models for Intelligent Buildings **337**
M. Trejo-Perea, G. Herrera-Ruiz, D. Vargas-Vázquez, P. Talamantes-Contreras, E. A. Rivas-Araiza, R. Luna-Rubio and G. J. Ríos-Moreno

Chapter 10 Performance of a Passive Aquarium's Conditioning System **353**
Majdi Hazami, Sami Kooli, Mariem Lazaar, Abdelhamid Farahat and Ali Belguith

Chapter 11 New Progress in Liquid Desiccant Cooling Systems: Adsorption Dehumidifier and Membrane Regenerator **373**
Xiu-Wei Li and Xiao-Song Zhang

Index **385**

PREFACE

Air conditioning refers to the cooling and dehumidification of indoor air for thermal comfort. This requires systems for heating, venting and air conditioning (HVAC). This book discusses the application and performance on thermal comfort, indoor air quality and energy efficiency of a variety of room air distribution modes. Efficient energy performance of HVAC systems due to active and passive control of energy consumption are also addressed. Furthermore, this book explores a passive energy system, called air-conditioning cupboard which exploit renewable energies as thermal sources for the purpose of saving energy. Additionally, the thermal supply for air conditioning systems can be carried out efficiently using reversible air pumps. Thus this book examines how to improve the design of these pumps, which have great potential, particularly in Mediterranean climates. Other chapters in this book explore the use of artificial neural networks (ANN) for air conditioning systems; the ventilation concepts utilising ambient energy from air ground and other renewable energy sources and different types of air-conditioning systems with good energy-saving potential, such as the liquid desiccant cooling system (LDCS) and solar-powered air-conditioning systems.

Chapter 2 - The move towards a de-carbonised world, driven partly by climate science and partly by the business opportunities it offers, will need the promotion of environmentally friendly alternatives, if an acceptable stabilisation level of atmospheric carbon dioxide is to be achieved. This requires the harnessing and use of natural resources that produce no air pollution or greenhouse gases and provides comfortable coexistence of human, livestock, and plants. This study reviews the energy-using technologies based on natural resources, which are available to and applicable in the farming industry. Integral concept for buildings with both excellent indoor environment control and sustainable environmental impact are reported in the present communication. Techniques considered are hybrid (controlled natural and mechanical) ventilation including night ventilation, thermo-active building mass systems with free cooling in a cooling tower, and air intake via ground heat exchangers. Special emphasis is put on ventilation concepts utilising ambient energy from air ground and other renewable energy sources, and on the interaction with heating and cooling. It has been observed that for both residential and office buildings, the electricity demand of ventilation systems is related to the overall demand of the building and the potential of photovoltaic systems and advanced co-generation units. The focus of the world's attention on environmental issues in recent years has stimulated response in many countries, which have led to a closer examination of energy conservation strategies for conventional fossil fuels. One way of reducing building energy consumption is to design buildings, which are more economical in their use of energy for heating, lighting, cooling, ventilation and hot water supply. Passive measures, particularly natural or hybrid ventilation rather than air-conditioning, can dramatically reduce primary energy consumption. However, exploitation of renewable energy in buildings and agricultural greenhouses can, also, significantly contribute towards reducing dependency on fossil fuels. This article describes various designs of low energy buildings. It also, outlines the effect of dense urban building nature on energy consumption, and its contribution to climate change. Measures, which would help to save energy in buildings, are also presented.

Chapter 3 - A result of the projected world energy shortage is an important effort in the last years towards the usage of solar energy for environmental control. This effort is receiving much attention in the engineering sciences literature. The design and implementation of an exclusively solar-powered air-conditioning system is described in this chapter. Its operation principle, the controller design and experimental results are presented. A variable structure fuzzy logic controller for this system and its advantages are investigated. Two DC motors are used to drive the generator pump and the feed pump of the solar air-conditioner. The first affects the temperature in the generator while the second the pressure in the power loop. Two different control schemes for the DC motors rotational speed adjustment are implemented and tested for the system considered initially as Single-Input/Single-Output (SISO): the first, is a pure fuzzy controller, its output being the control signal for the DC motor driver. The second scheme is a two-level controller. The lower level is a conventional PID controller, and the higher level is a fuzzy controller acting over the parameters of the low level controller. Step response of the two control loops are presented as experimental results. Next, the design and implementation of a Two-Input/Two-Output (TITO) variable structure fuzzy logic controller is considered. The difficulty of Multi-Input/Multi-Output (MIMO) systems control is how to overcome the coupling effects among each degree of freedom. According to the characteristics of the system's dynamics coupling, an appropriate coupling fuzzy controller (CFC) is incorporated into a traditional fuzzy controller (TFC) to compensate for the dynamic coupling among each degree of freedom. This control strategy can not only simplify the implementation problem of fuzzy control, but can also improve control performance. This mixed fuzzy controller (MFC) can effectively improve the coupling effects of the systems, and this control strategy is easy to design and implement. Simulation results from the implemented system are presented.

Chapter 4 - The application and performance on thermal comfort, indoor air quality and energy efficiency of a variety of room air distribution modes, namely, mixing ventilation, displacement ventilation, task/personalized ventilation, impinging jet ventilation, confluent jet ventilation and stratum ventilation, are discussed. There is a worldwide plea to reduce carbon dioxide emission. In response to such a call, there is a trend to adopt a higher indoor temperature in summer, particularly in East Asia. To promote such an idea, the public need to be convinced that such practice would not sacrifice indoor environmental quality, especially thermal comfort of the occupants. To implement such a measure, existing room air distribution modes are screened according to a set of criteria. Suitable room air distribution mode(s) that work under the unconventional conditions are identified. Design principles are discussed. Experimental and computational results show that with properly designed supply air velocity and volume, locations of supply and exhaust, the proposed stratum mode has the potential to maintain better thermal comfort with a smaller temperature difference between head and foot level, lower energy consumption, and better indoor air quality (IAQ) in the breathing zone. In addition, efficient energy performance of heating, ventilating and air-conditioning (HVAC) systems due to active and passive control of energy consumption are addressed.

Chapter 5 - This chapter investigated the use of solar energy for drying. The objective of the study was to study a simple batch dryer for agriculture products in order to improve the design and to determine the most effective variables influencing its performance. For this purpose, an indirect conventional solar dryer unit consisting of a solar air heater (i.e., inclined solar collector with corrugations) and a drying chamber (i.e., basically a batch dryer) was designed and built at the College of Engineering, Sultan Qaboos University (SQU) in Muscat (Oman). This unit can be used for drying various agricultural products like fruits and vegetables. In this chapter, the solar dryer unit was used to dry radish crops as the test samples. The dryer unit was investigated experimentally and its performance was evaluated under the climatic conditions of Oman (21° 00 N, 57° 00 E). To this effect, a series of experiments were performed on the developed solar dryer during summer conditions in Oman (between May 4 and June 8, 2008). The performance of the solar dryer was computed and expressed in terms of the moisture evaporation (crop mass during drying). Several influencing design parameters, involved in the dehydration process, that were expected to affect on this performance and on the drying time were tested and discussed. The design parameters studied were the number of glazing cover, the type of the absorber of the solar collector, the loading capacity, and the drying airflow rate. In terms of the number of glazing cover sheet, experiments were performed with a single glazing and a double glazing. With regard to the type of the absorber plate, vertical and horizontal corrugated absorber plates were used. The thermal performance of the solar dryer was also evaluated under no absorber plate conditions. In addition, the thermal performance of the solar dryer was evaluated under various load conditions and the unit was also tested under various drying air velocity. The temperature and relative humidity data during the solar drying were also recorded for each experiment. The comparison revealed that the performance of the solar dryer was improved remarkably with the use of the vertical corrugated duct collector and the drying time was consequently reduced. The results also affirmed that the performance increased with reducing the loading capacity. The other parameters were found to have a negligible effect on the performance.

Chapter 6 - In Tunisia, the buildings' space heating sector represents a major part of the total energy consumption budget. These issues have been increasingly prominent concerns since the energy crisis. Hence, interest has been growing to adopt renewable energies as viable sources of energy that offer a wide range of exceptional benefits for present and future applications, with an important degree of promise especially in the buildings sector. However, the management of the renewable energy sources for space air heating/cooling is usually not economically feasible compared with the traditional carriers. In this chapter, we present a passive energy system, called an air-conditioning cupboard which exploits renewable energies [hot water supplied from a solar collector (40–50°C) and cold groundwater (19°C)] as thermal sources, conceived and tested in our laboratory (LMTE, Tunisia).

To evaluate the air-conditioning cupboard efficiency, indoor experiments were carried out under varied Tunisian environmental conditions for several days. Results show that the air-heating system has a good thermal effectiveness (80 %). It permits us to maintain the temperature inside the experimented room at the range of [24–27°C] during the cold months and [20–23°C] during hot months. A theoretical model is employed for the sizing of the air-conditioning cupboard to obtain the required temperature values. This model allows also the determination of the air-cupboard conditioning thermal performances.

Chapter 7 - The complex system for air conditioning heat sources composed from a fuel cell system and renewable energy has a multi-purpose operation. So, the analysis method for operational planning of the complex system, when two or more pieces of energy equipment were introduced in an individual house, was developed. The cooperative control with two or more objective functions of the complex system was simulated using a genetic algorithm. In this Section, details of the proposed analysis method are explained, and a case study is described. An active energy device, a renewable energy device, and an unutilized energy device were connected by an energy network, and a preliminary survey of the system design required for cooperative operation was conducted. A genetic algorithm, which can analyze nonlinear problems and many variables at a time, was used for the software for the operational planning of the complex system developed in this study. As for the energy network to reduce energy costs and the environmental load, successive introductions are predicted.

Chapter 8 - In temperate climates, the thermal supply for air conditioning systems can be carried out efficiently using reversible heat pumps. In this chapter, once the potential efficiency improvement of these machines is analyzed, several alternatives to the traditional designs, proposed by ASHRAE, are undertaken. Results for performance simulations are presented. The comparison shows that new designs do, indeed, lead to a considerable improvement in energy efficiency. Finally, two new systems are shown, the heat pump with total energy recovery and the variant refrigerant volume system, VRV. These are heat pumps capable of exploiting both the heat emitted by the condenser and the one absorbed by the evaporator, achieving high energy efficiency levels.

Chapter 9 - The use of artificial neural networ (ANN) in several applications related to energy management in buildings has been increasing significantly in recent years. This work proposes and analyzes a load predictor for air conditioning systems in intelligent buildings. A linear autoregressive model with external input (ARX) is used to predict the indoor air temperature in a building. The accuracy of the load predictor can be improved by using a cascaded predictor architecture where the input variables for the ANN are: a) the output of a predicted temperature ARX model; b) historical hourly load data; c) day; and d) time. The load predictor uses a multi-layer perceptron (MLP) neural network, which is trained by Levenberg-Marquardt backpropagation (BP) algorithm. The performance of the load predictor was evaluated using real data obtained from a building located at the University of Queretaro, Mexico. The performance of the ANN model was evaluated by means of the analysis of variance (ANOVA). In addition, load values estimated by the ANN were compared, with regression-estimated and current values, using ANOVA and mean comparison procedures. The obtained results shows that the average of electric load estimated values for the selected ANN and the real data are very close, with a 95% of confidence level. Therefore, results from the ANN are significantly better than those obtained by conventional regression.

Chapter 10 - The objective of this work is to study the opportunity of exploiting seawater at a temperature of 18 °C, extracted from a well bored close to the museum (10 m) and close to the seashore (20 m), as a frigorific source for the cooling of SALAMMBO museum aquariums. Therefore an experimental prototype composed of an aquarium and a control basin containing a network of the capillary heat exchanger and a basin of seawater filtration is conceived. Experimental measurements were carried out under climatic conditions similar to those required by maritime species elevated in SALAMMBO museum's aquariums. A numerical study is also worked out. This numerical study allows the sizing of the capillary exchanger to be used by the passive cooling system in order to obtain the required temperature inside each aquarium. Theoretical analysis based on heat balance equations were testified to agree well with experimental results.

Chapter 11 - A liquid desiccant cooling system (LDCS) is a new type of air-conditioning system with good energy-saving potential. Its performance is dominated by dehumidification and regeneration processes. At present, few works have been done to propose a general principle for better dehumidifier design and most works about regeneration are only concentrated on the thermal regeneration method. For both aspects, new progress has been made and presented in this paper. On one hand, a new design method has been derived from the experiments: an adsorption dehumidifier, developed by integrating a solid desiccant with liquid dehumidifier, could greatly improve the dehumidification effects. On the other hand, a new regeneration style has been conceived: a membrane regenerator, which consists of many alternatively placed cation- and anion-exchange membranes, would regenerate the liquid desiccant in an electrodialysis way; while a solar photovoltaic generator provides electric power for fueling this process. This new regeneration method is immune from the adverse impact from outside high humidity, and it also has a pretty good performance, as well as the benefit that purified water can be obtained along with the regeneration process. These two developments can make LDCS more practical and competitive in the future market.

In: Air Conditioning Systems
Editors: T. Hästesko, O. Kiljunen, pp. 1-147

ISBN: 978-1-60741-555-8

Chapter 1

ALTERNATIVE REFRIGERATING, HEAT-PUMPING AND AIR-CONDITIONING SYSTEMS ON THE BASIS OF THE OPEN ABSORPTION CYCLE AND SOLAR ENERGY

***Alexander V. Doroshenko*[*1]*,**
***Leonid P. Kholpanov*[†2] *and Yury P. Kvurt* [2]**
[1] Odessa state academy of refrigeration, 1/3 Dvorianskaya st., Odessa, Ukraine, 65082,
[2] The Institute of Problems of Chemical Physics of RAS, Acad. Semenov av., 1, Chernogolovka, Russia, 142432

INTRODUCTION

Since 1992 at a number of conferences held under the aegis of UNO they have discussed the problem directly of humanity—the problem of global warming caused by the constantly increasing concentration of so-called greenhouse gases (GG) in the atmosphere [129, 130]. Burning fossil fuel as a source of releases of carbon dioxide, which is one of the main greenhouse gases, makes the greatest contribution to their continuous accumulation in the atmosphere. The greenhouse gases, the release of which is controlled by the UNO Convention, include methane; its source is also power engineering and decomposition of domestic and industrial waste. The UNO Convention (the climate convention) was adopted in 1992 in Rio de Janeiro at the UNO conference on the environmental protection and development and was devoted to adopting measures by the world community for smoothing the global warming caused by the increase of the GG concentration in the atmosphere. In December 1997 in Kyoto (Japan) at the third session of the conference of member-countries of the frame UNO Convention on climate change, the Kyoto Protocol was adopted which was ratified by participating countries (55 countries including Russia and Ukraine); it is these countries that provide 55% of global carbon dioxide releases [130].

* E- mail: al_dor@ukr.net
† E- mail: kholp@icp.ac.ru

Obligations to reduce the annual amount of GG releases are considered in the Kyoto Protocol from different viewpoints: participating countries undertook the obligation to reduce releases by the year 2000 by 6–8% (and refused from initial intentions to reduce releases by 15% because of their being unreal) and only 4 countries—Russia, Ukraine, Norway and New Zealand—can maintain releases at the level of 1990. It should be noted that it is much less than what is necessary for showing the rates of accumulating GG in the atmosphere [130]. Unfortunately, most of the countries have failed to reduce releases in recent years. Only Germany has a positive experience (due to the stopping of industry in the eastern zone), as well as Great Britain (coal was replaced by natural gas), France (atomic energy), in other industrially developed countries the releases continue to grow (for example, in Canada by 10%). Due to quite understandable reasons, releases were somewhat reduced in countries with a transitional economy (countries of the former USSR).

The main measures to slow the global warming include:

- increase of energy use efficiency;
- wide practical application of renewable energy sources;
- utilization of domestic and industrial waste (releases of methane);
- spreading of modern agricultural and forestry technologies.

The problem of global warming becomes decisive in perspective planning and developing traditional power engineering and hence in all energy-consuming branches of industry and agriculture without exception, changing and defining all the ideology of life support being made on the threshold of a new century. The main point in perspective planning power engineering will be determined by the use of energy-efficient technologies and the intensive development of alternative power engineering—the use of renewable energy sources. Refrigerating and air-conditioning machinery today consumes on average in industrially developed countries up to 30% of the total energy generated [20, 97-100].

Renewable energy resources (RER) will play an important role in the European energy structure in the near future. Till 2020 RER can become the only significant contribution to the supply of Europe with primary energy and will be able to provide over 50% of the world demand for energy till 2060 [70].

RERs are modern technologies which find all-round support from the public and have many advantages as compared to the usual energy sources used. Renewable energy resources make it possible to considerably reduce releases of CO_2 and other pollutants connected with the activities of the energy sector of the economy. Besides, they increase the safety of various sources and decrease the dependence on imports. The development of an industrial base for deliveries to the potentially large market will help revive the regions of Europe where industry is falling into decay. The employment potential for renewable power engineering is almost 5 times lager than that for the fossil fuel. It provides employment at the local level and can play an important role in the regional development at the expense of introducing profitable and stable sources of income in rural areas. Spreading of renewable energy resources enables them to become a means of developing distant regions and communication between them. Renewable energy resources involve:

- *Wind power engineering.* The cost of electric power generated by wind power stations has sharply reduced in recent years, in many cases being less than 0.04

eku/kWh. This method will become more acceptable for the public, if problems of visual and noise affects of such stations, located near inhabited areas, are solved. Wind power stations located near the seashore have commercial significance now; they have become more numerous in Denmark, and their development is planned in the United Kingdom and the Netherlands. It is wind power stations based on the shore that are preferable for the Ukraine, too, where wind potential is evidently insufficient on its main territory. 10% of the electric power for Europe can be generated by wind turbines which occupy no more of the Earth's surface than the island of Crete [36].

- *Bioenergetics.* The use of biomass makes a significant contribution to the supply of Australia and Denmark with energy and is very prospective for the Ukraine.
- *Water power engineering.* Water power stations make a significant contribution to the supply of Europe with electric power. Ecological effects and conflicts take place mostly in cases when large-scale systems are used. To avoid strong ecological influence, it is possible to design small water power stations. Though over 40% of a total potential of water power stations in Europe have already been used, there are many potential areas for placing small-sized water power station, especially in the Ukraine. There is also a potential for modernizing and restoring exiting systems.
- *Solar power engineering.* Passive and active solar heaters, solar thermal (hot water supply and heating) and photo-voltaic stations can make a considerable contribution to the energy structure of Europe. Photo-voltaic stations can provide 450000 MW power in the EEC, i.e. provide 16% of Europe's demand for electric power. The covering of faces and roofs of buildings by photo-voltaic batteries create for the whole of Europe a potential, which is estimated only for roofs in 500 TW with existing technologies [18]. It is necessary to attract architects for designing buildings with the use of both passive and active solar heaters; it provides the conformed approach to the development of industrial infrastructure in order to stimulate the market. Mass production in Europe should result in considerable reduction of the price due to saving in sizes and increasing of the scales of manufacture. But as it was proved, the photocell cost can become competitive in reaching peak values of electric power cost during the nearest ten years [68]. The photocell production causes ecological problems and only the technologies using silicon are considered to be acceptable from the viewpoint of ecological consequences. It is important to widen the sphere of practical use of solar energy, and first and foremost, it concerns the creation of solar refrigerating and air-conditioning equipment.
- *Wave power engineering.* Energy potential of wave power stations for the EEC approximately amounts to 155 TW (which makes 6.7% of the current production of electric power in the European Community.
- *Tide power engineering.* The affect of tide power stations on the environment is the problem of great importance, and in many cases it exceeds the potential benefit.
- *Geothermal power engineering.* Geothermal power stations can be used only when there is no considerable influence on the sensitivity of ecosystems and where it is possible to form closed cycles.

The global warming problems, as well as problems peculiar to vapour compression refrigerating and air-conditioning equipment, caused by the necessity of developing ozonenondepleting working bodies, have aroused a great, increasingly growing interest to potentialities of open absorption systems which operate at extremely low temperature differences and use solar energy as a heating source. Schematic solutions, configurations and designation of such systems are extremely various, just as the list of working bodies used in them (solid and liquid sorbents). The open cycle can be the basis of a new generation of refrigerating, heat pumping and air-conditioning systems which are wholly based on the use of renewable energy sources, such as solar energy, and provide a reduction of energy consumption simultaneously with ecological cleanliness of solutions used.

1. Development of Optimal Schematic Approaches and Selection of Working Parameters

1.1. General State of the Problem and Main Objectives of the Project

The number of papers devoted to studying the capabilities of the open absorption cycle in conformity with problems of air-conditioning and refrigerating systems (pre-dehumidifying the air, evaporative cooling the media, solar regeneration of the sorbent) is extremely numerous and is continuously increasing. It is caused by the search for fundamentally new approaches in connection with quickly aggravating interrelated problems of power engineering and ecology. The wide practical use of renewable energy sources is referred to only as some known basic measures on slowing global warming. The practical use of solar energy is most prospective for European countries, the Ukraine and Russia in particular. It is connected both with comparative simplicity of equipment and operation of solar plants, and with a great amount of coming solar energy. The idea of solar cooling and air-conditioning has been known for practical use since the 1890s. Various versions of such systems were created in the USSR in the 1960s in Tashkent. A solar cooler using LiBr solution was manufactured in Brizbene, Australia, in 1958. Later on (1966) a solar house with the solar cooler was made in Queensland. In the USA about 500 solar air-conditioners were installed in 1976, and they operated just on solar energy during 75–80% of the year—the rest of the year on electric power or liquid fuel [90]. The following years showed the increasingly growing interest for the capabilities of solar systems. Some conferences of IIR/IIF (Jerusalem, 1982 [125]; "Advances in the Refrigeration Systems, Food Technologies and Cold Chain", Sofia, Bulgaria [27, 29], 1998; Symposium Nantes '98 "Hygiene, Quality and Security in the Cold Chain and Air-Conditioning", 1998, Nantes, France; "EuroSun 98"— the second ISES—Europe Solar Congress, 1998, Portoroz, Slovenia; 20th International Congress of Refrigeration IIR/IIF, Sydney, 1999) paid great attention to this problem. The program of solar heating and cooling, which has existed since 1977, up to now was one of the first joint projects of the International Energy Agency (IEA). The last decade was especially active in this field in Japan [90, 101, 102, 114] and the USA [90, 97-100]. One of the most important national programs on renewable sources of energy is the Japanese Solar Project which is funded by the Japanese Ministry of International Trade and Industry (MITI) by means of the organization of new energy development (NEDO). The American Department of Energy (DOE) supports solar cooling on a wide basis. Attention is paid to the development of collectors, closed and open absorption systems. In 1996, over 110 reports were made at the conference devoted to absorption heat pumps in Montreal, Canada, and only one of them dealt with the use of solar energy (open system using zeolite). The situation is somewhat worse in the EEC, but here too the last years have been marked by growing interest for problems of solar cooling. An important conference devoted to solar energy systems, "EuroSun 96" was held in Freiburg, Germany, in 1996. Over 340 reports were submitted and only 5 of them were about solar cooling. In the work "Solar Energy for Building Air Conditioning" presented in Dresden, Germany, in 1994 50 participants, mainly from Germany, discussed 8 different papers balanced on the themes between open and closed systems. In the work "Solar-assisted Air Conditioning of Buildings using Low-grade Heat" presented in Freiburg, Germany, in 1995, more than 50 participants, mainly from Germany too, discussed 17 different papers: 5 researches were devoted to open systems and 5 to closed ones. However, 4 papers devoted to open systems were related to the commercial use, and only one to R&D, on closed systems—2 papers are connected with industrial application, and 3 with systems being developed. "Munich Discussion Meeting on Solar-assisted Cooling with Absorption Type Chillers" was held in Munich, Germany, in 1995. International experts in solar cooling with the use of sorption systems discussed the fundamental item of the solar problem to find out the most promising trends. 38 participants from 10 countries submitted 13 papers among which there were 4 papers devoted to the development of solar collectors, 3 to open cooling systems, and 5 to closed systems [90].

Principal capabilities both of open and closed systems were manufactured in one-, two- [52, 103] and three-stage version combinations of similar systems [58, 53] Open absorption systems operating at extremely low gradients of temperature and moisture content and at atmospheric pressure are more flexible in their operation, consume less power, and can ruin temperatures of the heating source from 60 to 100 °C [95, 96]. The idea of solar air-conditioning both for comfort and technological use [97, 101] is rather prospective, in particular, due to the existence of certain correlation between the insulation and the required level of cooling (the complex of thermal-humid air parameters for AACS), it occurs at the peak of energy consumption during the day time, and therefore can be especially advantageous. One should note an active prospective market for solar-system air-conditioning that is easy to operate, has low energy consumption, and causes no harmful ecological consequences [53, 98].

A classical scheme for open systems is the so-called Pennington cycle [5, 9, 20, 98], in which the fresh air entering the system is dehumidified in the absorber and then is cooled in the process of direct evaporation and passes into the room. The air from the room is cooled by evaporation, provides fresh air cooling in the heat exchanger and then, after additional heating, provides the regeneration process in the desorber. There are lots of versions of this scheme simple enough and well-known since the beginning of the century. As a rule, such systems include as the main components [4, 9, 19, 20, 23, 26, 29, 39, 41-42, 47, 52-53, 55-56, 57, 63, 66, 73-74, 91, 92-93, 97-100, 101-102, 105, 106, 108, 114, 121, 125, 126, 128, 131]: the absorber (adsorber) where the air flow is dehumidified, the evaporative cooler of the direct and indirect evaporative type, and the desorber (regenerator) of the direct or indirect type, as well as systems of heat-exchangers the necessity of which is due to low temperature gradients [41-42, 52-53, 97-100]. In dehumidifying the air, its moisture content is reduced, and hence, the values of the wet-bulb temperature and dew point which provides the possibility for deep cooling in the evaporative cooler. For air-conditioning systems it means a possibility of providing comfortable thermal-humid parameters with the use of evaporative cooling methods only, without using vapour-compression coolers. In conformity with solar air-conditioning systems the main schemes can be of a ventilation mode, VM, and a recirculation mode, RM, when a part of the air flow leaving the room is used to this or that degree for organizing the main air-conditioning process, or is simply mixed with the fresh air flow entering the room [20, 49]. Usually the amount of air recirculating in such a system is 10–20% [20, 23-24].

Solid sorbents (silica gel, zeolites, hydrides) [5, 9, 63, 66] and liquid sorbents on the basis of solutions LiBr-H_2O, LiCl-H_2O [4, 23, 26-27, 29, 30, 39, 41-42, 52-53, 63, 73-74, 89-90, 103, 126] are used with the number and variety of working media constantly increasing. To organize a continuos process in the case of using solid sorbents (solid sorption system), they use either switching adsorbents or drums rotating at a definite speed [9, 20, 66] whose sections are filled with the adsorbent, at constant and simultaneous pumping through different sectors of the drum the air flows being dehumidified and regenerated.

The adsorber is characterized by small overall dimensions and good characteristics of the process, but has high resistance to the movement of heat carriers and requires much higher temperatures of regeneration. In this respect the use of liquid sorbents (liquid sorption systems) is more preferable, but the influence of working substances on the main microclimate characteristics and the construction materials stability is of importance. An interesting solution for the rotational film absorber (absorption heat-pumping system) has been used in the work [69], where a mathematical model of the absorption process was developed for such a rotating absorber.

The solar regeneration of the absorbents in open systems can be direct [39, 101-102] and can occur in direct contact of the air flow and the absorbent film in the air solar collector, that is the absorbent regeneration here takes place under conditions of direct effect of solar radiation and the air flow. Here there is a danger of contaminating the absorbent and problematic character of providing heated air in amount needed for the regeneration [41-42]. Conclusions made in the work [42] with respect to advantages of systems with direct regeneration of the absorbent are not convincing and have not been confirmed in later researches. It should be noted that the comparative study made by these authors for systems with direct and indirect regeneration was based on the use of flat solar air collectors in both cases, which is not perspective for systems with indirect regeneration of the absorbent and is practically met nowhere else. The indirect regeneration provides the availability of the desorber with an external outstanding heat-exchanger or built-in heat-exchanger [52-53], to which the water heated in the heliosystem is supplied [73-74, 126]. In the absorber we can observe the opposite situation, cooling in required there. A cooling tower is most frequently introduced into the scheme for this purpose. The absorber with internal evaporative cooling is very appealing, however, its creation is connected with certain constructive difficulties [49, 97-100].

Along with apparatuses of direct evaporative cooling [39, 73-74] it is promising to use, as evaporative coolers, apparatuses of the indirect evaporative type [22-27, 29, 37, 52-53, 86, 94, 97-100, 126, 127], in particular, regenerative ones [24, 25, 66, 73-74]. In these apparatuses contact-free cooling of the main air-flow is achieved, that is cooling at the unchangeable moisture content, which is undoubtedly favorable for air-conditioning systems. The direct evaporative cooler [24, 37] can be used as an additional cooler after the indirect type cooler. Its use for the air-flow which has already been cooled and dry, does not caused any problem of an excess moisture in the room being conditioned.

The most important matters defining the future of alternative solar systems is the creation of highly efficient (absorber, desorber, evaporative coolers, heat exchangers). As the number of such apparatuses, included in the systems, is rather high, it will required considerable energy consumption for organizing the movement of heat carriers, that is the electrical power expenses for the operation of air fans and liquid pumps. From the thermodynamic viewpoint, allowing for small moving forces of processes characteristic of apparatuses; but it is, of course, connected with the increase of energy consumption and to some extent, it results in the depriciation of advantages of the principle used. Unfortunately, this problem is not considered in most of the works, which is evidently connected with today's theoretical and experimental level of developments. It is perspective to use apparatuses of the film type providing separate gas and liquid flows in the multi-channel ordered packing bed of such apparatuses [22-32, 86, 97-100] for alternative solar systems. The cross-flow scheme of the flows interaction in apparatuses [24, 25, 49, 73-74, 127, 134] makes it possible to obtain additional advantages, as it is characterized by slight air resistance. We can say, that it is promising to use in such systems apparatuses with movable quasi-liqnefied packing bed which has comparatively high air resistance, but capable to operate in a stable regime in complicated, particularly, highly polluted media [24, 134]. It should be supported that the field of such apparatuses practical use is systems of considerable capacity.

The possibility to use heliosystems with thermal solar collectors (SC) as external heating sources for providing solar energy regeneration [4, 23, 26-27, 29, 30, 39, 41-42, 52-53, 63, 73-74, 89-90, 103, 126] is of certain interest. These collectors can be nonevacuated flat collectors, evacuated flat or tube collectors, evacuated low-concentrating collectors, etc. They are rather expensive and today the development of solar energetics is under certain state support. Unfortunately, their efficiency decreases with the increase of the temperature achieved, and most of solar refrigeration technologies require the temperature level of about 100 °C. the situation is a little better for solar air-conditioning systems for which the temperature level of 60-100 °C can be sufficient [52-53, 97-100, 101-102]. It should be noted that the cheapest and most common type of the solar collector today – flat SC can provide only 50-65 °C and the operation on its basis even of the least fastidious in this sense, open solar system on the LiBr|H_2O solution is rather problematic. Solar alternative systems require the creation of a compensation mechanism, connected with the problem of natural variations of solar activity. For such systems it is important to create efficient heat energy accumulators. It is also advantageous in this to developed various combined systems providing the possibility of joint using different sources of low-grade heat along with solar energy, gas and liquid boilers [95-96, 97-100].

The commercial situation with solar refrigerating and conditioning systems [90] is rather uncertain. It is explained not only by today's stage of developing such new systems but the rushing change of priorities and contradictory estimations of the situation in this field. Solar alternative systems, as the available though rather limited experience of their practical use shows, [4, 101-102] can provide two-fold reduction of energy consumption [24, 101-102] as compared to vapour-compression coolers. They are undoubtedly deprived of ecological problems characteristic of traditional vapour-compression equipment (the global warming problem, ozone security) but they have serious disadvantages too – large overall dimensions, problematical character of solar regeneration only, hazards of corrosion effect on construction materials, the need for highly efficient energy accumulators. However, these drawbacks are characteristic of all alternative sources of energy without exception and the evaluation of prospects in this field must be based not on technical and economical comparison with traditional approaches, well-known today, but proceeding from long-term perspective concerns where the importance of solar refrigerating and conditioning systems must be decisive allowing for the available and constantly aggravating energy-ecology problems.

Main Conclusions

- it should be noted that the interest in capabilities of solar systems for air conditioning and media cooling is considerable and constantly growing. It is defined by the ecological cleanliness of these systems and low energy consumption. At the same time, these systems are at the stage of theoretical developments and experimental tests, and today we should speak about their prospective importance rather than a serious commercial aspect;
- along with various combined cycles the most prospective one is the open absorption cycle, based on the use of liquid and solid sorbents and the possibility of using solar energy as an external heating source;
- the preferable field of practical use of alternative systems based on the use of the open absorption cycle is solar air-conditioning, which is defined by a certain correlation between the insulation and the required level of cooling (the complex of heat-humidity air parameters for AACS), comparatively low as compared to refrigerating systems, temperatures of the sorbent regeneration and much wider prospective commercial demand for such systems [90], solar refrigerating systems require higher regeneration temperatures, they have a limited climatic zone of application (low moisture content of the outdoor air) [53];
- solar air-conditioning systems—AACS—on the basis of the open absorption cycle using liquid sorbents (absorbents) have obvious advantages over the systems using solid sorbents: a lower required temperature level of the sorbent regeneration, the construction design simplicity, lower energy expenditures on the organization of the heat carriers movement; solutions of LiBr are referred to the most widely spread types of absorbents today;
- solar AACS with direct and indirect absorbent regeneration have their advantages but it is indirect regeneration that makes it possible to exclude the possibility of the sorbent pollution and to lessen the overall dimensions of the solar system on the whole;
- the main problems of solar AACS, requiring the urgent solution to enter the market in the future are as follows:

  - ✓ the necessity to improve the solar heat receiver (heliosystem with solar collectors), increasing its efficiency and attaining temperatures of 100 °C and more;
  - ✓ the use of new ideas in the field of the cooling part of the system—evaporative cooler, which is of a special significance for open systems and more important than in the case of closed systems;
  - ✓ the creation of compact heat-and-mass transfer equipment for open absorption systems, unified for all its main components and providing the minimization of energy expenditures on organizing the heat carriers movement and high compactness. Obviously, it is the most important task, the successful solution of which will influence the development of this trend in the near future;

- ✓ the creation of combined systems for autonomous cooling (air conditioning) and heating, which are flexible and operative enough in control;
- ✓ modeling operation processes in the main links of alternative systems and creating fundamentals of the alternative solar system design.

1.2. The Development of New Approaches

The application of the open absorption cycle provides new possibilities for creating a prospective generation of engineering heat-and-cold supply systems—those of refrigeration, heat pumping and air conditioning. The cycle can operate at extremely small temperature differences, is ecologically clean and low power-intensive. Low-grade heat, natural gas or solar energy can serve as an external heating source. The solar energy source, in a practical sense, can be a heliosystem with flat solar collectors, i.e. the cheapest and most reliable type of heliosystem, developed by the authors [23, 26, 30] for hot water supply and including the necessary number of collectors and a tank-accumulator, depending on the required capacity [30, 118].

Versions of a schematic for an alternative system developed by the authors is given in figures 1.1–1.5 (in conformity with the problem of air-conditioning by AACS) on the basis of the open absorption cycle and solar regeneration of the absorbent. The scheme consists of two main parts: dehumidifying air (denoted ··········) and refrigerating (denoted – · – · – ·). In the dehumidifying part the heat needed for the absorbent regeneration is provided by the heliosystem with flat solar collectors 6 (7, 8, 9 are solar collectors, a tank-accumulator, an additional heating source, respectively), and cooling the absorber is provided by a fan cooling-tower 5. The given schemes includes as the main components absorber 3 (the air dehumidifier), desorber 4, designed for solar regeneration of the absorbent; combined evaporative cooler 1-2 and a system of regenerative heat exchangers 10-12, 17, whose necessity is caused by low temperature heads available. Air flow 13 (fresh outdoor air) being dehumidified in absorber 3 decreases moisture content x_g and dew point temperature t_{dp}, which provides a considerable potential of cooling in the evaporative cooler. The system includes fan cooling tower 5, cooling the absorbent at the inlet to the absorber (figures 1.1-1.4) or (figure 1.5) an absorber with internal evaporative cooling development by the authors is used.

As an evaporator is used the indirect evaporating cooling apparatus IEC [22, 23, 24, 26, 27, 29, 32, 94] developed by combined scheme in the form of the multi-channel packing bed with alternating «wet» (the auxiliary air flow and water recirculating through the apparatus) and «dry» channels (the main air flow cooled in the IEC at the unchangeable moisture content) (figures 1.1-1.5, figures 1.6-1.7 – position 1, IEC is a film cross-flow heat-and-mass transfer apparatus.). The amount of water evaporated water loop is replenish by feeding fresh water 15. As a results of evaporative cooling the water in “wet” channels, contact free cooling the main air flow in “dry” channels IEC is provided through a thin heat conducting wall which separates these channels. The wet bulb temperature at the inlet of IEC is a natural limit of evaporative cooling in a single-stage IEC.

Figure 1.2 shows the schematic diagram of two-stage indirect evaporative cooler. The authors had studied theoretically and experimentally [22, 24, 32, 81] possibilities of combined coolers IEC. The number of such stages should not exceed three, as the further growth of their number results in a slight increase of the cooler efficiency at the considerable growth of energy consumption for running the process. It is undoubtedly useful to include into schemes the coolers of regenerative heat exchangers (HEX 12 in figure 1.1 and HEX 12 and 17 in figure 1.3), in this case, however, one should take into account the growth of attendant energy consumption. Figure 1.1 shows a combined evaporative cooler as a part of the IEC (the first stage of cooling) and the DEC (the direct evaporative cooler as the second stage of cooling) – IEC/DEC. As it follows from previous studies of the authors [22, 23, 24, 32, 86] and American researches [37], such an evaporative cooler is promising for dry and hot climates. According to the authors' data, its efficiency is slightly lower than that of the two-stage evaporative cooler in figure 1.2 but the latter requires much more energy consumption for organizing the process. The regenerative scheme providing a high efficiency of the process is of a special interest, but it is distinguished by high power intensity. In figure 1.1 it is a joint operation of the indirect evaporative cooler IEC (1) and the heat exchanger HEX 12, and in figure 1.2 it is given for the interpretation of the two-stage evaporative cooler scheme. If we allow, that the regenerative scheme of the indirect evaporative cooler can potentially provide cooling up to the dew point temperature of the air coming to the evaporative cooler, then the level of cooling can be rather high, taking into account its preliminary dehumidification in the absorber. The combined evaporative cooler CEC can operate both by the ventilation and recirculation schemes, but in the latter case the flow 13 at the inlet to the IEC is the air leaving the room being conditioned.

The internal evaporative cooling of the absorber (figure 1.5) provides a high efficiency of the absorption process (3-6 times as much as that of the usual absorber [97, 100]), which makes it possible to considerably reduce the absorbent consumption and due to it to cut expenditures down needed for its regeneration and to increase the total C.O.P of the system by 30-35% [97, 100]. The main problem in designing such apparatuses is to separate liquid flows at the outlet from the absorber [97, 100]. It can be solved with the help of the indirect evaporative cooler design developed by the authors, in which packing bed components from closed channels thus providing the separate movement of air flows and the required tightness of channels (figure 1.8C).

Of a special interest for cooling and air-conditioning systems is the scheme given in figures 1.3 and 1.4, in which cooling tower 5 is used as the second stage of the combined evaporative cooler. The air dried in absorber 3 and having a low dew point temperature (the limit of evaporative cooling in the regenerative indirect evaporative cooler IEC – HEX 12) is cooled, while the moisture content is unchangeable, in the IEC and enters the cooling is provided. This water can be used in ventilated heat exchangers-cooler (18) installed directly in air-conditioned rooms (20) or refrigerating chambers. In this case the refrigerating unit can be outside air-conditioned rooms and the building. Cooling tower 5 is also a film cross-flow device, and the refrigerating unit CEC includes, along with CTW, two regenerative heat exchangers for air flows leaving IEC and CTW (figure 1.7).

Figures 1.8A-1.8B give the schematic description of the absorber (dehumidifier) and the desorber (solar regenerator of the absorbent) in some basic versions of the design. In figure 1.8A these apparatuses are shown as a version with placed outside (external) heat exchangers for cooling and heating the absorbent which corresponds to the schematic shown in figures 3.1–3.4. In figure 1.8B apparatuses are made in combination with appropriate heat exchangers, which corresponds to schemes in figures 1.1–1.5 the packing bed of these apparatuses has the tube-plate construction, previously developed and investigated by the authors in conformity with evaporative condensers of refrigerating plants [24, 81], and in these developments the longitudinally corrugated sheet with regular roughness of the surface is also used.

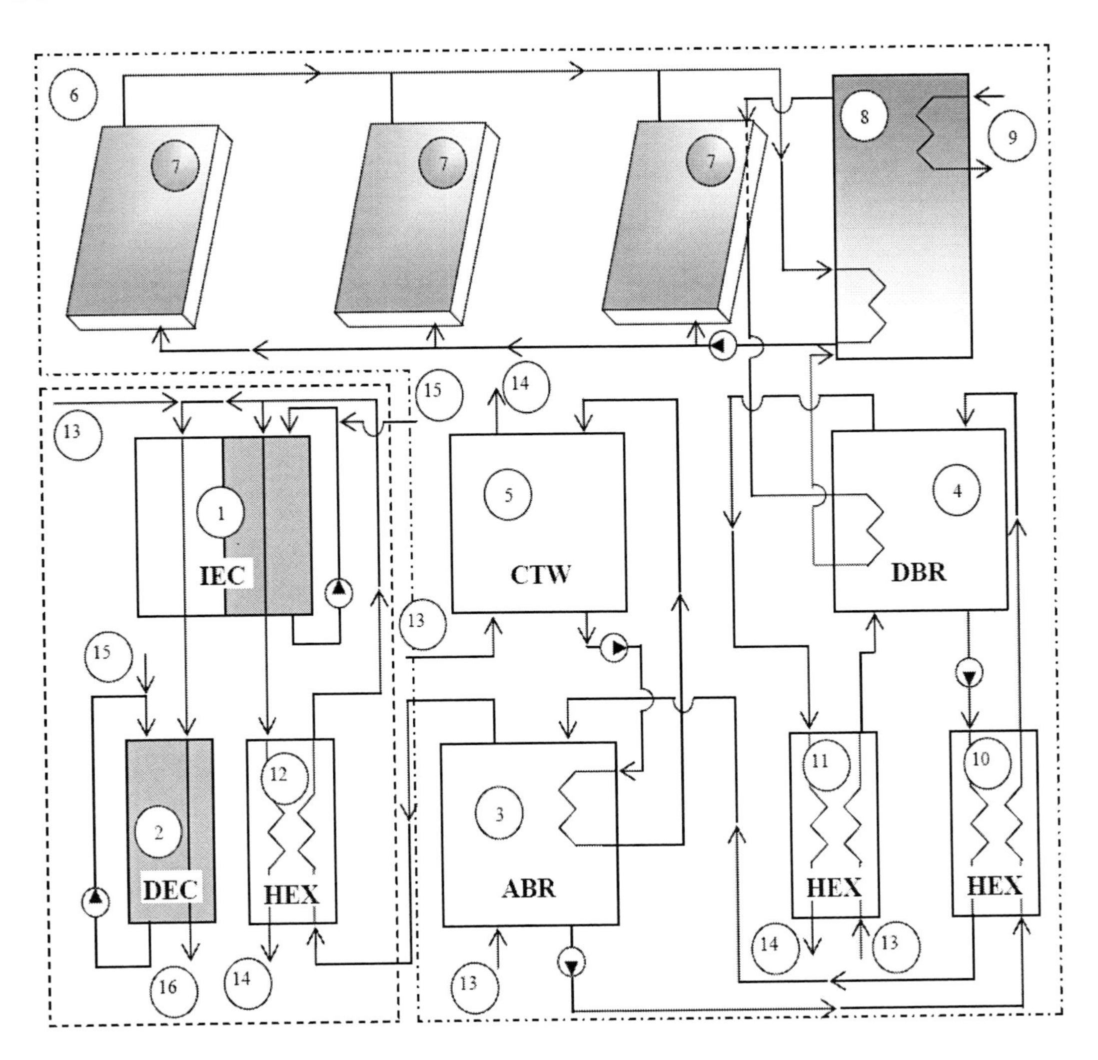

Figure 1.1. Schematic of the alternative air conditioning system (AACS) with heliosystem as an external source of heating. Nomenclature: 1 – indirect evaporative cooler; 2 – direct evaporative cooler; 3 – absorber; 4 –desorber; 5 – cooling tower; 6 – heliosystem (7, 8 – solar collector, tank-accumulator); 9 – additional heating source; 10, 11, 12, 17 – regenerative heat-exchangers; 13 – fresh (outdoor) air; 14 – release into the atmosphere; 15 – water (replenishment of the system or fresh water); 16 – air in the room; 18 – fan-coils; 19 – rooms air; 20 – room; - - - - - - -dehumidifying part; -.-.-.-.-.-.-refrigerating part (CEC).

The design of all the components of the schemes (DEC, IEC, CTW, ABR, DBR) is unified. They are manufactured in the form of film counter- or cross-flow apparatuses [24, 81], in which a longitudinally (in the direction of the liquid film flow) corrugated sheet of the packing bed with regular roughness on its surface as an intensification method is used as a part of the packing bed (figure 2.10). In this case the jet-film mode of liquid flowing (in the riffles of corrugation) and wet-dry mode of contacting gas and liquid flows are created. It provides the minimization of energy consumption and replenishment of the system with fresh water. The problems of such flows stability, relationship of heat-and-mass transfer surfaces, etc. have been studied by the authors theoretically and experimentally. The design of such units as the absorber, the desorber, and the cooling tower can also be executed in the form of apparatuses with a movable pseudoliquefied packing bed [24, 40, 134] developed and put into production by the authors in producing fan cooling towers, they proved to be good while operating in any contaminated media. This solution is preferable for large capacity systems.

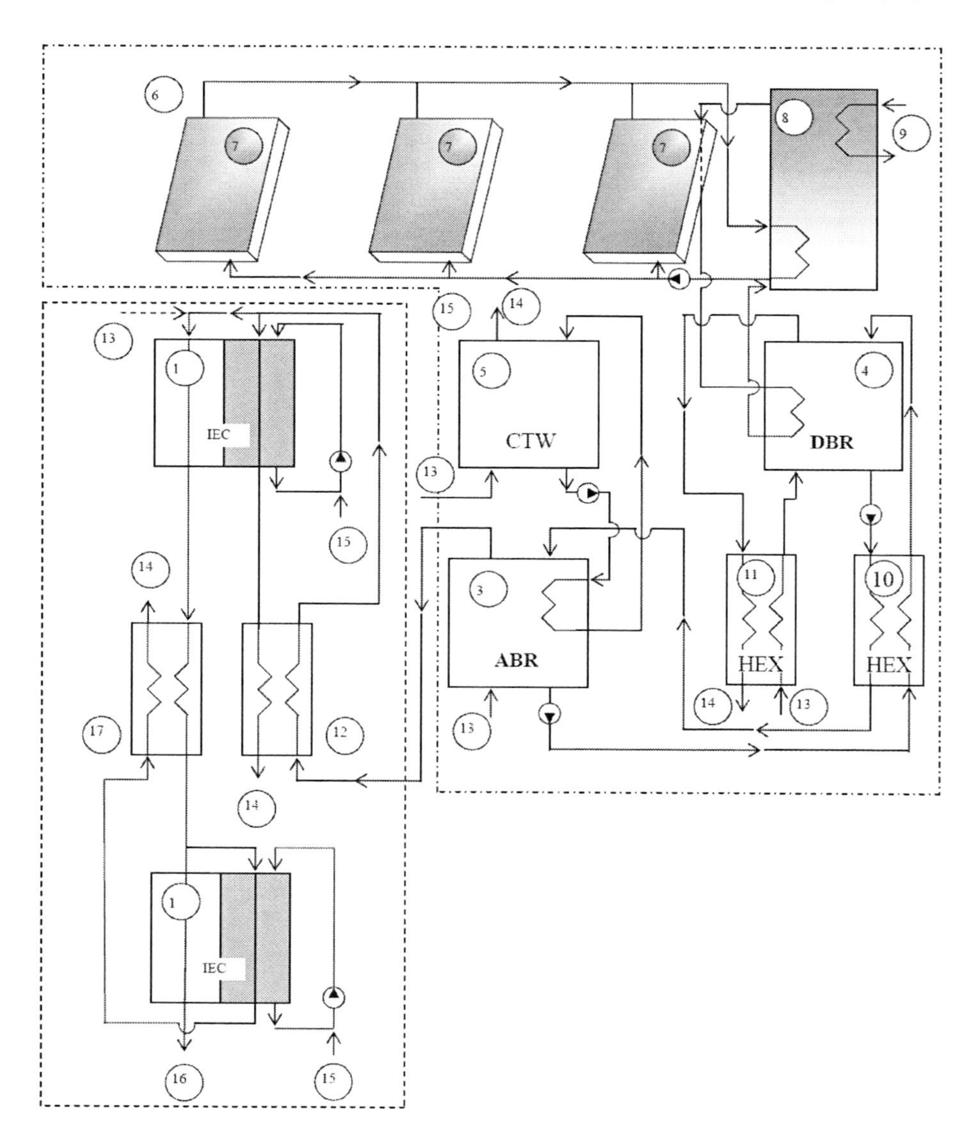

Figure 1.2. Variant of the AACS schematic from figure 1.1. Evaporative cooler in the form of a two-stage indirect-evaporative system (17 – additional regenerative heat-exchanger). Nomenclature is the same as in figure 1.1.

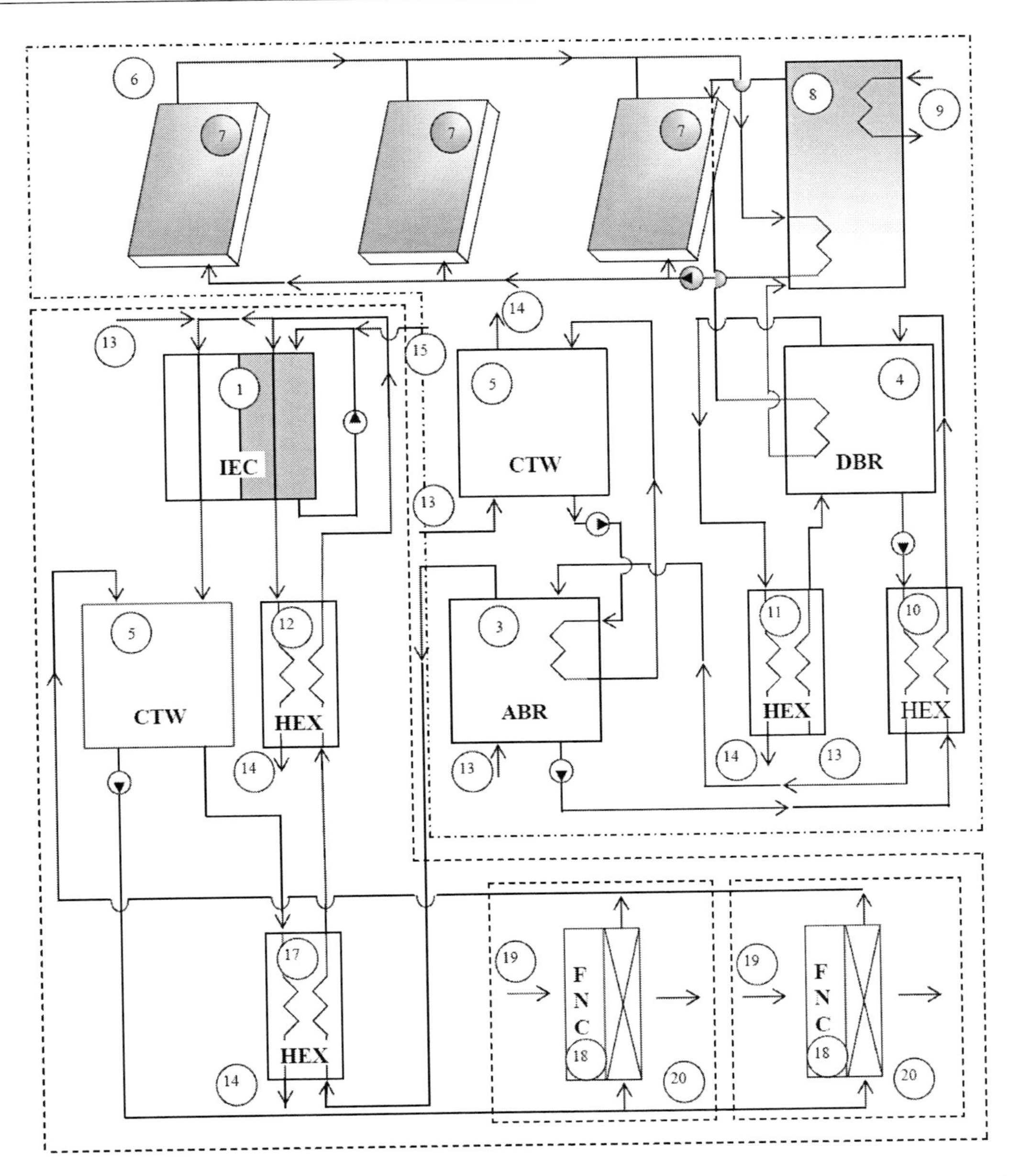

Figure 1.3. Schematic of the alternative air-conditioning system (AACS) with heliosystem (solar energy) as an external source of heating and feeding cold water to heat exchangers-coolers. Nomenclature is the same as in figure 1.1. Additional nomenclature: 17 – regenerative heat exchanger; 18 – ventilated heat exchanger-cooler; 19 – room air; 20 – conditioned room; - - - - - -unit of air cooling; -.-.-.-.-.-.-unit of evaporative cooling.

The heliosystem (figures 1.1-1.5) with flat solar collectors 7 is used as an external heating source during the absorbent regeneration. Such a system has been developed by authors (Research-Production firm "New Technologies") and proved to be good for hot water supply. Figure 2.27 gives a general view of the developed solar collector SC. The heat absorbing panel of SC (figure 2.26 is made in the form of a pipe register. Pipes have float-type fins made of corrosion-resistant aluminium alloy. The SC includes hydraulic collectors (2, 3), the housing of shaped aluminium. The special construction of fasteners provides a simple and reliable method of installing glass 6. Two SC modifications with the area of the heat absorber 1.1 and 2.0 m^2 are produced. One collector of SC-1.1 modification provides heating 80 litres of water up to 60-65 °C under the July conditions in Odessa, which is quite acceptable for the organization of the absorbent regeneration process. An additional heating source 9 is provided in the tank-accumulator 8 of the heliosystem (figures 1.1-1.5); it compensates natural variations of solar activity.

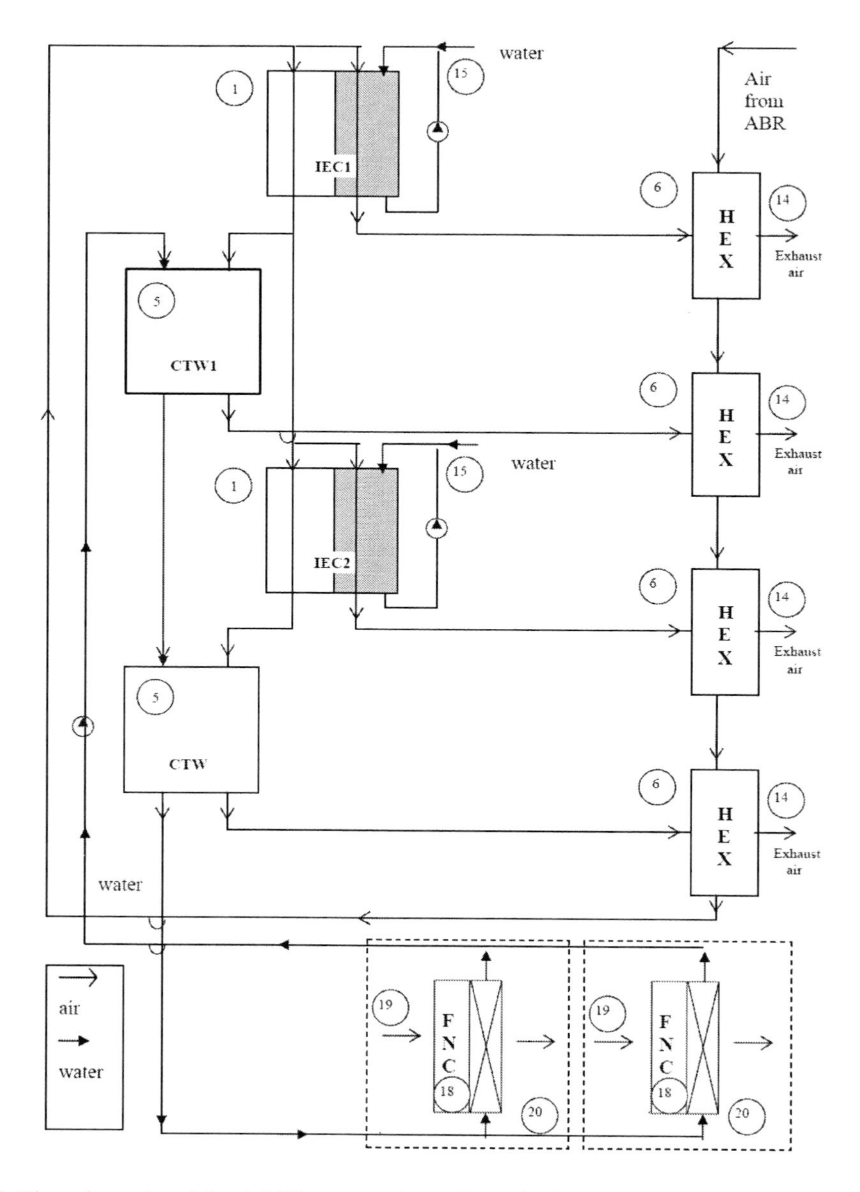

Figure 1.4. The schematic of the AACS evaporative unit on the basis of the two-stage combined system of IEC/CTW. Nomenclature is the same as in figure 1.1 and 1.3.

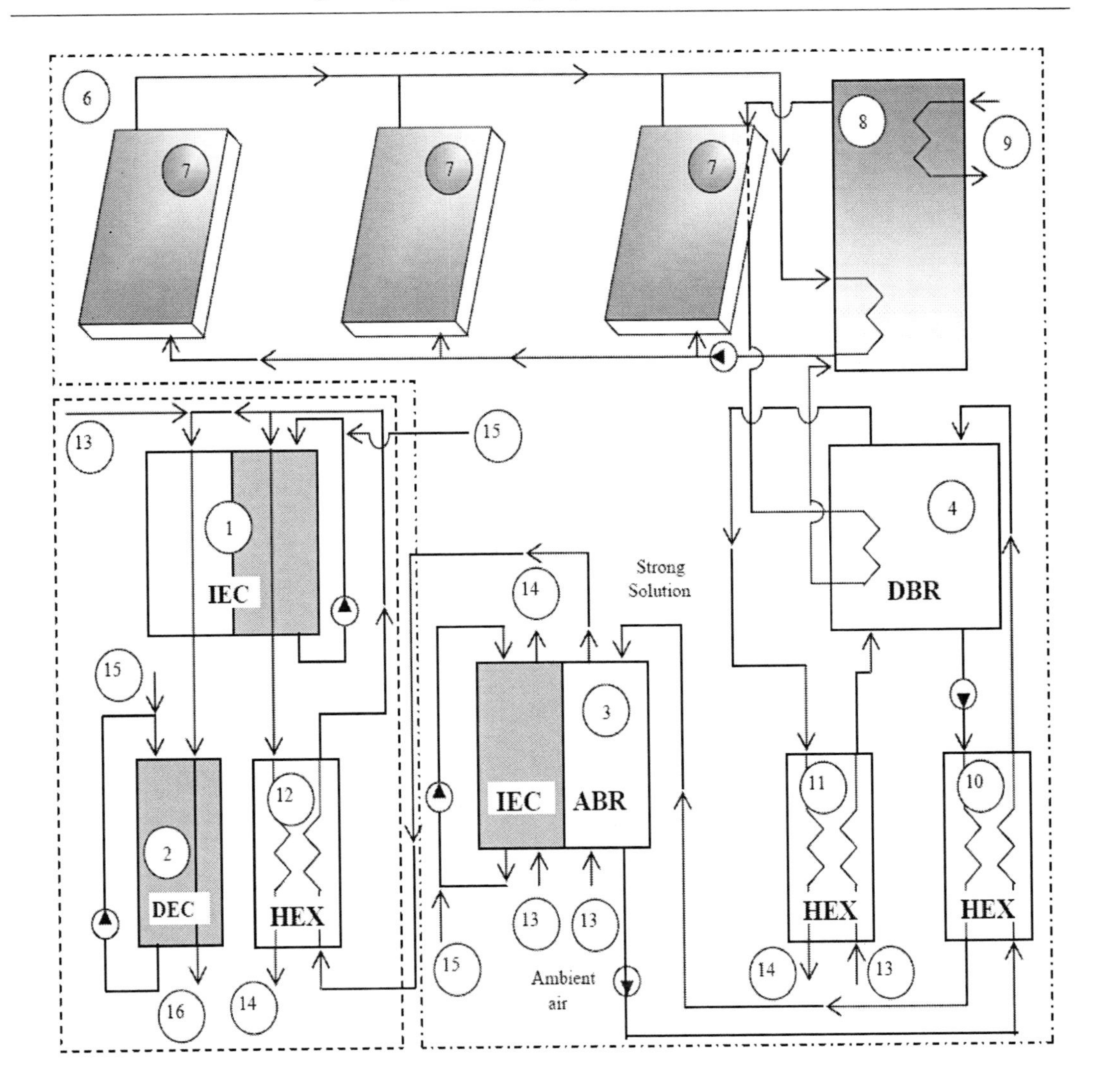

Figure 1.5. Schematic of the alternative air conditioning system (AACS) with heliosystem as an external source of heating and absorber with internal evaporative cooling. Nomenclature is the same as in figure 1.1.

To meet the own needs of the system for electric power (pumps, fans) it is possible to use solar energy of photo-voltaic stations (direct transformation of solar energy into electricity). In this case the solution is completed with renewable ecologically clean sources of energy. It should be noted, that the experience gained by the authors in the development, production and use of heat-and-mass transfer apparatuses, heliosystems and photo-voltaic transformers of solar energy can greatly speed up process of bringing into production alternative system under consideration. As these elements were brought into production for other problems in the form of standard-size series of various unit capacity, there are no fundamental problems in putting AACS into production (as well as refrigerating and heat pumping systems) of any capacity and configuration.

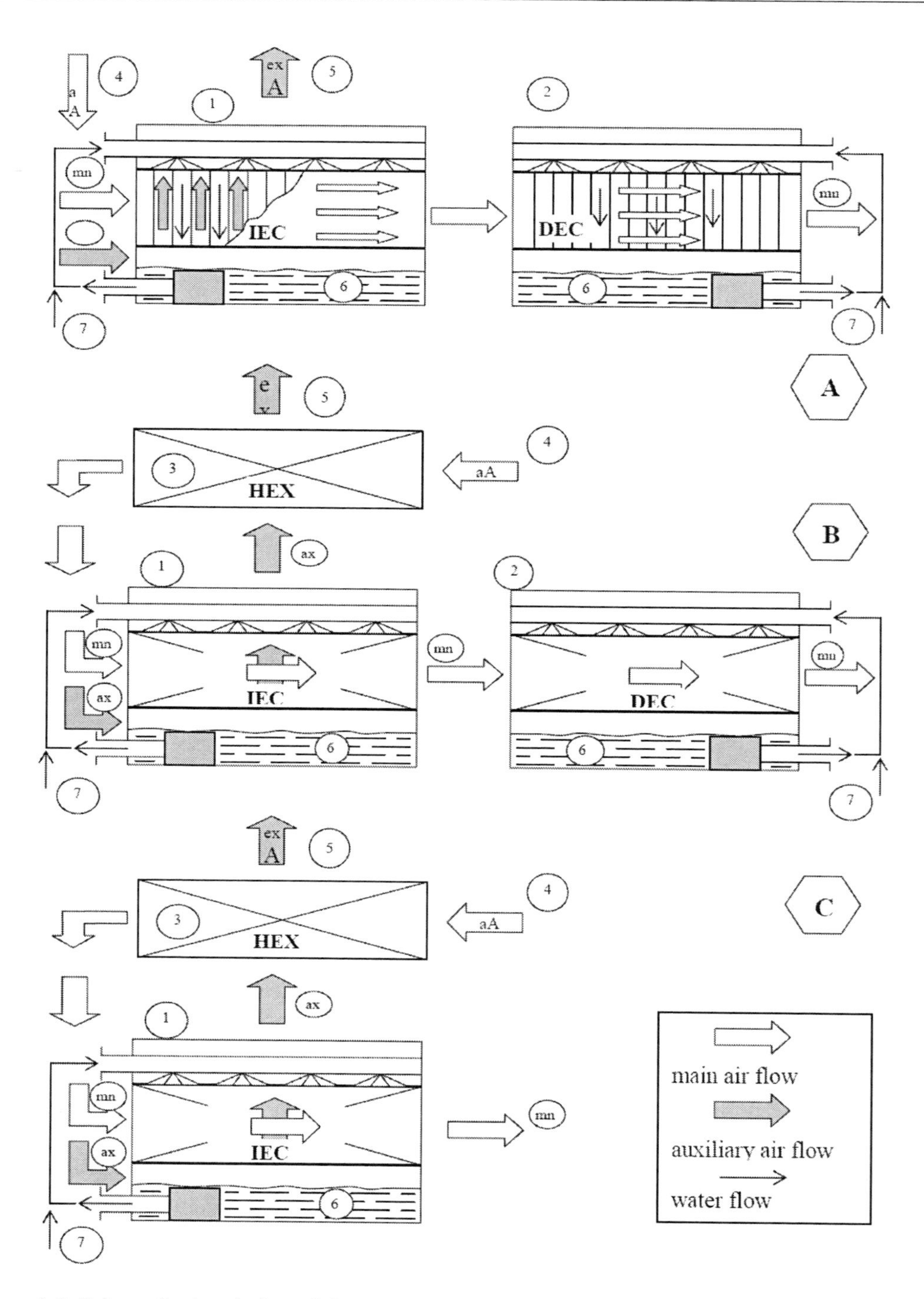

Figure 1.6. Schematic description of the combined evaporative cooler CEC on the basis of IEC/DEC (A, B) and IEC/R (C). Notation: 1 – indirect evaporative cooler IEC; 2 – direct evaporative cooler DEC; 3 – HEX; 4 – ambient air; 5 – exhaust air; 6 – water tank; 7 – water (replenishment of the system or fresh water).

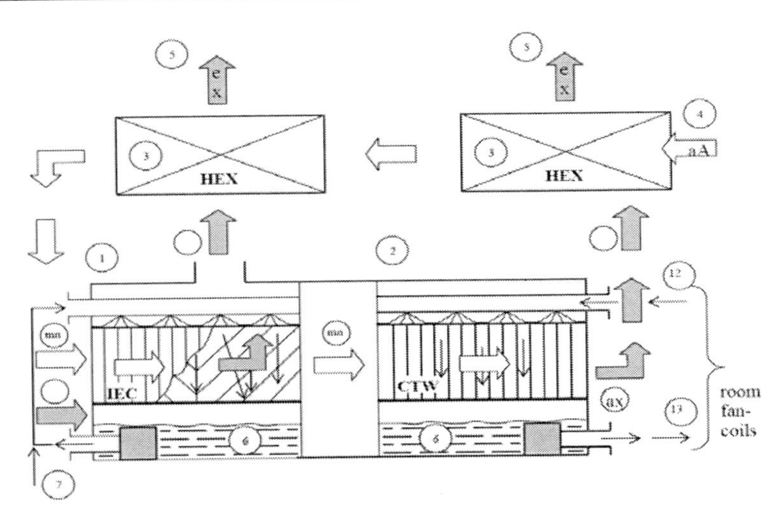

Figure 1.7. Schematic description of the combined evaporative cooler CEC on the basis of IEC|CTW. Nomenclature is the same as in figure 1.6. Additional notations: 12 ,13 – chilled water.

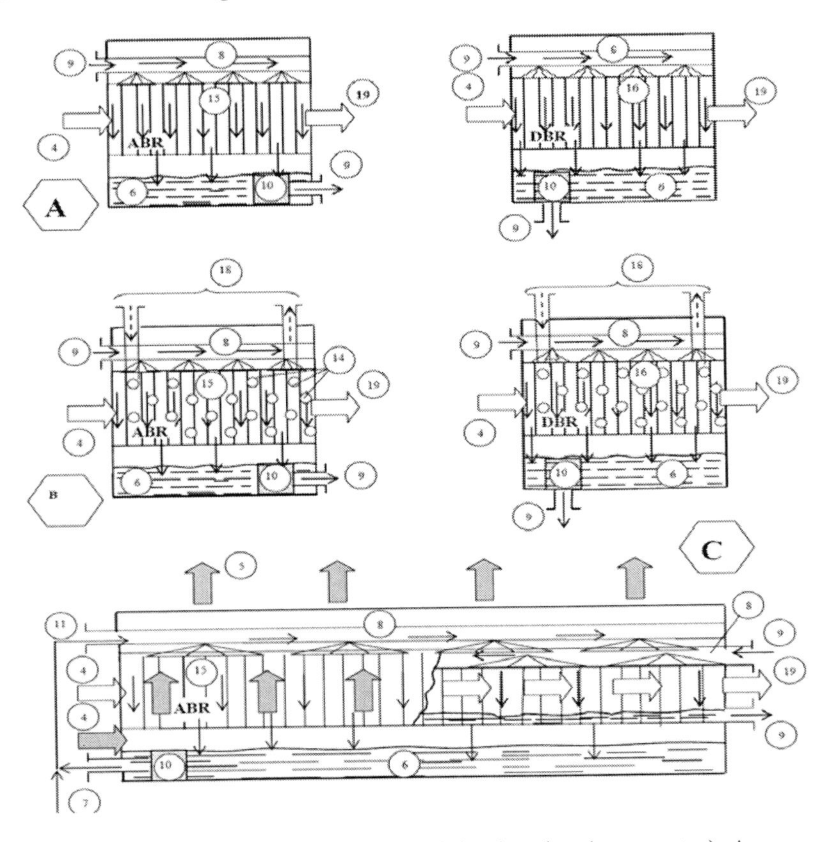

Figure 1.8. Schematic description of the absorber and the desorber (regenerator). A – apparatuses with placed outside heat exchangers, the cooler and the heater, respectively; B – apparatuses of the combined with heat-exchangers type; C – the absorber with internal evaporative cooling. Notations in figure 1.6, additional notations: 8 – the liquid distributor, 9 – the absorbent; 10 – the liquid pump; 11 – the recirculation water loop in the absorber; 14 – built-in heat-exchanger; 15 – absorber; 16 – desorber; 18 – the heat transfer medium (water) from the heliosystem; 19 – the dehumidified air flow.

2. SIMULATION OF WORKING PROCESSES IN THE ALTERNATIVE AIR-CONDITIONING SYSTEMS

2.1. The Execution of the General Thermodynamic Analysis

Thermodynamic efficiency of cycles is determined by their heat COP and degree of thermodynamic perfection θ

$$COP = \frac{Q_0}{Q_{DBR}}, \theta = \frac{COP}{(COP)_K}, \quad (2.1)$$

where

Q_0 – refrigerating capacity of the cycle;

Q_{DBR} – heat load of the regenerator (desorber);

$(COP)_K$ – heat coefficient of the Carnot cycle, drawn in the same temperature range for heat supply and heat removal.

$$(COP)_K = \frac{T_H - T_{aA}}{T_H} \cdot \frac{T_C}{T_{aA} - T_C}, \quad (2.2)$$

Refrigerating capacity and heat load of the regenerator are determined by the amount of moisture evaporated in the evaporator $(\Delta G_L)_{IEC}$ and absorbed in the regenerator $(\Delta G_L)_{DBR}$

$$Q_0 = (\Delta G_L)_{IEC} \cdot r_{IEC}, \quad Q_D = (\Delta G_L)_{DBR} \cdot r_{DBR}, \quad (2.3)$$

As

$$(\Delta G_W)_{IEC} = (x_{out} - x_{in})_{IEC} \cdot (G_g)_{IEC},$$
$$(\Delta G_W)_{DBR} = (x_{out} - x_{in})_{DBR} \cdot (G_g)_{DBR} \quad (2.4)$$

where

x_{out}, x_{in} – moisture content of the air at the inlet and outlet from the corresponding device.

As the moisture is not accumulated in the scheme under established working conditions

$$\left(x_{out} - x_{in}\right)_{IEC} = \left(x_{out} - x_{in}\right)_{ABR}. \tag{2.5}$$

In the single-flow (by the air flow) scheme and atmospheric the following conditions are observed in apparatuses:

$$\left(G_g\right)_{IEC} = \left(G_g\right)_{DBR} = G_g, \qquad r_{IEC} = r_{DBR} = r. \tag{2.6}$$

Then

$$COP = \frac{Q_0}{Q_{DBR}} = \frac{\left(x_{out} - x_{in}\right)_{IEC}}{\left(x_{out} - x_{in}\right)_{DBR}} = \frac{\left(x_{out} - x_{in}\right)_{IEC}}{\left(x_{out} - x_{in}\right)_{ABR}}. \tag{2.7}$$

For an idealized cycle at zero under-recuperation at the ends of heat-and-mass transfer apparatuses

$$COP_{id} = \frac{x_s\left(T_c\right) - x_{min}}{x_{aA} - x_{min}}, \tag{2.8}$$

where

$x_s\left(T_c\right)$ – moisture, content of the saturated air at the outlet from the evaporator;

x_{min} – minimum moisture content, corresponding to the partial pressure of water vapour in the strong absorbent solution;

x_{aA} – moisture content of the outdoor air, coming into the system.

Figures 2.1 – 2.2 shows the dependencies of COP and θ for an source temperature at different values of relative humidity of the outdoor air (it is accept that $T_C = 7$ °C, $T_{aA} = 30$ °C). The relationships are draw for relative humidity values of air corresponding to wet bulb temperature (dew point) which is higher than the required temperature of cooling in the cycle. The case of lower relative humidity was not considered, as the required reduction of temperature is realized in the usual evaporative cooler under these conditions. As one can see from figure 2.1 the temperature of the heating source of 60 °C and above provides the acceptable COP values in the idealize cycle practically for all values of relative humidity of the outdoor air.

In going to the real cycle, for the determination of moisture content values in corresponding points at the COP calculation it is necessary to introduce into calculations the value of heat efficiency η for apparatuses. For accepted in all devices the cross-flow scheme of movement η can be determined by F.Trefni equation with the accuracy sufficient for the analysis and engineering calculations [107, 124]:

$$\eta = \frac{1-\exp\left\{-S\cdot\left[1+A\cdot\left(1-2\cdot f_{\varphi}\right)\right]\right\}}{1+A\cdot\left(1-f_{\varphi}\right)-A\cdot f_{\varphi}\cdot\exp\left\{-S\cdot\left[1+A\cdot\left(1-2\cdot f_{\varphi}\right)\right]\right\}}, \tag{2.9}$$

where

f_{φ} – flow scheme coefficient for the accept scheme $f_{\varphi}=0.495$;

S – number of transfer units, $S = k/w$;

$A = \frac{w_{max}}{w_{min}}$;

w_{max}, w_{min} – minimum and maximum (for the given apparatus) total heat capacity of mass flow rate.

In calculating the moisture content of the air flow at the inlet and outlet from the absorber, desorber (regenerator) and evaporative cooler it is necessary to previously determine the temperature of each flow at the inlet and outlet from the apparatus for the determination of COP of real cycles.

For schematic solutions under consideration temperatures of each flow at the inlet into the apparatus T_1' and T_2' are usually known. The outlet temperatures of each flow T_1'' and T_2'' can be determined by:

$$T_1'' = T_1' - E_{\Phi}\cdot\left(T_1' - T_2'\right);$$
$$T_2'' = T_2' - E_Z\cdot\left(T_1' - T_2'\right), \tag{2.10}$$

where

E_{Φ} – efficiency of heating;

$E_{\Phi} = \frac{T_1' - T_1''}{T_1' - T_2'}$;

E_Z – efficiency of cooling;

$$E_Z = \frac{T_2'' - T_2'}{T_1' - T_2'}.$$

Calculations of moisture content values for determining COP and θ of the real cycle with real heat and hydraulic losses in heat-and-mass transfer apparatus for a wide range of temperatures of a heating source and relative humidity of outdoor air will be carried out at further stages of the work.

2.2. Selection of Working Substances Taking into Account Thermophysical Properties

To raise the efficiency of evaporative cooling process in AACS atmospheric air is predehumidified in the absorber. In case of high moisture content in the air at the system inlet its predehumidification is the only condition for the possible use of evaporative cooling to attain comfort air parameters. Aqueous solutions of calcium chloride, lithium chloride, lithium bromide as well as multicomponent solution on the basis of the above-mentioned substances are usually substances are usually used as absorbent in open system.

The solution of $CaCl_2$ is the cheapest, ecologically pure, well studied absorbent, still its application in AACS, as it will be shown below, is problematic.

The solution of LiCl is widely used as an absorbent. Its main advantages are bactericidal effect and harmlessness for people. The air treated by this solution is highly sterilized – in dehumidified air the amount of microorganisms can be reduced to 97%. It is characterized by the capability to absorb harmful smells; the possibility to be regenerated by low-grade heat (low-temperature water from heat-and-power stations, industrial waste, solar energy); the possibility of significant reducing moisture content in the air being dehumidified; a wide range of working parameters [136]. The disadvantage of the lithium chloride solution is the corrosive effect on metals, which requires the use of special covers of surfaces or the introduction of inhibitors into the solution. Brass, aluminium, tin, etc. are not affected by the solution.

Aqueous solution of lithium bromide (LiBr) [65, 136] has the best absorbing capability among the above-mentioned absorbents. Nevertheless, it is very agressive with respect to metals and other material and more expensive than the lithium chloride solution. The lithium bromide solution is most widely used as a working body in closed absorption systems. Data on the solution thermophysical properties in a wide range of parameters were obtained with the help of calculation equations given in papers [65, 136].

To achieve the required moving force in AACS the required concentration of LiBr should be 60–65%, i.e. the process line is quite near the crystallization line in the temperature range of 30–60 °C. In this connection, new working substances are being developed, which contain components increasing the solubility and decreasing the corrosive activity. $LiNO_3$, $ZnCl_2$, $CaBr_2$, LiI and other components are used as additives. The system $LiBr+LiNO_3+H_2O$ (4:1 $LiBr:LiNO_3$ in moles) has been recently used, which, in the authors opinion, [65], decreases the corrosive activity and improves the cycle characteristic as compared to the solution of lithium bromide. The authors [65] have measured the density, the viscosity and other thermophysical properties of the solution in a wide range of parameters.

The system $LiBr+ZnCl_2+CaBr_2+H_2O$ (1.0:1.0:0.13 $LiBr:ZnCl_2:CaBr_2$ in mass fraction respectively) is undoubtedly of interest as a working body, which equally with $LiBr+LiNO_3$, decreases the corrosive activity and improves the cycle characteristics as compared to the solution of LiBr, as the authors show [64]. The said system, as well as $LiBr+LiNO_3$ and LiBr, has a bactericidal effect and is harmless for people. Data on thermophysical properties of the solution in wide range of parameters are given in paper [64].

The saturated vapour pressure and the solubility of the five-component solution $H_2O+LiBr+LiI+LiNO_3+LiCl$ were studied in paper [85]. It is difficult to make a conclusion concerning the efficiency of this solution, as the results of measurments presented in the graphical form for the temperature range, we are interested in, have a very small scale. Moreover, there are no data on thermophysical properties of this solution.

This section gives the preliminary analysis of working substances on the basis of their thermophysical properties. Later on the solution will be designated: $H_2O+LiBr$ – LiBr; $H_2O+LiBr+LiNO_3$ – LiBr+; $H_2O+LiBr+ZnCl_2+CaBr_2$ – LiBr++.

2.2.1. Vapour Pressure

The vapour pressure p_s is the most important characteristic of working substances in AACS. To raise the system's efficiency the relationship $p_s(\zeta)$ should match the following requirements:

- low pressure at the absorption temperature (~30 °C);
- strong temperature dependence of the vapour, and, hence, high vapour pressure at desorption temperatures (~60 °C in using solar energy as a heating source);
- steep form of the crystallization line.

The last factor providing high reliability of using heat-and-mass transfer equipment in the vicinity of the crystallization line, as the probability of salt deposits on surfaces of the equipment is reduced when the conditions of using are disturbed.

As it was shown above, the use of the calcium chloride solution in open AACS is problematic. It, for instance, the parameters of the surrounding air t^1_{aA} = 30 °C, x^1_g = 10 g/kg, (a typical situation for regions with moderate climate), the final moisture content at the outlet from the absorber is x^2_g = 5 g/kg then partial pressure of the water vapour are equal to p^1_g = 1,6 kPa and p^2_g = 0,8 kPa, respectively. The surrounding temperature by the wet bubl is about 20 °C. With the reality of the heat-and-mass transfer processes in the cooling tower and heat-exchanger between water-strong solution taken into account, the solution temperature at the inlet of the absorber is not less than 25 °C. The minimal partial pressure (on the line of crystallization) over the solution surface at this temperature is 0.93 kPa. Even in such an almost ideal situation (in reality the absorbent temperature will be higher) the moving force of the process becomes negative, therefore this solution won't be considered as an absorbent below.

The vapour pressure of the LiBr++ solution at different temperatures and concentrations was calculated by Antoine formula [47, 65]:

$$\log p_s = \sum_{n=0}^{6} A_n \cdot \zeta^n + [1000/(T - 43.15)] \cdot \sum_{n=0}^{6} B_n \cdot \zeta^n , \qquad (2.11)$$

The vapour pressure of the LiBr+ solution at different temperatures and concentrations was calculated by Antoine formula [64-65] for temperature range 278.35 T$\leq$ 335.95 K:

$$\log p_s = \sum_{n=0}^{4} A_n \cdot \zeta^n + [1000/(T - 43.15)] \cdot \sum_{n=0}^{4} B_n \cdot \zeta^n . \qquad (2.12)$$

Figure (2.3) gives vapour pressures of aqueous solutions of LiCl(1), LiBr(2), LiBr+(3) and LiBr++(4) on two isotherms characteristic of open absorption cycle: 303.15 K (absorption) and 333.15 K (desorption). At the same pressures on isotherms 303.15 K solutions LiBr and LiBr+ are father from the crystallization line than solution LiCl, therefore their use is preferable from the viewpoint of reliability of the equipment operation in AACS at the same moving forces of absorption processes. The steeper line of crystallization is also an advantage of solution LiBr and LiBr+. At high concentrations (over 45%) the LiBr solution has a lower pressure than the LiBr+ solution (the difference is about 0.1 kPa near the line of crystallization), i.e. it has a better absorption capacity. At the temperature 333.15 K and ζ = 47% (concentration on the line of crystallization at T = 303.15 K) the vapour pressure over the LiCl solution is 1.73 kPa. At this temperature and the similar concentration the vapour pressure over the LiBr solution is 1.34 kPa, and over the LiBr+ solution is 1.8 kPa. The fact that under the same conditions the lower desorption temperature is required for the LiBr+ solution is of great importance when low grade sources of energy are used as a heating source. Low corrosive activity is also an advantage of this solution.

The solution of LiBr++ is also of interest as a working body. As it follows from figure 2.3, it is characterized by: high solubility (79.2%) at T = 303.15 K, low pressure (p = 0.01 kPa on the crystallization line at T = 303.15 K), almost vertical position of the crystallization line. It one assumes that the pressure of the LiBr++ solution at T = 303,15 K (point B_1) is the same as that of the LiBr+ solution on the crystallization line (point A_1), then the pressure of LiBr++ (point B_2) at desorbtion will be practically the same as that of the LiBr+ (point A_2) solution, however, the line B_1B_2 is rather far from the crystallization line. Unfortunately, it is difficult to use almost vertical position of the crystallization line for the solution LiBr++ in AACS with solar regeneration, as a too high temperature is required for the desorbtion near the line of crystallization. This region is interesting for deep dehumidifying air.

On the basis of analysis performed it is possible to make a conclusion about the perspectiveness of using solutions LiBr, LiBr+, LiBr++ as working bodies in suitable from the viewpoint of reliability from the heat-and-mass transfer equipment. The working interval of this solution concentrations is approximately 70-75%.

2.2.2. Heat conduction

There are no experimental data on heat conduction of solution LiBr+ and LiBr++, therefore the heat conduction of these solutions was calculated by the formula [136]:

$$\lambda = \lambda_0 \cdot \left(1 - \sum_{i=1}^{h} \beta_i \cdot \zeta_i\right), \quad (2.13)$$

where

λ_0 – is the heat conduction of water Wt/(m·K);

β_i – are coefficients being determined by the data [136];

ζ_i – is the concentration of the i-th component, kg of substance per 1 kg of the solution.

The heat conduction of water is approximated by the expression [136]:

$$\lambda_0 = 0.5545 + 0.00246 \cdot t + 0.00001184 \cdot t^2 \quad (2.14)$$

with the error of 0.01 Wt/(m·K) in the temperature interval from 0 to 100 °C. Coefficients β_i obtained by mathematical processing the experimental data for solutions under consideration are:

1) for LiBr $\beta = 0.16442$;
2) for $LiNO_3$ $\beta = 0.19961$; 3) for $ZnCl_2$
$\beta = 0.37108$; 4) for $CaBr_2$ $\beta = 0.41701$.

Paper [65] gives the investigations of the LiBr+ solution in which the relationship between LiBr and $LiNO_3$ was 4:1 in mole fractions. In this connection mole concentrations were converted into mass ones. In the mixture LiBr++ the relationship between components LiBr:$ZnCl_2$:$CaBr_2$ was 1.0:1.0:0.13 in mass fractions, so the concentration of each component was O.469×ζ, 0.469×ζ and 0.069×ζ.

2.2.3. Volumetric Heat Capacity

The heat flow, which is transferred by the working substance at the temperature difference in the outlet and inlet of the apparatus equal to ΔT is defined by the well-known equation:

$$Q = (\rho \cdot C_p) \cdot V \cdot \Delta T. \quad (2.15)$$

The heat capacity per unit volume or the product (ρC_p) is a complex characterizing the heat carrier "transport capability". The dependence of volumetric heat capacity on the concentration is shown in figure 2.4 for solutions LiBr, LiBr+ and LiBr++. In the range of working concentrations ρC_p of the LiBr+ solution is slightly higher than that of the LiBr solution and equals about 3400 kJ/(m^3·K) and greatly lower than that of the LiBr++ solution (volumetric heat capacity of the latter is equal to 5000 kJ/(m^3·K) at $\zeta = 72\%$). Thus from the viewpoint of "transport capability" LiBr++ is also the best heat carrier.

2.2.4. Viscosity

Being a thermophysical property viscosity is of a special interest, as it defines to a great extent the mode of the working body motion in the heat-and-mass transfer equipment and affects the pressure losses value. Overviscosity does not permit organizing the turbulent flow at rational transport expenditures, and the transition to the laminar mode results in decreasing the heat transfer coefficient, hence, in increasing the surface of apparatuses at the same heat load. For heat exchangers of the plate type the critical Reynold's number is equal to 200; the appropriate diameter of meshed-flow plates is 8 mm; the recommended flow rate is 0.25-0.8 m/s. Thus, the "critical" viscosity value is

$$\nu^* = \frac{w \cdot d_e}{Re^*} = 10^{-5} \quad m^2/c. \tag{2.16}$$

Figure 2.5 shows the dependence of kinematic viscosity of LiBr, LiBr+ and LiBr++ solutions, solutions LiBr and LiBr+ have practically the same viscosity and are below the "critical" line. When the concentration is 70%, the viscosity of the LiBr++ solution is less than that of the solutions LiBr and LiBr+. When the concentration is 75% on the isotherm 303.15 K the viscosity of the LiBr++ solution exceeds the "critical" value. The "critical" value of concentration ζ^* corresponding to the onset of the turbulent mode is 74%. Thus, from the viewpoint of viscosity the solution LiBr++ is most suitable, at the concentration of 70%, the working range of concentrations of this solution is from 70% to 74%.

2.2.5. Heat Transfer Factor

At turbulent flow in plate heat exchangers heat transfer is described by criterion equation [61]

$$Nu = 0.135 \cdot Re^{0.73} \cdot Pr^{0.43} \cdot (Pr_L / Pr_{wl})^{0.25} \tag{2.17}$$

On the basis of (2.17) it is possible to form a complex consisting of thermophysical properties which characterizes the intensity of heat transfer. This complex is called a heat transfer factor and has the form:

$$F_\alpha = 0.135 \cdot \lambda^{0.57} \cdot (\rho \cdot C_p)^{0.43} \cdot \nu^{0.48} \tag{2.18}$$

It is possible to determine the heat transfer coefficient with the help of F_α

$$\alpha = F_{\alpha} \cdot \frac{w^{0.73}}{d_e} \tag{2.19}$$

Figure 2.6. shows the dependence of F_{α} on the concentration for solutions LiBr, LiBr+ and LiBr++. In the crystallization line area the solution LiBr has a larger value of the heat transfer factor than the solution LiBr+, but both solutions have less, than at a solution LiBr++ F_{α} at ζ = 70%. At the concentration of 75% the LiBr++ solution has a approximately the same F_{α} as the LiBr+ solution. On the basis of the analysis made it is possible to conclude about the perspectiveness to use the solution LiBr++ from the viewpoint of the highest intensity of the heat transfer process in the concentration interval of 70-75%.

2.2.6. Pressure Drop Factor

The hydraulic resistance of "one run" of the plate apparatus can be calculated by similarity equation [61].

$$Eu_L = 1350 \cdot Re^{-0.25}. \tag{2.20}$$

As in case of the heat transfer process, let us form complex F_p characterizing the hydraulic resistance of the apparatus

$$F_p = 1350 \cdot \nu^{0.25} \cdot \rho. \tag{2.21}$$

Hydraulic losses are connected with the pressure drop factor * in the following way:

$$\Delta p = F_p \cdot d_e^{-0.25} \cdot w^{1.75}. \tag{2.22}$$

The concentration dependence of complex F_p for solutions LiBr, LiBr+ and LIBr++ is given in figure 2.7. Solutions LiBr and LiBr+ have practically the same values of F_p. At the concentration of 70% the LiBr++ solution has the same value of F_p as the LiBr+ solution on the line of crystallization, and at ζ = 75% F_p of LiBr++ is by 30% larger. Thus, from the viewpoint of the pressure drop the solution of LiBr++ at 70% has the same characteristics as LiBr and LiBr+ solutions, and at higher concentrations yields to them.

Conclusions: On the basis of the analysis made in this section it is possible to conclude that:

- the solution $CaCl_2$ is not suitable for use in AACS as an absorber due to too high vapour pressure values;
- among substances used as absorbents the most prospective from the viewpoint of thermophysical properties are aqueous solutions on the base of lithium bromide; in this case it is expedient to use additives reducing corrosive activity and increasing solubility ($LiNO_3$, $ZnCl_2$, $CaBr_2$ etc);
- it is expedient to use solutions LiBr, LiBr+ and LiBr++ for solving air-conditioning problems;
- the solution LiBr++ is the most preferable one from the viewpoint of heat-and-mass transfer characteristics, as well as the reliability of using (high solubility and relatively low corrosive activity). But at high concentrations it has a high temperature of regeneration. The working interval of concentrations is approximately 70–75%.

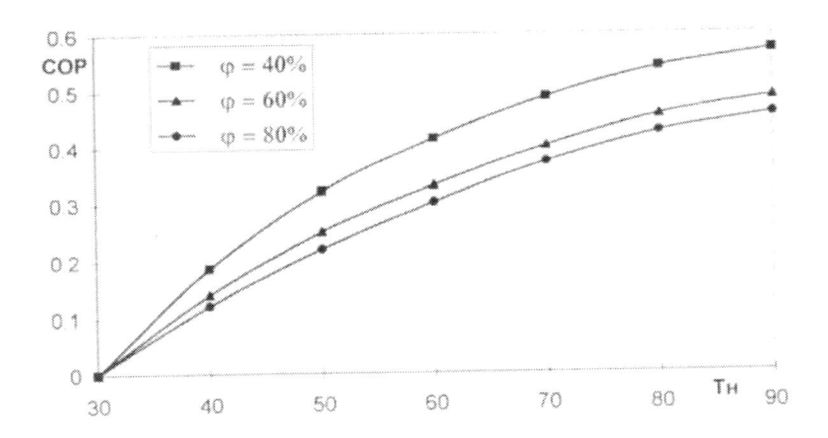

Figure 2.1. COP as a function of T_H for φ var.

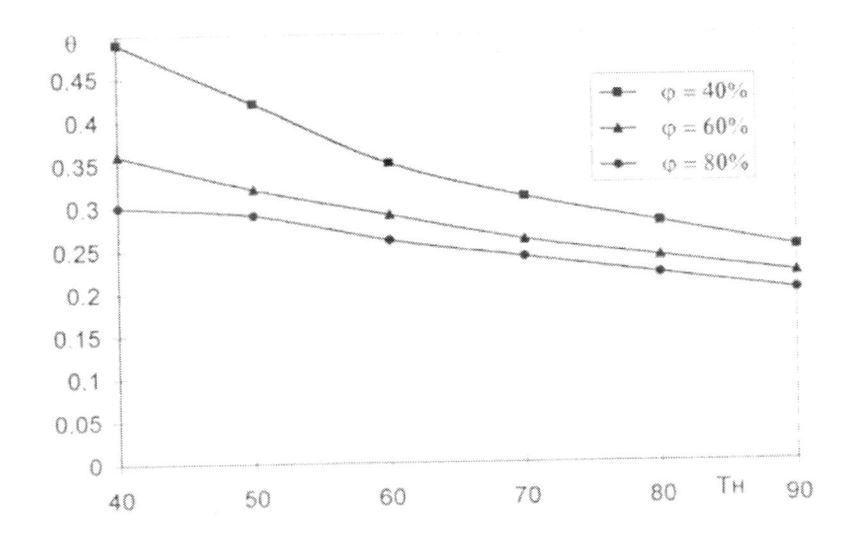

Figure 2.2. θ as a function of T_H for φ var.

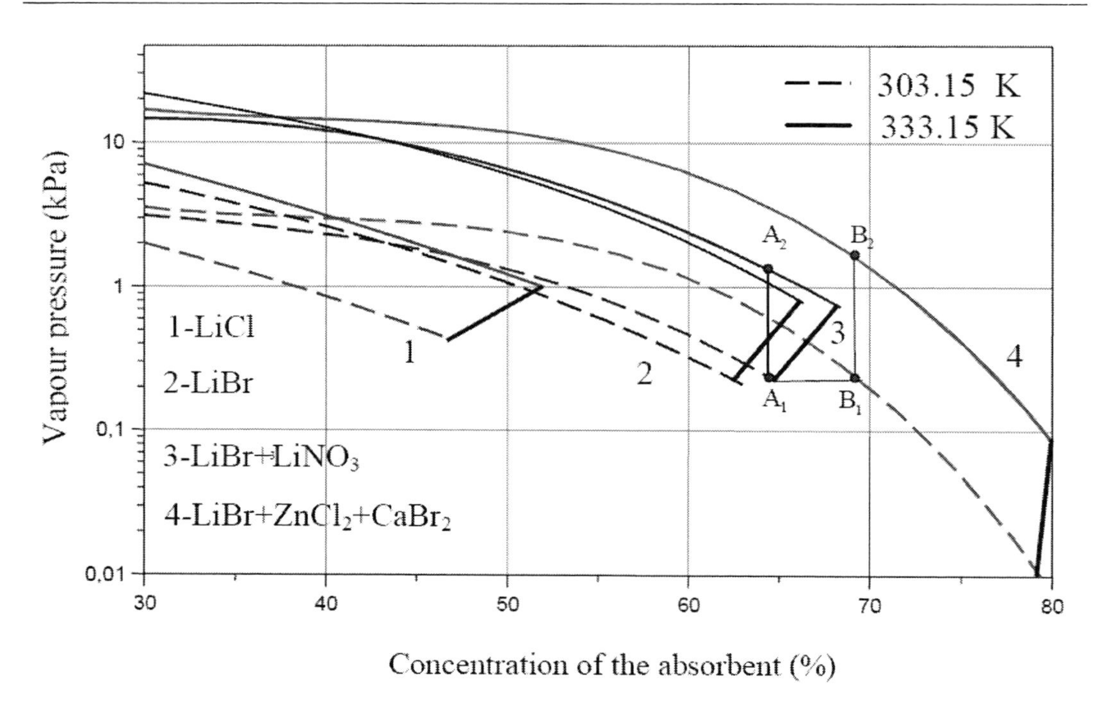

Figure 2.3. Vapour pressure over the absorbent surface. __________ crystallization line.

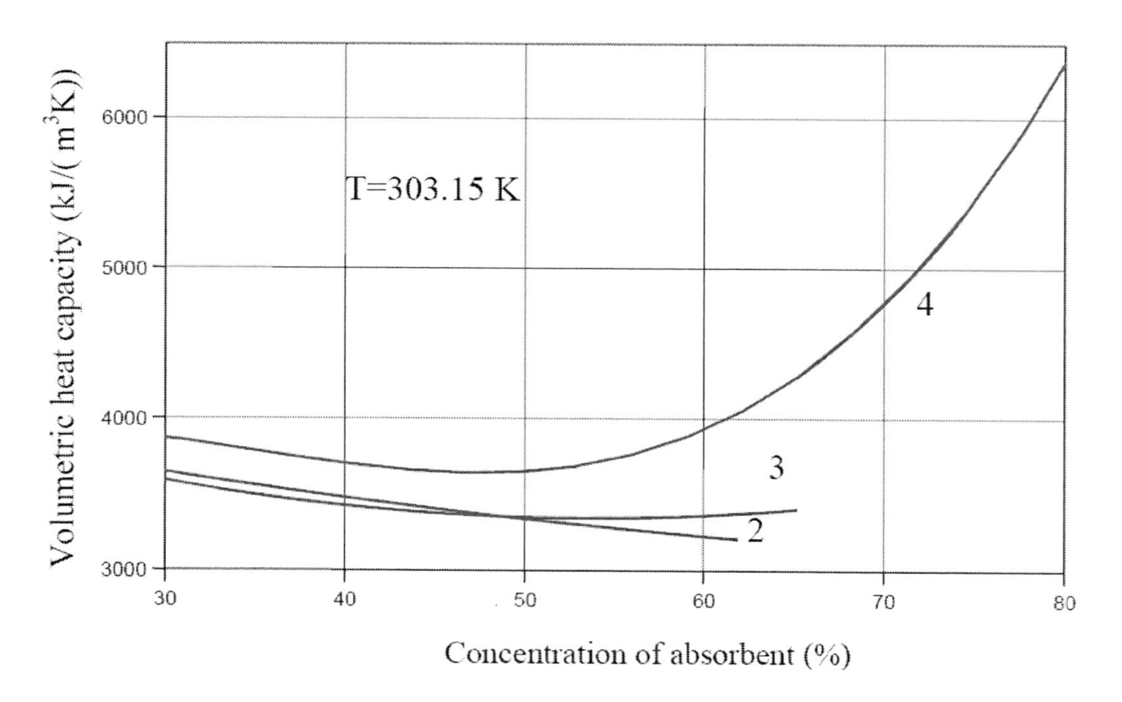

Figure 2.4. Volumetric heat capacity of absorbents. The nomenclature is the same as in figure 2.3.

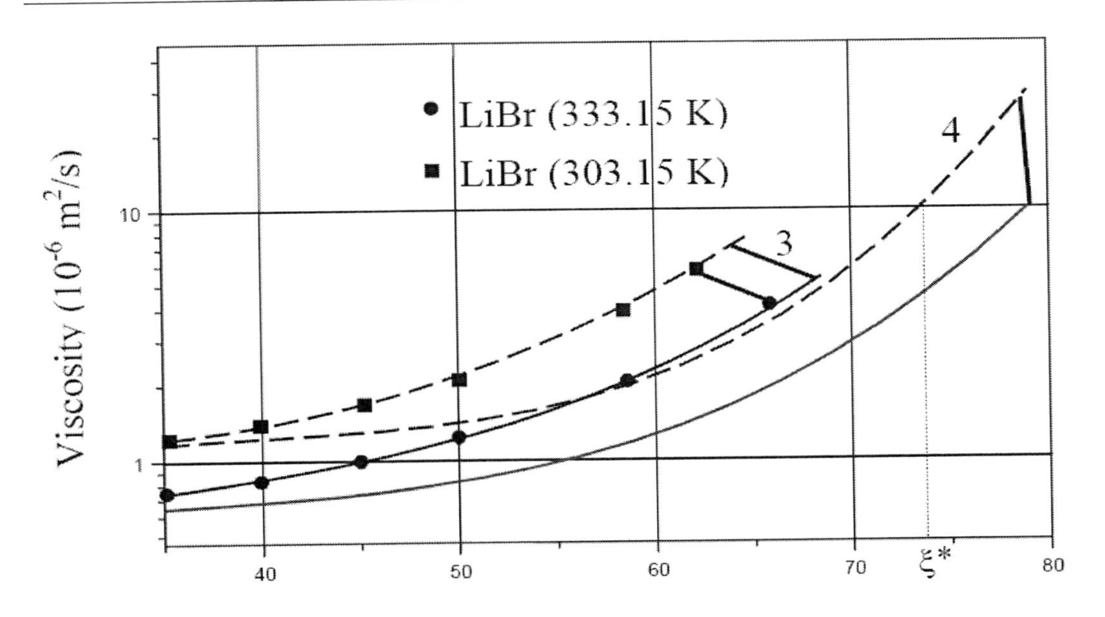

Concentration of the absorbent (%)

Figure 2.5. Kinematics viscosity of absorbents. The nomenclature is the same as in figure 2.3.

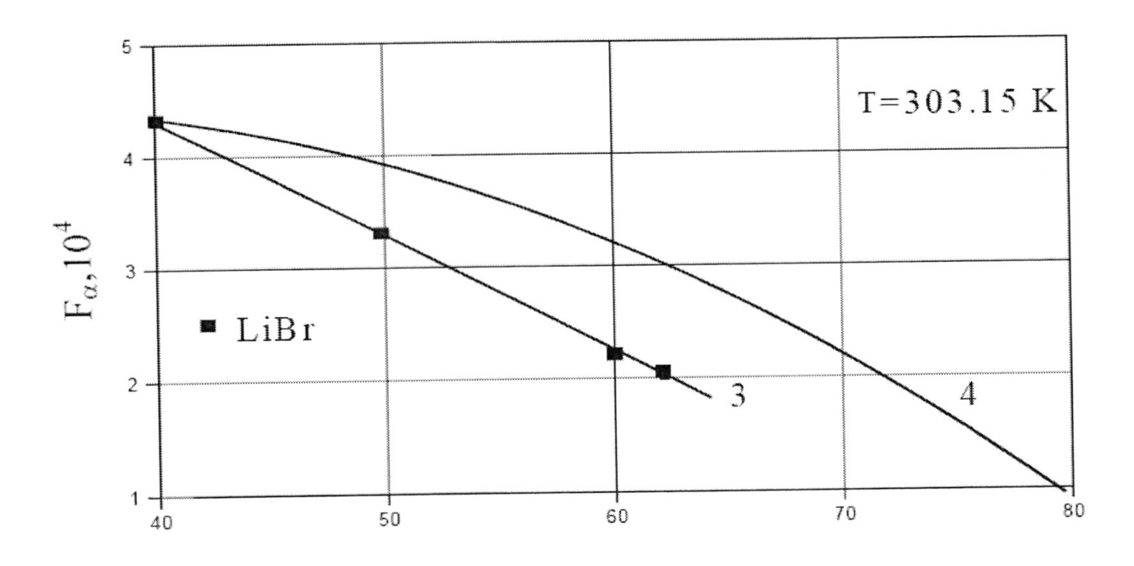

Concentration of the absorbent (%)

Figure 2.6. Heat transfer factor.

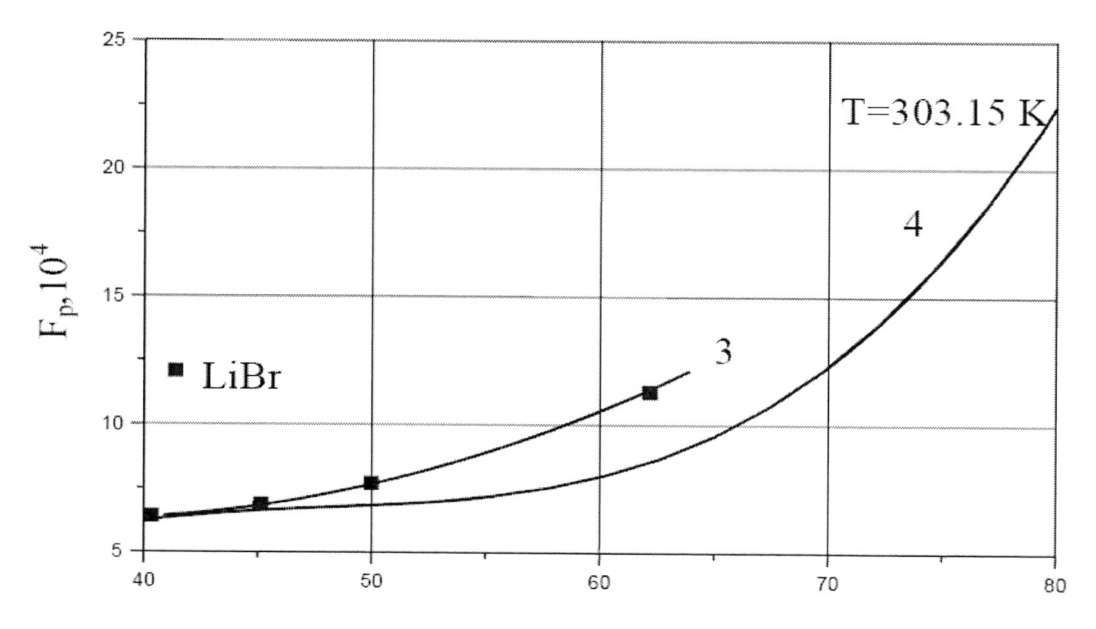

Concentration of the absorbent (%)

Figure 2.7. Pressure drop factor.

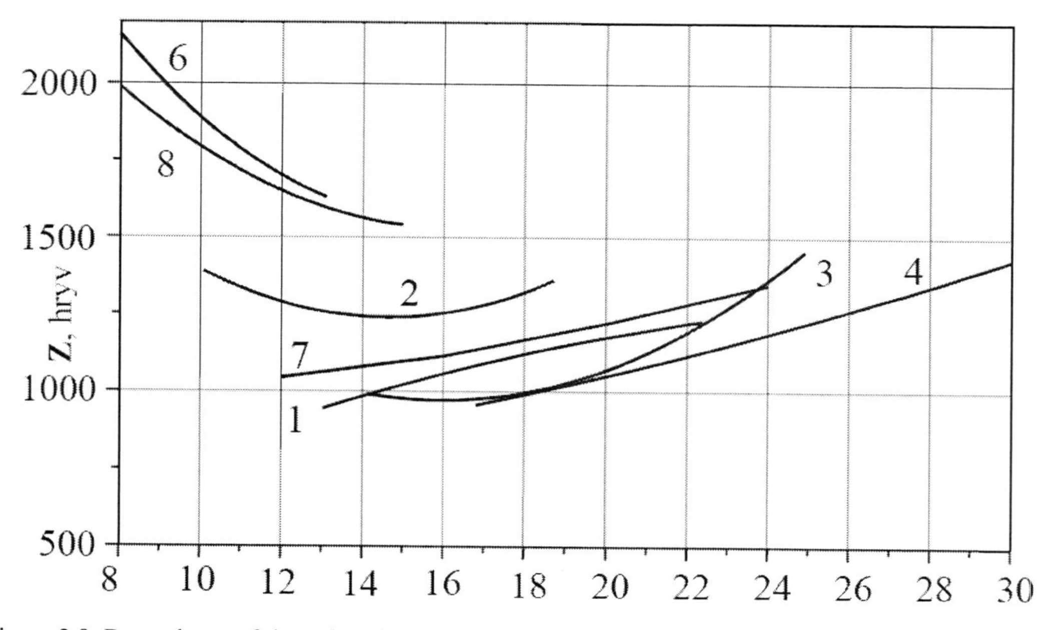

Figure 2.8. Dependence of the referred expenditures on the heat exchanger area. Plate type: 1-ПР-0.2; 2-ПР-0.3; 3-ПР-0.5Е; 4-ПР-0.5М; 6-Р11; 7-Р12; 8-Р13.

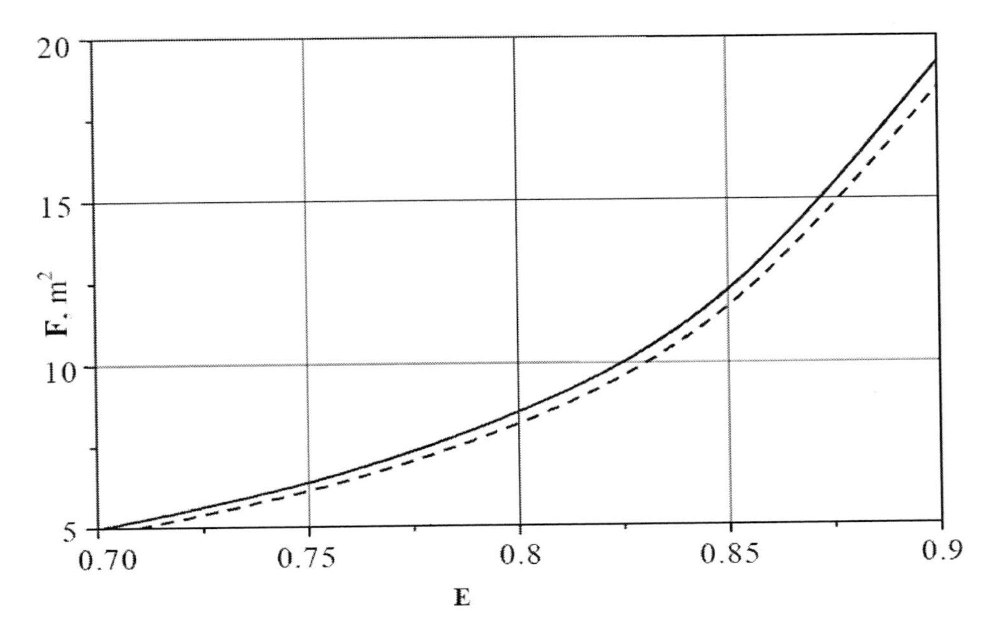

Figure 2.9. Dependence of the area of the heat exchanger with ПР-0.3 plates on its efficiency; ________ solution LiBr+; - - - - - - LiBr++.

2.3. Modelling of the Working Processes in the Main Components of Alternative Air-Conditioning and Cooling Systems (AACS, ARFS)

2.3.1. General Principles of Designing Heat-and-Mass Transfer Apparatuses for Alternative System

2.3.1.1. Main Heat-and-Mass Transfer Apparatuses of Alternative System

The low aerodynamic resistance during the transport of working substances (air and liquid flows) is the general requirement of heat-and-mass transfer apparatuses (HMA) for alternative systems, as the number of HMA and heat-exchangers (HEX) used in schemes is rather great. On the basis of the authors' long-time experience in the development (theory and experiment), production and maintenance of HMA [24, 81], the film-type HMA was chosen as the main universal solution for all HMA of alternative systems. It provides separate movement of gas and liquid flows with low aerodynamic resistance. The cross-current scheme of contact between gas and liquid was chosen as the most suitable when it is necessary to put together numerous HMA and HEX in a single unit of apparatus. The cross-current scheme provides a smaller number of flow turns nodes and lower Δp values as compared to the counter-current scheme of contacting, as it is characterized by higher values of maximum permissible velocity of the gas flow movement in the channels of the HMA packing-bed [82, 83] (loss of stability, liquid drops removal by the gas flow from the working zone).

The packing-bed of film cross-current HMA (figure 2.10) is formed by longitudinally corrugated (in the direction of the liquid film flow) thin-walled sheets of material, placed equidistant to each other. The surface regular roughness is used as the main method of intensifying processes of joint heat-and-mass transfer in HMA. Optimal values of roughness parameters are

$$k_{opt} = \frac{p}{e} = 8...14, \qquad (2.23)$$

where

p and e–are the spacing and the height of roughness fins, determined by the authors during theoretical and experimental studies of peculiarity of film flows of viscous liquid thin layers on surfaces of a complicated shape [10, 24, 27, 81, 82, 83].

Fins of roughness are evenly dispersed along the surface of longitudinally corrugated sheets of packing-bed (P and E are the spacing and the height of the main corrugation of pacing-bed sheets – figures 2.10, 2.20). Optimal values of the main corrugation parameters, densities of pacing-bed dimensions were determined by the authors. Within the framework of theoretical and experimental studies of the film two-phase flows stability problem the maximum permissible values of gas flow movement velocities in [10, 83] were determined. They were used in calculations below (Section 2.3.2).

The construction of all HMA, comprising the alternative systems under development, is unified (DBR, ABR, CTW, DEC). There is some difference in the construction of the IEC (figures 2.19 – 2.20), where the packing-bed is an alternation of «dry» and «wet» channels, designed for the movement of the main and auxiliary air flows, respectively. It is equally valid for the construction of DBR and ABR, shows in figure 1.8. The design scheme in figures 3.1 – 3.4 gives the simplified solution for these HMA with HEX 11 and 12 carried out. The variant of an absorber with internal evaporative cooling (figure 2.22), which was not discussed at this stage of the project, it of a special perspective interest. The problem in conformity with a film cross-flow heat-and-mass exchange apparatus, has been solved for the case when you take into account the real character of liquid film flowing along the longitudinally corrugated vertical surface forming the walls of «dry» and «wet» channels. In this case the film-jet flow with a characteristic distribution of "dry" and "wetted" sections of the surface (figure 2.20) is formed in "wet" channels of IEC and CTW. This circumstance must be taken into account in calculating the joint heat-and-mass transfer, and particularly in case of high heat conduction materials of the packing bed, as it occurs, for example, in IEC or in ABR with internal evaporative cooling (figure 2.22). These problems were studied by the authors in particular [24, 81, 82, 83], including the real character of the surface, the stability of film flowing at the interaction of laminar wave liquid film and the turbulent gas flow, the liquid film thickness, etc., in conformity with heat-and-mass transfer apparatuses included in alternative systems.

2.3.1.2. Heat Exchangers for Alternative Systems

Because of low temperature heads AACSs characterized by rather a large number of heat exchangers construction should provide an optimal combination of heat efficiency, comfort and reliability of using at low capital investments and operating costs. The use of plate heat transfer to take place between two liquids. It is due to their following advantages:

- high operating reliability at moderate pressure (up to 2 MPa) and temperatures (up to 150 °C);
- high value of the heat transfer surface per a volume unit;
- high degree of the liquid flow turbulization between plates. Even viscous liquids can be pumped through winding channels under turbulent conditions at low Reynolds numbers (for most types of plates the critical Re number value does not exceed 200). It results in the intensification of heat transfer between liquid flows and reduces the formation of most varieties of sediments to the minimum;
- simplicity of disassembly and assembly during manual cleaning the heat exchanger;
- variety of materials used to manufacture the plates. this circumstance makes it possible to avoid corrosion of the plates while using aggressive absorbents;
- light pressure losses. Coefficients of pressure loses in plate heat exchangers are much greater than those in pipe heat exchangers at the same Reynolds numbers [120]. However, the velocities of the flow between plates are much lower, and the length of plates, necessary to attain specified values of heat transfer units number, is much shorter than in pipes. Due to it pressure losses are usually lower under the same conditions than in flowing through pipes.

The construction of plate heat exchangers and geometry of plate surfaces are described in papers [61, 119, 120]. Tables 2.1 and 2.2 give the characteristics of most commonly types of plates.

Table 2.1. Characteristics of band-flow plates [7]

Characteristics of plates	P-5 (П-1)	P-11 (П-2)	P-12	P-13
Sizes: L×B, mm	800×225	1020×315	1170×420	890×280
Thickness of the wall, mm	1,2	1,25	1,3	1,25
Heat transfer surface, m^2	0,15	0,21	0,4	0,2
Mass, kg	1,4	3,2	5,6	3,0
Appropriate diameter of the channel, m	0,005	0,006	0,008	0,008
Cross-section area of the channel, m^2	0,0004	0,00075	0,0015	0,0008
Plate spacing, mm	2	3	4	3,5
Corrugation spacing along the flow, mm	23	22,5	22,5	23
Number of corrugations on the plate	7	7	7	7
Reduced lengh of the channel, m	0,7	0,8	1,0	0,8
Cross-section area of the angular hole, m^2	0,002	0,003	0,0045	0,002
Diameter of the connecting union, mm	50	50	76	50

While designing AACS it is convenient to define heat carrier temperatures at the outlet of the heat-exchanger, assuming its efficiency with the further determination of the heat exchange surface, and constructive parameters. The efficiency of the heat exchanger through which two heat carriers pass: heated (1) and cooled (2) is defined by the relation:

$$E = (G \cdot C_p)_1 \cdot (t_1^{out} - t_1^{in}) / (G \cdot C_p)_{min} \cdot (t_2^{in} - t_1^{in}) = \\ = (G \cdot C_p)_2 \cdot (t_2^{out} - t_2^{in}) \quad / (G \cdot C_p)_{min} \cdot (t_2^{in} - t_1^{in})' \tag{2.24}$$

where

$(G \cdot C_p)_{min}$ is the least of $(G \cdot C_p)_1$ and $(G \cdot C_p)_2$ values.

Table 2.2. Characteristic of net-flow plates [7]

Characteristics of plates	ПР-0,2	ПР-0,3	ПР-0,5E	ПР-0,5M
Sizes: L×B, mm	650×650	1370×300	1380×500	1380×550
Thickness of the wall, mm	1,2	1,0	1,0	1,0
Heat transfer surface, m²	0,2	0,3	0,5	0,5
Mass, kg	3,6	3,2	5,4	5,6
Appropriate diameter of the channel, m	0,0075	0,008	0,008	0,0096
Cross-section area of the channel, m²	0,0016	0,0011	0,0018	0,0024
Plate spacing, mm	3,8	4	4	5
Characteristics of plates	ПР-0,2	ПР-0,3	ПР-0,5E	ПР-0,5M
Corrugation spacing (mm): along the flow along the normal to corrugation	 20,8 18	 20,8 18	 18 16	 20,8 18
Number of corrugations on the plate	21	50	66	66
Reduced lengh of the channel, m	0,44	1,12	1,15	1,0
Cross-section area of the angular hole, m²	0,0082	0,0045	0,017	0,017
Diameter of the connecting union, mm	100	50	150	150
Slope of the corrugations to the vertical axis of symetry, °	60/30	60	60	60

The preliminary thermal and hydraulic calculations were carried out for heat exchangers with plates given in tables 2.1 and 2.2 for aqueous solutions of $LiBr+ZnCl_2+CaBr_2$, $LiBr+LiNO_3$ and LiBr when the absorbent concentration $\xi = 55\%$ and the mass flow rate of solutions G = 10000 kg/h for efficiency values from 0.7 to 0.9. The velocity of the heat carrier movement has been defined by the formula

$$\omega = \frac{G}{3600 \cdot \rho \cdot N \cdot f}, \tag{2.25}$$

where

N is the number of channels.

Let us find the Reynolds number changing, it necessary, the velocity to attain the turbulent conditions. The values of the Nusselt-number are calculated using the equation of similarity in a plate apparatus [61, 119]

$$Nu = 0.135 \cdot Re^{0.73} \cdot Pr^{0.43} \cdot \left(\frac{Pr}{Pr_{w1}} \right)^{0.25} . \tag{2.26}$$

The relations known were used to find coefficients of heat transfer from both heat carriers and the coefficient of heat exchange. Then the heat flow was defined

$$Q = G \cdot C_p \cdot \left(t^{out} - t^{in} \right) \tag{2.27}$$

and the total area of the apparatus too

$$F = \frac{Q}{K \cdot \Delta t} . \tag{2.28}$$

On the basis of performed calculations the number of plates was found and the lay-out diagram of the apparatus was made. The one pass pressure loss for plates of P-5 (П-1), P-11 (П-2), P-12 and P-13 marks was calculated by the equation of similarity [61]

$$Eu = 1850 \cdot Re^{-0.25} , \tag{2.29}$$

and for plates ПР-0.2, ПР-0.3, ПР-0.5E and ПР-0.5M

$$Eu = 1350 \cdot Re^{-0.25} . \tag{2.30}$$

The pressure loss in one pass was found by the relation

$$\Delta p = Eu \cdot \rho \cdot \omega^2 . \tag{2.31}$$

The technico-economic analysis and the selection of the optimal type of plates for the plate heat exchanger were made with the help of above-mentioned relations at the efficiency of E = 0.8 for solutions $LiBr+ZnCl_2+CaBr_2$. The cited expenditures were taken as a criterion of optimality. The calculation of the cited expenditures was carried out for some tentative prices for electric power, equipment, etc. With other prices quantitative results will change, the method of calculation and the qualitative consideration will be the same. The cited expenditures can be calculated by the following expression [51]

$$Z = K \cdot E_n + C_a + C_r + \tau \cdot C_e \cdot W_{pump} , \tag{2.32}$$

where

K is the cost of a heat transfer apparatus, hryv (K = 135·F, hryv.);

E_n is the normative coefficient of the capital outlays efficiency, (E_n = 0.15);

C_a I the part of the heat exchanger cost, allocated annually for its depreciation (C_a = 0.125·K);

C_r is the same for its current repair (C_r=0.45·K);

τ is the number of the heat exchanger operation hours a ear (τ = 8000);

C_e is the electric power cost (C_e=0.15 hryv/(kWt·h));

W_{pump} is the pump capacity depending on the pressure loss, mass flow rate and number of passes in the heat exchange apparatus ($W_{pump} = W_1 \cdot n$, where W_1 is the pump capacity at n = 1).

W_1 was calculated by the formula

$$W_1 = \frac{G \cdot \Delta p}{\rho \cdot 3600 \cdot \eta_{pump}}, \tag{2.33}$$

Where:

η_{pump} is the coefficient of performance of the pump (η_{pump}= 0.6).

The calculation results are given in figure 2.8. The representation of relation Z(F) in the form of continuous lines is to some extent tentative, as the heat exchange surface is a discrete value multiple of the area for one plate of the said type. Such a representation has been used to make the analysis visual and convenient. The use of ПР-0.2 plates results in expenditures indicated, which are beyond the limits of the figure. It follows from the figure that the use of ПР-0.5Е plates (curve 3) allows the minimization of the said expenditures while the surface of heat exchange is relatively small. Figure 2.9 gives the dependence of the heat exchanger surface on its efficiency for solutions of LiBr+ (weak and strong solution) and LiBr++. As it should be expected, according to ideas given in Section 2.2, at the same efficiency the solution of LiBr++ requires a smaller surface than LiBr+, though this difference is comparatively small (~6% at E = 0.8).

For heat transfer to make place between two gas flows it is expedient to plate-fin heat exchangers. The thermal and hydraulic calculations for these heat exchangers have been carried out by the method described in paper [6].

Conclusions

- The thermal and hydraulic calculations have been made for heat exchangers being components of AACS. On the basis of the technico-economic analysis performed the type of heat transfer plates corresponding to the minimum of the said expenditures has been selected.

- It has been shown, that the use of the LiBr++ solution results in smaller heat exchange surface as compared to the rest of the absorbents discussed in the report. This conclusion agrees with the results of the analysis of working bodies' thermophysical properties.

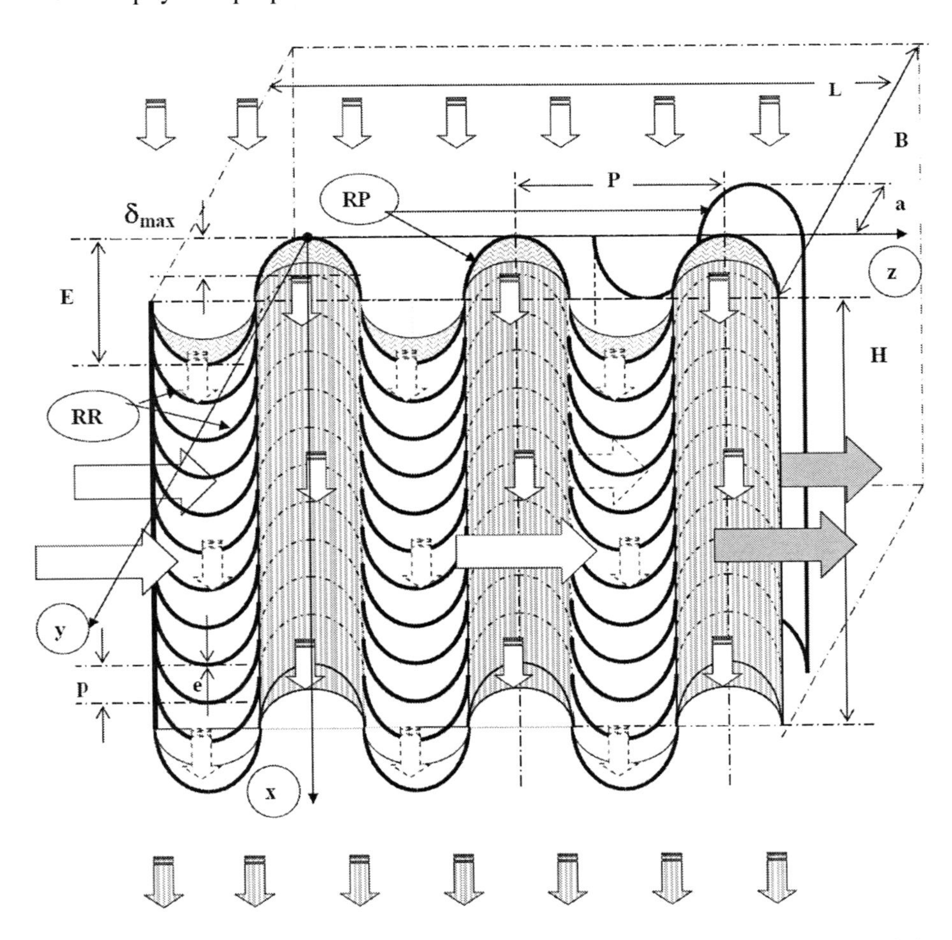

Figure 2.10. Schematic of gas and liquid cross flows (film-jet liquid flow) in channels of the regular packing bed (RP) of the head-and-mass exchanging apparatus (HMA), which is formed by longitudinally corrugated sheets (in the direction of the liquid flow) with regular surface roughness (RR) located equidistantly. Notation: ⇨ gas; ⇩ liquid; P, E are the pitch and the height of the main corrugation of sheets; p, e – the pitch and the height of the regular roughness of the sheet surfaces; H, L, B are the height, the length and the width of the packing bed of the heat exchanging apparatus; a is the distance between sheets; δ_{max} is the liquid film thickness.

2.3.2. Peculiar Features of Wave Film Flows of a Viscous Liquid Along the Longitudinally Corrugated Vertical Surface of Heat-and-Mass Transfer Apparatuses

2.3.2.1. Wave Flow of a Viscous Liquid Thin Layer Along the Flat Vertical Surface Having Regular Roughness

To calculate HMA it is necessary to know flow wave parameters as they are included in calculation formulas for the determination of mass transfer coefficients [10, 24, 77-79, 81].

Therefore the theoretical description of a liquid film flow along the surface with RR, when the interaction with a gas flow is taken into account, is timely. Packing beds of HMA under discussion (figure 2.10) consist of equally separated from each other flat or corrugated along the liquid flow sheets with counter flow RR made in the form of ordered bulges having specified configuration. The section of such a surface perpendicular to the plane, will be a flat line whose equation is a periodic function, moreover, a continuous one. It can be given in the form of a Fourier series [81]

$$f(z) = \frac{a_0}{2} + \sum_{n=1}^{\infty} \left(a_n \cdot \cos\frac{n\pi z}{e} + b_n \cdot \sin\frac{n\pi z}{e} \right),$$

where e is the function half-period;

$$a_0 = \frac{1}{e} \cdot \int_{-e}^{e} f(z) dz;$$

$$a_n = \frac{1}{e} \cdot \int_{-e}^{e} f(z) \cdot \cos\frac{n\pi z}{e} dz;\ b_n = \frac{1}{e} \cdot \int_{-e}^{e} f(z) \cdot \sin\frac{n\pi z}{e} dz;\ (n=1,2,3,\ldots).$$

The movement of a non-compressed liquid in the general case is described by the Navier-Stocks equations and that of continuity, connecting components of velocity and pressure vectors. When a thin layer of liquid is flowing down the flat vertical surface with RR by gravity (figure 2.10), these equations will take the form [81]:

$$\begin{cases} \frac{\partial w_x}{\partial t} + w_x \cdot \frac{\partial w_x}{\partial x} + w_z \cdot \frac{\partial w_x}{\partial z} = -\frac{1}{\rho} \cdot \frac{\partial P_L}{\partial x} + \nu \cdot \left(\frac{\partial^2 w_x}{\partial x^2} + \frac{\partial^2 w_x}{\partial z^2} \right), \\ \frac{\partial w_z}{\partial t} + w_x \cdot \frac{\partial w_z}{\partial x} + w_z \cdot \frac{\partial w_z}{\partial z} = -\frac{1}{\rho} \cdot \frac{\partial P_L}{\partial z} + \nu \cdot \left(\frac{\partial^2 w_z}{\partial x^2} + \frac{\partial^2 w_z}{\partial z^2} \right) + g, \\ \frac{\partial w_x}{\partial x} + \frac{\partial w_z}{\partial z} = 0, \end{cases} \quad (2.34)$$

where t – time.

Boundary condition:

1. Adhesion condition

$w_x = w_z = 0$; at x=f(z);

Kinematics condition on the free surface

$$w_x = \frac{\partial h}{\partial t} + w_z \cdot \frac{\partial h}{\partial z}, \text{ at } x{=}h(z,t);$$

Dynamic condition of tangential stresses

$$P_{n\tau} = \tau_r;\; -4 \cdot \mu \cdot \frac{h_z^{1}}{1+h_z^{12}} + \mu \cdot \frac{1-h_z^{12}}{1+h_z^{12}} \cdot \left(\frac{\partial w_z}{\partial x} + \frac{\partial w_x}{\partial z} \right) = \tau_g, \text{ at } x{=}h(z,t);$$

Dynamic condition of normal stresses equality

$$P_{nn} = P_{nn}^{1};$$

$$-P_L - 2 \cdot \mu \cdot \frac{1-h_z^{12}}{1+h_z^{12}} \cdot \frac{\partial w_z}{\partial z} - 2 \cdot \mu \cdot \frac{h_z^{1}}{1+h_z^{12}} \cdot \left(\frac{\partial w_z}{\partial x} + \frac{\partial w_x}{\partial z} \right) = -P_g + \frac{\sigma \cdot h_z^{11}}{\left(1+h_z^{12}\right)^{3/2}},$$

at x=h(z,t).

The film HMT the ranges of input parameters changes are such that $\varepsilon = e/\delta_N < 1,\quad \alpha = \delta_N / p < 1 \quad (\alpha^2 << I),\quad 1 < (\alpha \cdot Re_{fL}) \sim \alpha^2$. In this case the wave condition is defined only by stagnant waves, and running waves will be slight ripples spreading over the surface of stagnant waves.

The procedure of making dimensionless is used:

$$\bar{x} = \frac{x - f(z)}{\delta_N},\; f(z) = \max|f(z)| \cdot f_1(z) = e \cdot f_1(z),$$

$$\frac{f(z)}{\delta_N} = \frac{e}{\delta_N} \cdot f_1(z) = \varepsilon \cdot f_1(z),$$

$$k = \frac{p}{e},\; \bar{z} = \frac{z}{p},\; \bar{w}_z = \frac{w_z}{w_0},\; \bar{w}_x = \frac{w_x \cdot p}{\delta_N \cdot w_0},\; \bar{P}_L = \frac{P_L}{\rho \cdot w_0^2},\; \bar{P}_g = \frac{P_g}{\rho \cdot w_0^2},$$

$$\bar{h} = \frac{h}{\delta_N},\; \bar{\delta} = \frac{\delta}{\delta_N} = \frac{h-f}{\delta_N} = \bar{h} - \varepsilon \cdot f_1(z),\; \bar{t} = \frac{t \cdot w_0}{p},\; \bar{\tau} = \frac{\tau_r \cdot \delta_N}{\mu \cdot w_0},$$

$$Re_{fL} = \frac{w_0 \cdot \delta_N}{\nu},\; Fr = \frac{w_0^2}{g \cdot p},\; We = \frac{\rho \cdot p^2 \cdot w_0^2}{\sigma \cdot \delta_N}.$$

It resulted in obtaining the following equations with an accuracy to terms of the α^2 order:

$$\begin{cases} \dfrac{\partial P_L}{\partial x} = 0, \\ w_x \cdot \dfrac{\partial w_z}{\partial x} + w_z \cdot \dfrac{\partial w_z}{\partial z} = -\dfrac{\partial P_L}{\partial z} + \dfrac{1}{Fr} + \dfrac{1}{\alpha \cdot Re_{fL}} \cdot \dfrac{\partial^2 w_z}{\partial x^2}, \\ \dfrac{\partial w_x}{\partial x} + \dfrac{\partial w_z}{\partial z} = 0. \end{cases} \quad (2.35)$$

$$w_x = w_z = 0, \text{ at } x = \varepsilon \cdot f_1(z); \; w_x = \frac{\partial h}{\partial t} + w_z \cdot \frac{\partial h}{\partial z}, \quad \text{at } x = h(z,t);$$

$$\frac{\partial w_z}{\partial x} = \tau_g, \quad \text{at } x = h(z); \; -P_L = -P_g + \frac{1}{We} \cdot h''_z, \quad \text{at } x = h(z,t). \quad (2.36)$$

From the equation of continuity it follows: $w_x = -\int\limits_{\varepsilon f_1(z)}^{x} \frac{\partial w_z}{\partial z} dx$.

Then we shall have

$$\begin{cases} -\dfrac{\partial w_z}{\partial x} \cdot \int\limits_{\varepsilon f_1(z)}^{x} \dfrac{\partial w_z}{\partial z} dx + w_z \cdot \dfrac{\partial w_z}{\partial x} = \dfrac{1}{Fr} + \dfrac{1}{We} \cdot \dfrac{\partial^3 h}{\partial z^3} + \dfrac{1}{\alpha \cdot Re_{fL}} \cdot \dfrac{\partial^2 w_z}{\partial x^2}, \\ \dfrac{\partial}{\partial z} \int\limits_{\varepsilon f_1(z)}^{h} w_z dx = 0 \end{cases} \quad (2.37)$$

$$w_z = 0, \text{ at } x = \varepsilon \cdot f_1(z); \; \frac{\partial w_z}{\partial x} = \tau, \text{ at } x = h(z).$$

Let us substitute variables by formula $\xi = x - \varepsilon \cdot f_1(z)$ by averaging the firs equation by layer thickness (i.e. integrating both of its parts by ξ from 0 to δ and dividing by δ), then we shall get:

$$\begin{cases} -w_z\big|_{\xi=\delta}\cdot\int\limits_0^{\delta}\frac{\partial w_z}{\partial z}d\xi+2\cdot\int\limits_0^{\delta}w_z\cdot\frac{\partial w_z}{\partial z}d\xi=\frac{\delta}{Fr}+\frac{\delta}{We}\cdot\frac{\partial^3 h}{\partial z^3}+\frac{1}{\alpha\cdot Re_{fL}}\cdot\int\limits_0^{\delta}\frac{\partial^2 w_z}{\partial \xi^2}d\xi, \\ \frac{\partial}{\partial z}\int\limits_0^{\delta}w_z d\xi=0 \end{cases} \quad (2.38)$$

$$w_z=0,\ \text{at}\ \xi=0;\ \frac{\partial w_z}{\partial \xi}=\tau,\ \text{at}\ \xi=\delta(z).$$

According to [10, 81]:

$$\int\limits_0^{\delta}w_z d\xi = w\cdot\delta=1. \quad (2.39)$$

conditions (2.38 – 2.39) w_z it is necessary to approximate by the following quadratic function

$$w_z=\frac{-6+3\cdot\tau\cdot\delta^2}{4\cdot\delta^3}\cdot\xi^2+\frac{6-\tau\cdot\delta^2}{2\cdot\delta^2}\cdot\xi. \quad (2.40)$$

Substitution of (2.40) in the first equation of system (2.37) results in such an equation with respect to the dimensionless film thickness:

$$\delta^3\cdot\delta'''+\frac{48-2\cdot\tau\cdot\delta^2-\tau^2\cdot\delta^4}{40}\cdot We\cdot\delta'+\left(\frac{3\cdot We}{\alpha\cdot Re}+\varepsilon\cdot f_1'''\right)\cdot\delta^3-\frac{6-3\cdot\tau\cdot\delta^2}{2\cdot\alpha\cdot Re_{fL}}\cdot We=0. \quad (2.41)$$

Supposing that $\delta'=\delta_1,\ \delta''=\delta_1'=\delta_2$ we get

$$\begin{cases} \delta'=\delta_1, \\ \delta_1'=\delta_2, \\ \delta_2'=-\frac{48-2\cdot\tau\cdot\delta^2-\tau^2\cdot\delta^4}{40\cdot\delta^3}\cdot We\cdot\delta_1-\left(\frac{3\cdot We}{\alpha\cdot Re_{fL}}-\varepsilon\cdot f_1'''\right)+\frac{6-3\cdot\tau\cdot\delta^2}{2\cdot\alpha\cdot Re_{fL}\cdot\delta^3}\cdot We. \end{cases}$$

The periodic solution of this system has been found by the Kutta-Merson method at boundary conditions $\delta(0)=\delta(p),\ \delta_1(0)-0,\ \delta_2(0)=0$.

$$f(z) = \begin{cases} 0, & \text{if } 0 \le z \le (P-T)/2 \\ e \cdot \cos\dfrac{\pi \cdot (z - p/2)}{T}, & \text{if } (P-T)/2 < z < (P+T)/2 \\ 0, & \text{if } (P+T)/2 \le z \le P \end{cases} \tag{2.42}$$

From the obtained solution it follows that the availability of gas counter-flow with velocities up to 5-6 m/s does not practically affect the average film thickness, the flow is stable, the flooding of the apparatus will not take place.

2.3.2.2. The Flow of Viscous Liquid Thin Layer Along the Longitudinally-Corrugate Surface with Regular Roughness (RR)

It is known that if the packing bed component is a sheet longitudinally corrugated along the flow (figure 2.10), the liquid is stored and flows along riffles of corrugations, rather than covers the whole sheet, i.e. we have a wet-dry condition of contacting which, under the conditions of packing bed sheets heat-conduction, results in the necessity to take into account real transfer surfaces [24]. Along the riffle the corrugated sheet with RR (figure 2.10) a stream of liquid flows by gravity. In the set up flows of liquid and gas the weight force is balanced by the tangential stress on the channel wall and free surface of the liquid (the surface wettability effect and the force of the surface tension are not taken into consideration):

$$\rho \cdot g \cdot S_f = P_W \cdot \lambda_L \cdot \frac{\rho_L \cdot w_L^2}{2} + P_M \cdot \lambda_g \cdot \frac{\rho_g \cdot w_g^2}{2}. \tag{2.43}$$

Taking into account that $w_L = G_L / S_L$, $w_g = G_g / S_g$, we shall get:

$$\rho \cdot g \cdot S_f = P_W \cdot \lambda_L \cdot \frac{\rho_L \cdot G_L^2}{2 \cdot S_L^2} + P_M \cdot \lambda_g \cdot \frac{\rho_g \cdot G_g^2}{2 \cdot S_g^2}. \tag{2.44}$$

As $S_L << S_g$, then the total area between neighboring sheets by the length, equal to the period, can be approximately taken as S_g, i.e. $S_g \approx P \times H$. For the parabolic profile of corrugation its equation for a half-period has the form: $y = \left(8 \cdot E / P^2\right) \cdot x^2$ (as designated in Figure 2.10), for the simplification we assume that $(8 \cdot E)/P^2 = b$, then we get $y = b \cdot x^2$. The maximum thickness of the liquid in the channel is δ_{max}, then

$$S_L = 2 \cdot \left(\delta_{max} \cdot \sqrt{\frac{\delta_{max}}{b}} \cdot \int_0^{\sqrt{\frac{\delta_{max}}{b}}} b \cdot x^2 dx \right) = \frac{4}{3} \cdot \delta_{max} \cdot \sqrt{\frac{\delta_{max}}{b}} \, ; \; P_M = 2 \cdot \sqrt{\frac{\delta_{max}}{b}} \, ;$$

$$P_W = 2 \cdot \int_0^{\sqrt{\frac{\delta_{max}}{b}}} \sqrt{1 + 4 \cdot b^2 \cdot x^2} dx = -\frac{1}{2 \cdot b} - \left[\ln\left(2 \cdot \sqrt{b \cdot \delta_{max}} + \sqrt{1 + 4 \cdot b \cdot \delta_{max}}\right) + \right.$$
$$\left. + 2 \cdot \sqrt{b \cdot \delta_{max}} \cdot \sqrt{1 + 4 \cdot b \cdot \delta_{max}} \right]$$

Let us introduce $\nabla = 2 \cdot \sqrt{b \cdot \delta_{max}}$, then $S = \nabla^3 / \left(6 \cdot b^2\right)$

$$P_W = -\frac{1}{2 \cdot b} \cdot \left[\ln\left(\nabla + \sqrt{1 + \nabla^2} \right) + \nabla \cdot \sqrt{1 + \nabla^2} \right], \; P_M = -\frac{\nabla}{b} ;$$

$$\nabla^9 = 5.51 \cdot \lambda_L \cdot b^5 \cdot G_L^2 \cdot \left[\ln\left(\nabla + \sqrt{1 + \nabla^2} \right) + \nabla \cdot \sqrt{1 + \nabla^2} \right] + \quad . \quad (2.45)$$
$$+ 0.306 \cdot \lambda_g \cdot b \cdot \frac{\rho_g}{\rho} \cdot \left(\frac{G_g}{P \cdot H} \right)^2 \cdot \nabla^7$$

For *the laminar-wave jet flow* $\lambda_L = \frac{12.3}{Re}$, for the turbulent one $\lambda_L = 0.012 \cdot \left(\frac{e}{d_e} + \frac{68}{Re} \right)^{0.25}$, λ_g was experimentally found in paper [24]. Then (2.45) for the laminar-wave condition of the film flow will take the form:

$$\nabla^9 = 8.4719 \cdot \nu \cdot b^4 \cdot G_L \cdot \left[\ln\left(\nabla + \sqrt{1 + \nabla^2} \right) + \nabla \cdot \sqrt{1 + \nabla^2} \right]^2 + \quad , \quad (2.46)$$
$$+ 0.306 \cdot \lambda_g \cdot b \cdot \frac{\rho_g}{\rho} \cdot \left(\frac{G_g}{P \cdot H} \right)^2 \cdot \nabla^7$$

for *the turbulent flow*

$$\nabla^{9} = 5.56 \cdot 10^{-2} \cdot b^{4.75} \cdot G_L^2 \cdot \left[\frac{1.5 \cdot b^2 \cdot e}{\nabla^3} + \frac{17 \cdot \nu}{G_L}\right]^{0.25} \cdot \left[\ln\left(\nabla + \sqrt{1+\nabla^2}\right) + \nabla \cdot \sqrt{1+\nabla^2}\right]^{1.25} + \qquad (2.47)$$

$$+ 0.306 \cdot \lambda_g \cdot b \cdot \frac{\rho_g}{\rho} \cdot \left(\frac{G_g}{P \cdot H}\right)^2 \cdot \nabla^7$$

The solution of these equations with respect to ∇ enables us to calculate all the magnitudes of jet-film flow parameters, we are interested in, namely δ_{max}, P_W, P_M, F_M, F_H, $\alpha_{HMT} = F_M / F_H$ depending on the liquid and gas flow rates for any values of the main corrugation parameters of the surface with RR.

2.3.2.3. The Stability of Separated Two -Phase Flow in the Flat Channel of the HMTA with a Regular Packing Bed

In film HMA-s for optimal progress of transfer processes it is necessary to provide the condition of active hydrodynamic phase interaction. However, the loads on the part of the liquid and gas should not be so large that intensive drop carrying-over and «flooding» of the apparatus could take place. The available results of theoretical and experimental investigations of maximum loads are contradictory. The authors have offered a mathematical model for determining the stability of the two-phase flow in the vertical channel, over whose surface the liquid film is flowing, when the gas flow is available, and the conditions of stationary regime of heat-and-mass transfer. The result is generalized for the case of channels the surface of which has RR [24, 81]. As it is shown in paper [81] for long waves the contribution to forces of fluctuating tangential stress which strive to change the wave shape, is mach less than that of normal stress. It makes possible in investigating the stability of the liquid film surface to regard liquid and gas as ideal media and to use the theory of potential flows. Equations of continuity and Koshi-Lagrange integrals for liquid and gas have the following form:

$$\frac{\partial^2 \varphi_i}{\partial x^2} + \frac{\partial^2 \varphi_1}{\partial y^2} = 0, \qquad (2.48)$$

$$\frac{\partial \varphi_i}{\partial t} + \frac{1}{2} \cdot \left[\left(\frac{\partial \varphi_i}{\partial x}\right)^2 + \left(\frac{\partial \varphi_i}{\partial y}\right)^2\right] - g \cdot x + \frac{P_i}{\rho_i} = f(t). \qquad (2.49)$$

Here the i=1 index refers to liquid, 2 – to gas, $\varphi_i = \varphi_i(t, x, y)$ are potential of velocities, defined by equalities $w_{Li} = \partial\varphi_i / \partial x$; $w_{gi} = \partial\varphi_i / \partial y$.

Boundary conditions are:

$$\frac{\partial \varphi_i}{\partial y} = 0, \text{ at } y = 0; \quad \frac{\partial \varphi_2}{\partial y} = 0, \text{ at } y = r;$$

$$\frac{\partial \varphi_1}{\partial y} = \frac{\partial h}{\partial t} + w_L \cdot \frac{\partial h}{\partial x}, \quad \text{at } y = h(t,x);$$

$$\frac{\partial \varphi_2}{\partial y} = \frac{\partial h}{\partial t} + w_g \cdot \frac{\partial h}{\partial x}, \quad \text{at } y = h(t,x).$$

The dynamic condition with forces of surface tension taken into account is:

$$P_1 - P_2 = \frac{\sigma \cdot h''_{xx}}{\left(1 + h_x^{12}\right)^{3/2}}, \tag{2.50}$$

where

y=h(t,x)	–	is an equation of free liquid surface;
w_L, w_g	–	are average, by the flow rate, velocities of liquid and gas,

respectively.

Let the equation of the free film surface have the form

$$h(t,x) = \delta_N + \alpha \cdot e^{i \cdot k \cdot (x - c \cdot t)}, \quad c = c_1 + i \cdot c_2 \text{ (i is an imaginary unit).}$$

Supposing that the length of the mach wave (perturbation wave) is large, i.e. $k \cdot \delta_N << 1$, we get $(h'_x)^2 << 1$, and the boundary condition (2.50) will be written:

$$P_1 - P_2 = \sigma \cdot k^2 \cdot h, \quad \text{at } y = h(t,x). \tag{2.51}$$

Velocity will have the form:

$$\varphi_1(t,x,y) = w_L \cdot x + \Psi_1(y) \cdot e^{i \cdot k(x - c \cdot t)},$$

$$\varphi_2(t,x,y) = w_g \cdot x + \Psi_2(y) \cdot e^{i \cdot k(x - c \cdot t)}.$$

Substituting these expression in equations (2.48) – (2.49) and using boundary conditions (2.50) – (2.51) we shall get:

$$\varphi_1(t,x,y) = w_L \cdot x + i \cdot a \cdot (w_L - c) \cdot \frac{h \cdot k \cdot y}{s \cdot h \cdot k \cdot \delta_N} \cdot e^{i \cdot k(x - c \cdot t)},$$

$$\varphi_2(t,x,y) = w_g \cdot x + i \cdot a \cdot (w_g - c) \cdot \frac{- \cdot h \cdot k \cdot (z - y)}{s \cdot h \cdot k \cdot (z - \delta_N)} \cdot e^{i \cdot k \cdot (x - c \cdot t)}.$$

The dispersion relationship is obtained by excluding pressure P_i from (2.49, 2.51). With an accuracy to terms of the second-order infinitesimal it has the form:

$$P \cdot (w_L - -)^2 + G_L \cdot (w_g - c)^2 = \sigma \cdot k, \tag{2.52}$$

where

$$P = \rho_1 \cdot c \cdot th(k \cdot \delta_N),\ G_L = \rho_2 \cdot c \cdot th(k \cdot (z - \delta_N)).$$

Separating real and imaginary parts in (2.52) we shall get:

$$\begin{cases} P \cdot \left[(w_L - c_1)^2 - c_2^2\right] + G_L \cdot \left[(w_g - c_1)^2 - c_2^2\right] = \sigma \cdot k, \\ c_1 = (P \cdot w_L + G_L \cdot w_g)/(P + G_L). \end{cases} \tag{2.53}$$

Excluding c_1 from equations (2.53) we have:

$$c_2^2 = \left(\frac{P \cdot G_L}{P + G_L} \cdot (w_L - w_g) - \sigma \cdot k \right) \Big/ (P + G_L).$$

Hence, if the velocity of the gas w_g satisfies the condition

$$w_g > w_L + \left[(P + G_L) \cdot \sigma \cdot k / (P \cdot G_L)\right]^{1/2},$$

then $c_2 > 0$ and the surface of phase separation is unstable, long-wave perturbations will exponentially increase as time goes on. When $c_2^2 < 0$, the surface of the liquid will be stable. For the neutral curve ($c_2 = 0$) the maximum velocity of the gas, after which instability begins, is determined by the formula

$$w_g^* = w_L + \left[(P + G_L) \cdot \sigma \cdot k / (P \cdot G_L) \right]^{1/2}$$

For the counter flow, when $\frac{G_L}{P} << 1$, we get

$$w_g^* = \left[\frac{\sigma \cdot k}{\rho_2} \cdot th(k \cdot (r - \delta_N)) \right]^{1/2} - w_L .$$

Let us write the wave number in the form $k = \xi / \delta_N$, $(\xi = const)$ then

$$w_g^* = \left[\frac{\sigma \cdot \xi}{\rho_2 \cdot \delta_N} \cdot th\left(\xi \cdot \left(\frac{r}{\delta_N} - 1 \right) \right) \right]^{1/2} - w_L . \qquad (2.54)$$

Constants ξ in equation (2.54) is selected on the basis of the generalized criterion equation, defining maximum loafs of the gas and liquid phase with the counter-flow available and taking into account the length and the diameter of the channel. The equation is given in paper [24], it is a generalization of a lot of experiments and has the form:

$$w_g^* = 1.346 \cdot \frac{\nu^2}{d} \cdot Re_L^{0.38} \cdot We^{0.113} \cdot \left(\frac{\rho_2}{\rho_1} \right)^{0.513} \cdot \left(\frac{\mu_1}{\mu_2} \right)^{0.955} \cdot \left(\frac{d}{\delta_N} \right)^{1.628} \cdot f(H, d), \qquad (2.55)$$

$$f(H, d) = \frac{H}{d} \cdot (0.38 \cdot d - 0.015) + 0.07 \cdot d^{-0.8} \qquad (2.56)$$

The relative root-mean-square deviation of experimental data from relationship (2.55) does not exceed 4%. According to [24] we specify ξ in the following form:

$$\xi = \gamma \cdot We^{\alpha_1} \cdot \left(\frac{\mu_1}{\mu_2} \right)^{b_1}$$

where γ, α_1, b_1 have been found by the least square method. It yields:

$$\gamma = 0.01, \quad \alpha_1 = -0.5, \quad b_1 = -0.12 . \qquad (2.57)$$

Relationship (2.54) obtained corresponds to the flow in the vertical channel with smooth walls. For vertical channels with RR of the surface, the height of which is comparable with δ_L, in flowing stationary waves of a large amplitude are formed on the film surface. When hydrodynamic interaction of phase is great, wave amplitudes tend to increase as time goes on, which leads to the decrease of w_g^*. The correction for the surface RR is introduced in formula (2.55) in the form of the power dependence on the relative height of the roughness bulges:

$$w_{g,gRR}^* = w_g^* \cdot \left(1 - \alpha_2 \cdot \left(\frac{e}{p}\right)^{b_2}\right), \tag{2.58}$$

where w_g^* is defined by relationship (2.55) and a_2, b_2 are to be determined. These constants are found by generalizing experimental data from paper [24, 81]. It yields

$$a_2 = 0.88, \quad b_2 = 0.572. \tag{2.59}$$

Relationships (2.55) – (2.58) are valid in the following range of varying parameters:

$$20 < Re_L < 100;\ 0.3\,m < H < 0.85\,m;\ 16\,mm < d_e < 24\,mm;\ 0 < \frac{e}{p} < 0.2.$$

The deviation from experimental data does not exceed ±10.1 %.

2.3.2.4. Experimental Study of Film Flows in Heat-and-Mass Transfer Apparatuses

In numerous experimental works devoted to the study of peculiar features of liquid film flowing along vertical surfaces they have mainly analyzed the flow through tubes or along the flat smooth surface. The number of works where they studied the flows along the flat surface with regular roughness (RR) is scarce, and only some aspects of flowing along the longitudinally corrugated surface have been considered conformably to the special features of the condensation process. But it is longitudinally corrugated sheets that found wide application in designing film HMTA. The flow along such surfaces is characterized by some distinguishing features.

Well-known methods of experimental measuring the liquid film thickness, $\overline{\delta}_L$, can be divided into two groups. *Methods of measuring local thickness* δ_L: the method of touching or electrocontact method; photoshady method, method of light flux absorption; method of measuring electrical capacitance between the sprayed wall and the sensor; method of electrical resistance or local conductance; method of measuring radioactive traces in the liquid film, etc. *Methods of measuring average thickness of the liquid film* $\overline{\delta}_L$: the method of instant power supply cut-off; method of consecutive weighing the dry and moistened part; neutron diagnostics method; modified method of electrical capacitance; method of electrical resistance or conductance.

The electrical conductance method is most widely used in studying two-phase systems [10, 87, 88]. It makes possible to study a lot of parameters (wave characteristics, gas content nature of flowing); it can be used in channels of complicated configurations; it has a good reproducibility of results obtained; it can be used with the gas flow both available and not available; it is suitable for local measurements when there are many sensors, in this case the flow does not get perturbed. To obtained the information about the distribution of δ_L on the channel perimeter and the value of the wetted surface the method of local electrical conductance (the ratio of the electrode size by the flow D to the distance between the electrodes is about 1) is used; to measure the average thickness $\overline{\delta}_L$ the method of integral electrical conductance (l/D >> 1) is used. The idea of the method is to measure the ohmic resistance of the liquid film by means of several electrodes located on the surface of sheets in such a way that the bulge should not be formed (flush with). The authors [10, 24, 81, 87] used tap water the natural is sufficient for making measurements. The alternating current is used to eliminate the polarization effect. The value $\overline{\delta}_L$, on the part being studied, was defined on the basis of the film resistance R_f and that of the liquid column R_f^*, which responds to complete filling the experimental channel with the liquid. Proceeding from the Ohm's law, $\overline{\delta}_L = D' \cdot R_f^* / R_f$, where D' is the channel width taken from the conditions: $D' >> \delta_f^*$ (δ_f^* is the limiting value of δ_L) and $D \leq L$ (L is the length of the channel part being measured). Then R_f and R_f^* will be of approximately the same order, that is the slope $R_a = f \cdot \frac{1}{\sqrt{\omega}}$ will be practically constant. Thus it is possible to replace

$$\left(\frac{R_f^*}{R_f}\right)_{f\to\infty} \to \left(\frac{R_{a,f}^*}{R_{a,f}}\right)_f$$

The working frequently has been chosen as equal to 5 kHz, referring to the linear part of the relationship through the whole range of changes in $\overline{\delta}_L$. Two schemes of placing electrodes on the sheets have been used. In the first one three electrodes (strap electrodes of aluminium foil 5 mm wide, 2 mm thick) are placed perpendicular to the film flow, which makes it possible to avoid leak flows. But in defining $\overline{\delta}_L$, especially in case of steam-line flow around the fin of regular roughness, an error appears, which is connected with the change of equality of experimental parts lengths arising in defining R_f and R_f^*. This equality is guaranteed by the scheme of two electrodes located along the flow but this scheme is not suitable for studying flows along the longitudinally corrugated sheet. To check the method an experiment has been carried out in the channel with the clearance of varying height, wholly filled with a liquid. This is equivalent to plotting the calibration curve.

The stand – a one-channel model (figure 2.11) provides the possibility to conduct research both with the gas flow net available and available. Here 1 is the liquid distributor, 2 is the fastening frame; 3 is a sheet of the pacing bed (backing) with the dimensions of the part being measured 0.25×0.5 m; 4 is the water rotameters block PC; 5 is electrodes; 6 is the liquid receiver; 7, 8, 9 are the filter, the pump, the tank; 10 is the heater; 11 is the receiver; 12 is the air rotameter PC; 13 is the air distributor; 14 is the observation panel; 15 is the contact thermometer. The width of the working channel is 15 mm. The observation panel is made of plexiglas's. To distribute the liquid film uniformly and to smooth pulsations a band of thin capillary-porous material (phlizeline) is placed in the upper part of the sheet. The same bands 5 mm wide are placed along the edges of the sheet to prevent the film taking-off at low loads. The investigation has been conducted on vertical flat and longitudinally corrugated sheets (figure 2.11B). Flat sheets were made of plexiglas's: 1 is smooth sheets (a sheet without regular roughness); 2 is a sheet with regular roughness in the form of horizontal bulges uniformly distributed along the surface of the sheet (the bulges have a rectangular form with rounded edges) – figure 2.11B.a. Height of bulges (of 1 mm) is accepted on the basis of the recommendations [24]; width at the basis of 1 mm. The quantity k= p/e was changed discretely in experiments (k = 2.5; 4.5; 6.5; 8.5; 11; 20; 36.5).

Longitudinally corrugated sheets were formed from epoxy resin (to exclude the samples electrical conduction) in some versions: 1 – without regular roughness; 2 – with fins of roughness – figure 2.11B.b. The sheet main corrugation parameters are follows: P =10 mm, E = 3.8 mm, those of roughness are: k = 12.5 at e =1 mm. The values of $\overline{\delta}_L$ were averaged by the channel width. In papers [10, 24] it is shown by the "searching needle" method that the maximum increase of the surface at the expense of thickening at the side walls does not exceed 0.7%, therefore such averaging is grounded. The investigation has been carried out both without the gas flow and with it, in the latter case special attention was paid to the problems of stability. The former case corresponds to the absence of hydrodynamic phase interaction (it is characteristic of the cross flow scheme).

Figures 2.12 – 2.13 give the main results. The transition from the laminar to the turbulent flow region (Re_L^*) depends on the availability of regular roughness and the magnitude k. The availability of two-dimensional roughness accelerates the transition, reducing the threshold value of Re_L^*. It follows from the stability theory – the surface roughness causes in the laminar flow perturbations which are added to the perturbations available due to a certain degree of turbulence in the external flow. The latter accelerates the flow mode transformation. For a smooth sheet $Re_L^* \approx 1650$. On a smooth sheets and sheets with k > 20 one can observe a slightly noticeable irregular wave flow. At c k < 20 a monotonous and periodical wave flow is formed, in this case the initial part length becomes less with the decrease of k.

At the same time the film becomes thicker (k = 20 → 4) with subsequent thinning at k < 4. The minimum value of Re_L^*, corresponds to k = 8-14. The further reduction of k results in the growth of Re_L^*, thus the relationship $Re_L^*(k)$ is described by a complicated curve. The type of the relationship is explained by the fact that at sufficiently low k fins of roughness are practically linked and conditions of film flows approach to those characteristic of the smooth sheet without regular roughness. The film thickening takes place which is connected with the growing influence of capillary forces on the liquid delay. It was shown by Japanese researchers [57, 82] who noted, that in this case the evaporative cooling efficiency could be even lower than that of the flow along the smooth sheet. According to $k = p/e$ three peculiar regions of liquid film flow can be distinguished:

1. k = 8-14. A stable regular wave mode of flow with the predomination of standing waves is notable; the initial part of the wave formation is practically missing: the splash formation at flowing around the fin is minimum:
2. k < 8. The flow laminarization takes place.; the values of $\overline{\delta}_L$ and Re_L^* increase; at sufficiently low k the situation is practically the same as in flowing along the smooth sheet;
3. k > 14. The regular wave mode of flowing is changed; the initial part of the wave formation increases; the splash formation grows; at sufficiently high k the situation is the same as in flowing along the smooth sheet.

Thus, the range of values k = 8-14, where there are optimal conditions for revealing roughness and maximum intensity of transfer processes in the liquid film is provided, is of special concern.

Experimental results are given in the form:

$$\overline{\delta}_L = x_1 \cdot \sqrt{\frac{3 \cdot \nu_L^2}{4 \cdot g}} \cdot Re_L^{x_2} \cdot k^{x_3} \cdot \exp(x \cdot_4 k) \cdot \sin\alpha^{x_5} \qquad (2.60)$$

The parameters $x_1 \ldots x_5$ were found by minimizing the efficiency function

$$\varphi = \sum_{i=1}^{n} \left(\overline{\delta}_{L,ex}^{i} - \delta_{L,cal}^{i} \right)^2, \qquad (2.61)$$

with the help of the search method by a polyhedron under deformation (Neldor-Mid), where n is the number of experimental points. The following expression are obtained:

for the laminar wave region

$$\overline{\delta}_L = 0.83 \cdot 10^4 \cdot \sqrt[3]{\frac{3 \cdot \nu_L^2}{4 \cdot g}} \cdot Re_L^{0.251} \cdot k^{-0.655} \cdot \exp\left(0.22 \cdot 10^{-1} \cdot k\right) \cdot \sin\gamma^{-0.48} \quad (2.62)$$

(the average error is 9%; $Re_L = 280 - 1260$; k = 2.25-40)

for the turbulent region

$$\overline{\delta}_L = 0.81 \cdot 10^3 \cdot \sqrt[3]{\frac{3 \cdot \nu_L^2}{4 \cdot g}} \cdot Re_L^{0.56} \cdot k^{-0.401} \cdot \exp\left(1 \cdot 10^{-2} \cdot k\right) \cdot \sin\gamma^{-0.327} \quad (2.63)$$

(the average error is 4%; $Re_L = 720 - 2100$; k = 2.25-40)

The boundary of flow modes existence is defined by the expression

$$Re_L^* = -3.36 \cdot 10^2 \cdot k^{0.661} \cdot \exp(-0.039 \cdot k) + 1650 \quad (2.64)$$

(the average error is 12.1%). The comparison of calculated (see 2.3.2.1 – 2.3.2.2) and experimental values of δ_L shows that they are in good agreement. The gas flow influence on the quantity δ_L was not registered in the range $\omega_g = 1.5 - 7.0$ m/s.

Flow Along the Vertical Longitudinally Corrugated Sheet with Regular Roughness

The transition from a flat to longitudinally corrugated sheet means the transition from frontally flowing film to jet-film flow in channel cavities with alternating wetted parts (figure 2.13). With the predominant abundance of corrugated sheets in film HMA-s such a flow is traditionally considered as the film one. The problem of correlation of heat-and-mass transfer surfaces appears to be important, especially in case of high heat conduction of the material the sheet is made of. While analyzing experimental data instead of the average $\overline{\delta}_L$ notion, we used the idea of the maximum film thickness in cavities of channels. In calculating $\overline{\delta}_L$, the wetted perimeter and the wetted surface quantity we supposed that the tree surface of the jet was smooth with some curving at the boundaries. The latter increases the wetted surface quantity by 10% at an average (figure 2.13B). It can be seen that the introduction of regular roughness results in film thickening (figure 2.12). The retaining of optimal range $k_{opt} = 8 - 14$ has been established for the flow along the longitudinally corrugated sheet too. The regular wave mode of the flow with the predomination of standing waves is notable. The following expression for the flow along the longitudinally sheet have been obtained:

$$\overline{\delta}_{L,L} = 0.198 \cdot 10^4 \cdot \sqrt[3]{\frac{3 \cdot \nu_L^2}{4 \cdot g}} \cdot Re_L^{0.404} \cdot k^{-0.591} \cdot \exp\left(1.44 \cdot 10^{-2} \cdot k\right) \cdot \left(\frac{P}{2 \cdot E}\right)^{0.549} \cdot \cos\alpha^{-0.249}, (2.65)$$

$$\overline{\delta}_{L,T} = 0.178 \cdot 10^3 \cdot \sqrt[3]{\frac{3 \cdot \nu_L^2}{4 \cdot g}} \cdot Re_L^{0.698} \cdot k^{-0.327} \cdot \exp\left(0.33 \cdot 10^{-2} \cdot k\right) \cdot \left(\frac{P}{2 \cdot E}\right)^{0.602} \cdot \cos\alpha^{-0.219} \quad (2.66)$$

The design and experimental values of $\overline{\delta}_L$ for the longitudinally corrugated sheet are in good agreement. It should be noted that for the smooth longitudinally corrugated sheet the magnitude Re_L^* coincides with the magnitude generally accepted for the film flow along the smooth flat surface (≈ 1600, figure 2.12). The shift of Re_L^* accounting for the surface the flat sheet with regular roughness registered by us. The value $a_{HMT} = F_M / F_H$ has been obtained by calculating experimental data. The control was exercised by the method of local electrical conduction in spot electrodes (d =0.5 mm, n is the number of electrodes equal to 15-30) were located around the channel perimeter in some of its sections by the height. The value of the wetted perimeter was registered by shorting a pair of extreme electrodes with the liquid film. On the average, the surface correlation $a_{HMT} = F_M / F_H$ is equal to 0.3-0.5 and it was later used in calculating transport coefficient (figure 2.13A). The design and experimental values of a_{HMT} are in good agreement. The influence of regular roughness fins on the value of the channel wetted perimeter was not taken into account in calculations (figure 2.13B).

The longitudinally corrugated sheet provides a film-jet mode of flowing with the alternation of dry and moistened parts (wet-dry mode of heat-exchange). In this case the former, under conditions of high heat-conduction of the packing bed material, served as finning of moistened parts, to some extent compensating the loss of totally transported heat in the system due to the reduction of F_M. This problem is specially dealt with in the paper [10] which shows the attainment of considerable water saving for the replenishment of the system with a slight loss of refrigerating capacity in conformity with evaporative coolers. Unlike the solution [10] our sheet construction simplifies the distribution of the liquid film, providing the same effect. The gas flow influence was studied on the counter-flow and cross-flow schemes of contacting. For the counter-flow the absence of this influence was registered up to the values of $w_g \approx 6.5$ m/s, when the flooding phenomenon is established. For the cross-flow the visualization model was used. The high corrugation is characterized by the picture which corresponds on the whole to the case of one-phase liquid flow. The shift of the latter in the direction of the gas movement is not height at all. In reducing the corrugation height (E ≈ 2.2 mm) at high velocities ($w_g > 10$ м/c) the film is shifted to the peak of the longitudinal corrugation and partly dispersed; drops are gone with the gas flow. For E ≈ 1.5 mm this phenomenon appears even at $w_g \approx 7 - 8$ m/s. These results confirm high stability of the organized film-jet flow along the surface of the longitudinally corrugated sheet and justify the choice of just the cross flow scheme of gas and liquid contacts in HMTA.

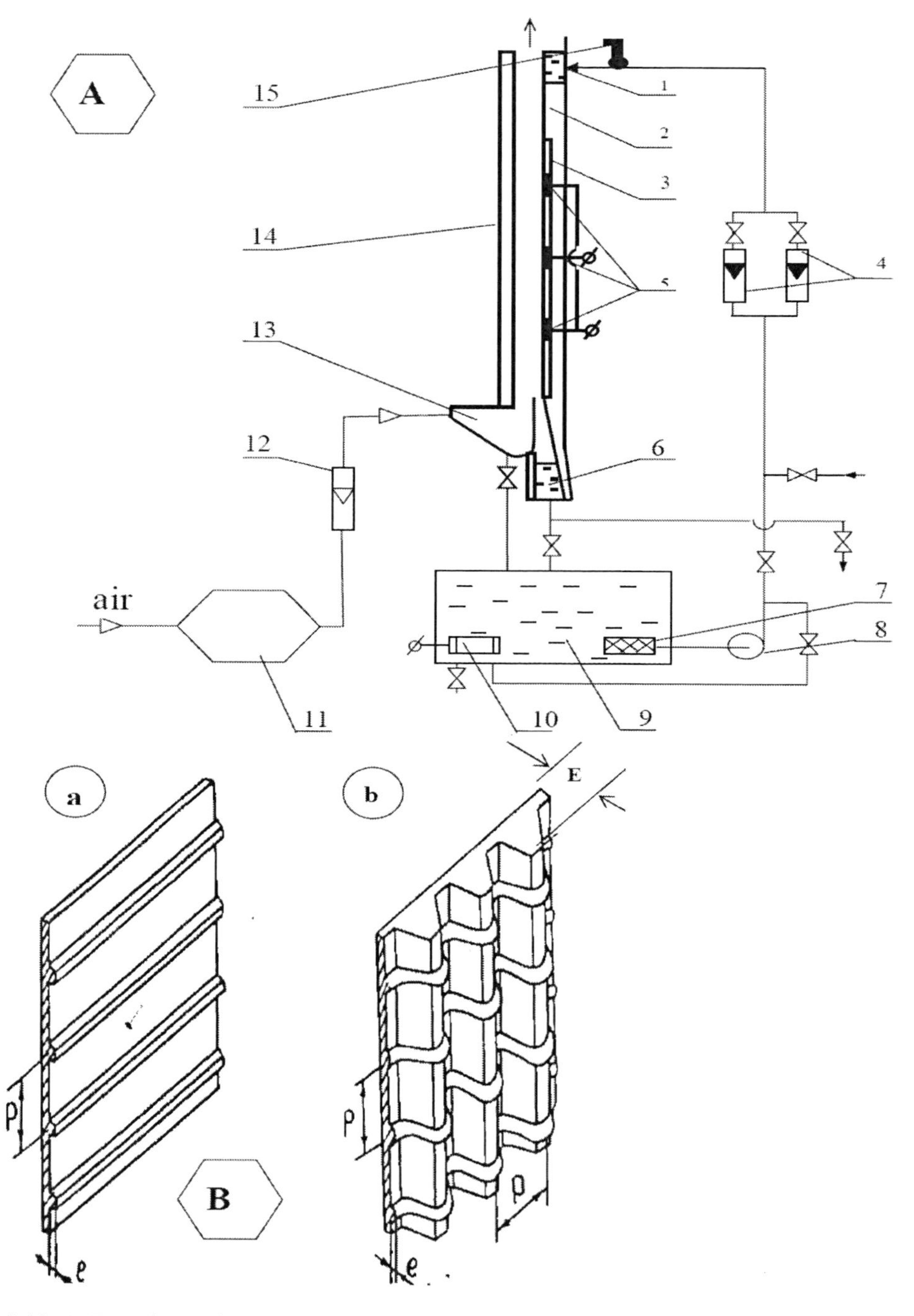

Figure 2.11. A. Experimental stand (the single channel model) for studying film flows. Nomenclature: 1 is the liquid distributor; 2 is the fastening frame; 3 is the packing bed sheet; 4 is the water rotameters block; 5 is electrodes; 6 is the liquid collector; 7, 8, 9 are the filter, the pump, the tank; 10 is the heater; 11 is the receiver; 12 is the air rotameter; 13 is the air distributor; 14 is the vision panel; 15 is the contact thermometer. B. The types of investigated sheets: a is the flat sheet with regular roughness RR; b is the longitudinally corrugated sheet with RR.

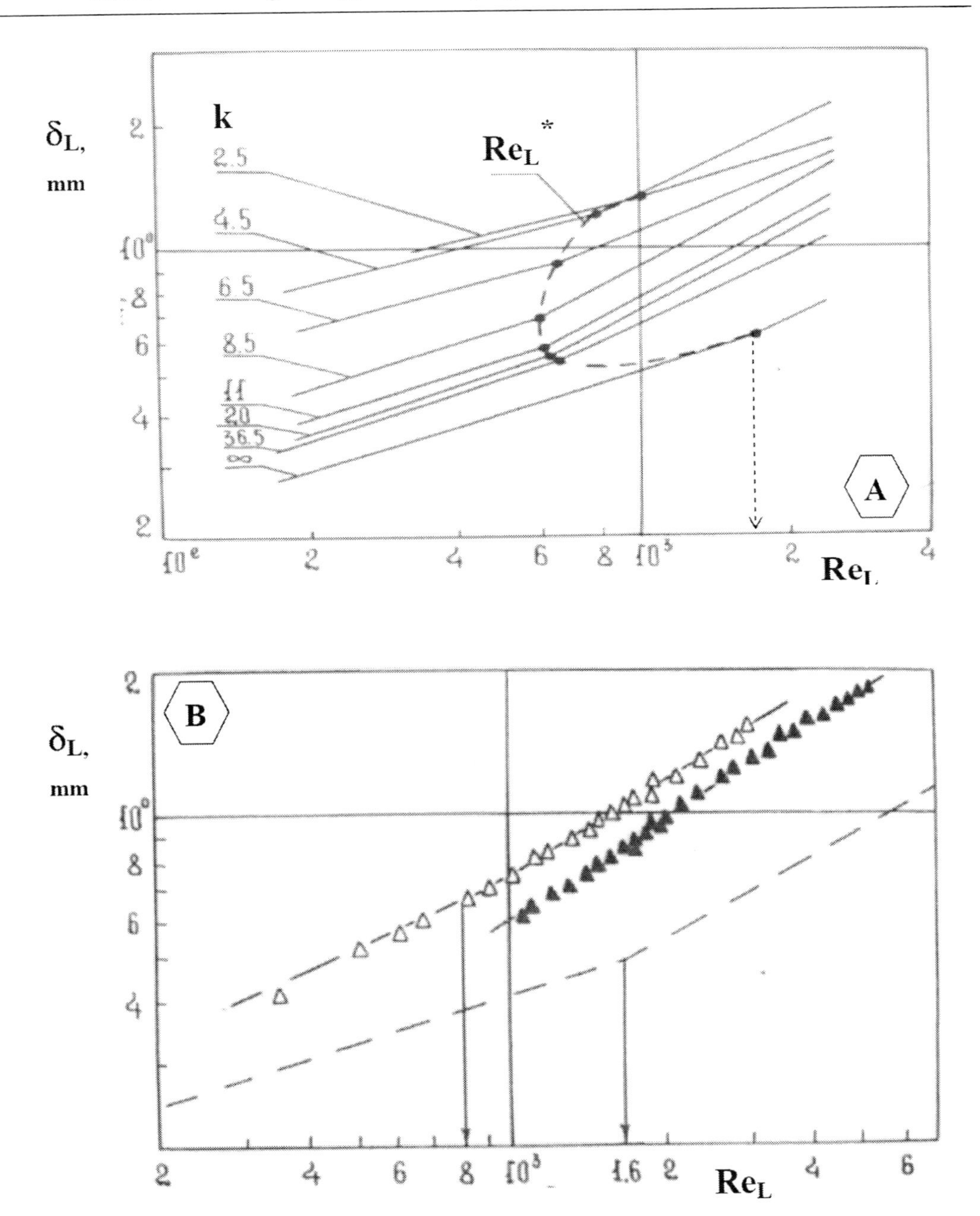

Figure 2.12. Results of experimental studying film flows in heat exchanging apparatuses. A is the relationship δ_L (Re_L, k) for a flat sheet with the surface regular roughness (ϑ is the smooth sheet); B is the influence of the longitudinal corrugation of the sheet: ▲ – the flat sheet with RR, Δ – the longitudinally corrugated sheet with RR, (_ _ _) is the smooth flat sheet (k = 8.5).

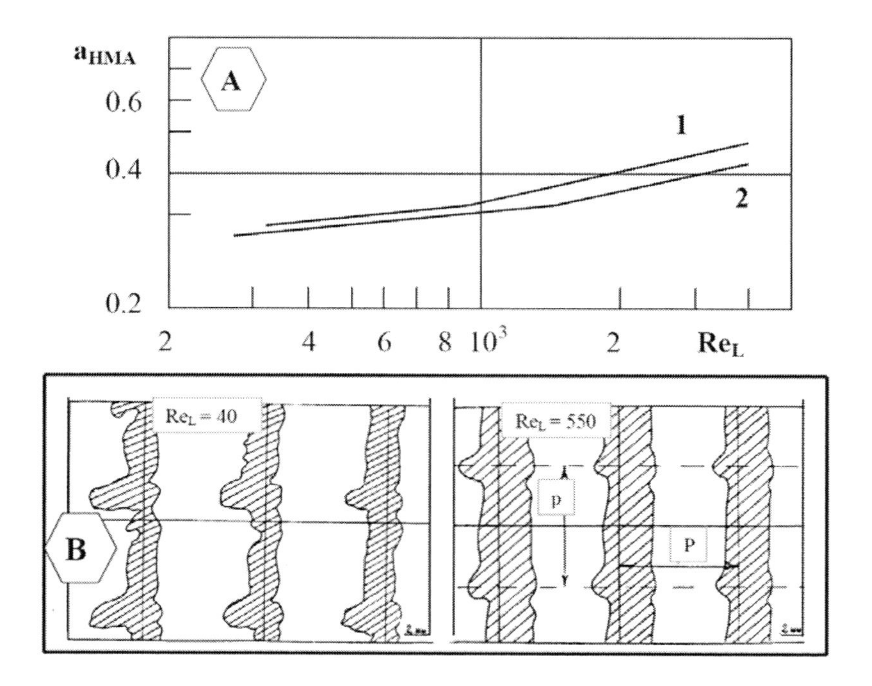

Figure 2.13. Results of experimental studies of film flows in heat-and-mass transfer apparatuses. A is the value of relations for surfaces of heat-and-mass transfer in heat-and-mass transfer apparatuses with longitudinally corrugated sheets of the packing bed: 1 – without the surface regular roughness (RR), 2 – with the surface regular roughness (k = 12.5, P = 10); B is the influence of Re_L on the value of the wetted surface of the longitudinally corrugated sheet.

2.3.3. Experimental Study of Working Characteristics of Cross-Flow Film HMTA

2.3.3.1. Experimental Equipment and Research Program

The stand (figures 2.14.A and 2.15) consist of the main part and systems of preparing the flows. The main part of the stand is the working (4), water distributing (3) and water catching (5) (multi-sectional) chambers and two chambers for measuring air flow parameters. All chambers are made of thick-walled transparent plexiglass. The water distributor is lamellar (the upper ends of the packing bed sheets are protruded into the water distributing chamber, and layers of capillary-porous material are arranged between and over them). The upper layer of foamed plastic serves for the uniform distribution and filtration of the liquid. The water-catcher consists of 6 pockets, the last of which is located outside the module in the direction of the gas movement. It provides differentiated measuring of the liquid flow-rate and its longitudinal (for the gas) drift with the air flow. Frame pockets the liquid passes to the unit of measuring the local flow-rates (12) with temperature sensitive elements placed in its measuring chambers. The chamber's dimension are 460×400× ×180 (width) mm. The chamber capacity is up to 2000 m^3/h for the gas. The air was supplied by a centrifugal fan, installed lower in the flow which made it possible to take into account the heating in the fan and high stability and uniformity of the distribution. The flow rate control was achieved by even changing the voltage, supplied to the electric motor. The air electric heaters are located in front of the fan and they switched on the two-position temperature regulator (the controlling accuracy is 0.3 °C). Flow meters (7, 8) of the collector type were used. To control the relative humidity the by-pass line with flow meter 8 is provided. The heat load was controlled by the one-phase regulator of the voltage; the liquid temperature – by the two-position temperature controller – with the accuracy of 0.1 °C. The liquid flow meter is a block of rotameters of PC-5 and PC-7 types (11).

The preference was mainly given to the study of conditions of evaporative water cooling (CTW), processes of direct (DEC) and indirect (IEC) evaporative cooling at the cross-flow scheme of contacting. The program was formed by the step planning method and consist of the following sections: the investigation of the working characteristics in a wide range of the layer density values; the determination of the optimal d_e (equivalent diameter of the channel) magnitude; the study of the influence of regular roughness parameters (RR); the study of the influence of the module height and length in the gas flow movement (cross-flow); the study of the effect of the nonuniformity of distributing contacting flows on the working characteristics.

On the whole, five types of packing beds with stacks of longitudinally corrugated sheets were studied (Table 2.3): the longitudinally corrugated sheet without regular roughness; the longitudinally corrugated sheet with regular roughness (RR) of the surface; in this case regular roughness fins are located horizontally (along the air flow movement) – the “flat double sheet” type – figure 2.14B.b; the same sheet with RR is placed at an angle to the running-on flow – the “straight oblique riffle” type – figure 2.14B.c; the sheet with oblique-angle double corrugation, when both channels themselves and fins of surface regular roughness are located at an angle to the vertical – the “double-oblique row” type – figure 2.14B.d;

- the sheet with horizontal main channels and the slope of RR fins towards the running-on flow (in the table is not shown).

The magnitude of the channels equivalent diameter ranged from 12 to 30 mm; the construction surface of the packing-bed in the layer volume unit was in the range of 170-200 m^2/m^3; the slope angle of the main packing-bed channel is 0-90°; the slope angle of the regular roughness fins to the horizontal is 0-30°; the roughness parameter value k = 10-14 ($k = p/e = 12.5$ in the majority of experiments).

2.3.3.2. Hydrodynamic Characteristics

The investigation was carried out within the range of values $d_e = 12.0 - 29.9$ m (calculated by the middle section of the channel) at k = 12.5 which is among the optimal values of $k_{opt} = 8 - 14$. We studied the influence of the surface regular roughness, the arrangement of regular roughness fins and the main channels with respect to the running-on air flow, the main corrugation parameters P and E, on the transfer intensity and energy consumption.

Table 2.3. Geometric characteristics of investigated cross-flow apparatuses

№	Packing bed modification	Packing bed type and material	Scheme of laying out riffles and RR: ——— RP, - - - RR	$d_э$, mm	Specific surface of packing bed a_x, m^2/m^3	Corrugation geometry		Regular roughness (RR) geometry		Riffle slope β	RR slope α
						K = P/E	P, mm	k = p/e	p		
I	1.1	Aluminium foil, corrugation without RR	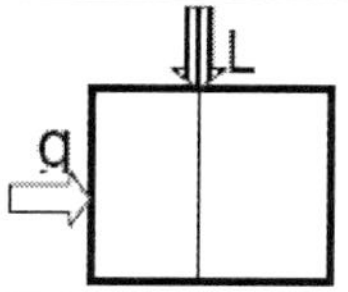	18.5	210	2.5	10			90	
	1.2			21.4	182	2.86	10			90	
	1.3			23.3	166	3.5	14			90	
II	2.1	Aluminium foil, "direct double riffle" figure 2.14.Bb,	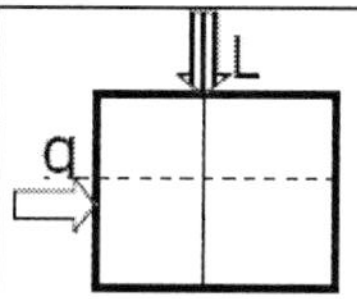	15.7	236	2.86	10	12.5	10	90	0
	2.2			19.3	197.2	2.5	10	10	10	90	0
	2.3			22.3	175	3.5	14	14	14	90	0
III	3.1	Aluminium foil, "direct oblique riffle" figure 2.14.Bc,	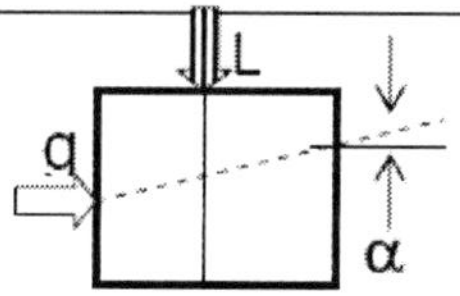	20.6	185	2.5	10	10	10	90	15
	3.2			21.8	175	5.0	10	10	10	90	15
	3.3			22.3	175	3.5	14	14	14	90	30
IV	4.1	Aluminium foil, "double oblique riffle" figure 2.14.Bd	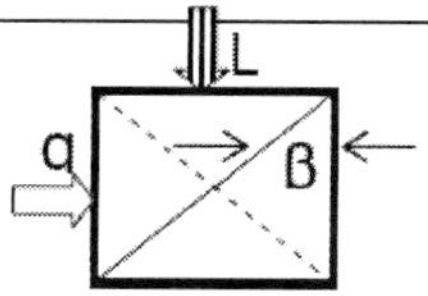	21.5	177.3	2.5	10	10	10	75	75

The transition to the cross-sectional scheme provides the reduction of the Δp level as compared to the counter-flow and the possibility to increase the loads considerably. The introduction of regular roughness for the vertical arrangement of the main corrugation does not practically affect the value Δp in contrast to the counter-flow both for one- and two-phase flows. The hydrodynamic interaction of phases is missing in the whole range of gas and liquid flow-rates. The traditional phenomenon of flooding for the cross-flow scheme is not available at all, up to values $w_g \approx 10-12$ m/s and at large w_g it is replaced by the phenomenon of the longitudinal liquid drift which results in its unfavorable redistribution in the packing bed volume and removal from the layer. The influence on the main corrugation parameters P and E was studied. The increase of E (2-5.3 mm (P = idem)) results in the growth of Δp; at $E^* \leq 2.2$ mm the longitudinal drift of the liquid begins from w_L >10.0 m/s. The decrease of p (14 →10 mm) results in the drop of Δp. In both cases the lines of the gas flowing become more flatted in channels of the regular packing bed, which reduces transfer intensity.

The channel slope within the limits of 75-90° does not practically affect the amount of resistance, but the situation with vertically and horizontally arranged channel is different. The latter is characterized by the abrupt change of the longitudinal drift of the liquid already at w_g = 2.5 m/s (the transformation of the cross-flow scheme into the parallel current one) and by the increase of the spray density influence on Δp. The vertical arrangement of the main channel is preferable: the film-jet liquid flow in its cavities is characterized by high stability in case of the cross-flow.

2.3.3.3. Heat-and-Mass Transfer in the System

The maximum increase of the transfer intensity was obtained for the "straight oblique riffle" (figures 2.16 and 2.17), and one can notice the favorable liquid distribution over all the surface of the sheet in the module. The influence of the slope angle of regular roughness fins to the running-on flow is illustrated by the data in figure 2.17. The angle α = 15° is optimal, which corresponds to results of visual observations of the liquid distribution. Unlike the counter-flow, where regular roughness fins are arranged horizontally, for the cross-flow scheme their optimal arrangement has been concretized for the first time. The absence of liquid on sheets in the lower part of the inlet section, noted in literature is practically not available, which provides the reduction of specific consumption of materials of the module by 15-20%. One can note the particular form of the relationship $K_h = f(q_L)$ for α = 15° which shows the absence of the longitudinal redistribution of the liquid. Considerable influence of the main channel slope was not detected and its vertical arrangement is preferable (figure 2.17B). It should be noted that RP of the "oblique double riffle" (figure 2.14B.a) type which is very popular abroad not provide any advantages over the vertical arrangement of the channel, according to the results of the present investigation. The version with the horizontal arrangement of the channel presents quite a different picture. In the velocity range $w_g \leq 3.5$ m/s low energy consumption, but with the further increase of w_g the longitudinal drift of the liquid sharply increases. With the decrease of q_L the critical value w_g^* increases and for q_L =4 m^3/(m^2·h) amounts to 1.5-5.0 m/s. Such an arrangement of the module is favorable under the conditions of thermal-moist air treatment (evaporative coolers IEC, DEC). It should be noted that with the growth of q_L, K_h is unusually decreasing, which is explained just by the phase interaction. The sheet with "double corrugation" should be recommended for the version with a vertical channel – for evaporative cooling of the liquid (CTW); with a horizontal channel – for thermal-moist treatment of the gas (IEC, DEC), the optimal arrangement of the regular roughness fins proves to be identical.

The influence of the main corrugation parameters is shown in figure 2.16B. The increase of E (2-5 mm) results in the increase of K_h, it is registered at two values of P (10 and 14 mm). The optimal range $K = \frac{P}{2 \cdot E}$ = 1.4-2.0 is defined by the influence of this parameters on the distribution of flow lines in the channel bent in the direction of the gas movement. This range of K values is recommended for both schemes of contacting.

The intensifying affect of regular roughness fins under the conditions of the cross-flow scheme is revealed in the reduction of thermal resistance in both phase [10, 24, 81] as compared to the smooth (background) sheet, and the increase of $K_h = f(w_g)$ is practically provided at the expense of reducing the thermal resistance of the phase R_g, which is connected with the lack of hydrodynamic phase interaction. The intensification mechanism can be explained on the basis of conclusions from papers [124], experiments with the sublimation of naphthaline under the conditions of the cross-flow scheme of flowing in channels with corrugated walls (the channel from and the distance between walls in our investigation and the works [124] are practically identical). Curvilinearity of the air flow lines and the appearance of secondary flows overlaying the main one are noted. The separation of flow on the lee side of the sheet results in abrupt reducing the transfer velocity, which increases in the same abrupt manner in the range of repeated addition of the flow. And it can be noted that the transfer intensity on the concave surface is much greater, than on convex one. It is important to show that in our case the places of appearing repeated flows are located just in the range of the film-jet flow on the convex walls of the sheet. The analysis of the specific energy consumption and the attainable degree of cooling confirms advantages of the sheet "straight oblique riffle" for the cross-flow scheme.

The investigations were carried out in the range d_e = 12.0-29.9 mm (k = 12.5) for the sheet of the "straight oblique riffle" type. With the increases of d_e the transfer intensity decreases. Experimental data allow the concretization of the range $d_{e.opt}$ = 20-25 mm, i.e. it is displaced to the side of higher values of d_e. The energy consumption for obtaining the specified value of the process efficiency E_L is much lower under the conditions of the cross-flow scheme than that of the counter-flow. The problem of the relations between main overall dimensions of cross-flow HMTA (H is the height, L is the length in the direction of the air flow, B is the width) was only raised in literature. The optimal formula of the main dimensions relation according to the results of this investigation is: H = B; the magnitude L is limited by the admissible magnitude of pressure losses in the air flow (but the ban features). Besides, one should be guided by the optimal range of working loads while choosing values H and B.

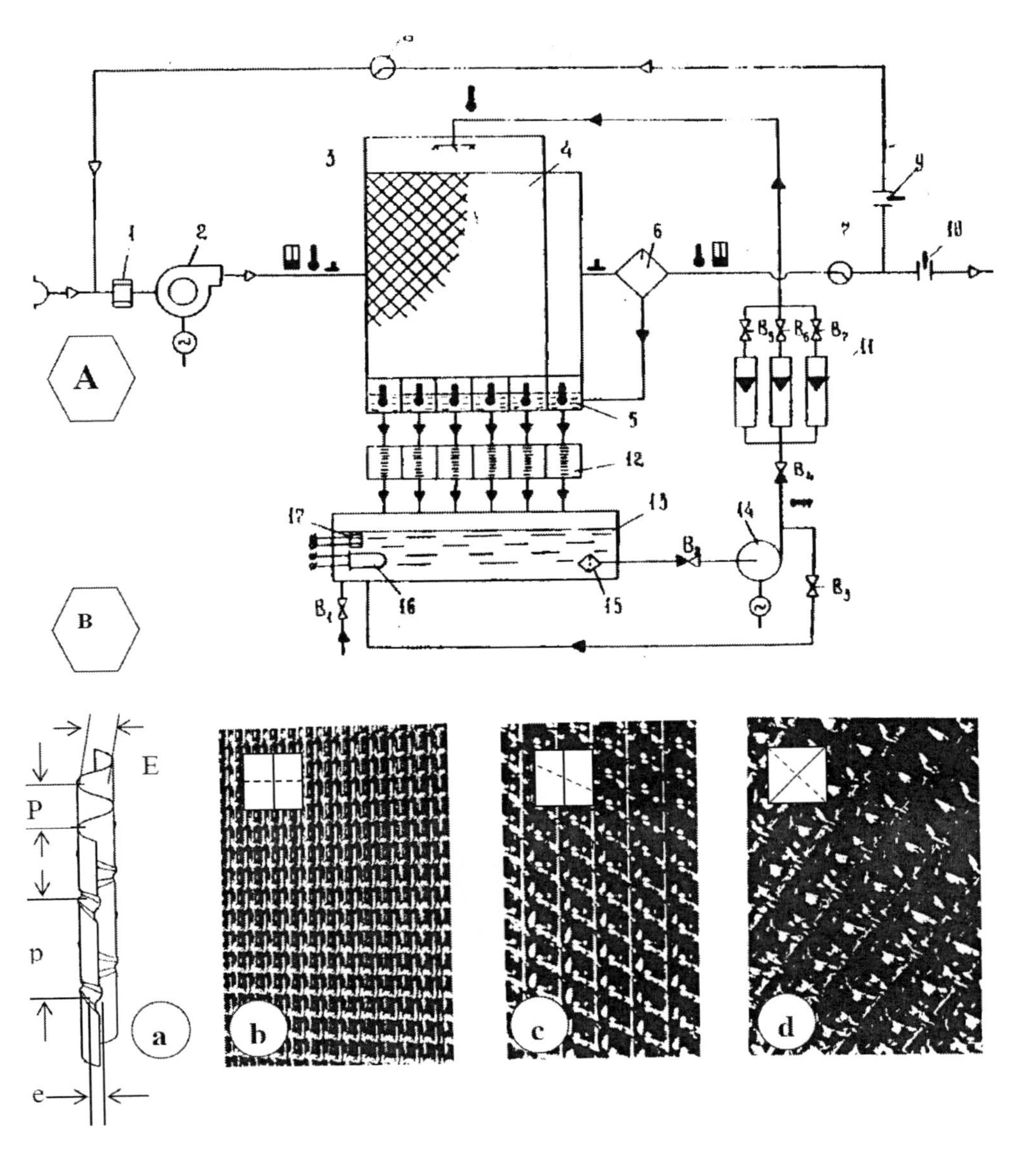

Figure 2.14. A. The experimental stand for investigating characteristics of cross-flow film heat-and-mass transfer apparatuses: 1 – electrical heater; 2 – fan; 3 – water distributing chamber; 4 – working chamber; 5 – accumulating sections; 6 – separator of drop moisture; 7, 8 – air flow-meters; 9, 10 – air flow-rate controls; 11 – water rotameters; 12 – sectional liquid flow-rate meter; 13 – water tank; 14 – pump; 15 – filter; 16 water heater; 17 – liquid temperature control. B – main types of the studied sheets of the packing bed: a – scheme of the packing bed sheet; b – straight double riffle; c – "straight oblique riffle"; d – "double oblique rifle" (____ – arrangement of the main channel, _ _ _ _ – arrangement of the regular roughness fin).

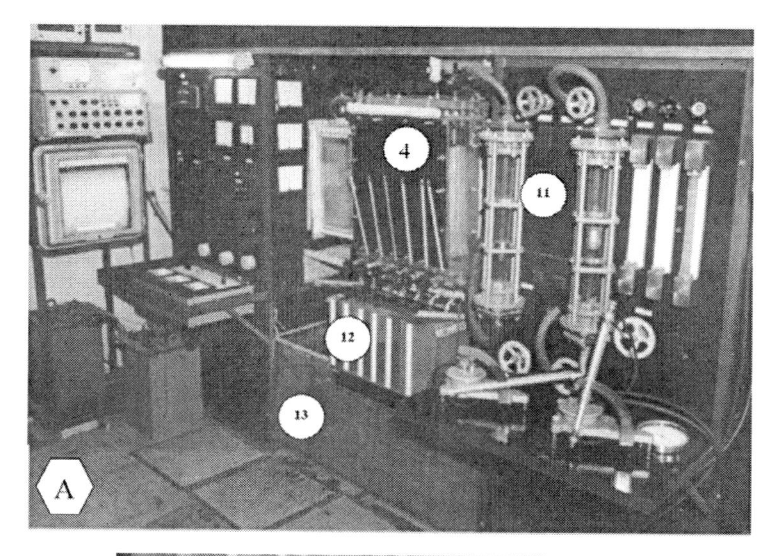

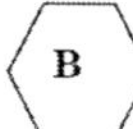

Figure 2.15. Pictures of the experimental stand (A) for investigating working characteristics of cross-flow film heat-and-mass transfer apparatuses and the working chamber (B) with a pack of beds "direct double riffle" installed in it. The nomenclature is as in figure 2.14.

The film-jet contacting provides the reduction of water consumption for feeding the system by decreasing the evaporation surface. As unsprayed parts of the heat transfer sheet surface play the role of fins in wetted parts the decreases of refrigerating capacity of the system on the whole is insignificant. It was first discussed in paper [35, see also the patent of France N2409481, F28F25/02, 1979], where special grooves were made on sheets for the liquid to flow. On the average, for the load range under the investigation it was possible to lower the water consumption for the evaporative process (feeding the system) by 30-40% while the refrigerating capacity being reduced only by 10-15% (IEC, DEC). The wet-dry principle of contacting, laid in the basis of developed HMTA, provides the possibility of using them in autonomous systems, and its combination with regular roughness of the surface makes it possible to control the relation between heat-and-mass transfer surface and the distribution of phase thermal resistances in the system.

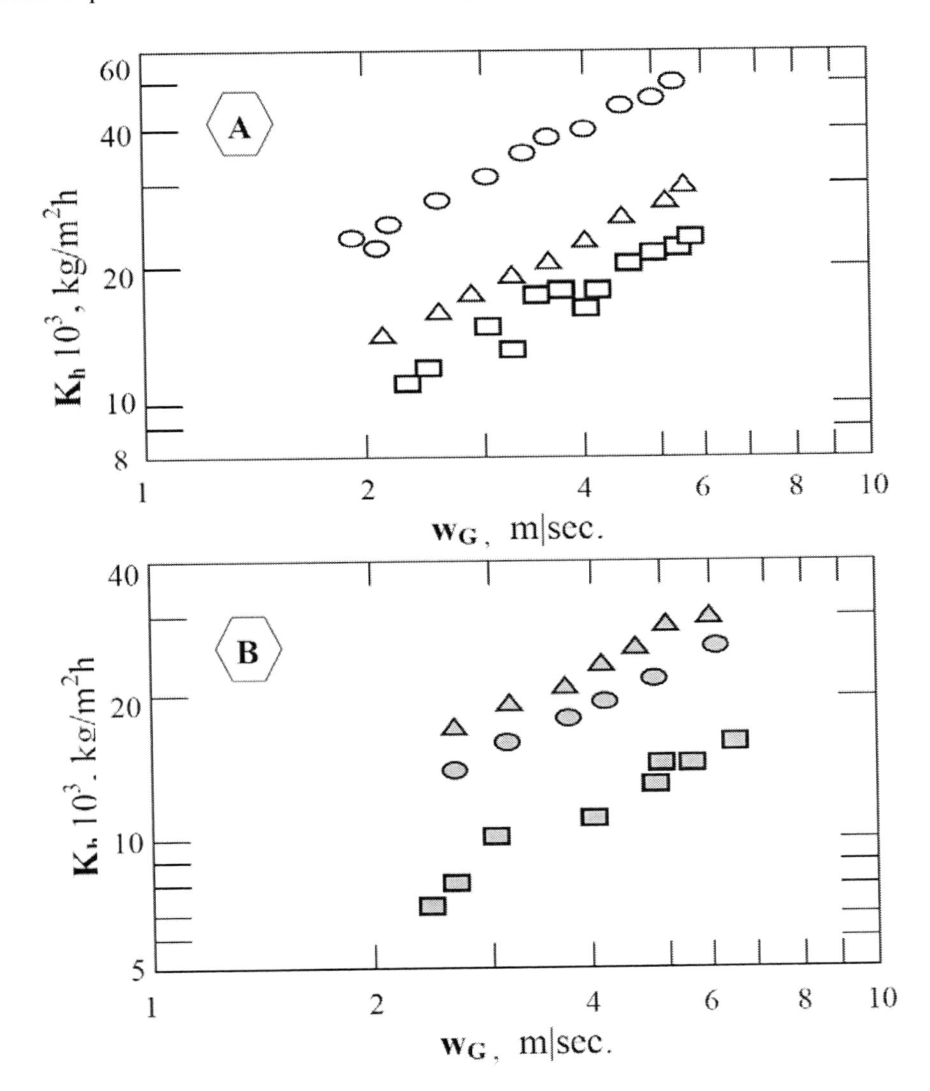

Figure 2.16. Results of experimental studies of working characteristics of cross-flow film heat-and-mass transfer apparatuses. A is the influence of regular roughness on the processes intensity: □ is the corrugated packing bed without RR; Δ is the corrugated packing bed "direct double riffle"; ○ is the packing bed "direct oblique riffle"; B is the influence of the main corrugation parameters: ▲ – E = 3 mm, ● – E = 4 mm, ■ – E = 2 mm. (P = 10мм.)

The vertical arrangement of the main channel is recommended for multi-channel HMTA with regular roughness. For the cross-flow – the "direct oblique riffle" with an inclined fin of regular roughness (α = 15°). All cases are characterized by the film-jet laminar-wave (or transitional) liquid flow at the turbulent gas flow. The wet-dry principle of containing makes it possible to decrease the amount of the liquid needed for feeding the system when the decrease of the refrigerating capacity is insignificant.

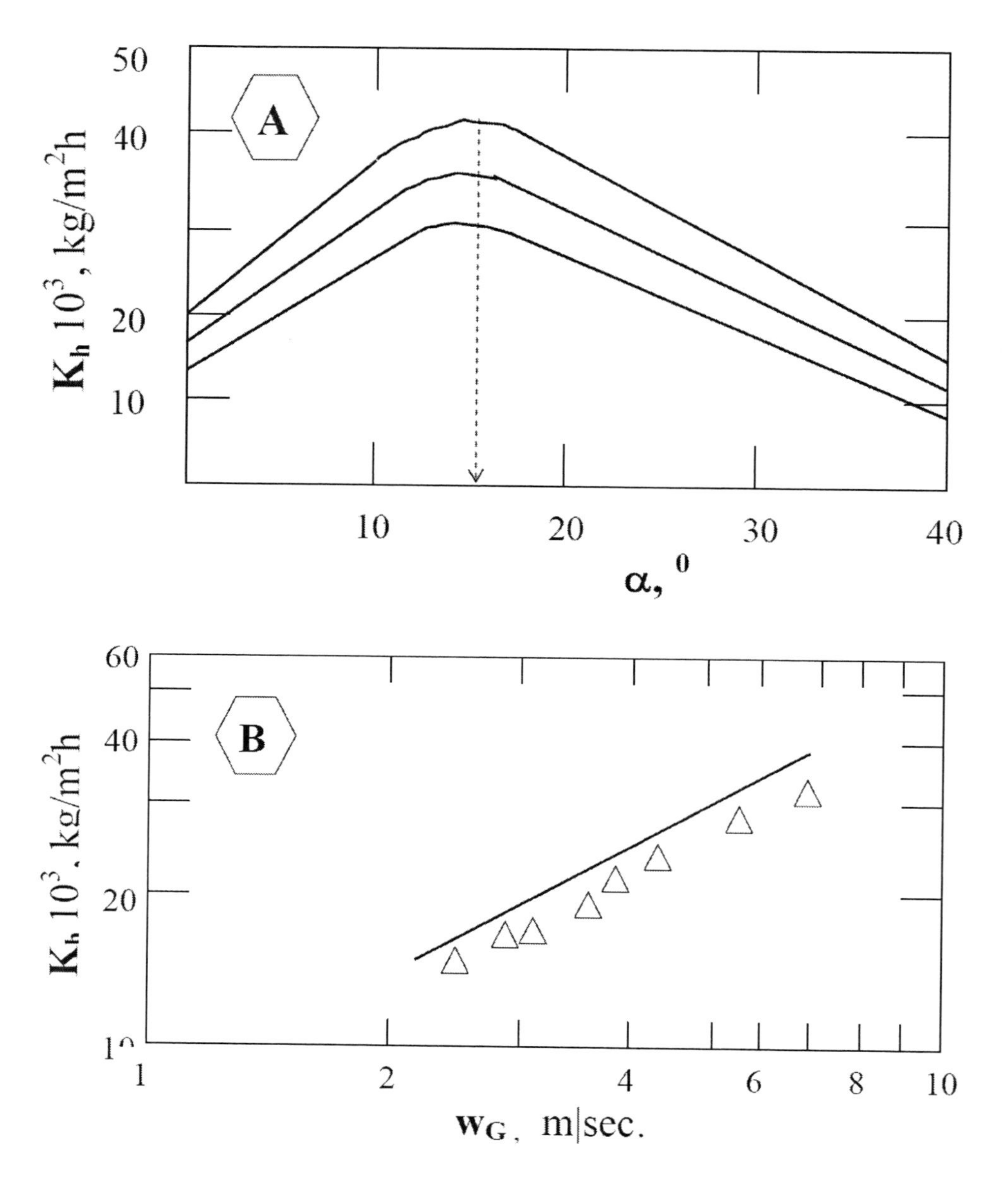

Figure 2.17. Results of experimental studies of working characteristics of cross-flow film heat-and-mass transfer apparatuses. A is the slope of the surface regular roughness (RR) to the run-on gas flow (packing bed sheet "direct oblique riffle"): 1 – w_g = 3 m|sec.; 2 – w_g = 4 m|sec.; 3 – w_g = 5 m|sec. B is the influence of the slope of the main packing bed sheet (main channel) corrugation to the run-on gas flow: ________ vertical channel ("direct double riffle" – figure 2.14Bb); Δ is the packing bed sheet "oblique double riffle" – figure 2.14Bd.

The cross-flow scheme has some advantages over the counter-flow: the range of working loads is much wider; the energy consumption is reduced at higher efficiency of the process; the height of HMTA is lowered; it is possible to install a fan outside the wet air flow and, if necessary, to reverse it. The joint arrangement of some heat-and-mass transfer apparatuses into one unit of an alternative cooler is simplified to a considerable extent. The possibility to use the cross-flow scheme at high loads without notable redistribution (a horizontal drift) is decisive for the use of it in HMTA for alternative air-conditioning and refrigeration systems, especially in designing absorbers and desorbers.

2.3.4. Simulation of Working Processes Sin the Combined Evaporative Cooler (CEC) on the Basis of IEC/DEC (IEC/CTW)

2.3.4.1. Direct Evaporative Cooling (DEC/CTW)

The cooling part of AACS is based on the use of the combined evaporative cooler (CEC) including indirect evaporative cooler (IEC) as the first stage and direct evaporative cooler (DEC) or cooling tower (CTW) as the second one: IEC/DEC (figure 1.6, 2.19) or IEC/CTW (figure 1.7). All apparatuses are of the film-type crosscurrent with multi-channel packing-bed of the structure made of longitudinally corrugate sheets (in the direction of liquid film flowing) with regular roughness of the surface as the basic method of heat-and-mass transfer processes intensification (figure 2.10).

Simulation of the processes in DEC/CTW – figure 2.18, CTW – was carried out in conformity with flat-and-parallel packing bed and cross-current scheme of phase contact and after it in conformity with more complicated situation with longitudinally corrugated sheets.

The heat-and mass transfer taken place on the free water surface, contacting with air, if their temperatures are different and the partial pressure of water vapour at the interface as well as in the air space are also different. For the evaporation (condensation) of the amount of moisture and supplied heat from interface F_s:

$$g_s = \beta_p \cdot (p_s - p) \cdot F_s, \tag{2.67}$$

$$Q_s = g_s \cdot r = r \cdot \beta_p \cdot (p_s - p) \cdot F_s. \tag{2.68}$$

This amount of heat can be provided, if there are no other sources, only convective mode from the side of gas and liquid:

$$Q_g = \alpha_g \cdot (t_s - t_g) \cdot F_s, \tag{2.69}$$

$$Q_L = \alpha_L \cdot (t_L - t_s) \cdot F_s. \tag{2.70}$$

At the phase of boundary the heat balance is maintained:

$$Q_{\Sigma} = Q_g + Q_s,$$
$$\alpha_L \cdot (t_L - t_s) \cdot F_s = \alpha_g \cdot (t_s - t_g) \cdot F_s + r \cdot \beta_p \cdot (p_s - p) \cdot F_s. \tag{2.71}$$

The moving force of mass-transfer on the free surface of the liquid is the difference of partial pressure $\Delta p = p_s - p$, at $\Delta p > 0$ the liquid evaporates, and hence, the air gets wet, when $\Delta p < 0$ the moisture condenses or, as it happens with liquid absorbents, condensation, absorption and dissolving of moisture in the solution takes place, hence, the air is dehumidified. Let us consider the laminar (laminar-and-wave) liquid film flowing on the vertical surface of the flat-and-parallel packing-bed in contact with the turbulent gas flow under the conditions of phase cross-current (it is this mode of flows movement that is characteristic of film heat- and –mass transfer apparatuses in real ranges of changing gas and liquid working loads). For elementary liquid volume $\Delta v = \Delta x \cdot \Delta z \cdot \delta$, the amount of heat lost in the time unit

$$dQ_L = -c_L \cdot G_L \cdot \Delta t, \tag{2.72}$$

where

$G_L = \rho_L \cdot \delta \cdot \Delta z \cdot v_L$ – is the mass flow rate of the liquid (v_L is the average by the flow rate velocity of the liquid).

On the other hand, the amount of heat removed by convection on the free surface $\Delta F_s = \Delta x \cdot \Delta z$ is equal to:

$$dQ_L = \alpha_L \cdot (t_L - t_s) \cdot \Delta F_s \tag{2.73}$$

and the equation of the heat balance for liquid volume Δv has the form:

$$-c_L \cdot g_L \cdot \Delta t = \alpha_L \cdot (t_L - t_s) \cdot \Delta x$$

where

$$g_L = \rho_L \cdot \delta \cdot v_L = \frac{G_L^m}{2 \cdot n \cdot L}$$

– mass flow rate of the liquid referred to the unit of the packing-bed sheet width, kg/(m.s)

G_L^m – mass flow rate of the liquid in a film apparatus, kg/s;

n – number of packing-bed sheets;

L – packing-bed size in the direction of the air movement, m.

Taking the relationship (2.71) into consideration and going to the limit $\Delta x \to 0$, well obtain:

$$\frac{\partial t_L}{\partial x} = a_1 \cdot \left(t_g - t_L\right) + b_1 \cdot \left(p - p_s\right), \tag{2.74}$$

where

$$a_1 = \frac{\alpha_g}{c_L \cdot g_L}, \qquad b_1 = \frac{\beta_p \cdot r}{c_L \cdot g_L}.$$

Let us consider the air movement in the flat-and-parallel pacing bed between two surface wetted by liquid film running off. We distinguish the elementary volume of air $\Delta v = \Delta x \cdot \Delta z \cdot b$ («b» is the distance between sheets). Then from the element of area $\Delta F_s = \Delta x \cdot \Delta z$ the free surface of liquid film

$$\Delta g_s = \beta_p \cdot \left(p_s - p\right) \cdot \Delta F_s \tag{2.75}$$

will evaporate. The increase of moisture content in the gas volume Δv_g will be

$$\Delta g_g = G_g \cdot \Delta x_p, \tag{2.76}$$

where

$G_g = \rho_g \cdot b \cdot v_g \cdot \Delta x$ – mass flow rate of the air in the channel cross-section $\Delta F_s^g = b \cdot \Delta x$;

$$x_p = 0.622 \cdot \frac{p}{p_g - p} = \frac{0.622}{p_g} \cdot p \quad (\text{as} \quad p_g >> p). \tag{2.77}$$

The material balance equation $\Delta g_g = 2 \cdot \Delta g_s$ at $\Delta z \to 0$ with equation (2.77) taken into account has the form

$$\frac{\partial p}{\partial z} = 3.2154 \cdot \frac{\beta_p \cdot p_g}{g_g} \cdot \left(p_s - p\right), \tag{2.78}$$

Where:

$$g_g = \rho_g \cdot v_g \cdot b = \frac{G_g^m}{n \cdot H}$$

– mass flow rate of the air reduced to the unit of the channel height, kg/(m.s);

G_g^m – total mass flow rate of the air of the film apparatus, kg/s;

n – number of sprayed channel of the packing-bed ;

H – packing-bed height, m.

As the result of heat-and-mass transfer for the air volume of Δv_g, it is enthalpy changes

$$\Delta Q_L = G_g \cdot \Delta h, \tag{2.79}$$

Where:

$h = c_g + r_0 \cdot x_p + c_p \cdot t_g \cdot x_p$ – specific enthalpy of wet air;

$r_0 = 2500 \quad kJ/kg$ – heat of vaporization at 0 °C.

Then equation (2.79) will take the form

$$dQ_L = G_g \left(c_v \cdot \Delta t_g + r_t \cdot \Delta x_p \right), \tag{2.80}$$

Where:

$$c_v = c_g + c_p \cdot x_p; \qquad r_t = r_0 + c_p \cdot t_g.$$

On the other hand, the amount of heat removed from the free surface of water $\Delta S_L = \Delta x \cdot \Delta z$ at convection and evaporation, is determined by the relationship

$$dQ_g + dQ_r = \left[\alpha_g \cdot \left(t_L - t_g \right) + r \cdot \beta_p \cdot \left(p_s - p \right) \right] \cdot \Delta x \cdot \Delta z. \tag{2.81}$$

Hence, the heat balance equation for the volume of the air under consideration is

$$dQ_L = 2 \cdot \left(dQ_g + dQ \right)$$

or going to the limit of $\Delta z \rightarrow 0$:

$$g_g \cdot \left(c_V \cdot \frac{\partial t_g}{\partial z} + r_t \cdot \frac{\partial x_p}{\partial z} \right) = 2 \cdot \alpha_L \cdot \left(t_L - t_g \right) + 2 \cdot r \cdot \beta_p \cdot \left(p_s - p \right) \quad (2.82)$$

$$\frac{\partial x_p}{\partial z} = \frac{0.622}{p_g} \cdot \frac{\partial p}{\partial z} = \frac{2 \cdot \beta_p}{G_g} \cdot \left(p_s - p \right),$$

$$g_g \cdot c_V \cdot \frac{\partial t_g}{\partial z} = 2 \cdot \alpha_g \cdot \left(t_L - t_g \right) + 2 \cdot \beta_p \cdot \Delta r \cdot \left(p_s - p \right), \quad (2.83)$$

Where: $\Delta r = r - r_t$.

As the numerical analysis shows for characteristic values defining the evaporation (condensation) of the water vapour

$$\frac{\beta_p \cdot \Delta p \cdot \Delta r}{\alpha_g \cdot \Delta t_g} << 1. \quad (2.84)$$

Hence, the gas temperature change can be found by the equation:

$$\frac{\partial t_g}{\partial z} = a_2 \cdot \left(t_L - t_g \right), \quad a_2 = \frac{2 \cdot \alpha_g}{c_V \cdot g_g}. \quad (2.85)$$

The mathematical model of evaporative cooling (condensation) processes in DEC/CTW at cross phase current is given by the system of equations:

$$\begin{cases} \dfrac{dt_L}{dx} = a_1 \cdot \left(t_g - t_L \right) + b_1 \cdot \left(p_g - p^* \right); \\ \dfrac{dt_g}{dz} = a_2 \cdot \left(t_L - t_g \right); \quad \dfrac{dp_g}{dz} = b_2 \cdot \left(p^* - p_g \right) \end{cases} \quad (2.86)$$

and boundary conditions

$$\text{at } x = 0, \quad t_L = t_L^0; \text{ at } z = 0, \quad t_g = t_g^0, \quad p_g = p_g^0. \quad (2.87)$$

The partial pressure value for saturated vapour on the free surface depends on the temperature of the liquid:

$$p^* = p^*(t_L). \tag{2.88}$$

Coefficients of heat-and-mass transfer for the flat-and-parallel packing-bed were found by relationships (2.33)

$$Nu_g = 0.023 \cdot Re_g^{0.8} \cdot Pr^{0.4}, \quad Sh = 0.95 \cdot Nu_g. \tag{2.89}$$

The boundary problem (2.86) – (2.87) is solved by the method of finite differences. Equations (2.86) are approximated by the following difference scheme:

$$\begin{cases} t_L^{i+1,k} = \left[1 - a_1 \cdot \Delta x \cdot t_L^{i,k}\right] + a_1 \Delta x \cdot t_g^{i,k} + b_1 \cdot \Delta x \cdot p_g^{i,k} - b_1 \cdot \Delta x \cdot p^*\left(t_L^{i,k}\right); \\ t_g^{i,k+1} = \left(1 - a_2 \cdot \Delta z \cdot t_g^{i,k}\right) + a_2 \cdot \Delta z \cdot t_L^{i,k}; \\ p_g^{i,k+1} = \left(1 - b_2 \cdot \Delta z \cdot p_g^{i,k}\right) + b_2 \cdot \Delta z \cdot p^*\left(t_L^{i,k}\right). \end{cases} \tag{2.90}$$

For boundary nodal points

$$t_L^{0.k} = t_L^0, \quad t_g^{i.0} = t_g, \quad p_g^{i.0} = p_g^0. \tag{2.91}$$

Liquid flowing on the corrugated surfaces has some features caused by the fact that because of forces of surface tension the liquid is accumulated in cavities a curved surface and the movement of the liquid in such a packing-bed is of a regular-jet nature, and effect of gas flow doesn't practically result in the removal of the liquid from the apparatus. On the basis of theoretical and experimental researches in jet flows on corrugated surfaces with regular roughness (P, E are corrugation parameters; p, e – regular roughness parameters) the authors obtained (section 2.3.2):

$$\delta_0 = 0.7986 \cdot G_{Lv}^{0.3277}, \tag{2.92}$$

Where:

G_{Lv} – the volume flow rate of the liquid in a jet, cm^3/s (at 8 mm≤ P ≤ ≤12 mm, 3 mm≤ E ≤4 mm; 8 mm≤ p ≤10 mm, 0.6 mm≤ e ≤0.8 mm.)

The value of free liquid surface (ℓ_1), as well as perimeters of wetted (ℓ_0) and dry (ℓ_2) sections of the corrugated surface are determined by the relations (figure 2.20):

$$\begin{cases} \ell_1 = p \cdot \sqrt{\dfrac{\delta_0}{2 \cdot E}}; \quad \left(\Delta = \dfrac{4}{p} \cdot \sqrt{2 \cdot \delta_0 \cdot E}, \quad \Delta_1 = \dfrac{4 \cdot E}{p}\right) \\ \ell_0 = \dfrac{p^2}{16 \cdot E} \cdot \left[\ln\left(\Delta + \sqrt{1+\Delta^2} + \Delta_1 \cdot \sqrt{1+\Delta^2}\right)\right] \\ \ell_2 = \dfrac{p^2}{32 \cdot E} \cdot \left[\ln\left(\Delta_1 + \sqrt{1+\Delta_1^2}\right) + \Delta_1 \cdot \sqrt{1+\Delta_1^2}\right] - \dfrac{\ell_0}{2} \end{cases} \tag{2.93}$$

Processes of the evaporative cooling (condensation) in DEC/CTW with corrugated components having regular roughness are again defined by the problem (2.86) – (2.88), in this case coefficients a_1, a_2, b_1, b_2 in Equations (2.86) take into account dry and wetted sections of the surface and are calculated by formulas:

$$a_1 = \frac{\alpha_L \cdot \ell_1 + k_\alpha \cdot \ell_0 + K}{c_L \cdot g_L}, \quad b_1 = \frac{\beta_p \cdot r \cdot \ell_1}{c_L \cdot g_L},$$

$$a_2 = \frac{2 \cdot \left(\alpha_L \cdot \ell_1 + k_\alpha \cdot \ell_0 + 2 \cdot K\right)}{p \cdot c_v \cdot g_g}, \quad b_2 = 3.2154 \cdot \frac{\beta_p \cdot p_g}{p \cdot g_g}, \tag{2.94}$$

$$k_\alpha = 1 \Big/ \left(\frac{1}{\alpha_s} + \frac{\delta}{\lambda_L} + \frac{1}{\alpha_L}\right),$$

Where:

k_α	–	heat transfer coefficient;
K		takes into account the efficiency of the fin BD of the corrugated surface.

Heat transfer coefficient are determined by the Kader formulas, usually used when the liquid (gas) is flowing through the channels with large bulges of the walls roughness [24, 81].

2.3.4.2. Simulation of Working Processes in IEC

The evaporative cooling process in the IEC is described by a system of equations (Figures 2.19 – 2.20):

$$\begin{cases} \dfrac{dt_w}{dx} = a_1 \cdot (t_{AX} - t_w) + b_1 \cdot (p - p^*) + c_1 \cdot (t_{mn} - t_w); \\ \dfrac{dt_{AX}}{dz} = a_2 \cdot (t_w - t_{AX}); \quad \dfrac{dp_{AX}}{dz} = b_2 \cdot (p^* - p); \\ \dfrac{dt_{mn}}{dz} = c_2 \cdot (t_w - t_{mn}). \end{cases} \tag{2.95}$$

$$\text{at } x = 0, \quad t_w = t_w^0; \text{ at } z = 0, \quad t_{AX} = t_{AX}^0, \quad p = p^0, \quad t_{mn} = t_{mn}^0. \tag{2.96}$$

Here constants $a_1, a_2, b_1, b_2, c_1, c_2$ take into account dry and sprayed sections of corrugated surface with regular roughness (2.92), (2.93), and heat exchange coefficients are determined by Kader formulas [24, 81]. The solution of equations (2,95) was carried out by the method of finite difference set:

$$\begin{cases} t_w^{i+1,k} = [1 - (a_1 + c_1) \cdot \Delta x] \cdot t_w^{i,r} + (a_1 \cdot t_{AX}^{i,k} - b_1 \cdot p^{i,k} + b_1 \cdot p^{i,k^*} + c_1 \cdot t_{mn}^{i,k}) \cdot \Delta x, \\ t_{AX}^{i,k+1} = (1 - a_2 \cdot \Delta z) \cdot t_{AX}^{i,k} + a_2 \cdot t_w^{i,k} \cdot \Delta z, \\ p^{i,k+1} = c_2 \cdot (1 - b_2 \cdot \Delta z) \cdot p^{i,k} + b_2 \cdot p^{i,k^*} \cdot \Delta z, \\ t_0^{i,k+1} = (1 - c_2 \cdot \Delta z) \cdot t_{mn}^{i,k} + c_2 \cdot t_w^{i,k} \cdot \Delta z, \end{cases} \tag{2.97}$$

Where:

i, k – define nodal points on x and z coordinates.

For boundary nodal points they define parameters of all the channels in the IEC both for single – multistage scheme. System (2.86) – (2.95) is easily transformed for the use with DEC (for the IEC/DEC or IEC/CTW scheme).

Discussed are the processes of regenerative indirect evaporative cooling of air (IEC/R) during which the whole air-flow is first cooled without contracting the liquid in «dry» channels and only then a portion of the flow (an auxiliary air flow) is passed to "wetted" channels of the packing bed, where the water film is cooled by evaporation in direct contact with the air flow. "Dry" and "wetted" channels alternate multichannel packing bed with heat conducting walls of high heat conductivity made of a thin-walled aluminium sheet. Thus, unlike the IEC scheme, shown in figure 1.6, the flows are separated not at the inlet of the apparatus but at the outlet of it, and a portion of the flow returns to "wetted" channels of the apparatus packing bed. The limit of cooling in IEC/R is the dew point temperature of the air to be cooled. The combination of such a regenerative IEC with the heat-exchanger HEX 13 (figures 1.1 – 1.5) without DEC is one of its versions. It is this version of CEC that was used in AACS designs (Section 3.3). Mathematical models developed make it possible to calculate temperature and humidity parameters (fields) in "dry" and "wetted" channels of air evaporative coolers and to optimize the cooling process allowing for the minimization of the energy consumption for its development.

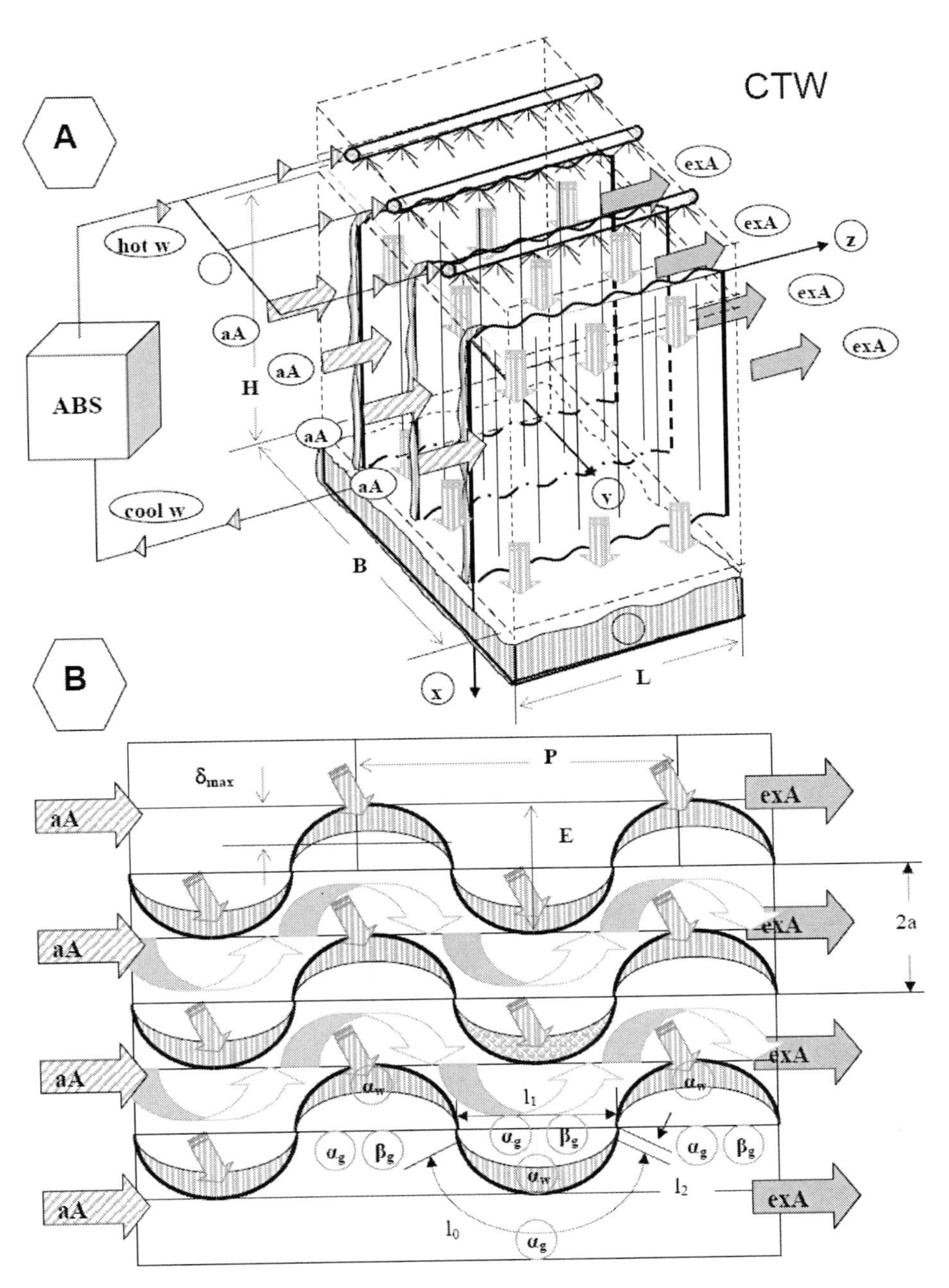

Figure 2.18. Schematic of the fan cross-flow-tower (A) and the scheme of gas liquid flows (B) in the channels of the regular packing bed (RP) of the heat-and- mass exchanging apparatus. Notations are given in figure 2.10. aA entering air flow; exA discharged flow; water.

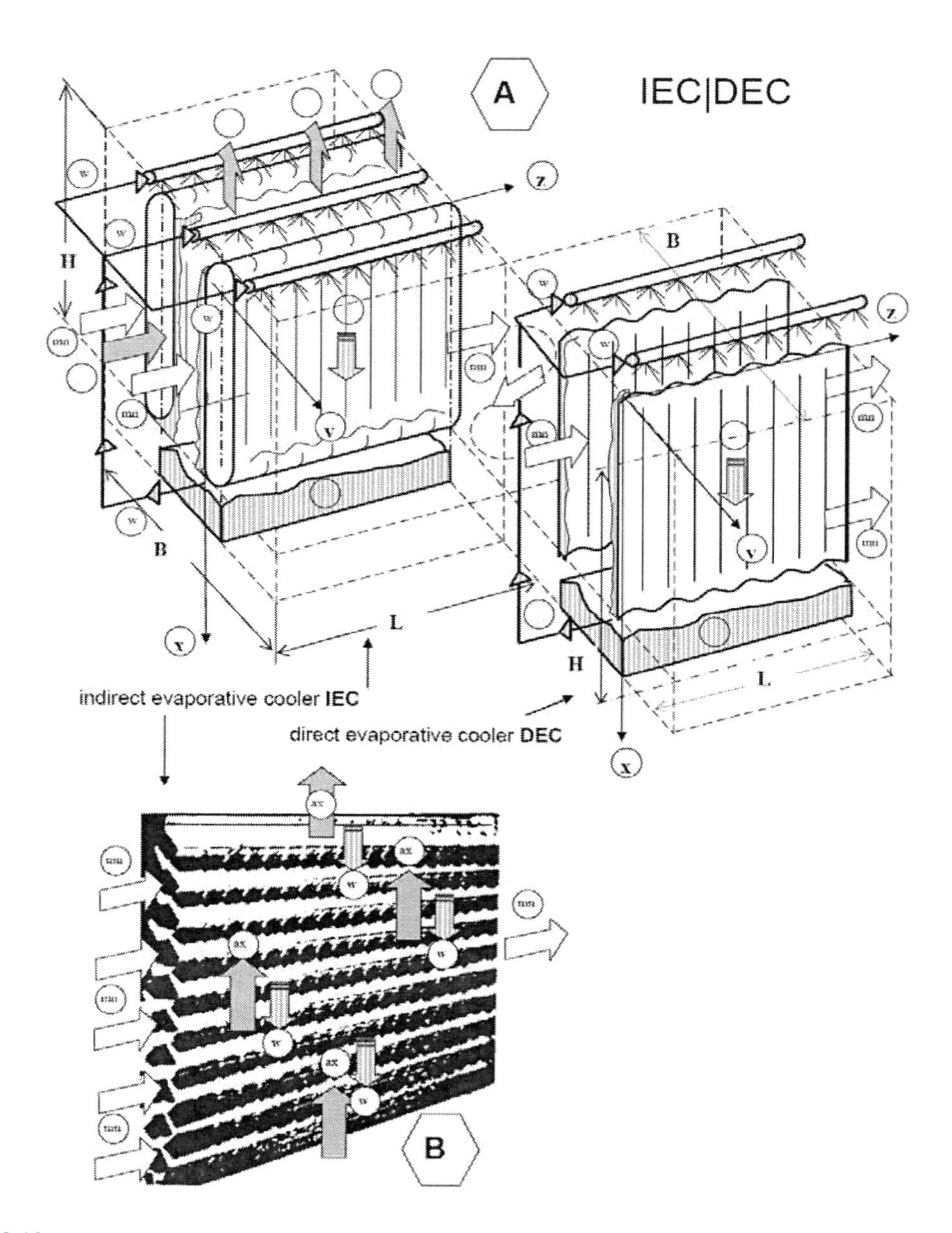

Figure 2.19. A. Evaporative unit of the ASAC. Schematic of the combined evaporative system on the IEC|DEC basis. Notations are given in the figure 2.10. B. Picture of the heat exchanging component – "wafer" – channels for the movement of the main air flow in IEC. Additional notations: ⇨ the main air flow; ➡ the auxiliary air flow; ——▷ the water flow. ⇨

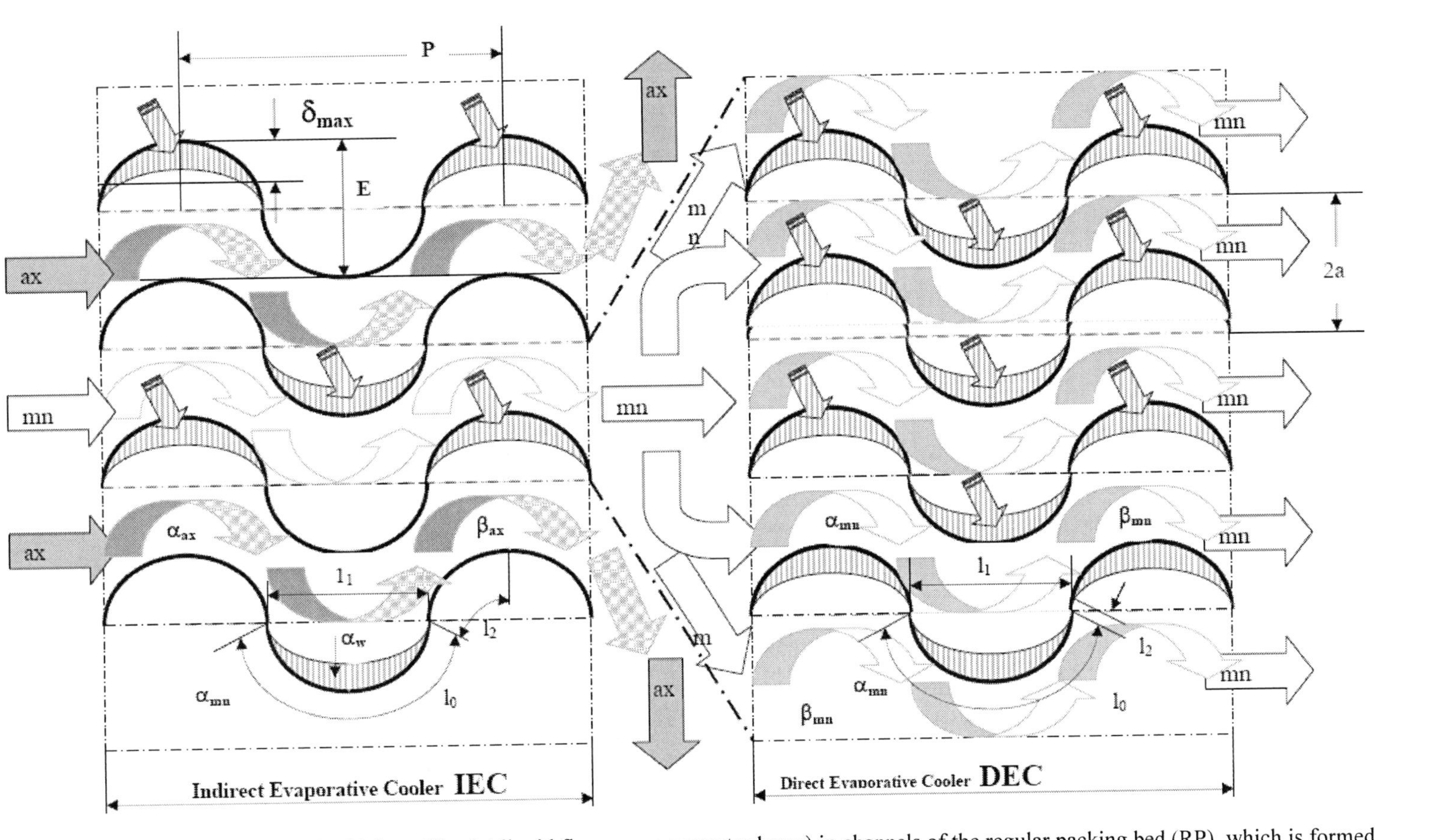

Figure 2.20. Schematic of gas and liquid flow (film-jet liquid flow, cross-current scheme) in channels of the regular packing bed (RP), which is formed by longitudinally corrugated sheets. Evaporative unit of AACS, the combined IEC|DEC cooler. Notation are given in figure 2.19. Additional notations:

the main air flow; the auxiliary air flow; the water flow.

2.3.5. Simulation of Working Processes in the Air-Solution Contactor (Pre-Dehumidification of Air by Liquid Absorbent and Solar Regeneration of Absorbent)

The drying part of AACSs including the dried (absorber) and the regenerator (desorber) is schematically shown in figure 1.8A in the form of film cross flow heat-and-mass transfer apparatuses (A) and apparatuses of combined with heat exchangers types – the cooler and the heater (Figure 1.8B). In the latter case heat-and-mass transfer apparatuses – the absorber and the desorber – have the packing bed in the form of tube-plate surfaces, which was developed by the project authors in conformity with air condensers of refrigerating plants [24]. The construction in figure 1.8A does not practically differ in any way from the cooling tower CTW described above, or the direct evaporative cooler DEC. Figures 1.8C and 2.22 show the absorber with internal evaporative cooling. The packing bed design of such an absorber is similar to that of the apparatus for indirect evaporative cooling. In the inner space of heat exchanging components-wafers the air flow being dehumidified moves in contact with the absorbent film streaming down the internal surfaces; in the interwafer space there moves the air flow and the water film streaming down the outer surfaces which removes the absorption heat, when the evaporative cooling takes place, through a thin heat conducting wall of heat-exchanging components (figure 2.22)

The design of the air dehumidifier (absorber) and the regenerator of solution (desorber) is carried out on the basis of the developed mathematical model of absorption (desorption) processes. Aqueous solutions of salts are used as absorbents for dehumidifying the air. Aqueous solutions of LiCl and LiBr are most widely used for dehumidifying the air, as well as aqueous solutions of salts with additives (inhibitors) which reduce the corrosive effects of the given solutions on construction materials (see 2.2). Mathematical model of absorption (desorption) processes under the conditions of cross phase-contacts is presented by a system of equations

$$\begin{cases} \dfrac{\partial t_s}{\partial x} = a_1 \cdot (t_g - t_s) + b_1 \cdot (p_g - p^{**}); \\ \dfrac{\partial t_g}{\partial z} = a_2 \cdot (t_s - t_g); \quad \dfrac{\partial p_g}{\partial z} = b_2 \cdot (p^{**} - p_g) \end{cases} \tag{2.98}$$

and boundary conditions

$$\text{at } x = 0, \quad t_s = t_s^0, \quad \xi = \xi^0 \quad \text{(initial concentration of the solution)},$$
$$\text{at } z = 0, \quad t_g = t_g^0, \quad p_g = p_g^0. \tag{2.99}$$

Significant distinction between the problem under consideration (2.98), (2.99) and that of evaporative liquid cooling (2.86), (2.87) lies in the fact that problem (2.98), (2.99) is mainly non-linear, as coefficients a_1, a_2, b_1, b_2, as well as p_s are function of the solution temperature and concentration. Coefficients of heat transfer in this case are determined by formulas as before [24, 81]. The problem (2.98), (2.99) is solved by the finite differences method. In designing the desorber the same equations (2.98) are used, but the solution temperature at the inlet to the apparatus is selected by the iteration method so that the concentration of the solution should reach the needed value ξ^0 at the outlet from the apparatus. The design programs for air-solution contactor (absorber – desorber) include equations approximating thermophysical properties of working liquids (solution) in the required ranges of working parameters (temperatures and concentrations) and thus the programs provide the possibility to calculate the working processes in the AACS irrelative of types of absorbents used.

2.3.6. Modeling the Processes in a Heliosystem with Flat Solar Collectors (An External Heating Source of AACS) and in a Tank-Accumulator

In developed AACS schemes a heliosystem with flat SCs is used as an external heating source. The heliosystem usually includes a SC system, a thermally insulated tank-accumulator, a system of downcomimg and lifting pipelines. The circulation of the heat carrier is realized by the natural way due to density differences in downcoming and lifting parts of the heliosystem, or by the forced one with the help of a pump (figures 2.33, 2.25). This section describes the methods of thermal and hydraulic designs developed by the authors for both types of heliosystems.

2.3.6.1. Heliosystem with Natural Circulation of the Heat Carrier

Natural circulation of the heat carrier is used in comparatively small heliosystem. Despite the simplicity of these systems design as compared to systems with forced circulation, methods of theoretical investigations are more complicated because of the necessity to conduct both thermal and hydraulic designs simultaneously, as the heat-carrier flow rate is the value to be determined. Some additional difficulties occur because of the nonstationarity of processes taking place in heliosystem components.

The heliosystem design has been made on the basis of a mathematical model accounting for the most important characteristics of thermal and hydrodynamic processes occurring in the system on the whole and its every component, at the same time mathematical description provided for the conjugation of all heliosystem components. The model being developed is presented by nonstationary equations of energy, movement and continuity for all components of the plant with the relevant initial and boundary conditions which provide conjugation of these components. The design process can be conditionally divided into some stages: thermal design of the SC; thermal design of TA together with connecting pipelines; determination of the heat carrier flow rate at its natural circulation.

A mathematical SC model has been developed (figures 2.26 – 2.27) which had some modifications and additions as compared to know solutions [3, 17, 21]:

- the offered model is based on the two-dimensional concept of heat processes allowing for the material heat capacity;
- the longitudinal heat conduction of the absorber plate has been taken into consideration in the direction of the liquid movement;
- the effect of cooling the absorber by the liquid coming from the TA was taken into account.

The transparent coating temperature $T_s = T_s(\tau, x, y)$ is found by the equation:

$$\delta_s \cdot \rho_s \cdot c_s \cdot \frac{\partial T_s}{\partial \tau} = U_{ps} \cdot (T_p - T_s) + U_{s0} \cdot (T_s - T_0). \tag{2.100}$$

In calculating U (the coefficient of heat losses) we took into account the plate and glass radiation, free convection between the plate and the glass, the heat conduction through the thermal insulation, etc. according to the methods given in [17].

The plate temperature is defined by the equation

$$\delta_p \cdot \rho_p \cdot c_p \cdot \frac{\partial T_p}{\partial \tau} = \lambda_p \cdot \delta_p \cdot \left(\frac{\partial^2 T_p}{\partial x^2} + \frac{\partial^2 T_p}{\partial y^2} \right) + \left(J - U_i \cdot (T_p - T_0)\right) + U_{ps} \cdot (T_s - T_p), \tag{2.101}$$

where $U_i = \dfrac{\lambda_i}{\delta_i}$.

The initial condition of equations (2.100) and (2.101) has the form

at $\tau = 0, \quad T_p = T_s = T_0$.

Boundary conditions for the plate range $0 \le x \le w/2, \quad 0 \le y \le L_{sc}$ can be presented in the following way:

at $x = \dfrac{w}{2}, \quad \dfrac{\partial T_p}{\partial x} = 0$,

$$\text{at } x = 0, \quad 2 \cdot \lambda_p \cdot \delta_p \cdot \frac{\partial T_p}{\partial x} = \alpha_L \cdot \pi \cdot d_L \cdot (T_p - T_L). \tag{2.102}$$

Here one can suppose that the heat flow is uniformly distributed around the pipe perimeter

at $y = 0, \quad \dfrac{\partial T_p}{\partial y} = 0$,

$$\text{at } y = L_{sc}, \quad \frac{\partial T_p}{\partial y} = 0. \tag{2.103}$$

Temperature fields $T_s = T_s(\tau, x, y)$ and $T_p = T_p(\tau, x, y)$ are determined in the range:

$0 \le \tau \le \tau_s$, (τ_s is the light day duration)

$$0 \le x \le \frac{w}{2}, \quad 0 \le y \le L_{sc}. \tag{2.104}$$

The typical SC construction is defined by inequality $L_{sc} >> W/2$, therefore it seems possible to average the temperature field by the coordinate x, i.e. to reduce the problem to a one-dimensional one, supposing

$$\overline{T}_p(\tau,y) = \frac{2}{w} \cdot \int_0^{w/2} T_p(\tau,x,y)dx, \; \overline{T}_s(\tau,y) = \frac{2}{w} \cdot \int_0^{w/2} T_s(\tau,x,y)dx,$$

In heliosystems with the natural circulation the laminar mode of the liquid flow is usually ound, for which the following relationship is valid: $\alpha_L = 4.36 \cdot \frac{\lambda_L}{d_L}$.

Taking into account this relationship and the above boundary conditions one can write equations (2.100) and (2.101) in the form:

$$\delta_s \cdot \rho_s \cdot c_s \cdot \frac{\partial T_s}{\partial \tau} = U_{ps} \cdot (T_p - T_s) + U_{s0} \cdot (T_s - T_0), \tag{2.105}$$

$$\delta_p \cdot \rho_p \cdot c_p \cdot \frac{\partial T_p}{\partial \tau} = \lambda_p \cdot \delta_p \cdot \frac{\partial^2 T_p}{\partial y^2} - 13.7 \cdot \frac{\lambda_L}{w} \cdot (T_p - T_L) + \\ + (J - U_i \cdot (T_p - T_o)) + U_{ps} \cdot (T_s - T_p). \tag{2.106}$$

Averaging signs in these formulas are omitted. The system of equations (2.105) and (2.106) was solved under the following conditions:

initial conditions

$$\text{at } \tau = 0, \quad T_s = T_p = T_0, \tag{2.107}$$

boundary conditions

$$\text{at y=0 and at } y = L_{sc}, \quad \frac{\partial T_p}{\partial y} = 0 \tag{2.108}$$

The temperature distribution in the liquid flowing through SC pipes is defined by the differential equation of convection

$$\rho_L \cdot c_L \cdot \frac{\pi \cdot d_L^2}{4} \cdot \frac{\partial T_L}{\partial \tau} + c_L \cdot \frac{G_L}{n} \cdot \frac{\partial T_L}{\partial y} = 13.7 \cdot \lambda_L \cdot \left(T_p - T_L\right). \qquad (2.109)$$

The thermal SC design is to solve equations (2.105), (2.106) and (2.109) at the following initial and boundary conditions:

$$\text{at } \tau = 0, \quad T_s = T_p = T_L = T_0;$$

$$\text{at } y = 0, \quad \frac{\partial T_p}{\partial y} = 0, \quad T_L = T_2\left(L_2\right);$$

$$\text{at } y = L_{sc}, \quad \frac{\partial T_p}{\partial y} = 0 \qquad (2.110)$$

The mathematical TA model (figure 2.26) takes into account the effect of mixing stratified liquid layers as a result of convection when the hot heat carrier is supplied. The energy conservation equation for the TA has the form:

$$\rho_L \cdot c_L \cdot f_T \cdot \frac{\partial T_T}{\partial \tau} + c_L \cdot G_L \cdot \frac{\partial T_T}{\partial z} = \lambda_L \cdot f_T \cdot \frac{\partial^2 T_T}{\partial z^2} + U_{T0} \cdot P'_T \cdot \left(T_1 - T_0\right). \qquad (2.111)$$

The initial conditions is: at $\tau = 0, \quad T_T = T_0$.

The boundary conditions is: at $z = 0, \quad T_T = T_2\left(\tau, L_2\right)$,

$$\text{at } z = H_T, \quad \lambda_L \cdot \frac{\partial T_T}{\partial z} = U_{T0} \cdot \left(T_T - T_0\right). \qquad (2.112)$$

The liquid temperature in the lifting pipelines $T_2 = T_2\left(\tau, y\right)$ is defined by equation

$$\rho_L \cdot c_L \cdot \frac{\pi \cdot d_2^2}{4} \cdot L_2 \cdot \frac{\partial T_2}{\partial \tau} + c_L \cdot G_L \cdot L_2 \cdot \frac{\partial T_2}{\partial y} = U_{20} \cdot \pi \cdot d_2 \cdot L_2 \cdot \left(T_2 - T_0\right). \qquad (2.113)$$

The initial conditions is: at $\tau = 0, \quad T_2 = T_0$.

The boundary conditions is: at $y = 0, \quad T_2 = T_L\left(\tau, L_{sc}\right)$. (2.114)

The liquid temperature in the downcomimg pipeline $T_1 = T_1\left(\tau, y\right)$ is defined by equation

$$\rho_L \cdot c_L \cdot \frac{\pi \cdot d_1^2}{4} \cdot L_1 \cdot \frac{\partial T_1}{\partial \tau} + c_L \cdot G_L \cdot L_1 \cdot \frac{\partial T_1}{\partial y} = U_{10} \cdot \pi \cdot d_1 \cdot L_1 \cdot \left(T_1 - T_0\right). \quad (2.115)$$

The initial condition is: at $\tau = 0, \quad T_1 = T_0$.

The boundary conditions is: at $y = 0, \quad T_1 = T_T\left(\tau, H_T\right)$. (2.116)

The determine the liquid flow rate around the close loop we use the Navier-Stokes equation for a one-dimensional liquid flow through the pipe [59]

$$\frac{\partial \omega}{\partial \tau} + \omega \cdot \frac{\partial \omega}{\partial \xi} = -\frac{1}{\rho_L} \cdot \frac{\partial p}{\partial \xi} + \omega \cdot \frac{\partial^2 \omega}{\partial \zeta^2} + g_\xi,$$

or

$$\rho_L \cdot \left(\frac{\partial \omega}{\partial \tau} + \omega \cdot \frac{\partial \omega}{\partial \xi}\right) = -\frac{\partial p}{\partial \xi} + \mu \cdot \frac{\partial^2 \omega}{\partial \zeta^2} + \rho \cdot g_\xi. \quad (2.117)$$

Averaging equation (2.18) by the cross-section of the pipe we shall obtain

$$\rho_L \cdot \left(\frac{\partial \overline{\omega}}{\partial \tau} + \overline{\omega} \cdot \frac{\partial \overline{\omega}}{\partial \xi}\right) = -\frac{\partial \overline{p}}{\partial \xi} + \tau_{st} + \rho \cdot g_\xi, \quad (2.118)$$

Where:

$\overline{\omega} = \overline{\omega}(\tau, \xi), \quad \overline{p} = p(\tau, \xi)$ are the cross-section average velocity and pressure of the liquid; τ_{st} is the tangential stress on the pipe wall.

Here it is taken into account that on the pipe axis ($\zeta = 0$)

$$\tau = \mu \cdot \left.\frac{\partial \omega}{\partial \zeta}\right|_{\zeta=0} = 0,$$

and on the pipe wall ($\zeta = d/2$)

$$\mu \cdot \left.\frac{\partial \omega}{\partial \zeta}\right|_{\zeta=\frac{d}{2}} = \tau_{st}.$$

The sign of averaging is further omitted. As the liquid motion as a result of natural convection takes place in the closed loop of the hydraulic system of the plant, we shall integrate equation (2.118) by this loop:

$$\oint \rho_L \cdot \frac{\partial \omega}{\partial \tau} d\xi + \oint \rho_L \cdot \omega \cdot \frac{\partial \omega}{\partial \xi} d\xi = -\oint \frac{\partial p}{\partial \xi} d\xi + \oint \tau_{st}(\xi) d\xi + \oint \rho_L \cdot g_\xi \cdot \frac{\partial \omega}{\partial \tau} d\xi \cdot \qquad (2.19)$$

Let us analyse the values in the right part of equation (2.119). It is possible to show that [118]

$$\oint \tau_{st}(\xi) d\xi = -\rho_L \cdot g \cdot \sum_i h_i ,$$

Where: h_i is specific head losses for overcoming all forces of resistance on i-th part of the hydraulic loop.

Let us consider the integral .

$$\oint \left[\rho_L \cdot g_\xi - \frac{\partial p}{\partial \xi} \right] d\xi$$

As it is know, the difference between the volume force and the pressure force can be taken as equal to $g \cdot (\rho - \rho_0)$ at the natural convection of the liquid, thus

$$\oint \left[\rho_L \cdot g_\xi - \frac{\partial p}{\partial \xi} \right] d\xi = -\oint g \cdot (\rho_L - \rho_0) \cdot \sin \varphi_\xi d\xi = -g \cdot \oint (\rho_L - \rho_0) \cdot \sin \varphi_\xi d\xi ,$$

where φ_ξ is the slope angle of the loop $d\xi$ with respect to the horizontal.

Let us analyse the values in the left part of equation (2.119). As the liquid velocity at the natural circulation is low, the velocity component $\oint \rho_L \cdot \omega \cdot \frac{\partial \omega}{\partial \xi} d\xi = \rho \cdot \oint \frac{\partial}{\partial \xi} \left(\omega^2 / 2 \right) d\xi$ can be neglected. Let us represent the first component of the left part in the form:

$$\oint \rho_L \cdot \frac{\partial \omega}{\partial \tau} d\xi = \sum_i \rho_i \cdot \frac{\partial \omega_i}{\partial \tau} \cdot L_i \cdot$$

As a result of it, equation (2.119) will have a form more convenient for practical calculations

$$\sum_i \frac{\partial \omega_i}{\partial \tau} \cdot L_i = -g \cdot \oint \frac{1}{\rho_0} \cdot (\rho_L - \rho_0) \cdot \sin \varphi_\xi d\xi - g \cdot \sum_i h_i \cdot \qquad (2.120)$$

Writing down the equation of continuity for all heliosystem components (SC, TA, of the downcoming and lifting pipelines) and allowing for the relation $\frac{H_T}{f_T} << \frac{L_1}{f_1}$ we can write the first component in the form:

$$\sum \frac{\partial \omega_i}{\partial \tau} \cdot L_i = \frac{L}{\rho_L} \cdot \left(\frac{L_{sc}}{n \cdot f_{sc}} + \frac{L_1}{f_1} + \frac{L_2}{f_2} \right) \cdot \frac{\partial G_L}{\partial \tau} \cdot$$

Introducing the volume expansion coefficient, we will write the first component in the right part of equation (2.120) in the form:

$$\frac{1}{\rho_0} \cdot \oint (\rho_L - \rho_0) \cdot \sin\varphi d\xi = \beta' \cdot \int_0^{L_{sc}} [T_L(y) - T_2(L_2)] \cdot \sin\varphi_{sc} dy - \beta' \cdot \oint [T_T(y) - T_1(L_1)] dz \cdot \qquad (2.121)$$

Where:

φ_{sc} is the slope angle of the SC; the temperature difference along the length of connecting pipelines is neglected.

The last integral in the right part of equation (2.121) is the moving force for the free convection in the TA resulting in the formation of closed liquid currents inside the tank; therefore, the said integral can be neglected.

Writing down the expression for hydraulic losses, neglecting local resistances as compared to the resistance along the length, and allowing for the fact that for the laminar mode of flowing, which, as a rule, takes place at natural circulation, the coefficient of resistance along the length is defined by the relation: $\lambda' = \frac{64}{Re}$. The second component in the right part of formula (2.121) can be written down in the form:

$$g \cdot \sum h_i = \frac{32 \cdot \nu_L}{\rho_L} \cdot G_L \cdot \left(\frac{L_{sc}}{n \cdot d_{sc}^2 \cdot f_{sc}} + \frac{L_1}{f_1 \cdot d_1^2} + \frac{L_2}{f_2 \cdot d_2^2} \right) \cdot \qquad (2.122)$$

Here it is taken into account, that $\frac{H_T}{d_T^2 \cdot f_T} << \frac{L_1}{d_1^2 \cdot f_1}$ hence, the equation of the liquid movement will have the form:

$$\left(\frac{L_{sc}}{n \cdot f_{sc}} + \frac{L_1}{f_1} + \frac{L_2}{f_2} \right) \frac{\partial G_L}{\partial \tau} = \frac{\rho_L g \beta'}{2} \int_0^{L_{sc}} [T_L(y) - T_2(L_2)] dy - \qquad (2.123)$$

$$- 32 \cdot \nu_L \cdot G_L \cdot \left(\frac{L_{sc}}{n \cdot d_{sc}^2 \cdot f_{sc}} + \frac{L_1}{f_1 \cdot d_1^2} + \frac{L_2}{f_2 \cdot d_2^2} \right).$$

The initial condition is: at $\tau = 0, \quad G_L = 0$.

Thus, the heliosystem with the natural circulation is defined by the following system of equations, initial and boundary conditions:

For The Solar Collector

$$
\begin{cases}
\delta_s \cdot c_s \cdot \rho_s \cdot \dfrac{\partial T_s}{\partial \tau} = U_{ps} \cdot (T_p - T_s) + U_{s0} \cdot (T_s - T_0), \\
\delta_p \cdot c_p \cdot \rho_p \cdot \dfrac{\partial T_p}{\partial \tau} = \lambda_p \cdot \delta_p \cdot \dfrac{\partial^2 T_p}{\partial y^2} - 13.7 \cdot \dfrac{\lambda_L}{w} \cdot (T_p - T_L) + \\
\quad + [J - U_i \cdot (T_p - T_0)] + U_{ps} \cdot (T_s - T_p), \\
c_L \cdot \rho_L \cdot \dfrac{\pi \cdot d_1^2}{4} \cdot \dfrac{\partial T_L}{\partial \tau} + c_L \cdot \dfrac{G_L}{n} \cdot \dfrac{\partial T_L}{\partial y} = 13.7 \cdot \lambda_L \cdot (T_p - T_L), \\
\text{at } \tau = 0,\ T_s = T_p = T_L = T_0; \\
\text{at } y = 0, \dfrac{\partial T_p}{\partial y} = 0,\ T_L = T_2(L_2); \\
\text{at } y = L_{sc}, \dfrac{\partial T_p}{\partial y} = 0.
\end{cases}
\qquad (2.124)
$$

For the Tank-Accumulator (TA) in Downcoming and Lifting Pipelines

$$
\begin{cases}
c_L \cdot \rho_L \cdot f_T \cdot \dfrac{\partial T_T}{\partial \tau} + c_L \cdot G_L \cdot \dfrac{\partial T_T}{\partial z} = \lambda_L \cdot f_T \cdot \dfrac{\partial^2 T_T}{\partial z^2} + U_{T0} \cdot P'_T \cdot (T_0 - T_T), \\
c_L \cdot \rho_L \cdot \dfrac{\pi d_2^2}{4} \cdot L_2 \cdot \dfrac{\partial T_2}{\partial \tau} + c_L \cdot G_L \cdot L_2 \cdot \dfrac{\partial T_2}{\partial y} = U_{20} \cdot \pi \cdot d_2 \cdot L_2 \cdot (T_2 - T_0), \\
c_L \cdot \rho_L \cdot \dfrac{\pi d_1^2}{4} \cdot L_1 \cdot \dfrac{\partial T_1}{\partial \tau} + c_L \cdot G_L \cdot L_1 \cdot \dfrac{\partial T_1}{\partial y} = U_{10} \cdot \pi \cdot d_1 \cdot L_1 \cdot (T_1 - T_0), \\
\text{at } \tau = 0,\ T_1 = T_T = T_2 = T_0; \\
\text{at } y = 0,\ T_1 = T_L(L_{sc}),\ T_2 = T_T(H_T); \\
\text{at } z = 0,\ T_T = T_1(L_1); \\
\text{at } z = H,\ \lambda_L \cdot \dfrac{\partial T_T}{\partial z} = U_{T0} \cdot (T_0 - T_T).
\end{cases}
\qquad (2.125)
$$

The equation of the movement is

$$\begin{cases} \left(\dfrac{L_{sc}}{n\cdot f_{sc}}+\dfrac{L_1}{f_1}+\dfrac{L_2}{f_2}\right)\cdot\dfrac{\partial G_L}{\partial \tau}=\dfrac{\rho_L\cdot g\cdot \beta'}{2}\cdot\left[\int\limits_0^{L_{sc}} T_L(y)dy - T_2(L_2)\cdot L_{sc}\right]- \\ -32\cdot \nu_L\cdot G_L\cdot\left(\dfrac{L_{sc}}{n\cdot d_{sc}^2\cdot f_{sc}}+\dfrac{L_1}{f_1\cdot d_1^2}+\dfrac{L_2}{f_2\cdot d_2^2}\right); \\ \text{at}\quad \tau=0,\ G_L=0. \end{cases} \qquad (2.126)$$

The calculation was done by the method of finite differences using the implicit scheme of Krank-Nikolson which is absolutely stable. The heat design program for a heliosystem was developed which makes it possible to determine the heat carrier temperature in all components of the plant at any time. Figure 2.24 gives the comparison of our temperature calculation results for the TA of the helioplant with the results of approximate calculations [17]. The design was carried out for a helioplant with the natural circulation of the heat carrier having 12 SCs of 12 m^2 total area and the volume of the TA 1 m^3. The solar radiation intensity and the ambient temperature were taken from the Reference-book on Climate for Odessa in July. It follows from Figures that the account of factors, mentioned in the beginning of the section, in the model results in the temperature increase in the lower part of the TA, and, hence, at the inlet to the SC, which decreases the design capacity of the plant. The temperature in the upper part of the TA is lower than that in reference [17] because of which the average liquid temperature in the TA decreased. Such behavior agrees quite will with physical conceptions of the influence of changes occurring in the model on heliosystem characteristics.

2.3.6.2. Heliosystem with Forced Circulation of the Heat Carrier

Mathematical model of the heliosystem with forced circulation is simplified as compared to the above-mentioned one since the flow rate of the heat carrier is not the parameter to be defined. With the increase of the flow rate the COP of the heliosystem increases, but at the same time more energy is consumed for the movement of the heat carrier. In practice they usually select the flow rate proceeding from the relation $G\cdot c_p / F_{sc}\cdot U =$ 2-4, which corresponds to the efficiency of SC equal to 86 – 90 % of the maximum possible one under given outer conditions [17]. In modeling heliosystems with forced circulation two approaches are usually used. One is based on large-scale averaging the characteristics of heliosystems proceeding from accumulated experimental data. In this case the results are obtained rather easily, however because of low accuracy they can be used only for approximate evaluation of heat capacity of the heliosystem. The other approach, which is being developed by the authors of the project, is based on solving differential balance equations for heliosystems and makes it possible to predict accurately enough time dependencies of heat characteristics of heliosystems allowing for their construction peculiarities and varying climatic conditions. Both approaches are analysed below.

Normative method of calculating heliosystems. COP of the heliosystem can be calculated by formula [17]

$$\eta = 0.8 \cdot \left\{ \theta - 9 \cdot U \cdot \frac{\left[0.5 \cdot \left(t^{out} - t^{in}\right)\right] - t_0}{\sum_i J_i} \right\}, \tag{2.127}$$

where 0.8 is a coefficient taking into account the dust content and shadowing; the summation is done by hours of the light day.

The solar radiation intensity in the solar collector plan is founded by expression:

$$J = J_S \cdot k_S + J_D \cdot k_D.$$

Coefficient k_S is defined by Tables [17], and k_D is calculated by the relation

$$k_D = \cos^2 \beta/2,$$

where β is the slope of the SC plane to the horizon (30° for the latitude of Odessa).

The data for J_S, J_D and t_0 during every hour of the current month are given in the construction norms and rules CN&R 2.04.05-91 ("Heating, ventilation and air-conditioning", State Construction Committee of the USSR, Moscow, 1997). The required area of the SC is defined by formula

$$F_{sc} = \frac{Q'}{\eta \cdot \sum_i J_i}, \tag{2.128}$$

where Q' is the required 24-hour heat load for the warmest month when there is a duplicate source of heat in the heliosystem.

The volume of tank-accumulator is defined by empirical formula

$$V_{TA} = (0.06 - 0.08) \cdot F_{sc}, \text{ m}^3/\text{m}^2. \tag{2.129}$$

The largest values of the volume in this formula refer to the IV climatic zone. To find a value of COP seasonal η_{seas} depending on F_{sc} and V_{TA} one can use the nomograph given in [17]. The amount of heat energy generated by the heliosystem during a season (April-September) is equal to

$$Q_p = \eta_{seas} \cdot Q_{seas},$$

where Q_{seas} is the amount of heat falling on the SC of the heliosystem during a season.

Mathematical model of the heliosystem. The calculation procedure described above is of an approximate character and can be used for orientation evaluation of its efficiency. The mathematical model offered is to a large extent deprived of disadvantages, characteristic of the normative procedure, and makes it possible to obtain more realistic results. While deriving equations which describe processes of heat exchange in the heliosystem it was assumed:

- processes in the heliosystem are of a quasi-steady nature. It allows the use of characteristics, obtained under stationary conditions, in calculations. The legitimacy of this assumption is stipulated for a slow change of the solar radiation intensity and the temperature of surroundings;
- heat losses through pipe walls into surroundings were not taken into account. As the calculations and the experience in using heliosystems show, these losses are insignificant due to small surface of pipes and the heat insulation available;
- the temperature separation (stratification) of the heat carrier in TA was not taken into account. With the forced circulation the velocity of the heat carrier is greater than that with the natural one, which results in stirring the liquid. The assumption of complete stirring leads to under-estimated values of heat capacity, providing in such away a calculation store of generated heat energy.
- The heat balance equation for the heliosystem has the form

$$J \cdot \eta \cdot F_{sc} d\tau = M_{TA} \cdot c_p dt + k_{TA} \cdot F_{TA} \cdot (t - t_0) d\tau \quad , \qquad (2.130)$$

where t is the current temperature of the heat carrier in TA;

M_{TA} is the mass of the heat carrier in TA.

Equation (2.130) can't be solved with respect to the unknown temperature as ηdepends on t. Let us the known expression for COP of SC

$$\eta = 0.8 \cdot \left\{ \theta - U \cdot \frac{\left[0.5 \cdot (t^{out} - t) - t_0\right]}{J} \right\} . \qquad (2.131)$$

From the heat balance equation of the heliosystem

$$J \cdot \eta \cdot F_{sc} = G \cdot c_p \cdot \left(t^{out} - t\right) \qquad (2.132)$$

we shall find temperature t^{out} introducing the idea of specific flow rate $g = G/F_{sc}$

$$t^{out} = t + \frac{J \cdot \eta}{g \cdot c_p} . \qquad (2.133)$$

From expressions (2.132) and (2.133) we obtained

$$t^{out} = t \cdot \left(\frac{g \cdot c_p - 0.4 \cdot U}{g \cdot c_p + 0.4 \cdot U} \right) + 0.8 \cdot \left(\frac{J \cdot \theta + U \cdot t_0}{g \cdot c_p + 0.4 \cdot U} \right). \tag{2.134}$$

Thus, the problem of determining the time dependence of temperature in TA is reduced to solving the nonlinear differential equation of the first order

$$\frac{dt}{d\tau} = \frac{J(n,\tau) \cdot \eta \cdot F_{SC}}{M_{TA} \cdot c_p} - \frac{k_{TA} \cdot F_{TA} \cdot [t - t_0(n,\tau)]}{M_{TA} \cdot c_p} \tag{2.135}$$

allowing for expressions (2.131) and (2.134). In equation (2.34) n means the number of the month for which the calculation is done. The equation was solved by the method of Runge-Kutta with the time period of 1 hour. The calculations given with the time period twice as much showed a slight deviations of results.

In using the model the following factors were taken into account:

- the dependence of the heat loss coefficient on temperature. It is usually assumed that U = const. However, as the calculations show, with the increase of the absorbing plane temperature from 50 °C to 100 °C the heat loss coefficient increases by 30 %;
- the influence of the wind velocity on the heat loss coefficient. The increase of the wind velocity from 0 to 5 m/s increases the heat loss coefficient by 15 – 20 %, therefore the calculations were done allowing fore a real wind load for the area where AACS is located;
- the dependence of the penetration capacity of the transparent insulation from the angle of incidence of solar rays on the SC plane. It is necessary to take this factor into account, as the glass will reflect practically total radiation falling on its surface, if the said angle is over 60°, [17].

Using the described procedures the calculation was done of the heliosystem for hot water supply of a 50-flat house situated in Odessa. The total area of SC was 236 m^2, the hot water temperature – 50 °C. The seasonal generation of heat calculated by an approximated method amounted to 310 GJ, and the heat obtained by means of the developed mathematical model was 275 GJ. Such a discrepancy is caused both by a greater physical correctness of the approach offered by the authors and the account of the above-mentioned factors.

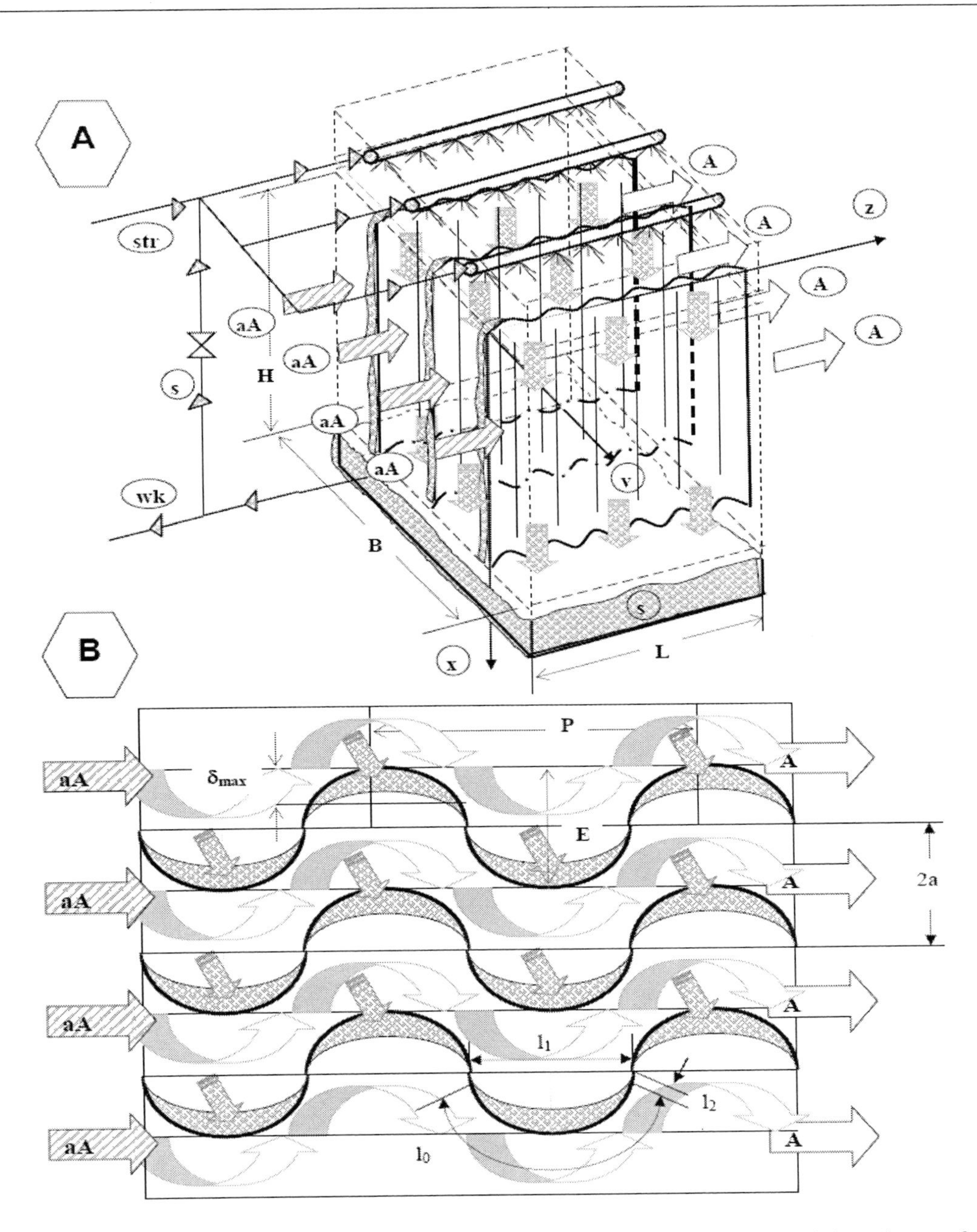

Figure 2.21. Dehumidifying unit of the AACS. Schematic of the absorber (A) and the scheme of gas and liquid flows (B) *of the air flow being dried and the absorbent* – in the channels of the regular packing bed (RP). Nomenclature is the same as in figure 2.19 In addition to the above [aA arrow] entering air flow; [A arrow] dried flow; [arrow] absorbent.

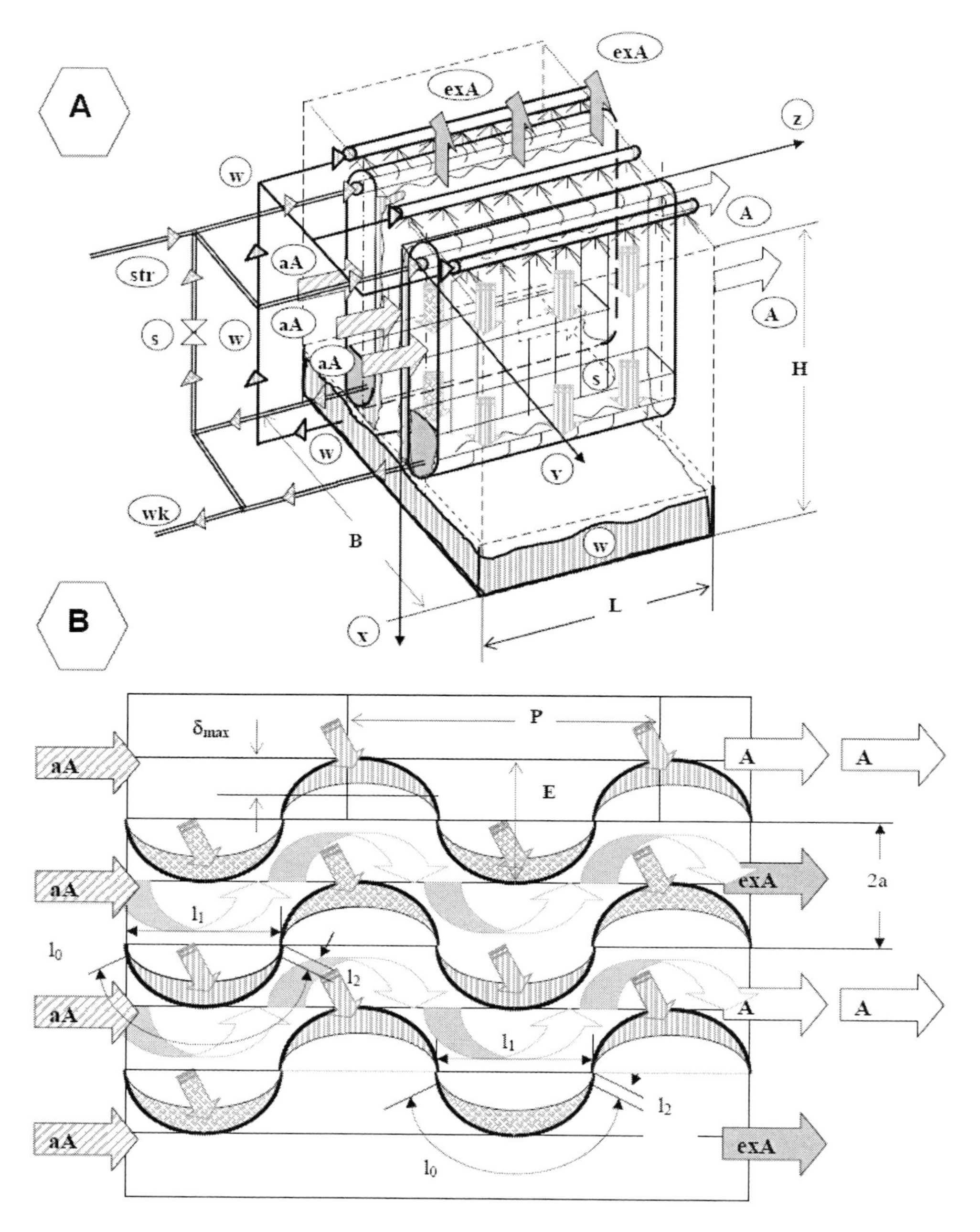

Figure 2.22. Dehumidifying unit of the ASAC. Schematic of the absorber (A) with internal evaporative cooling and the scheme of gas, water and absorbent flows (B) in channels of HMA packing bed.

Nomenclature is the same as in figure 2.19: aA inlet of air; A outlet of the dried air flow; exA outlet of the auxiliary air flow; absorbent; water.

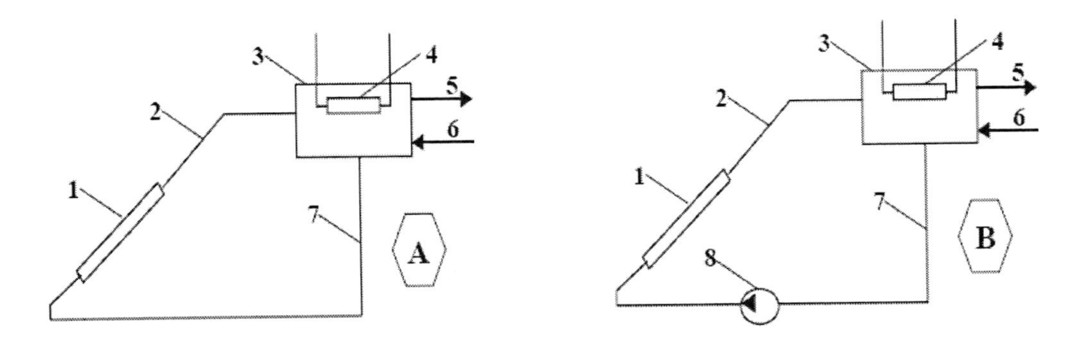

Figure 2.23. Schematics of heliosystems. A – heliosystem with natural circulation of heat carrier, B – heliosystem with forced circulation of heat carrier. 1-SC, 2-lifting pipe-line, 3-tank-accumulator, 4-duplicate source of heat, 5-hot water extraction for the consumer, 6-cold water replenishment, 7-lowering pipe-line, 8-water pump.

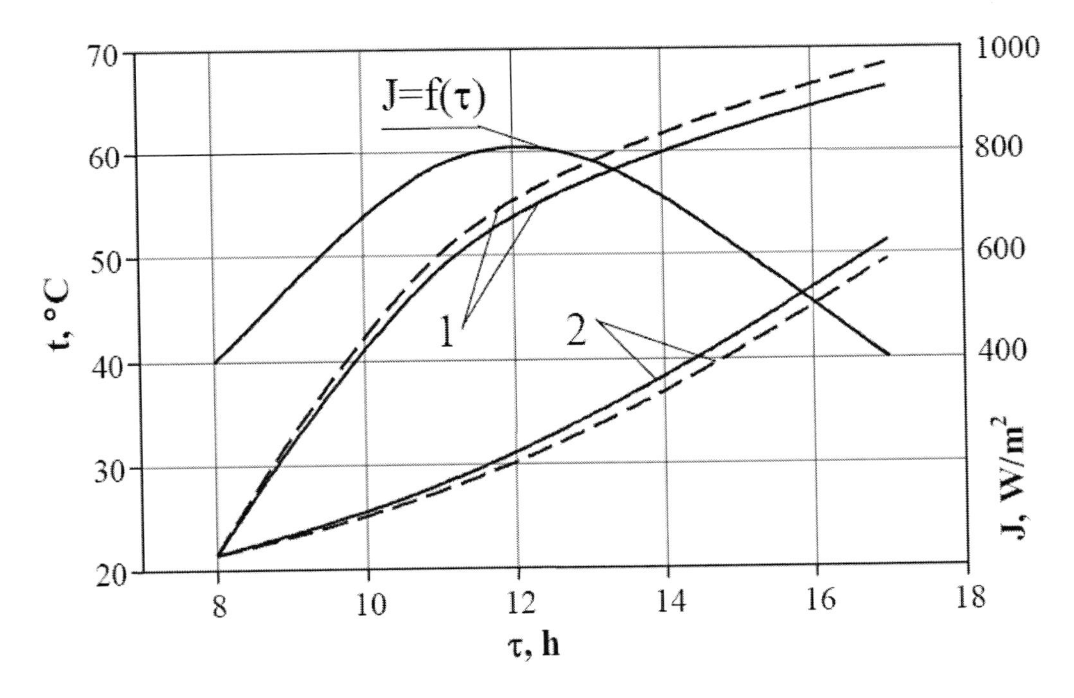

Figure 2.24. Water temperature in the tank-accumulator. 1-upper part of tank-accumulator, 2-lower part of tank-accumulator. - - - - - results of approximated calculations; ———— calculation by developed mathematical model.

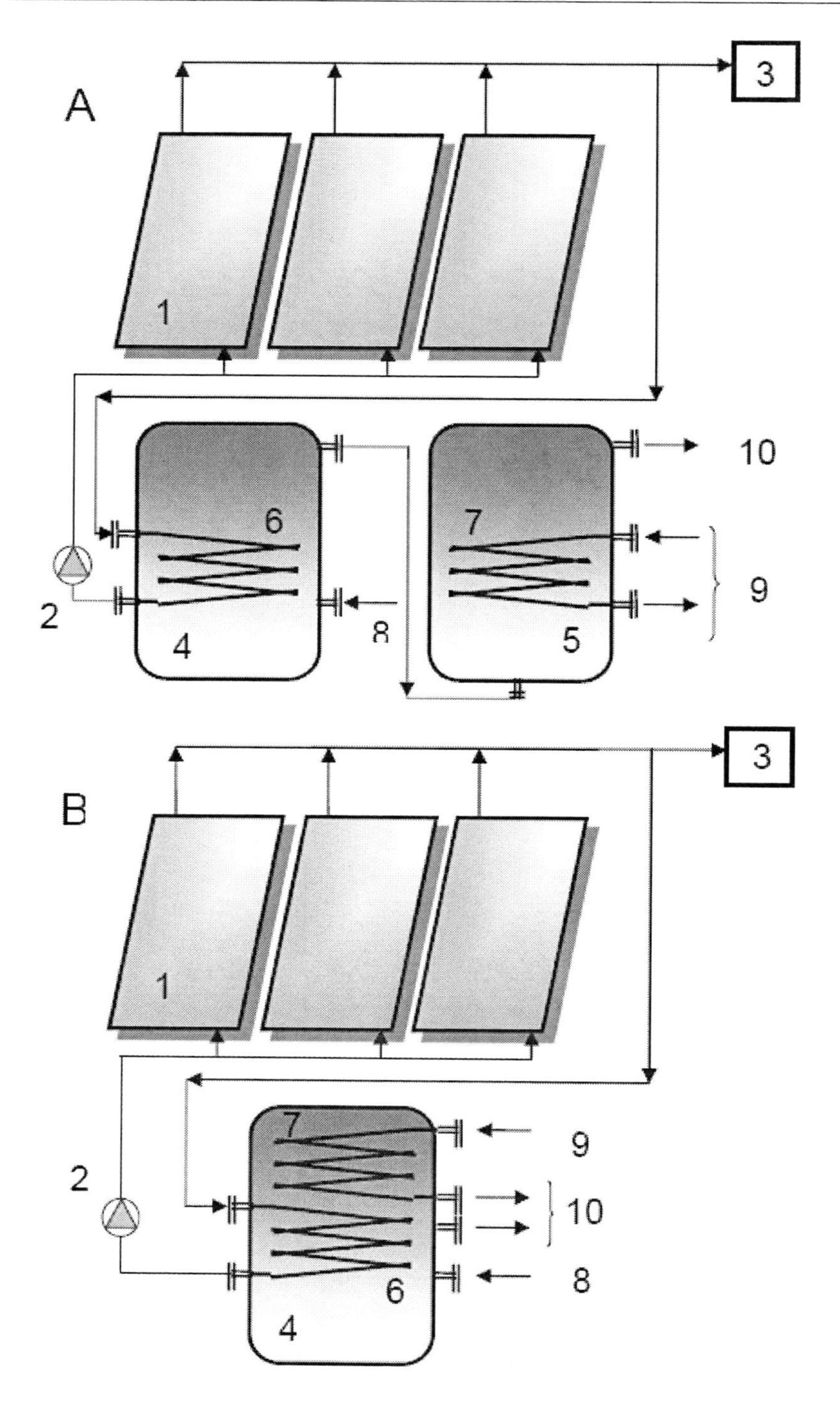

Figure 2.25. The heliosystem – the external heating source, providing solar regeneration of the absorbent in an alternative system; A - Two circuit heat supply system with two tanks- accumulators; B - Two circuit heat supply system with a combined tank- accumulator. Notation: 1 – solar collectors SC; 2 – pumps; 3 – expansion tank; 4 – tank-accumulator; 5 – tank for hot water; 6, 7 – heat exchangers; 8 – cold water replenishment; 9 – additional source of heat; 10 – water for the heat supply system.

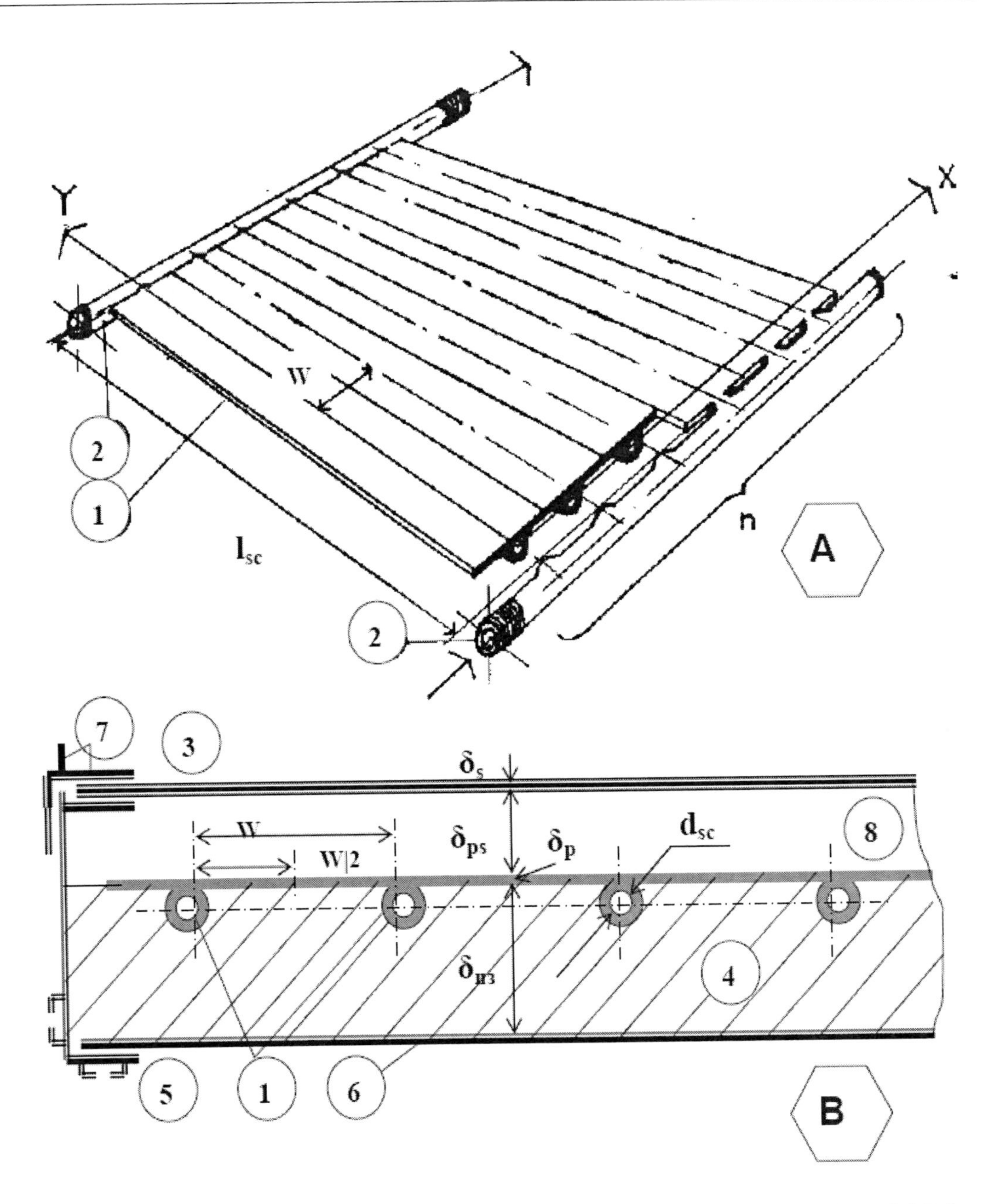

Figure 2.26. On modeling working processes in a flat solar collector. A is the absorber of SC (heat receiver), B is the cross section of the SC absorber. Nomenclature: 1 – the pipe register SC; 2 – collector pipes; 3 – glass (transparent insulation); 4 – heat insulation; 5 – the collector housing; 6 – the bottom; 7 – the clamping device (fastening the glass); 8 – the air gap.

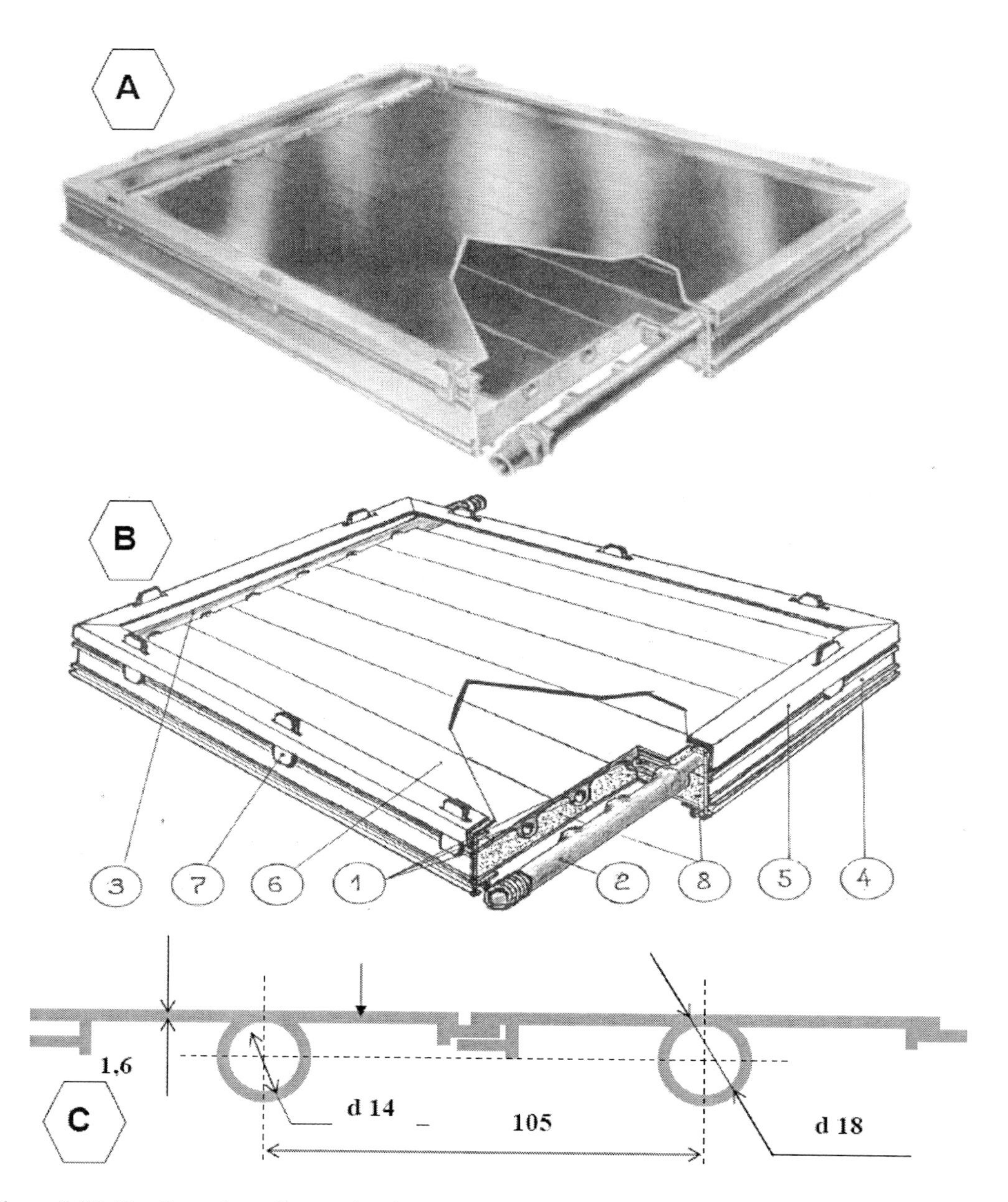

Figure 2.27. The flat solar collector developed by the SPF "New Technologies" produced since 1991. A is a picture of the CK-1.1 modification collector. B is the general view of the collector. Notation: 1 – pipe register; 2, 3 – collector pipes; 4 – SC housing; 5 – fastening angle; 6 – glass; 7 – fastener; 8 – thermal insulation. C is the heat receiver in the form of a pipe register with fins.

3. ENGINEERING SUPPLEMENT (TECHNIKO-ECONOMICAL) CHARACTERISTICS OF A PILOT PLANT

3.1. Modeling of Working Processes in Alternative Refrigerating and Air-Conditioning Systems (The Design Procedure)

The main design scheme is given in figure 3.1, it shows the numbers of design points (parameters of flows). The simplified design of heat-and-mass transfer apparatuses is given – of the absorber and the desorber—with outstanding heat-exchangers HEX 11 and 12. The scheme is oriented to the AACS and includes a combined evaporative cooler IEC/R or IEC/DEC in the cooling part. In figure 3.2 this scheme is given with the use of cross-flow heat-and-mass transfer apparatuses (the absorber, desorber, cooling tower, indirect and direct evaporative coolers), which corresponds to the calculations carried out and executed developments of alternative air-conditioning systems. The cross-flow scheme corresponds to all other schematic developments of the AACS given in figure 3.3 – 3.4 and 3.23.

The algorithm of alternative systems design is built in the following sequence:

- the calculation of the air dehumidification process in the absorber and further the process of cooling the air in the evaporative cooler IEC/DEC, the determination of heat-moisture parameters of the main air flow leaving the evaporative cooler (of the temperature and the relative humidity) and coming to the conditioned room; the determination of the AACS refrigerating capacity;
- the calculation of processes in the dehumidification part of the AACS (in the absorber and the desorber), the determination of the cooling water temperature and flow rate (the calculation of the process in the cooling tower) and the temperature and capacity of the heating source;
- the calculation of the heliosystem with the determination of the solar collectors type and the required surface of heat receivers.

1. We assume the type, concentration ξ (Section 2.2) and the temperature of the absorbent t_s^8 (in the first approximation) at the inlet to the absorber. All geometrical characteristics of heat-and-mass transfer apparatuses are selected on the basis of the investigation carried out in the work (Section 2.3). The rate flow of the air entering the conditioned room is defined by the area and the height of this room, and the required multiplicity of air exchange in the room k (CN&R 2.04.05-91, "Heating, ventilation and air-conditioning", State Construction Committee of the USSR, Moscow, 1997). Let us select the working value of air movement velocity in channels of crossflow heat-and-mass transfer apparatuses (Section 2.3.3) which enables the calculation of the main overall dimensions of the packing bed of these apparatuses and their air resistanse magnitude. The design parameters of the outside air (temperature and moisture content) are defined for a chosen city or region allowing for the existing norms (construction norms and rules CN&R 2.04.05-91 "Heating, ventilation and air-conditioning", State Construction Committee of the USSR, Moscow, 1997).

- Solving the systems of equations (2.98) with boundary conditions (2.99) which describe heat-and-mass exchange in the absorber, we shall get concentration ξ and temperature t_s of the absorbent at the outlet from the absorber. Aqueous solutions of LiBr, LiBr+ and LiBr++ were considered as working bodies. Thermophysical properties of the solutions were defined on the basis of the literature data and formulas given in Section 2.2. Then let us calculate the main parameters of the air flow leaving the absorber – temperature t^2 and moisture content x^2.
- Then let us calculate the evaporative cooling processes in the cooling part of the plant. In the main, the generative indirect evaporative cooler IEC/R is used here; it includes the indirect evaporative cooler IEC proper and the heat exchanger HEX 13 (figure 3.3). The air flow rates on the process line 2–3 and on the line 5–7 amount to G_g and $G_g/2$, respectively. The heat balance equation at

$$G_g \cdot c \cdot \left(t^2 - t^3\right) = \frac{G_g}{2} \cdot c \cdot \left(t^7 - t^5\right), \text{ at } E_{HEX} = \left(t^7 - t^5\right) / \left(t^2 - t^5\right).$$

Let us define the air temperature values which are

$$t^5 = 2.5 \cdot t^3 - 1.5 \cdot t^2, \qquad (3.1)$$

$$t^7 = 0.5 \cdot \left(t^3 + t^2\right). \qquad (3.2)$$

for E_{HEX} = 0.8. All geometrical characteristics of IEC are chosen on the basis of the investigation carried out (Section 2.3). On solving the differential problem (2.95) – (2.96) by the specified value of air parameters at the inlet to the IEC (t^3 and x^3), we shall find the temperature of the auxiliary air flow at the outlet from the cooler:

$$t^5 = IEC\left(t^3\right), \qquad (3.3)$$

when the temperature of the water recirculating through the IEC reaches the steady state (the main condition of stabilizing the process in the indirect evaporative cooler). Thus, the heat calculation of IEC/R is reduced to a repetitive procedure with the use of equation (3.1) and operator equation (3.3). We assume the initial value of the auxiliary air flow temperature $t^{3,0}$ and solve the problem (2.95) – (2.96) of defining the temperature of the auxiliary air flow t^5 under the condition of the steady temperature of the water recirculating through the IEC t_w. It also defined in the course of the repetitive procedure by the variation of t_w at the inlet to the apparatus till the water temperature values at the inlet to and outlet from the apparatus become the same. As a result

$$t^{5,1} = IEC\left(t^{3,0}\right), \tag{3.4}$$

from (3.1) follows $t^{3,1} = 0.4 \cdot t^{5,1} + 0.6 \cdot t^{2}$. Then, having chosen $t^{3,0} = t^{3,1}$ as an initial value of the auxiliary air flow temperature, we continue the repetitive procedure till the subsequent value of $t^{3,k}$ differs little from the previous one (with the required accuracy of calculating this value). Finally, on solving the initial value problem Koshi (2.95) – (2.96) we define the temperature of the main air flow t^{4} at the outlet from IEC. The moisture content of the main air flow in this case does not change in cooling and is equal to x^{2}. Then, if it is necessary, we calculate the process in the direct evaporative cooler DEC [by solving the system of equations (2.86) – (2.87)], to which the main air flow is directed for additional cooling.

2. The calculation of processes in the dehumidifying part of the AACS (in the absorber and the desorber). We perform the heat calculation of the desorber, for which we define the absorbent temperature t_{s}^{11} in the course of the repetitive procedure; this temperature provides the restoration of the initial concentration of the absorbent ξ at the specified parameters of the outside air. This repetitive process corresponds to the solution of the differential problem (2.98) – (2.99), describing the process of heat-and-mass exchange in the desorber. Then, we define the temperature of the strong absorbent solution at the outlet from the desorber.

- The absorbent temperature values found make it possible to turn to the calculation of heat-exchangers HEX 10, HEX 11 and HEX 12. The calculation of heat carriers temperature are done by equation (2.24), Section 2.3.1.2. The area of the heat transferring surface is calculated by relations (2.25) – (2.88). On the basis of results obtained during the calculations the value of $E_{HEX} = 0.8$ was chosen, which corresponds to rather a wide range of changing the temperature of working bodies for an acceptable surface of heat-exchange.
- Let us do the heat calculation of cooling tower CTW. Geometrical characteristics of the apparatus packing bed, density of the layer and the relationship between the flow rates of water and air are chosen on the basis of the investigation performed (Section 2.3). By solving the system of equations (2.86) – (2.87), describing the processes of heat-and-mass exchange in the direct evaporative cooler, we obtain the water temperature at the outlet from the cooling tower

$$t_{w}^{16,1} = CTW\left(t_{w}^{15}\right) \tag{3.5}$$

in the first approximation. From equation (2.24, HEX 11) we shall find a new value of the absorbent temperature t_{s} at the inlet to the absorber. Then in the course of the repetitive procedure we define the final value of the absorbent temperature at the inlet to the absorber t_s^8 with the required accuracy.

3. The heating source temperatures t_w^{17} and t_w^{18} (heat-exchanger HEX 12) were defined as a result of the heat calculation of AACS. At the known flow rate of the heat carrier (it was assumed to be equal to the absorbent flow rate) we found the heat load of the heliosystem Q and by formulas (2.128) and (2.129) calculated the area of SC (solar collector) and the volume of the tank-accumulator in the first approximation. The type of the SC is defined by the level of required temperatures and was chosen in accordance with the data from table 3.1. In calculating the COP of the heliosystem (the expression (2.127)) we used the values of the reduced optical COP θ and the reduced coefficient of heat losses U, also given in the table. For the flat SC with single glazing produced by RPF "New Technologies", the data have been obtained as a result of full-scale tests [30, 118], for others the data have been taken from literature sources [17, 21, 90].

Table 3.1. Characteristics of the main types of SCs

Type of solar collector (SC)	Working temperature, °C	θ	U, W/(m^2×°C)
Flat SC with single glazing without selective coating of the heat receiver surface	60-70	0.73	5.6
The same with double glazing	60-80	0.65	4.5
Flat SC with single glazing and selective coating of the heat receiver surface	70-90	0.75	3.5
Evacuated glass tube SC	100-150	0.7	2.0

Depending on the required area of SC the type of the heliosystem was chosen by the mode of the heat carrier movement. For small heliosystems it is possible to use the scheme with natural circulation of the heat carrier. To define the time dependence of the temperature in the TA equations (2.124) – (2.126) have been solved. For heliosystems of higher capacity the scheme with forced circulation was used, and the problem was reduced to numerical solving differential equation (2.135). If the water temperature in the TA did not reach the required value, the correction of the total area (number) of SC was done, and the calculation was repeated.

3.2. Analysis of the Results Obtained

On the basis of the developed procedure for calculating processes in alternative systems the calculations were carried out with respect to problems of air-conditioning and cooling the media. The main results of calculations for AACS are given in tables 3.2 – 3.3 in the form of tabulated data by the basic parameters for three types of absorbents $LiBr+H_2O$, $LiBr+LiNO_3+H_2O$ and $LiBr+ZnCl_2+CaBr_2+H_2O$ (further on LiBr, LiBr+ and LiBr++, respectively). The main part of calculations is done conformably to LiBr++.

Calculations of AACS were done for a wide range of possible climatic conditions and operating parameters:

- x_g^1 = 7-25 g/kg of dry air (for the range of values x_g^1 = 7-13 g/kg of dry air calculations were carried out in conformity with a combined evaporative system CEC, Table 3.4; for the range of values x_g^1 = 13-25 g/kg of dry air the calculation results are given in table 3.6);
- t_g^1 = 20-45 °C;
- t_w = 27-45 °C;
- for each of three types of absorbents in changing ξ_{LiBr} = 45-50 %, ξ_{LiBr+} = 45-60 % and ξ_{LiBr++} = 60-75 %;
- the relationship of flow rates of gas and liquid for CTW is assumed, on the basis of the previous investigations, to be equal to unit [24, 81], for other apparatuses (IEC, ABR, DBR) the optimal value of this relationship has been defined by calculations and is given below.

We studied the influence of initial parameters of climatic condition: the moisture content and the temperature of outdoor air; the type and the concentration of the absorbent solution; the relationship of gas and liquid flow rates through the apparatus; temperatures of the heating source and the cooling medium; the required level of cooling on capabilities and working characteristics of AACS. The requirements of thermal-moist air parameters in a conditioned room are of decisive importance for air-conditioning system. Systems of air-conditioning should provide favorable microclimate in rooms. The main characteristics of a microclimate air are: the temperature in specific zones of rooms, humidity and hygienic state of inner air. Temperature and humidity, their maximum calculated values, fluctuations and changes during 24 hours and within a year are the most important factors influencing the people in the room. The velocity of air moving in the room is of some importance. The main task of AACS is to supply the conditioned room with the required amount of fresh air having the temperature and humidity which provide comfort heat and moisture parameters for residential and public premises. According to standards of the former USSR [Construction norms and rules CN&R 2.04.05-91 “Heating, ventilation and air-conditioning” the State Construction Committee of the USSR, Moscow, 1997] the zone of *optimal* heat and moisture parameters for a warm (summer) period of the year is limited by isotherms of 20 and 25 °C and by lines φ = 30% and 60 % (figure 3.7, the comfort parameters zone is marked with grey color). The standard specifies an extended comfort zone with *admissible* parameters: isotherms of 18 and 28 °C,

Table 3.2. The calculation performances of AACS on the basis of the open absorption cycle (nomenclature – figures 3.1, 3.2)

N/N	Ambient Air				Dehumidifying Part: Absorber							Dehumidifying Part: Desorber							Refrigerating Part: IEC					
	G_g 10^{-3}	t_g^1	x_g^1	φ_g^1	G_s 10^{-3}	ξ_8	ξ_9	t_s^8	t_s^9	x_g^2	t_g^2	G_s 10^{-3}	ξ_{11}	ξ_{12}	t_s^{11}	t_s^{12}	t_g^{14}	x_g^{14}	G_{mn} 10^{-3}	t_g^4	x_g^4	G_{ax} 10^{-3}	t_g^5	x_g^5
	kg/h	°C	g/kg	%	kg/h	%	%	°C	°C	g/kg	°C	kg/h	%	%	°C	°C	°C	g/kg	kg/h	°C	g/kg	kg/h	°C	g/kg
LiBr + H_2O system; dependence $f(x_g^1)$																								
1	10	30	13.3	47	10	55	54.7	30	36	7.7	31.8	10	54.7	55	65.6	53	48.4	18.9	5	18.4	7.7	5	18.3	11.5
2	10	30	20	70	10	55	54.5	30	39.9	10.0	32.2	10	54.5	55	83.2	59.8	57.0	29.9	5	21.2	10.0	5	21.0	13.9
3	10	30	25	87	10	55	54.3	30	42.9	11.9	34.1	10	54.3	55	87.7	60.9	58.1	38.0	5	23.0	11.9	5	22.7	15.8
LiBr + $LiNO_3$ + H_2O system; dependence $f(x_g^1)$																								
4	10	30	13.3	47	10	55	54.7	30	34.7	8.45	31.5	10	54.7	55	60.5	48.3	45.1	18.3	5	19.1	8.5	5	19.0	12.5
5	10	30	20.0	70	10	55	54.5	30	38.6	10.9	32.8	10	54.5	55	72.7	53.9	51.1	29.1	5	21.3	10.9	5	21.0	14.7
6	10	30	25.0	87	10	55	54.3	30	41.4	12.8	33.7	10	54.3	55	82.5	58.8	55.9	37.1	5	23.8	12.8	5	23.5	16.6
LiBr + $LiNO_3$ + H_2O system; dependence $f(\xi)$																								
7	10	30	13.3	47	10	45	44.9	30	31	30.3	12.2	10	44.9	45	36.0	35.6	34.6	14.8	5	22.4	12.2	5	22.0	15.5
8	10	30	13.3	47	10	55	54.7	30	34.7	31.5	8.4	10	54.7	55	60.5	48.3	45.1	18.3	5	19.1	8.5	5	19.0	12.5
9	10	30	13.3	47	10	60	59.6	30	36.7	32.1	6.7	10	59.6	60	74.7	56.4	52.8	19.9	5	17.4	6.7	5	17.1	10.7

Note: 1. The table does not show the cooling process 2–3 of air (mn) in HEX 13, this calculation is performed on the basis of efficiency value E_{HEX}=0.8. The process 13–8 of cooling the absorbent solution in the heat-exchanger HEX 11 is not shown. 2. Value t_s^{11} defines the required value of the heating source temperature on the basis of value E_{HEX}=0.8 (HEX 12).

Table 3.3. The calculation performances of AACS on the basis of the open absorption cycle (nomenclature – figures 3.1 and 3.2)

N / N	Ambient Air				Dehumidifying Part																		Refrigerating Part					
					Absorber									Desorber									IEC					
	G_g 10^{-3}	t_g^1	x_g^1	φ_g^1	G_s 10^{-3}	G_g^{ab} 10^{-3}	ξ_8	ξ_9	t_s^8	t_s^9	t_w^{16}	x_g^2	t_g^2	G_s 10^{-3}	G_g^d 10^{-3}		ξ_{11}	t_s^{11}	t_w^{17}	t_s^{12}	t_g^{14}	x_g^{14}	G_{mi} 10^{-3}	t_g^4	x_g^4	G_{ax} 10^{-3}	G_w^d 10^{-3}	t_g^5
	kg/h	°C	g/kg	%	kg/h	kg/h	%	%	°C	°C	°C	g/kg	°C	kg/h	kg/h	%	%	°C	°C	°C	°C	g/kg	kg/h	°C	g/kg	kg/h	kg/h	°C
$LiBr + ZnCl_2 + CaBr_2 + H_2O$ system																												
Dependence $f(x_g^1)$																												
1	10	30	13.5	50	5	10	60	59.5	26	34.1	23.7	9.2	30.6	5	10	59.5	60	55.1	59.4	39.5	40.4	17.4	5	22.7	9.2	5	2	22.6
2	10	30	17	64	5	10	60	59.5	33	43.7	30.8	12.1	34.2	5	10	59.5	60	60.0	64.7	42.1	42.3	21.0	5	26.0	12.1	5	2	25.8
3	10	30	20	85	5	10	63	62.1	34	38.5	31.1	12.3	36.2	5	10	62.1	63	76.3	83.5	48.6	48.2	25.6	5	27.0	12.3	5	2	26.8
4	10	30	25	92	5	10	65	63.4	40	51.0	37.4	14.6	40.7	5	10	63.4	65	90.6	99.7	54.7	53.5	30.2	5	28.0	14.6	5	2	27.8
Dependence $f(t_g^1)$																												
5	10	25	15	75	5	10	60	59.4	29	34.5	26.9	10.3	29.7	5	10	59.4	60	60.4	65.9	39.2		19.6	5	22.8	10.3	5	0.5	22.7
6	10	30	15	56	5	10	60	59.6	32	36.8	30.6	11.3	33.1	5	10	59.6	60	56.6	60.9	40.4		18.7	5	25.0	11.3	5	0.5	24.8
7	10	35	15	43	5	10	60	59.8	38	39.6	37.0	13.1	37.6	5	10	59.8	60	50.5	52.8	41.5		16.9	5	28.0	13.1	5	0.5	27.9
9	10	40	15	33	5	10	60	59.9	43	41.9	42.7	14.9	41.5	5	10	59.9	60	43.7	44.1	42.0		15.1	5	30.1	14.9	5	0.5	30.0
Dependence $f(\xi_s)$																												
9	10	30	13.5	50	10	10	60	59.8	30	32.5	27.5	10.6	30.6	10	10	59.8	60	50.2	55.0	43.9	37.7	16.4	5	23.4	10.6	5	5	23.3
10	10	30	13.5	50	10	10	65	64.7	30	34.0	27.5	8.9	30.9	10	10	64.7	65	65.7	78.0	55.9	43.9	18.1	5	22.5	8.9	5	5	22.3
11	10	30	13.5	50	10	10	70	69.6	30	34.9	27.5	7.7	31.2	10	10	69.6	70	84.3	90.0	71.8	51.5	19.3	5	21.9	7.7	5	5	21.8
12	10	30	13.5	50	10	10	75	74.5	30	34.9	27.5	7.1	31.1	10	10	74.5	75	99.0	108	89.3	59.0	21.0	5	21.1	7.1	5	5	21.0

N/N	Ambient Air				Dehumidifying Part																		Refrigerating Part					
					Absorber									Desorber									IEC					
	G_g 10^{-3}	t_g^1	x_g^1	φ_g^1	G_s 10^{-3}	G_g^{ab} 10^{-3}	ξ_s	ξ_0	t_s^8	t_s^9	t_w^{16}	x_g^2	t_g^2	G_s 10^{-3}	G_g^d 10^{-3}		ξ_1	t_s^{11}	t_w^{17}	t_s^{12}	t_g^{14}	x_g^{14}	G_m 10^{-3}	t_g^4	x_g^4	G_{ax} 10^{-3}	G_w^d 10^{-3}	t_g^5
	kg/h	°C	g/kg	%	kg/h	kg/h	%	%	°C	°C	°C	g/kg	°C	kg/h	kg/h	%	%	°C	°C	°C	°C	g/kg	kg/h	°C	g/kg	kg/h	kg/h	°C
Dependence $f(t_w)$																												
13	10	30	15	56	5	10	65	64.1	30	40.3	27.0	3	33.6	5	10	64.1	65	76.1	83.3	48.8	48.8	21.6	5	23.8	3	5	0.5	23.6
14	10	30	15	56	5	10	65	64.2	35	42.0	32.9	1	35.7	5	10	64.2	65	73.3	79.6	50.1	50.5	20.9	5	25.2	1	5	0.5	25.0
15	10	30	15	56	5	10	65	64.3	40	43.5	38.7	9	37.7	5	10	64.3	65	70.3	75.4	51.3	52.0	20.1	5	26.6	9	5	0.5	26.4
16	10	30	15	56	5	10	65	64.5	45	44.9	44.7	0.9	39.6	5	10	64.5	65	66.8	70.8	52.4	53.4	19.1	5	28.0	0.9	5	0.5	27.9
Dependence $f(G_{ax}/G_w)$ (the influence of flow rates relationship in IEC)																												
17	10	30	16	60	5	10	60	59.4	30	32.6	28.0	32.7	11,4	5	10	59.4	60	59.9	64.9	41.8	41.8	20.6	5	25.0	11,4	5	5.0	24.9
18	10	30	16	60	5	10	60	59.4	30	32.6	28.0	32.7	11,4	5	10	59.4	60	59.9	64.9	41.8	41.8	20.6	5	24.9	11,4	5	2.0	24.8
19	10	30	16	60	5	10	60	59.4	30	32.6	28.0	32.7	11,4	5	10	59.4	60	59.9	64.9	41.8	41.8	20.6	5	24.8	11,4	5	1.0	24.7
20	10	30	16	60	5	10	60	59.4	30	32.6	28.0	32.7	11,4	5	10	59.4	60	59.9	64.9	41.8	41.8	20.6	5	24.8	11,4	5	0.5	24.7

Note: The dependence f(ξ) has been obtained with packaging density of the apparatus packing bed which is defined by the distance between sheets of the packing bed equal to b = 0.016 m and with the relationship of flows G_g/G_s=1.0; all other dependencies have been obtained for b = 0.01 m and the relationship of flows G_g/G_s = 2.0.

on the left side the curve φ = 30 %, on the right side the broken line with typical points (t = 27 °C, φ = 60 %; t = 26 °C, φ = 65 %; t = 25 °C, φ = 70 %; t = 24 °C, φ = 75 %) – figure 3.7. According to American Association of engineers on heating and ventilation ASHRAE 55 – 56 [ASHRAE 1989 Fundamentals Handbook (SI)] the comfort zone is limited for the summer period by the isotherms t = 10 and 24 °C and the lines φ = 30% and 70%, that is, the zone is even wider than it is shown above. In the basic calculations we oriented ourselves to the zone of admissible comfort air parameters by CN&R 2.04.05-91.

Calculations of combined evaporative coolers CEC were done allowing for new parameters of pre-dehumidified air in the open absorption system. The efficiency of multistage evaporative coolers (multistage IEC) is equal for the three-stage scheme of IEC E_{db} = 0.9-0.95 (is calculated with the orientation to the dew point temperature), but the flow rate of the main air flow with such a design of IEC. (figure 1.2 shows a two-stage scheme of IEC in the composition of AACS) is continuously (from stage to stage) decreasing which shows the growth of specific power consumption. For the combined scheme (IEC/DEC) the efficiency value, under comparable conditions, is lower and amounts to E_{db} = 0.84 but the power consumption is much less, which completely corresponds to the authors'experimental data [22 – 27, 94] and makes it possible to consider such a scheme to be more preferable for being included in AACS. As it is seen from the analysis of processes in AACS for a wide variety of climatic conditions, it is quite enough to use IEC/R or a combined scheme IEC/DEC in the cooling part to enter the zone of comfort heat-moisture parameters, and only in rare cases, at high outdoor air temperatures one has to use the double-stage scheme IEC/IEC.

Figures 3.5 – 3.6 show the characteristic calculation distribution of the basic gas and liquid flows parameters in the crossflow scheme of the interaction for the absorber, desorber, and indirect evaporative cooler for some typical calculation points. One can note a considerable separation of values for temperatures and moisture contents in the air flow leaving the desorber. The character of distributing the temperature of water recirculating through IEC at the outlet from the apparatus is of some interest, and the mean value of this temperature in the cycle is unchangeable and equal to the temperature of water at the inlet to IEC.

Figures 3.7 – 3.10 illustrate on the H-X diagram for moist air the course of working processes in the alternative air-conditioning system for the ventilation scheme, when the whole air flow being cooled comes into the conditioned room (ventilation mode, VM). The diagrams show the main processes taking place in AACS: dehumidification of the air in the absorber, followed by the increase of its temperature: its cooling in IEC (shown is only the process of the main air flow cooling which proceeds with unchangeable moisture content); the process of changing the air flow state in the desorber (the absorbent regeneration process). It is quite obvious, that even at the highest values of moisture content in air (x_g^1=25g/kg) the alternative system is capable to provide obtaining comfort parameters without using vapour-compression cooling. With the growth of moisture content the temperature of the heating source providing the absorbent regeneration increases too. The temperature increase of the water cooling the absorber results in the decrease of the air dehumidification process efficiency and the growth of the heating source temperature. In this case, of course, the energy expenditures on the organization of processes in cooling tower decrease too (the growth of t_w corresponds for instance to the decrease of the air flow rate in the cooling tower) and the possibility is evident for optimizing the process of evaporative cooling the water in the cooling tower which serves the absorber. It is also possible to use an auxiliary air flow leaving IEC for cooling the absorbent in the heat exchanger HEX 11. Of course, one should take into account the resistance, in the first place, aerodynamic drag, of all service lines in the scheme. The increase of the absorbent concentration intensifies the process of dehumidifying the air but causes the increase of the heating source temperature.

The processes taking place in IEC/R are shown in figure3.17 in more details, particularly, the figure shows the process in the auxiliary air flow. It also shows conventionally the state of water recirculating through the apparatus.

The coefficient of performance (COP) for AACS was calculated by formula $COP = Q_{IEC}/Q_{DBR}$, i.e. as the ratio of cooling capacity to heat obtained by the absorbent from the heliosystem. Figure 3.11 shows the dependence of COP on the surrounding air temperature at the constant moisture content. The character of this dependence is defined by a considerable reduction of Q_{DBR} and comparatively week growth of Q_{IEC} with increasing t_g^1. As it follows from figure 3.12, it is possible to note monotonous decreasing COP with the increase of the moisture content of the surrounding air. Such a behavior agrees both with general physical considerations and the results of the work [52], in which they calculated COP of the idealized cycle as the ratio of changes of moisture content in the evaporative cooler and the absorber.

We can make the following conclusions:

- for the whole discussed range of initial air parameters (x_g^1 = 13-25 g/kg; t_g^1 = 20-45 °C) the pre-dehumidification of air makes it possible to reduce its moisture content to the value x_g^1 < 13 g/kg, which, in its turn, gives the possibility to enter the zone of comfort air parameters by means of evaporative cooling;
- when x_g^1 < 13 g/kg of dry air, it is quite enough to use CEC in the composition of IEC/R or IEC/DEC and there is no need for the dehumidifying part of AACS; when x_g^1 > 13 g/kg it is impossible to provide entering the zone of comfort air parameters without the dehumidifying part of AACS only by means of evaporative cooling;
- the configuration of the cooling part of CEC in the composition of AACS, as it can be seen from the calculations carried out, can be the combination of IEC and regenerative heat exchanger HEX – 13, and only in case of a dipper dehumidification of air it is necessary to use CEC in the form IEC/DEC or the double-stage evaporative system IEC/DEC;
- the comparative estimation of working bodies-absorbents – LiBr and LiBr+ and LiBr++ showed their suitability for solving the problem of air-conditioning, and LiBr+ provides some reduction of the required temperature of the heating source at a lower extent of dehumidifying the air. Taking into account general considerations (see Section 2.2 of the given report) we may prefer LiBr++, which has, besides everything, lower corrosive activity;

- the analysis of parameters varying in calculations showed that the moisture content of the air (x_g^1)was important rather than the outdoor air temperature value (t_g^1); the increase of the absorbent solution concentration greatly influences the results of the dehumidification process;
- the optimal value of the relationship between working flows of the gas and the liquid in the apparatus VR is equal to unit for the cooling tower, to 2.0 for the absorber and the desorber [for example, the change of G_g/G_s from 1.0 to 2.0 results in a slight growth of the air moisture content at the outlet from the absorber (from 9.7 to 10.0 g/kg for initial parameters of the air flow being dried t_g^1 = 30 °C and x_g^1 = 13.5 g/kg) and the increase of the absorbent temperature at the inlet to the desorber from 50.5 °C to 53.9 °C which justifies, of course, the double-fold reduction of the absorbent flow-rate)]; for apparatuses of direct and indirect evaporative cooling – 10.0 which completely corresponds to experimental investigation conducted by the authors previously [22 – 27, 81, 94].

3.3. Description and Technico-Economic Characteristics of a Pilot Plant

The general requirement to heat-and-mass transfer apparatuses (HMA and HEX) for alternative systems is their small overall dimensions [22-32, 97-100] due to high intensity of heat-and-mass exchange processes at low energy expenditure for the transportation of working substances (air and liquid flows), as HMA and heat exchangers HEX, used in systems, are numerous. The main universal development for all HMA of alternative systems has been realized through the film type apparatuses which provide a separate movement of gas and liquid flows in packing bed channels at low air resistance and the cross-flow scheme of contacting gas and liquid flows as the most acceptable one if it is necessary to arrange numerous HMA and HEX in a single block. The cross-flow scheme provides minimum turnings of flows and lower air resistance as compared to the counter-flow scheme, because it is characterized by higher values of extreme velocities of gas flow movement in packing bed channels HMA [24, 81].

The packing bed of cross-flow HMA is universal for all apparatuses use the developed multi-channel packing bed of a regular structure, it is formed by longitudinally corrugated thin sheets equally distant from each other. Regular roughness of the surface is used as the main method of intensifying processes of joint heat-and-mass exchange in HMA. Optimal values of roughness parameters are $k_{opt} = p/e = 8 \ldots 14$, where p and e are the pitch and the height of roughness riffles. Roughness riffles are uniformly dispersed on the surface of the packing bed longitudinally corrugated sheets. Optimal values of the main corrugation parameters (P and E are the pitch and the height of the main corrugation of the packing bed sheets), of the packing bed layer density (the distance between sheets in a stack, the magnitude of the equivalent diameter of the apparatus packing bed channels) and the main overall dimensions of the packing bed have also been defined above (Section 2.3.2 and 2.3.3). Within the farmers of theoretical and experimental studies of the problem of two-phase film flows stability the extreme values of the gas flow velocities in HMA were defined. In developing heat-and-mass transfer apparatuses comprising alternative systems, on the basis of the researches carried out it was assumed: the value of the equivalent diameter of packing bed channels 25 mm, the pitch and the height of the main corrugation of the packing bed sheets P = 14 mm, E = 6 mm the magnitude of the parameter of the sheet surface regular roughness k = 12.5; the work range of the air flow velocity in the channels of the packing bed 5-7 m/s. the design of the packing beds for all HMA comprising the alternative systems under developing the unified (desorber DBR, absorber ABR, cooling tower CTW, direct evaporative cooler DEC).

Figures 3.13 –3.15 shows the schematic description of combined evaporative coolers CEC: on the basis of IEC/R (figure 3.13); on the basis of joint arranging evaporative coolers of direct and indirect types IEC/DEC (figure 3.14), evaporative cooler and the cooling tower IEC/CTW (figure 3.15). All these coolers are based on the use of only the evaporation principle and are intended for the operation in dry and hot climate. The main component of these approaches is the device of indirect evaporative cooling IEC, the packing bed of which is the alternation of dry and wetted channels designed for the movement of the main and auxiliary air flows respectively. Longitudinally corrugated parts of the packing bed form closed components-wafers in the inner space of which moves the main air flow, cooled contactlessly at the unchangeable moisture content. The auxiliary air flow moves in the space between closed components (wafers) arranged vertically and uniformly. This flow is perpendicular to the main one and counter-flow with respect to the water film streaming down the outer surfaces of closed components. In this case evaporative cooling of water removes heat from the main air flow. The apparatus has a vessel for water 6, recirculating through its packing bed with the help of a pump 10. The water distributor 8 is a system of perforated tubes uniformly dispersed over closed components-wafers. Natural losses of water during its evaporation are compensated by automatic replenishment of the system with fresh water – 14. The division of air flows into the main and auxiliary flows from the one which enters the apparatus, occurs automatically and is defined by air resistances of the packing bed and air service lines. It can be regulated by a shutter of the main flow air-duct. The movement of air flows is provided by a single fan located at the entrance of air into the CEC. The auxiliary air flow first enters the space of the apparatus under the packing bed and then into the packing bed of the apparatus. In the CEC version, shown in figure 3.13, a regenerative variant of the cooler is represented. It comprises IEC and a heat exchanger, where the air to be cooled is precooled by the cold and wetted auxiliary air flow which is discharged into the surroundings. The cooler based on the joint arrangement of evaporative direct and indirect coolers IEC/DEC (figure 3.14) includes the DEC apparatus providing additional cooling of the main air flow which leaves the IEC. The cooling process takes place at the direct interaction of the air and the water film flowing down the surfaces of the packing bed sheets is directed on the isoenthalpy line. The water recirculating through the DEC apparatus takes the temperature equal to the wet bulb temperature of the air coming into the a DEC. The air is cooled and rewetted which provides an additional possibility to regulate it heat-moisture parameters. The design of the DEC apparatus is simpler than that of the indirect evaporative cooler. The packing bed is a set of longitudinally corrugated sheets equally distant from each other on both sides of which the water film streams.

The cooler based on the joint arrangement of the indirect evaporative cooler and the cooling tower IEC/CTW (figure 3.15) includes a cooling tower 2, which provides evaporative cooling of water by the main air flow leaving the IEC. As the moisture content of the air is not changed while the air is being cooled in the IEC, its dew point temperature is reduced which provides deeper cooling of water in the cooling tower. Cold water (13) comes into the system of water-air heat exchanger. The cooling tower design is similar to the DEC apparatus. A separator for drop-like moisture made in the form of louvers is provided at the outlet from the cooling tower. The system includes two heat exchangers where the precooling of the air entering the CEC is provided by air flows leaving IEC and CTW. Figure 3.16 shows the general view of the air-conditioning system based on the IEC/CTW system. The CEC unit is located outside the system of water-air heat exchangers placed directly in rooms being served. The general analysis of principal capabilities of CEC coolers is given in Table 3.4 and on the wet air diagram, figure 3.17. Calculated air parameters of the table are given for different cities of the world for climatic conditions when $x_g < 13$ g/kg (zone A). The diagram shows the working processes, occurring in the cooler, on the example of two most specific points with parameters $t_g = 40$ °C, $x_g = 8$ g/kg и $t_g = 45$ °C, $x_g = 10$ g/kg , and it can be seen that an additional stage of cooling DEC is necessary only at high initial air temperatures ($t_g >$ 40 °C). The diagram shows the processes of changing the main air flow state (3–4) and those of the auxiliary flow (3–5) in the indirect evaporative cooler IEC (the state of water recirculating through the apparatus and having constant temperature – figures 3.5-3.6 is shown by the arbitrary point 21 in the diagram; the dashed line indicates the direction of the process line in the "wet" part of cooler). Lines 2–3 and 5–7 are the processes taking place in the heat exchanger. It is evident, that at $x_g > 13$ g/kg using only evaporative methods it is impossible to provide necessary comfort parameters of the air.

The schematic description of alternative air conditioning systems AACS on the basis of the open absorption cycle and solar regeneration of the absorbent is given in figures 3.18 – 3.20. Figure 3.18 shows the simplified version of AACS for the absorber and desorber (figure 1.8A) with outstanding heat exchangers of cooling and heating the absorbent. The above mentioned main versions of evaporative coolers are included in alternative systems as an evaporative part. In this case it is a version of the cooler on the basis of IEC/R. The design of the absorber and desorber is similar to that of the cooling tower described above. The apparatuses are supplied with a separator of drop-like moisture in the form of louvers. The layout diagram shows the heat exchanger of preheating the air entering the desorber and the heat exchanger of flows of the hot strong and cold weak solutions of the absorbent. The diagram shows two fans providing the movement of all air flows.

In figure 3.19 the layout diagram of the AACS is shown in the variant of combining heat exchangers of heating and cooling with the desorber and the absorber, respectively (figure 1.8B). The packing bed of these apparatuses is a tube-plate structure, and longitudinally corrugated sheets are used as plates like in all other apparatuses. The author of the project developed such apparatuses before as evaporative condensers of refrigerating plants, and they proved to be good [24, 81]. Such a development makes it possible to avoid additional heat exchangers and to reduce the total air resistance to the movement of heat carriers.

All the above-mentioned AACS are represented in the version of ventilation schemes (ventilation mode, VM), when all the cooled air flow comes into the room. The recirculation apparatus (recirculation mode, RM – figure 3.23) is more economic, when the amount of the fresh air entering the system can be about 20% [20, 24]. Various versions of similar schematic approaches have been discussed in Section 1.1 of the

Table 3.4. Design air parameters for an AACS of the evaporative type (IEC/R) ($x_g^1 < 13$ г/кг)

N/N	Country	City	Design air parameters (summer air-conditioning)				A combined evaporative cooler						
							The main air flow				The auxiliary air flow		
			$G_g \times 10^{-3}$	t_g^1	x_g^1	φ_g^1	$G_{mn} \times 10^{-3}$	t_g^4	x_g^4	t_g^6	$G_{ax} \times 10^{-3}$	$G_w \times 10^{-3}$	t_g^5
			kg/h	°C	g/kg	%	kg/h	°C	g/kg	°C	kg/h	kg/h	°C
1	Germany	Hamburg	10	27.8	9.76	42	5	22.31	9.76		5	0.5	22.30
2	Germany	Berlin	10	32.8	10.53	34	5	28.10	10.53		5	0.5	28.00
3	Kenya	Nairobi	10	28.3	8.62	37	5	21.95	8.62		5	0.5	21.93
4	Syria	Damascus	10	40.6	9.91	21	5	27.77	9.91		5	0.5	27.70
5	Libya	Djarabub	10	45.0	10.15	17	5	30.13	10.15	25.07	5	0.5	30.09
6	Iraq	Bagdad Mosul	10	46.1	8.20	13	5	30.00	8.20	23.90	5	0.5	29.50
7	Iran	Teheran	10	37.8	10.18	25	5	26.10	10.18		5	0.5	26.07
8	Hungary	Budapest	10	34.4	10.78	32	5	25.33	10.78		5	0.5	25.30
9	Australia	Adelaide	10	37.8	7.71	19	5	26.13	7.71		5	0.5	26.10
10	Australia	Canberra	10	36.1	9.30	25	5	25.50	9.30		5	0.5	25.42
11	Algeria	Colon Beshar	10	40.6	8.95	19	5	27.00	8.95		5	0.5	26.90

N / N	Country	City	Design air parameters (summer air-conditioning)				A combined evaporative cooler						
							The main air flow				The auxiliary air flow		
			$G_g \times 10^{-3}$	t_g^1	x_g^1	φ_g^1	$G_{mn} \times 10^{-3}$	t_g^4	x_g^4	t_g^6	$G_{ax} \times 10^{-3}$	$G_w \times 10^{-3}$	t_g^5
			kg/h	°C	g/kg	%	kg/h	°C	g/kg	°C	kg/h	kg/h	°C
12	England	London	10	27.8	9.73	42	5	22.28	9.73		5	0.5	22.26
13	USA	San Francisco	10	29.4	8.74	42	5	22.80	8.74		5	0.5	22.72
14	Israel	Elat	10	43.9	11.05	23	5	29.00	11.05	24.93	5	0.5	28.90

Notes: design air parameters (summer air-conditioning) are adopted according to Construction Norms and Rules CN&R, 2.04.05-91 "Heating, ventilation and Air-conditioning", State Construction Committee of the USSR, Moscow, 1997, ASHRAE 1989 Fundamentals Handbook (SI); for cities with a design air temperature over 40 °C a combined evaporative system IEC/DEC is used, and there is a column for the temperature t_g^6 (calculation scheme in figure 3.1) in the table.

present report. As a rule, the air leaving the room is cooled in the direct evaporative cooler DEC and then can be used, for example, for precooling the air flow which has been dehumidified in the absorber and enters the cooler, as is shown in figure 3.33, or is directed for the recirculation in the room.

In figure 3.20 the layout diagram of the ARFS is shown in the variant of the evaporative cooler on the basis of IEC/CTW and the cold water supply to heat exchangers-coolers. This development makes it possible to avoid the system of branched cost of construction work and the plant on the whole.

Table 3.5 gives the design technico-economical characteristics of a pilot plant AACS (the layout diagram is the same as in figure 3.18). All the calculating have been carried out for the flow rate of the air being dehumidified G_g = 10000 kg/h when the correlation $VR_{ABR}\left(VR_{DBR}\right)$ is equal to unit, in this case overall dimensions of these apparatuses are the same (H×B×L = 650×650×1000 mm) and the velocity of the air flow in packing bed channels is w_g = 5.8-6.8 m/s, which is quite acceptable from the view point of the flow stability.

The required capacity of the pilot plant has been defined owing to the necessity of providing comfort air parameters in rooms, like offices, of the area *200 m^2* situated in the city of Odessa, for the summer period of air-conditioning. In this case

- design air parameters in the city of Odessa in the summer period of air-conditioning amount to t = 28.6 °C; h = 62 kJ/kg; φ = 52% ("B" parameters, Construction Norms and Rules CN&R, 2.04.05-91 "Heating, ventilation and air-conditioning", State Construction Committee of the USSR, Moscow, 1997).
- optimal characteristics of the micro climate (comfort air parameters) for an office amount to t = 23-25 °C; φ = 40-60% (CN&R, 2.04.05-91). Let us choose from this range of parameters the values of t = 25 °C, φ = 50%.
- the required temperature of the air fed to room through the cooler AACS at the working temperature difference of 4 °C (the difference between the air temperature in the room and that of the air coming from the air cooler) will be t = 21 °C; h = 46 kJ/kg; x = 10.5 g/kg. These parameters, as it can be seen from results obtained by us, are quite possible to be provided by AACS. In this case, the air leaving the absorber will have the parameters t = 29 °C; x = 10.5 g/kg. The required content x = 10.5 g/kg is achieved in the single-stage IEC/R.
- the capacity of the forced-exhaust ventilation at the required multiplicity of the air exchange in the room k = 4-7 (CN&R, 2.04.05-91) will amount to

$$G = k \cdot F \cdot H = 5.5 \cdot 200 \cdot 4.5 = 4950 \text{ m}^3/\text{h}$$

Table 3.5. Technical-and-economical characteristics of a pilot plant of AACS (alternative air-conditioning system; figure 3.18)

	Characteristics	Magnitude		
I	• Refrigerating capacity • Air capacity • Heat required for regeneration	**Q**, 28.6 kW (VM), 14.3 kW (RM) **G**, m^3/h 5000 **F**, m^2 200		
II	Main overall dimension	H_{CEC}, mm 2000	B_{CEC}, mm 1000	L_{CEC}, mm 3000
III	Structural composition of the pilot AACS (Figure 3.18): 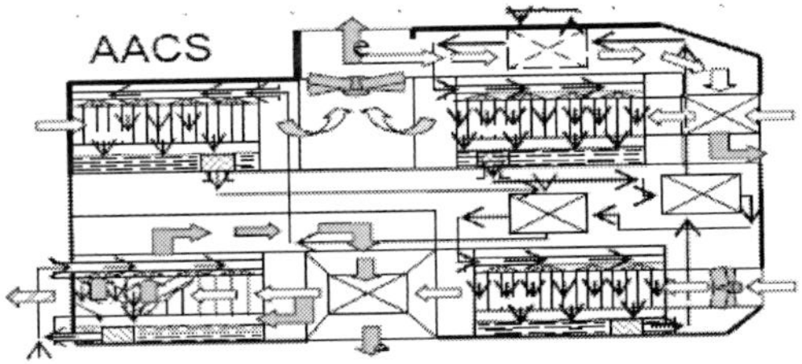 AACS	• absorber ABR • desorber DBR • indirect evaporative cooler IEC • direct evaporative cooler DEC • cooling tower CTW • heat exchangers: 2 – air-air 3 – liquid-liquid 2 • fan 2 • liquid pumps		
IV	Absorbent: • composition • concentration (working range)	$LiBr + ZnCl_2 + CaBr_2 + H_2O$ 65-75%		
V	Regime parameters: • relationship of working flows in the dehumidifying loop (air/absorbent):	absorber 2.0	desorber 2.0	
	• relationship of working flows in the cooling loop: – main (mn) / auxiliary (ax) air flows in IEC – – auxiliary air flow and recirculating water in IEC	 1.0 10.0		
VI	Heliosystem (external heating source) • total required area of collectors	$F_{\Sigma sc} = 150 m^2$		

Table 3.5. (Continued)

N / N	HEAT-AND-MASS TRANSFER APPARATUS HMA	MAIN CHARACTERISTICS						
		Overall dimensions of HMA			Packing bed of HMA			
		H, mm	B, mm	L, mm	P, mm	E, mm	d_e	k= =p/e
1	Absorber ABR	650	650	1000	14	6	25	12.5
2	Desorber DBR	650	650	1000	14	6	25	12.5
3	Indirect evaporative cooler IEC	650	650	1000	14	6	25	12.5
4	Direct evaporative cooler DEC	450	450	500	14	6	25	12.5
5	Cooling tower CTW (the cooling loop of the absorber)	650	650	1000	14	6	25	12.5
6	Cooling tower CTW	450	450	500	14	6	25	12.5

Note:

The direct evaporative cooler DEC can be not available in the scheme AACS, which is defined by initial calculation parameters of the air flow;

Geometrical characteristics of the “wet” and “dry” parts of the indirect evaporative cooler IEC are identical;

The absorbent concentration (the operating range) and the temperature level of regeneration depend on the type of the absorbent and the required degree of dehumidifying the air in the absorber.

(it was assumed in the design that k = 5.5, F = 200 m^2 – the aria of an office space, H – the height of the space),

- the refrigerating capacity of the pilot plant is $Q = G \cdot c \cdot \Delta \cdot h$ = 28.6 kW in ventilation mode (VM; the flow rate of air is taken equal to 5000 m^3/h); in the recirculation mode (RM), when the amount of fresh air fed to the room is only 20%, Q = 14.3 kW (figure 3.23, the design scheme and the general view of the recirculation system AACS).

Figures 3.21A and 3.21B shows the space layout diagrams of AACS in the form of separate units of cooling, and figures 3.22 and 3.24 shows the general view of alternative air-conditioning systems including the very unit of cooling 1, heliosystem 3 located on the roof of the building and including a tank-heat accumulator 4 with a duplicate source of heating 5, a system of air ducts of force ventilation 6 and exhaust one 7, and for version of AACS with a cooling tower – a system of water-air heat exchangers being ventilated 2 which are located in rooms of the building.

The general analysis of the principle capabilities of alternative air-conditioning systems is given in Table 3.6 and diagram H-X of the wet air, figure 3.25. The design parameters of the air in the table are given for different cities of the world for climatic conditions when $x_g > 13$ g/kg (zone B). The absorbent LiBr++ is used at the concentration ξ = 70%. The course of processes in AACS is shown on the example of typical points with calculation parameters t_g = 21.1 °C, x_g = 11.99 g/kg (Bogota); t_g = 45.6 °C, x_g = 22.7 g/kg (Port-Said); t_g = 32.2 °C, x_g = 24.96 g/kg (Duala); t_g = 35 °C, x_g = 18.47 g/kg (Tel Aviv). The diagram shows only the processes of dehumidifying the air in the absorber and the following evaporative cooling of the air in IEC, it is seen in this case that the need for an additional stage of cooling (DEC and IEC) appears only at high initial temperatures of air ($t_g > 40$ °C) – Port-Said. The alternative solar system of air-conditioning is capable to provide the obtaining of comfort air parameters for any climatic conditions of the world and in the form_of an evaporative cooler it is optimal to use a single-stage regenerative indirect evaporative cooler IEC/R. As compared to traditional vapour – compression systems of air – conditioning the alternative system AACS provides considerable redaction of energy consumption (30-60%), which is confirmed by not numerous data of using similar plants [24, 101, 102].

The general analysis of the principle capabilities of alternative cooling systems (the layout diagram of the alternative cooling system AACS on the basis of the open absorbtion cycle with a heliosystem used as a heating source and chilled water supply to heat exchangers-coolers (room fan coils) is given in figure 3.20; the calculating scheme is given in figure 3.4) is given in Table 3.7 and diagram H-X of the wet air, figure 3.26. The design parameters of the air in the table are given for particular points with design parameters – t_g = 25 °C, x_g = 10 g/kg and t_g = 28 °C, x_g = 20 g/kg (the processes of air dehumidification in the absorber **1–2**, of the 2-stage evaporative cooling in IEC – **2–4**; and of water cooling in the cooling tower are shown; the temperature of the water cooled is given in the form of a point in the line φ = 100%). The temperature of the cooled water is 8.5-11 °C, in this case the required temperature of the heating source is in the range of

Table 3.6. Design air parameters for a pilot plant of AACS to be used in different cities of the world ($x_g^1 > 13$ г/кг)

N / N	Country	City	Design air parameters (summer air-conditioning)				Dehumidifying Part: Absorber / Desorber					Air being dehu-midified		Refrigerating Part			
			$G_g \times 10^{-3}$	t_g^1	x_g^1	φ_g^1	$G_s \times 10^{-3}$	$G_g \times 10^{-3}$	ξ_s	t_w^{16}	t_w^{17}	x_g^2	t_g^2	$G_{mn} \times 10^{-3}$	t_g^4	x_g^4	t_g^4
			kg/h	°C	g/kg	%	kg/h	kg/h	%	°C	°C	g/kg	°C	kg/h	°C	g/kg	°C
1	Italy	Naples	10	32.2	13.0	42	5	10	70		105.7	6.9	36.7	5	26.3	6.9	
2	Italy	Milan	10	33.9	12.3	38	5	10	70		105.2	7.4	34.9	5	25.4	7.4	
3	Colombia	Medellin	10	32.2	13.5	45	5	10	70		106.6	7.1	33.7	5	25.1	7.1	
4	Colombia	Bogota	10	21.1	12.0	77	5	10	70		101.4	8.0	27.1	5	18.3	8.0	
5	Chile	Valparaiso	10	31.1	15.6	55	5	10	70		112.1	11.1	32.3	5	25.0	11.1	
6	Uganda	Kampala	10	31.7	15.2	55	5	10	70		111.5	11.8	33.1	5	33.1	11.8	
7	Tangiers	Tangiers	10	32.8	14.6	47	5	10	70		113.7	9.2	37.5	5	27.3	9.2	
8	USA	Washington	10	35.0	16.6	47	5	10	70		110.1	11.7	36.8	5	27.9	11.7	
9	USA	New-York	10	35.0	14.1	40	5	10	70		109.9	9.7	37.2	5	27.5	9.7	
10	Pakistan	Karachi	10	37.8	19.5	47	5	10	70		114.2	12.1	42.1	5	28.1	12.1	
11	Japan	Tokyo	10	32.8	19.6	62	5	10	70		115.4	12.2	38.5	5	26.9	12.2	

N / N	Country	City	Design air parameters (summer air-conditioning)				Dehumidifying Part							Refrigerating Part			
							Absorber / Desorber										
			$G_g \times 10^{-3}$	t_g^1	x_g^1	φ_g^1	$G_s \times 10^{-3}$	$G_g \times 10^{-3}$	ξ_s	t_w^{16}	t_w^{17}	Air being dehu-midified		$G_{mn} \times 10^{-3}$	t_g^4	x_g^4	t_g^4
												x_g^2	t_g^2				
			kg/h	°C	g/kg	%	kg/h	kg/h	%	°C	°C	g/kg	°C	kg/h	°C	g/kg	°C
12	Israel	Tel Aviv	10	35.0	18.5	52	5	10	70		111.0	8.3	40.4	5	27.2	8.3	
13	Cameroun	Douala	10	32.2	25.0	81	5	10	70		120.9	10.6	41.7	5	28.8	10.6	
14	Sudan	Port-Said	10	45.6	22.7	36	5	10	70		112.2	10.8	48.6	5	32.1	10.8	25.0
15	Noeway	Bergen	10	23.9	12.3	66	5	10	70					5			
16	Spain	Madrid	10	32.2	13.5	45	5	10	70		103.4	7.1	33.7	5	25.4	7.1	
17	Burma	Rangoon	10	32.2	24.3	77	5	10	70		119.5	10.3	48.0	5	29.2	10.3	
18	Haiti	Port-au-Prince	10	32.2	20.1	66	5	10	70		117.0	12.0	38.1	5	27.0	12.0	
19	Japan	Okinawa	10	32.2	24.9	78	5	10	70		120.7	10.5	41.7	5	28.9	10.5	
20	New Zealand	Oakland	10	27.8	16.0	68	5	10	70		104.9	11.9	30.2	5	24.2	11.9	

Notes: design air parameters (summer air-conditioning) are adopted according to Construction Norms and Rules CN&R, 2.04.05-91 "Heating, ventilation and Air-conditioning", State Construction Committee of the USSR, Moscow, 1997, ASHRAE 1989 Fundamentals Handbook (SI)

Table 3.7. Calculation results of the main parameters for the alternative refrigerating system

N / N	Design air parameters				Dehumidifying Part							Refrigerating Part						
												IEC 1			IEC 2			CT W
	$G_g \times 10^{-3}$	t_g^1	x_g^1	φ_g^1	$G_g \times 10^{-3}$	$G_s \times 10^{-3}$	ξ_s	t_w^{16}	t_w^{17}	air being dehumidified t_g^2	air being dehumidified x_g^2	$G_{mn} \times 10^{-3}$	t_g^4	x_g^4	$G_{mn} \times 10^{-3}$	t_g^4	x_g^4	t_w
	kg/h	°C	g/kg	%	kg/h	kg/h	%	°C	°C	°C	g/kg	kg/g	°C	k/kg	kg/h	°C	g/kg	°C
1	10	28	12	50	10	5	65	25,75	76,20	32,68	6,796	5	22,55	6,796	2,5	16,42	6,796	14,7
2	10	28	12	50	10	5	70	24,37	98,76	31,55	4,996	5	20,87	4,996	2,5	14,20	4,996	12,1
3	10	28	12	50	10	5	75	23,53	122,64	31,50	4,130	5	20,33	4,130	2,5	13,20	4,130	10,8
4	10	25	10	53	10	5	65	24,10	72,44	29,56	5,606	5	20,13	5,606	2,5	14,36	5,606	12,8
5	10	25	10	53	10	5	70	22,82	94,42	28,45	4,168	5	18,63	4,168	2,5	12,39	4,168	10,3
6	10	25	10	53	10	5	75	22,02	121,89	28,41	3,482	5	18,22	3,482	2,5	11,60	3,482	8,5

Notes: a two-stage evaporative cooler with further cooling of water in the cooling tower is used in the cooling part of the plant.

t_{17} = 95-122 °C. The use of two-stage cooler IEC/DEC greatly increases energy expenditures for the organization of the process, as from the second evaporative stage only ¼ of air which had been dehumidified is carried out. The possibility of operation of such a cooling system is limited by the range of comparatively low values of moisture content of the outdoor air $x_g <$ 10-15 g/kg.

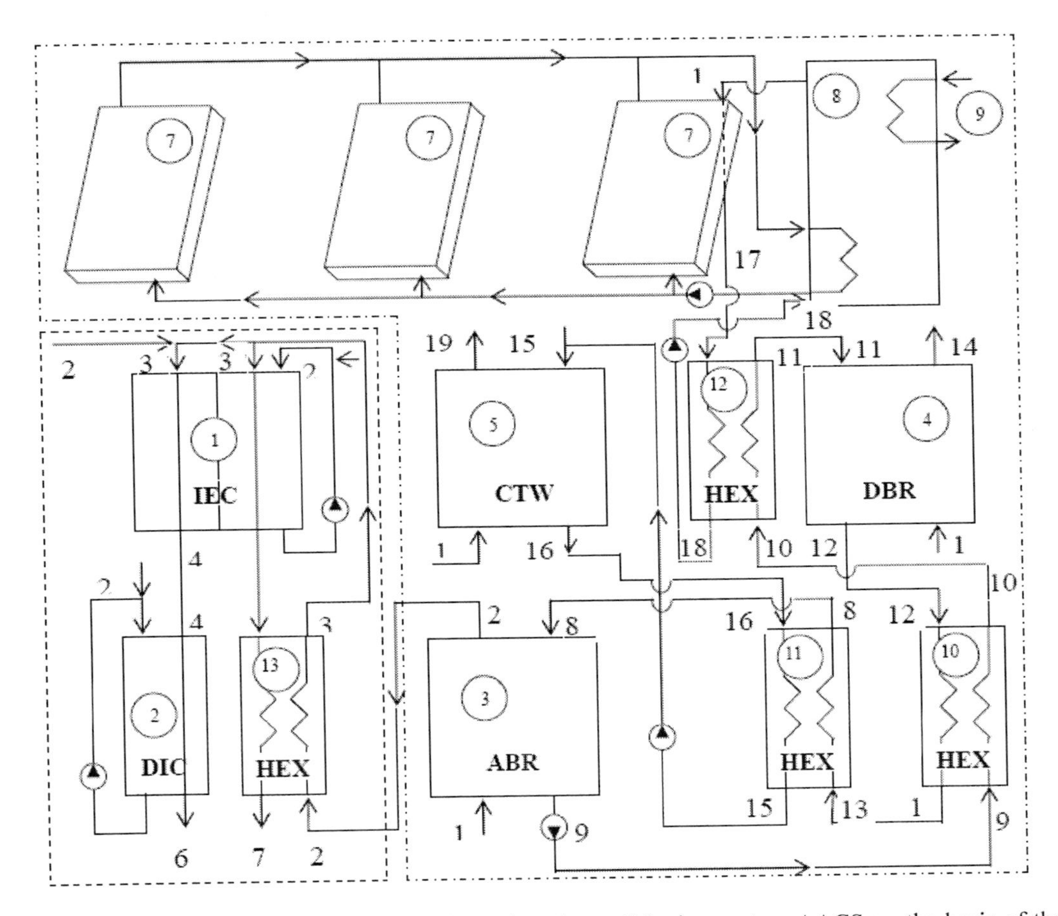

Figure 3.1. Schematic description of the alternative air-conditioning system AACS on the basis of the open absorption cycle with external (outstanding) heat exchangers HEX 11 (cooler) and HEX 12 (heater) as the design scheme. Nomenclature: (1)⇨(13) are heat-and-mass transfer apparatus (in figure 1.1); 1 – 22 – are flow parameters (numbers of design points).

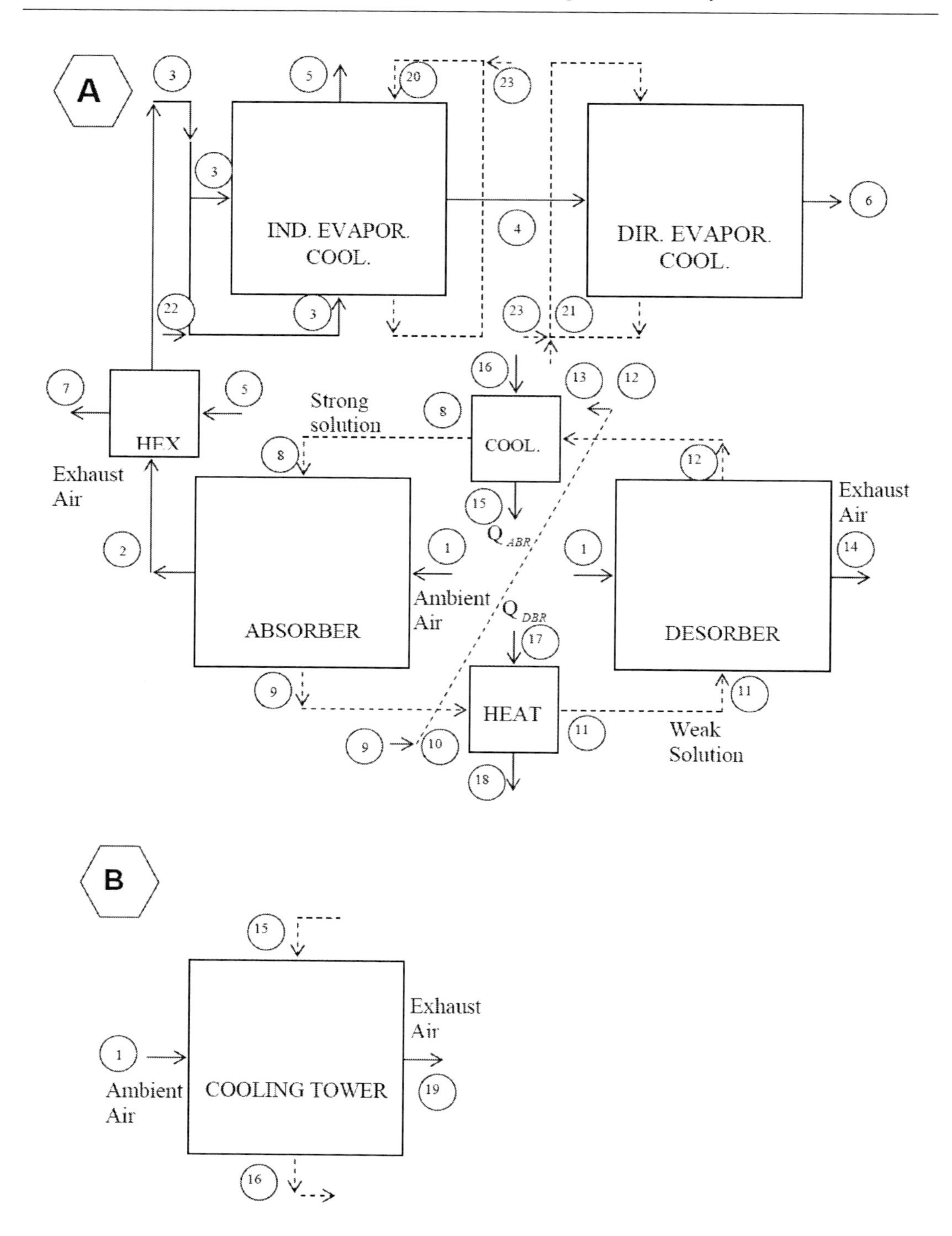

Figure 3.2. Sketch of the AACS – the dehumidifying unit and the evaporative cooling unit (A), and the cooling tower (B) in the form of film cross-flow Heat-and-mass transfer apparatuses. Nomenclature is the same as in figure 3.1.

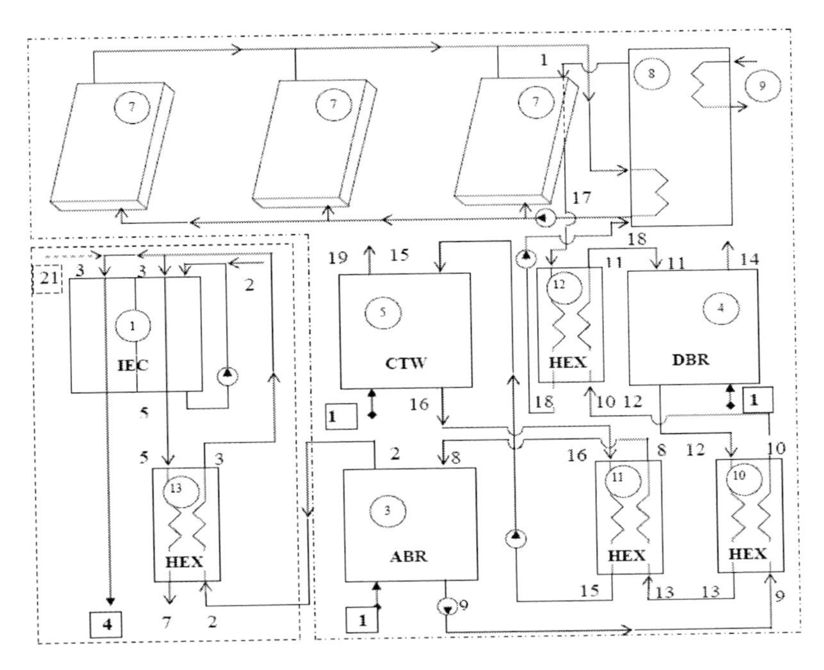

Figure 3.3. The design scheme of the alternative air-conditioning system AACS on the basis of the regenerative evaporative cooler IEC/R. Nomenclature is the same as in figure 3.1.

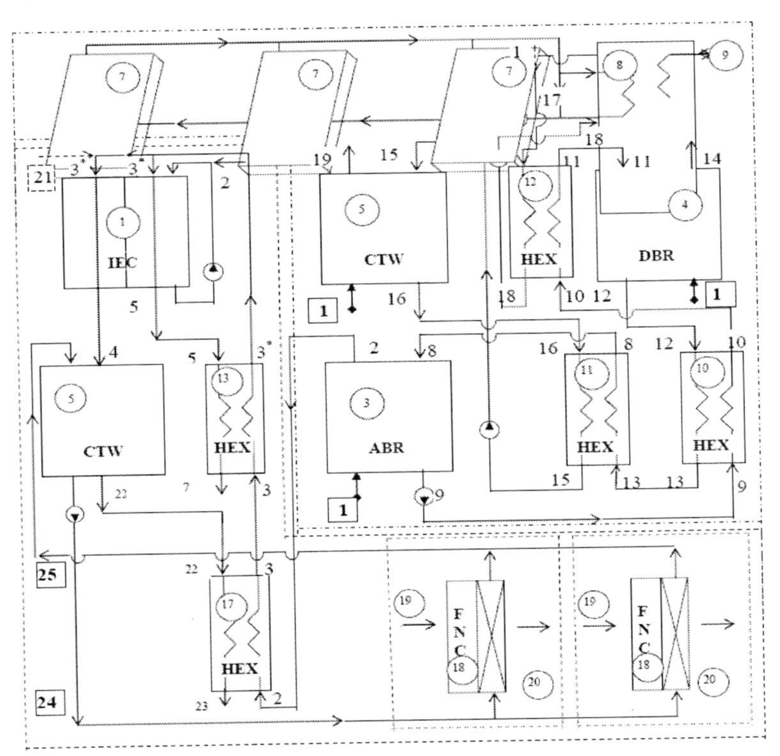

Figure 3.4. The design scheme of the alternative system on the basis of the open absorption cycle and cooled water supply to heat exchangers-coolers. Nomenclature is the same as in figure 3.1

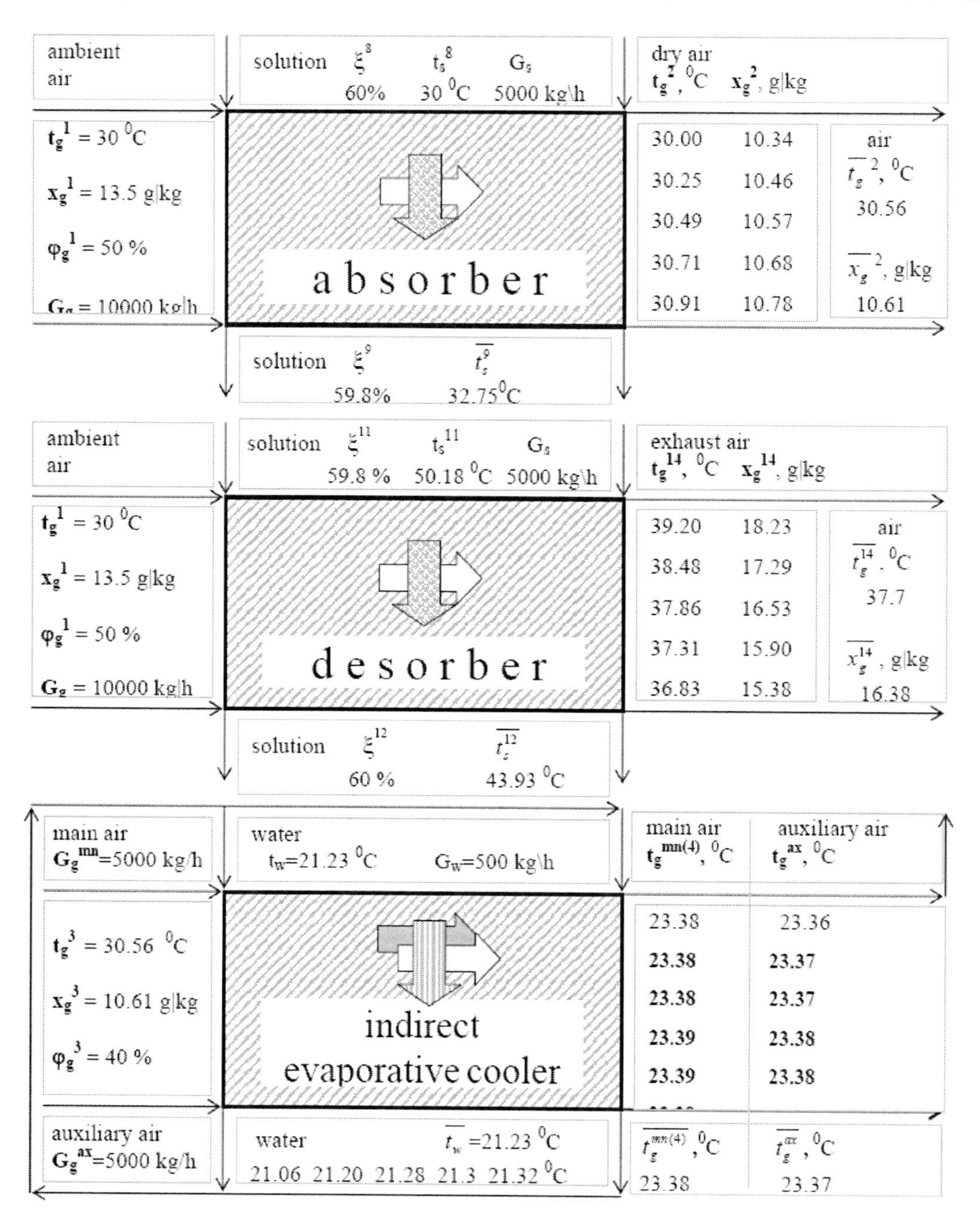

Figure 3.5. Characteristic distribution of flow parameters (calculation) with the cross-flow scheme of gas and liquid flow interaction (ξ_{LiBr++} = 60%).

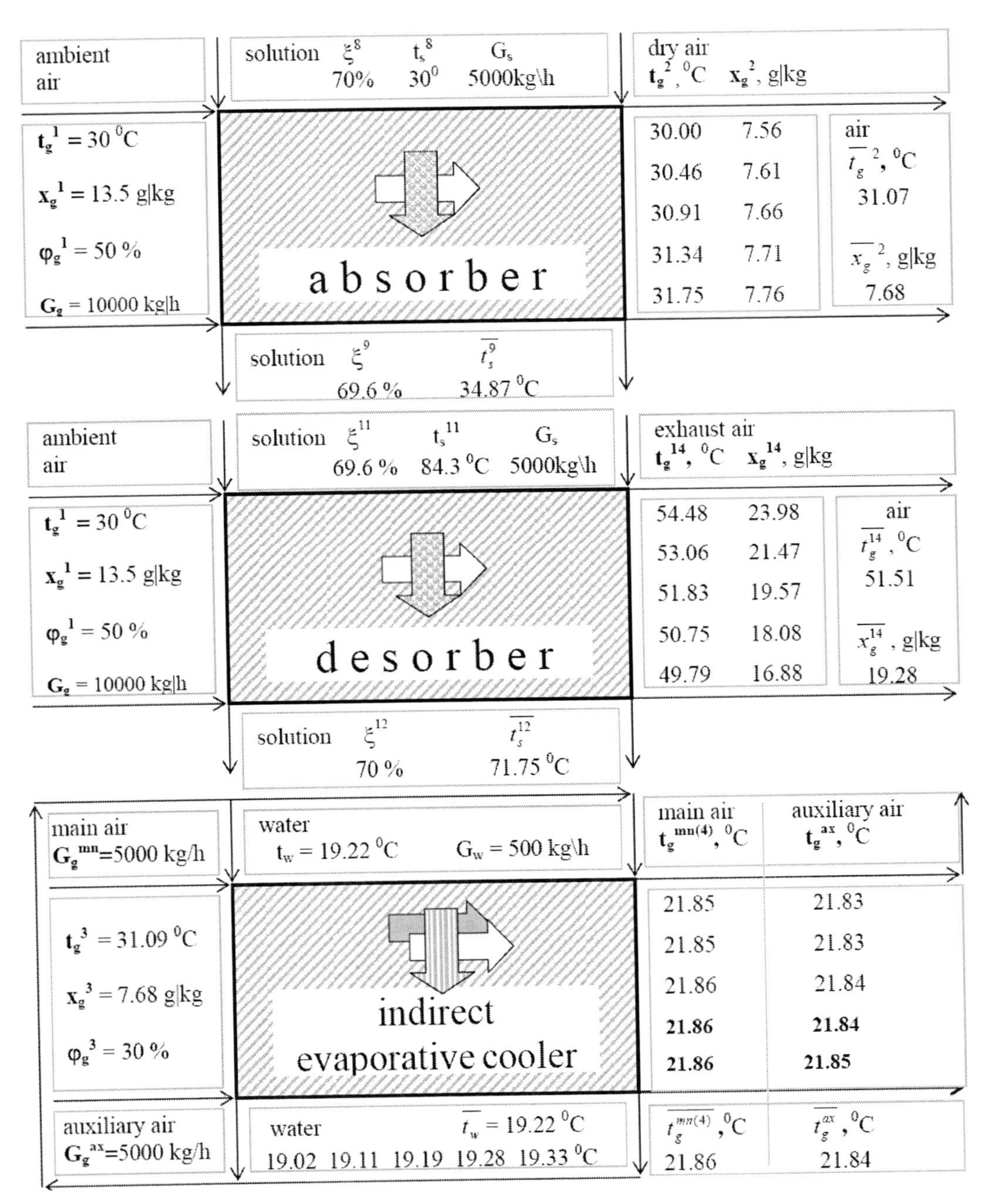

Figure 3.6. Characteristic distribution of flow parameters (calculation) with the cross-flow scheme of gas and liquid flow interaction (ξ_{LiBr++} =70%).

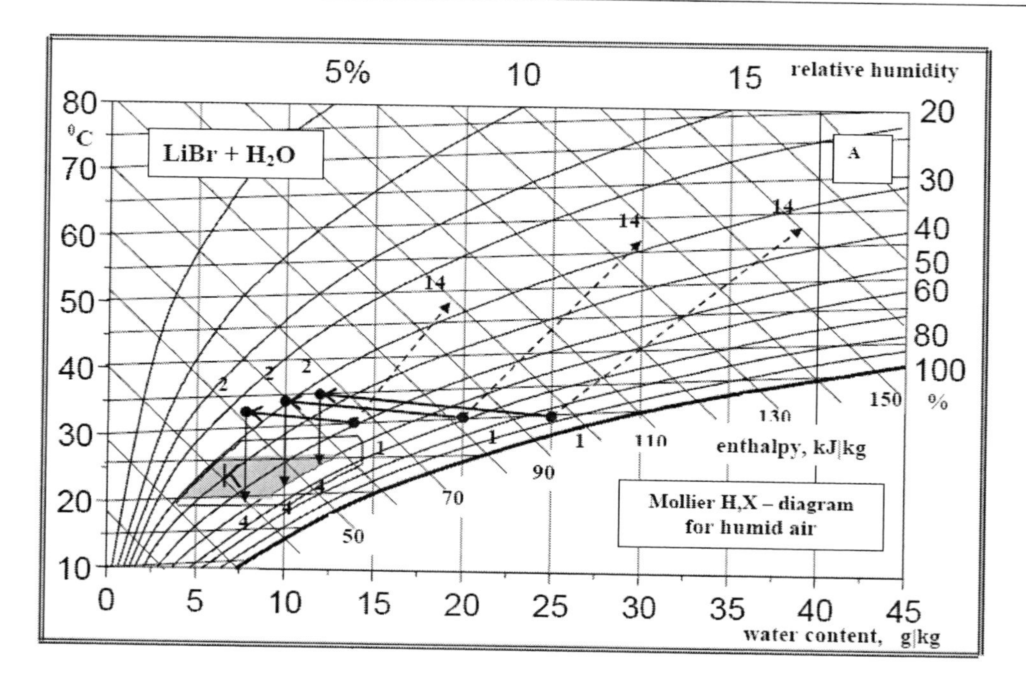

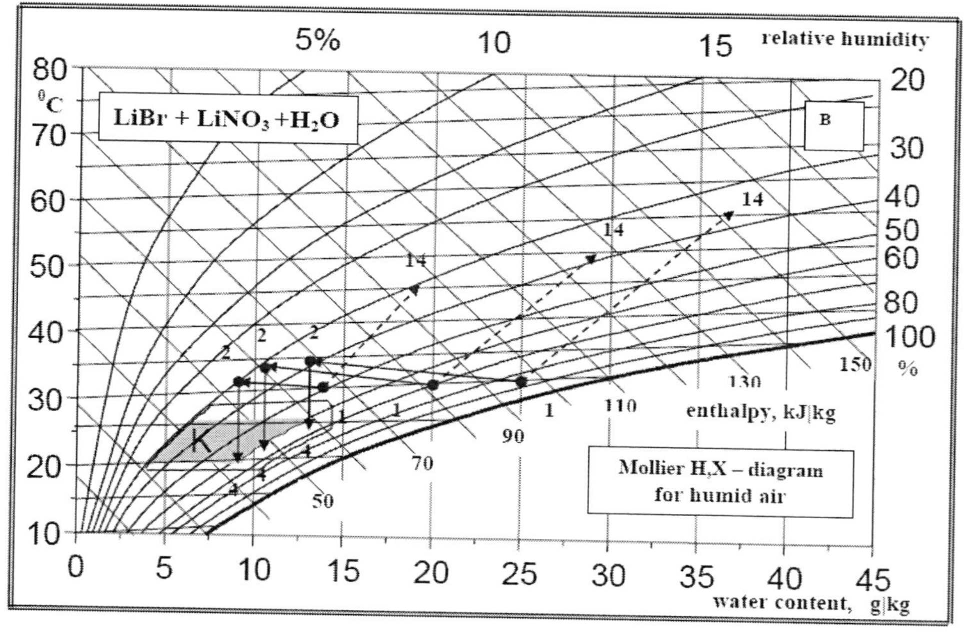

Figure 3.7. The representation of working process in AACS on H,X-diagram for wet air. (influence of the moisture content of outside air). Nomenclature: K is the zone of comfort heat-humidity parameters; 1–2 is the dehumidification of the air in the absorber; 2–4 is the air cooling in IEC (assimilation process in the space being conditioned is not shown); 1–14 – changing the state of the air flow during the absorbent regeneration in the desorber.

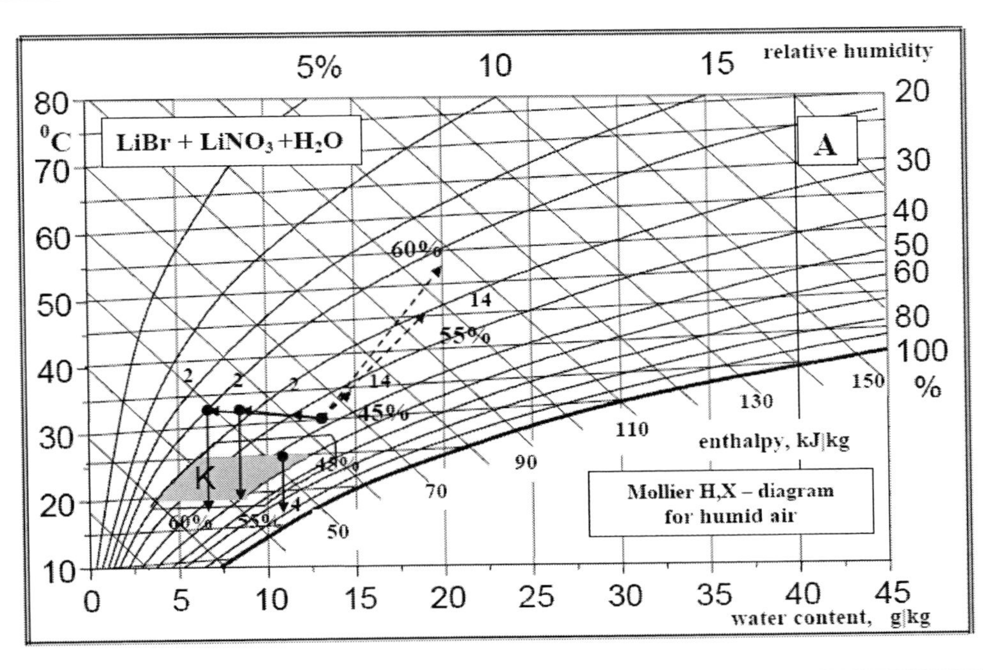

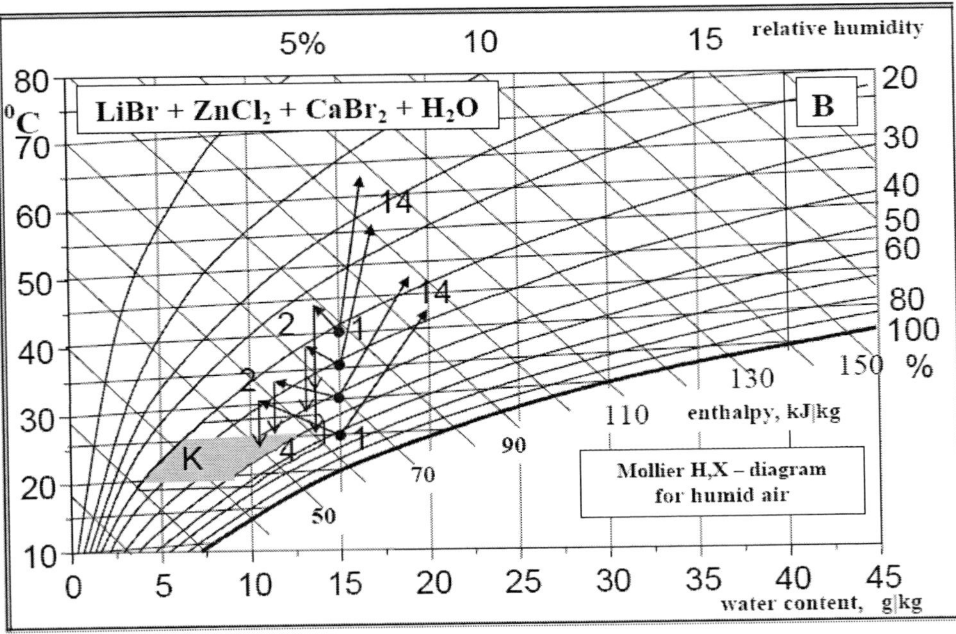

Figure 3.8. The representation of working processes in AACS on the H,X-diagram of wet air. A. The absorbent is LiBr + $LiNO_3$ +H_2O. The influence of the absorbent concentration on the working processes in the system. B. The absorbent is LiBr + $ZnCl_2$ + $CaBr_2$ + H_2O. The influence of the outdoor air temperature t_g = 25-40 °C, at X_g = 15 g/kg and ξ_s = 60%. Nomenclature is the same as in figure 3.7.

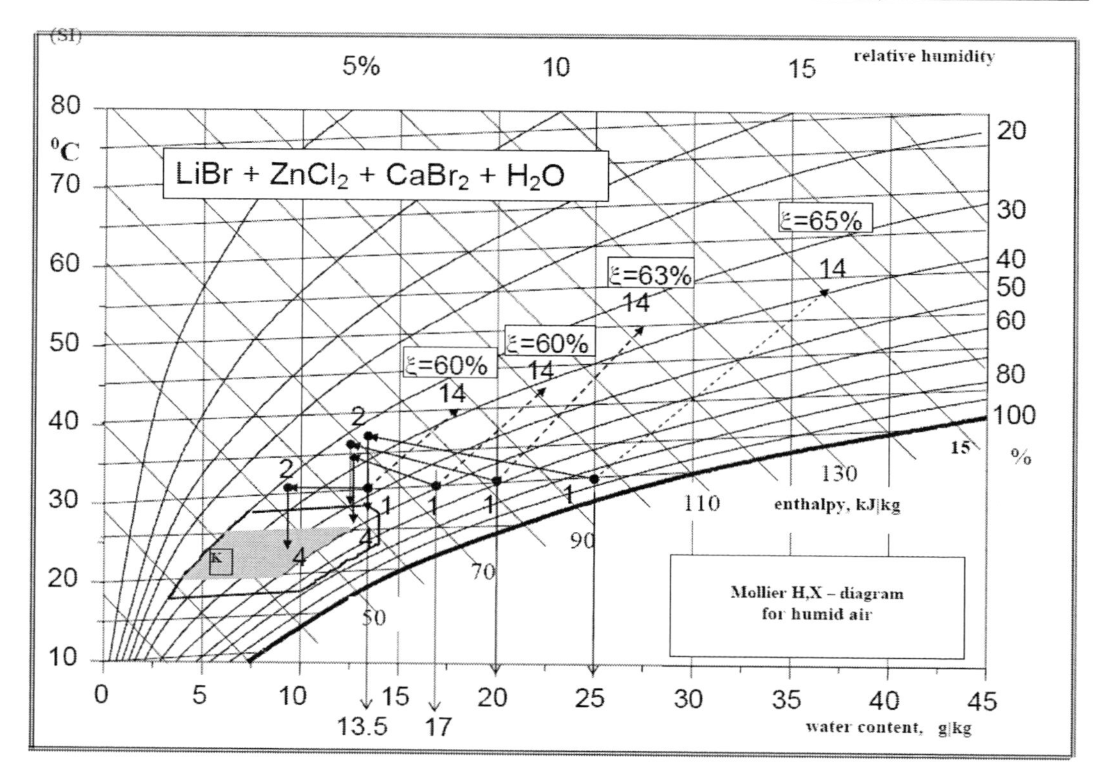

Figure 3.9. Working processes in the alternative system. Influence of the moisture content of the outside sir. Nomenclature is the same as in figure 3.7.

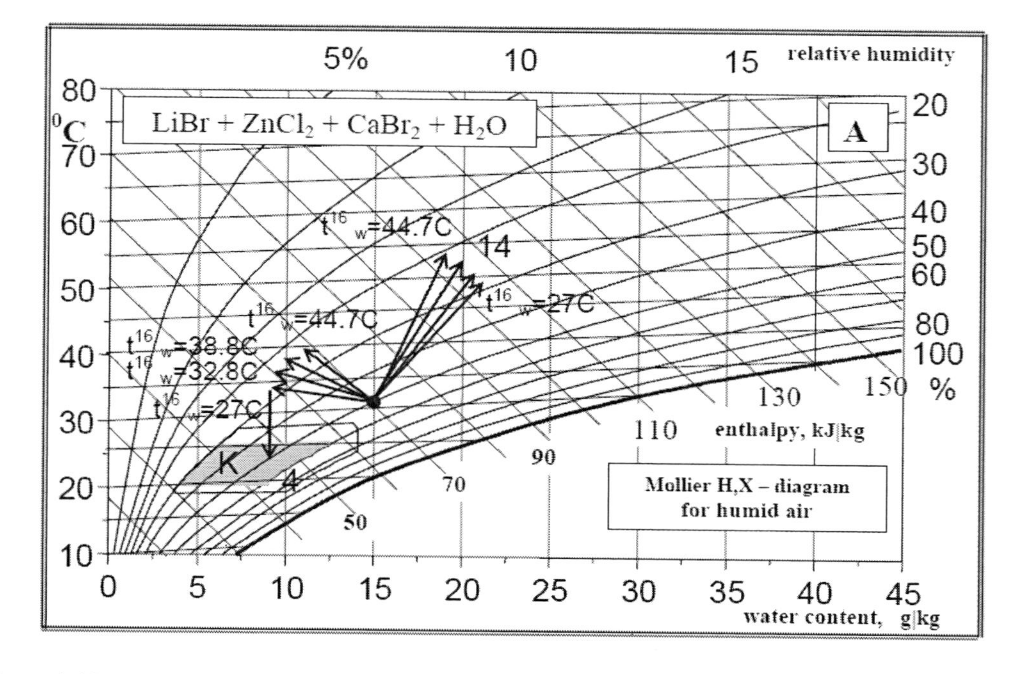

Figure 3.10. Continued on next page.

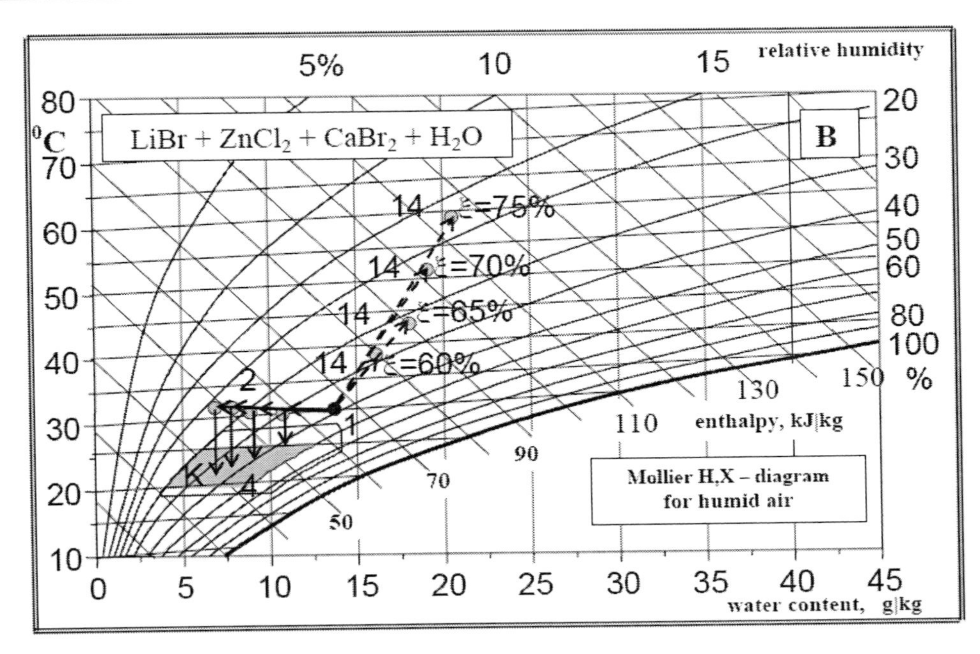

Figure 3.10. The representation of working processes in AACS on H,X-diagram of wet air. The absorbent is $LiBr + ZnCl_2 + CaBr_2 + H_2O$. A is the influence of the cooling liquid temperature t_w ; B is the influence of the absorbent concentration on the working processes in the system. Nomenclature is the same as in figure 3.7.

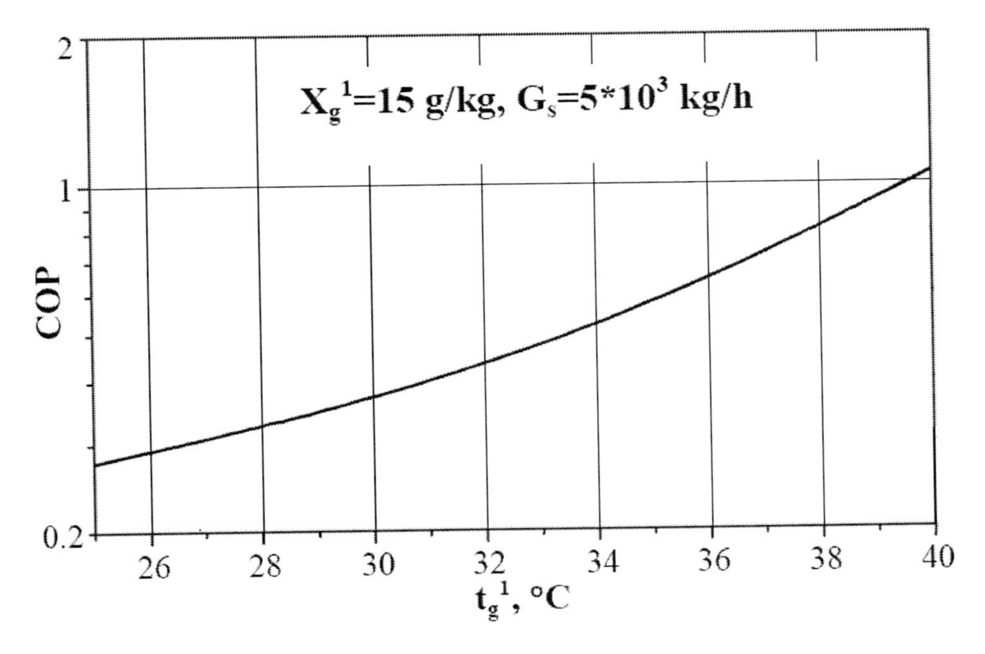

Figure 3.11. Dependence of COP on the temperature of ambient air.

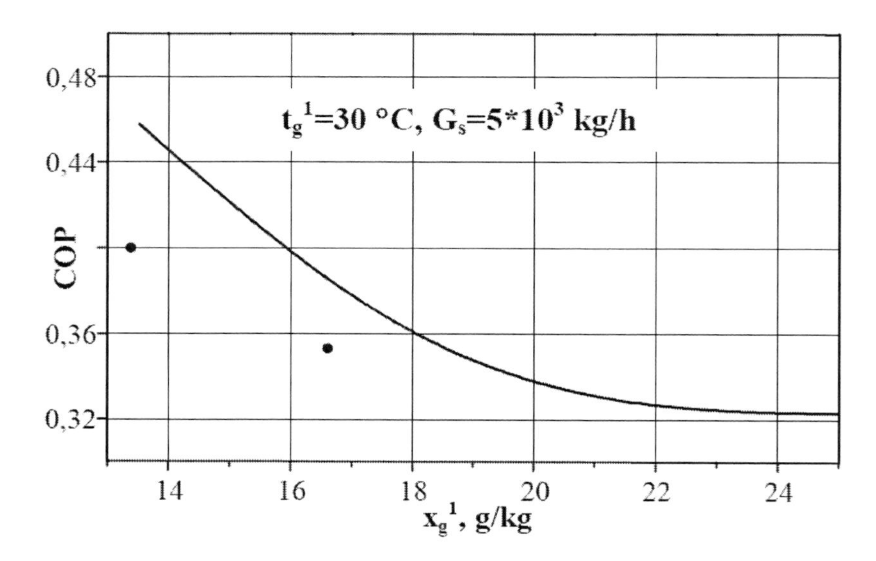

Figure 3.12. Dependence of COP on the moisture content of ambient air. • data of the work [52] at t_g^1 = 29.4 °C for the solution LiBr.

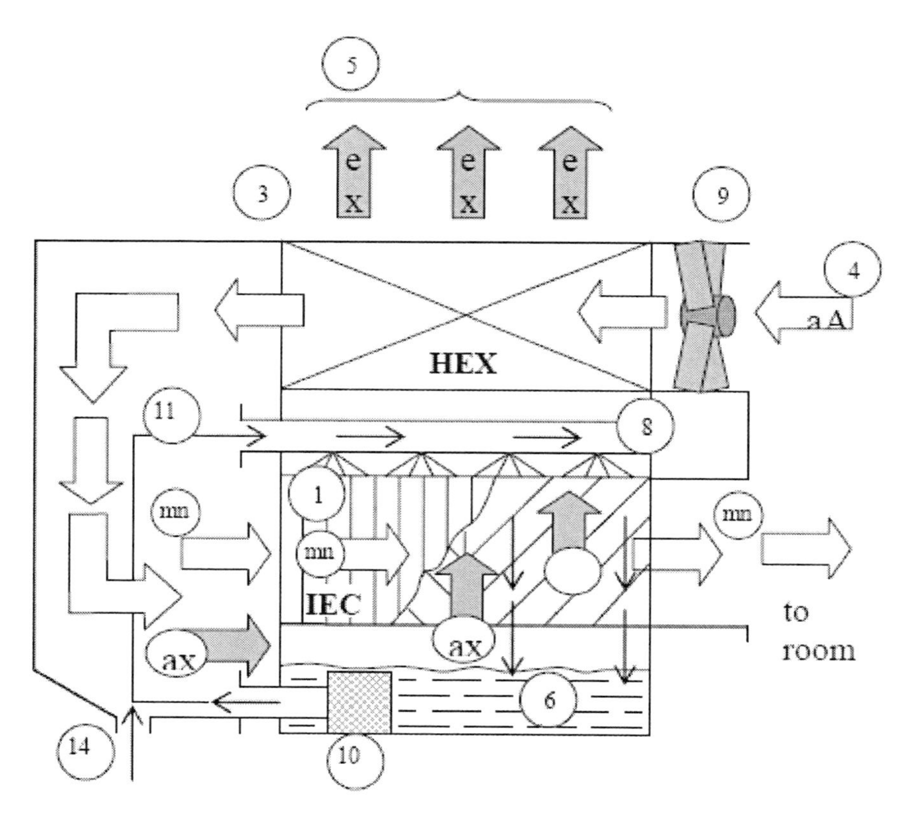

Figure 3.13. Schematic description of the combined evaporative cooler CEC on the basis of IEC/R. Nomenclature: 1 – indirect evaporative cooler IEC; 3 – heat exchanger HEX; 4 – entering (outdoor) air; 5 – air flow discharged into the atmosphere; 6 – water tank; 8 – liquid distributors; 9 – fan; 10 – water pump; 11 – recirculation water loop in IEC; 14 – replenishment of the system with water.

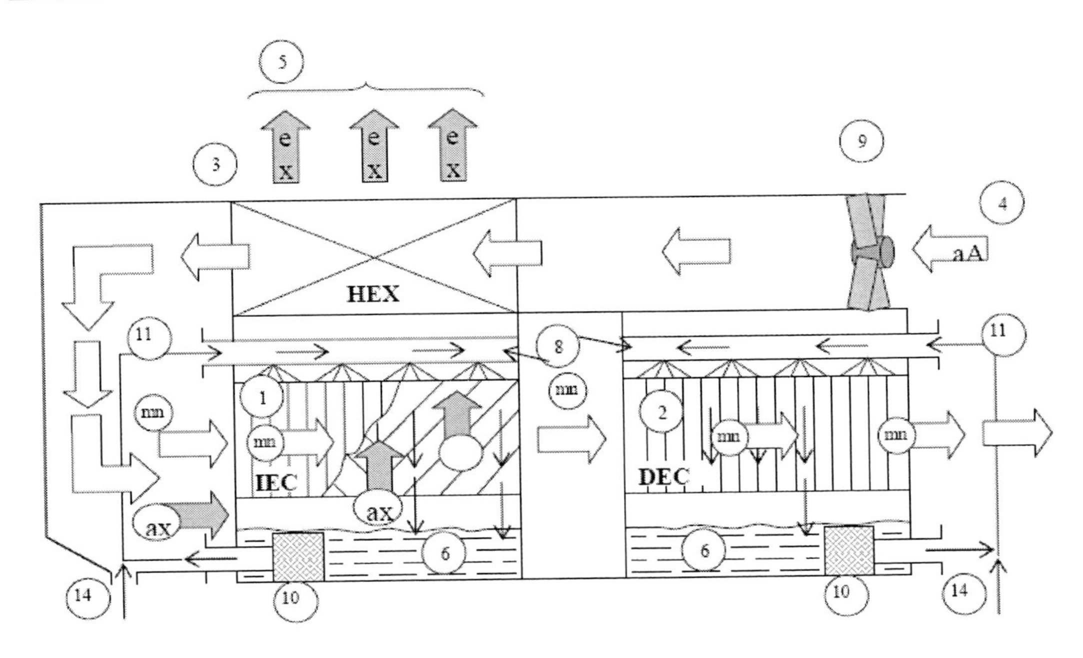

Figure 3.14. Schematic description of the combined evaporative cooler CEC on the basis of IEC|DEC. Nomenclature is the same as in figure 3.13 (2 – direct evaporative cooler DEC).

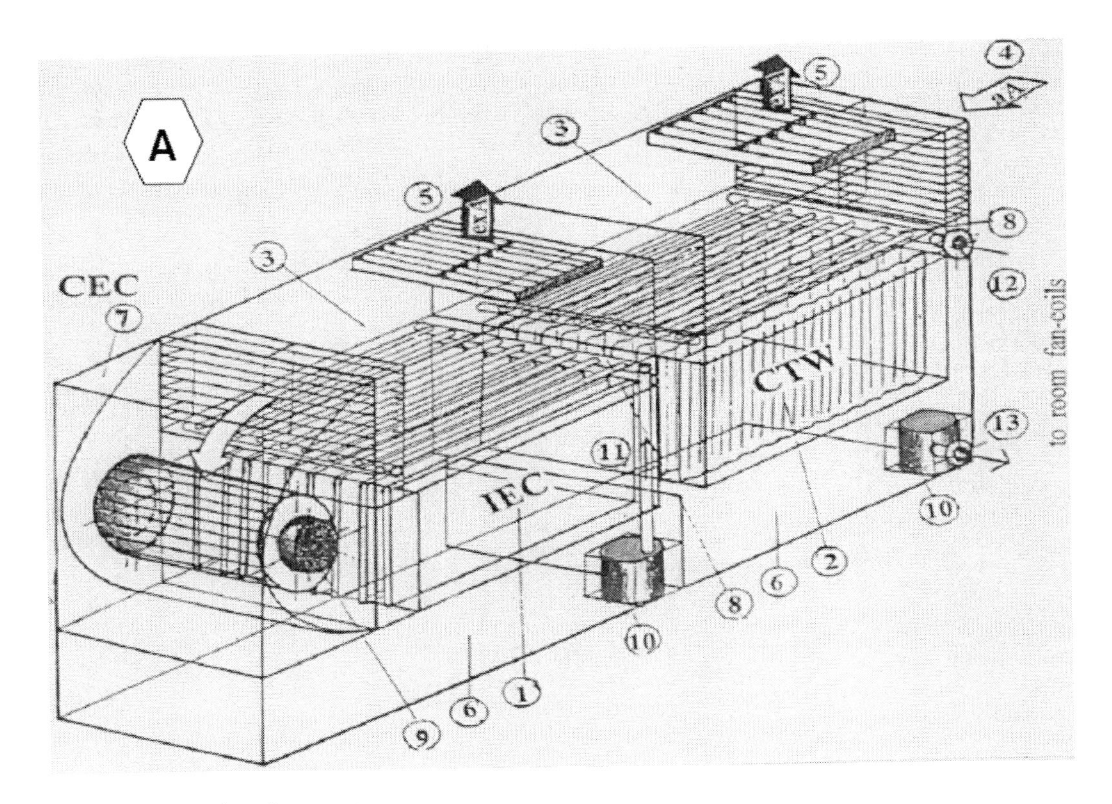

Figure 3.15. Continued on next page.

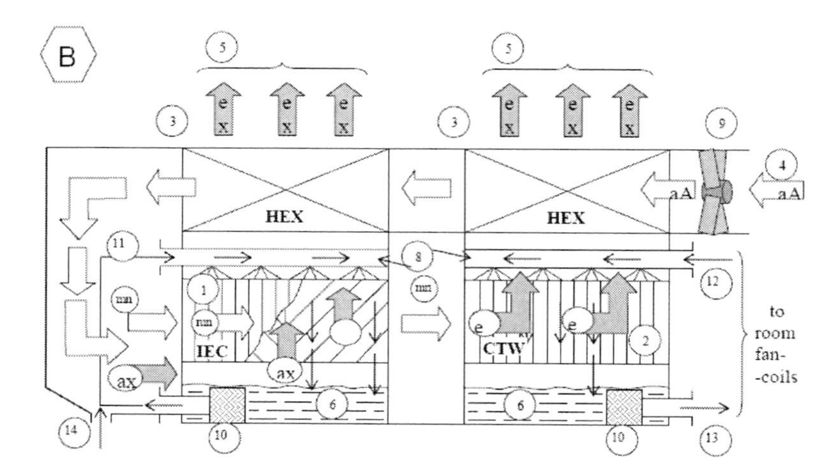

Figure 3.15. Schematic description of the combined evaporative cooler CEC on the basis of IEC/CTW. Notation: 1 – indirect evaporative cooler IEC; 2 –cooling tower CTW; 3 – heat exchanger HEX; 4 – entering (outdoor) air; 5 – air flow discharged into the atmosphere; 6 – water tank; 7 – combined evaporative cooler CEC; 8 – liquid distributors; 9 – fan; 10 – water pumps; 11 – recirculation water loop in IEC; 12, 13 – chilled water; 14 – replenishment of the system with water.

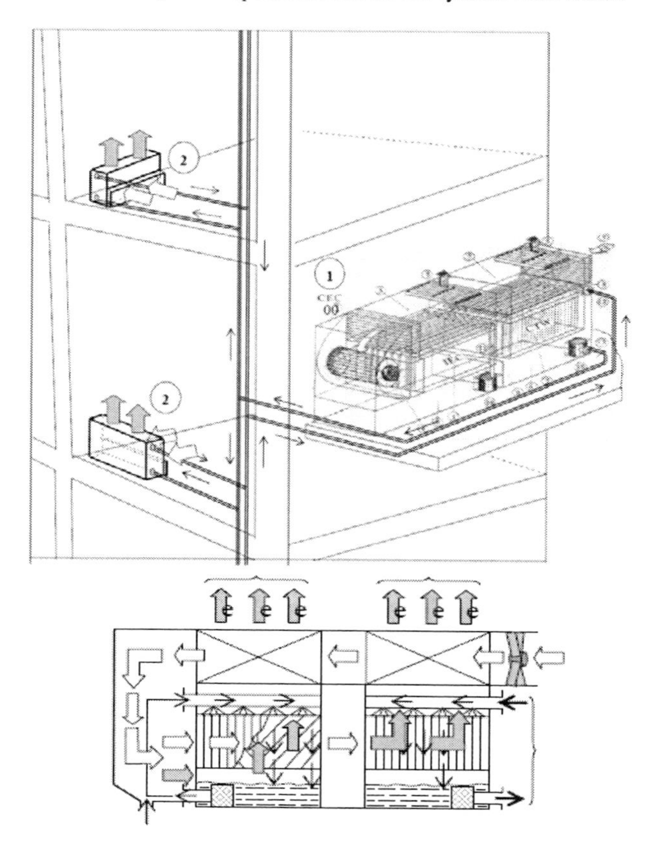

Figure 3.16. General view of the alternative air-conditioning system (AACS) with the combined evaporative air cooler CEC (1), located outside the building, and the system of water-air heat exchangers (2), located directly inside the rooms being served. All the nomenclatures shown in the CEC unit, are the same as in figure 3.15.

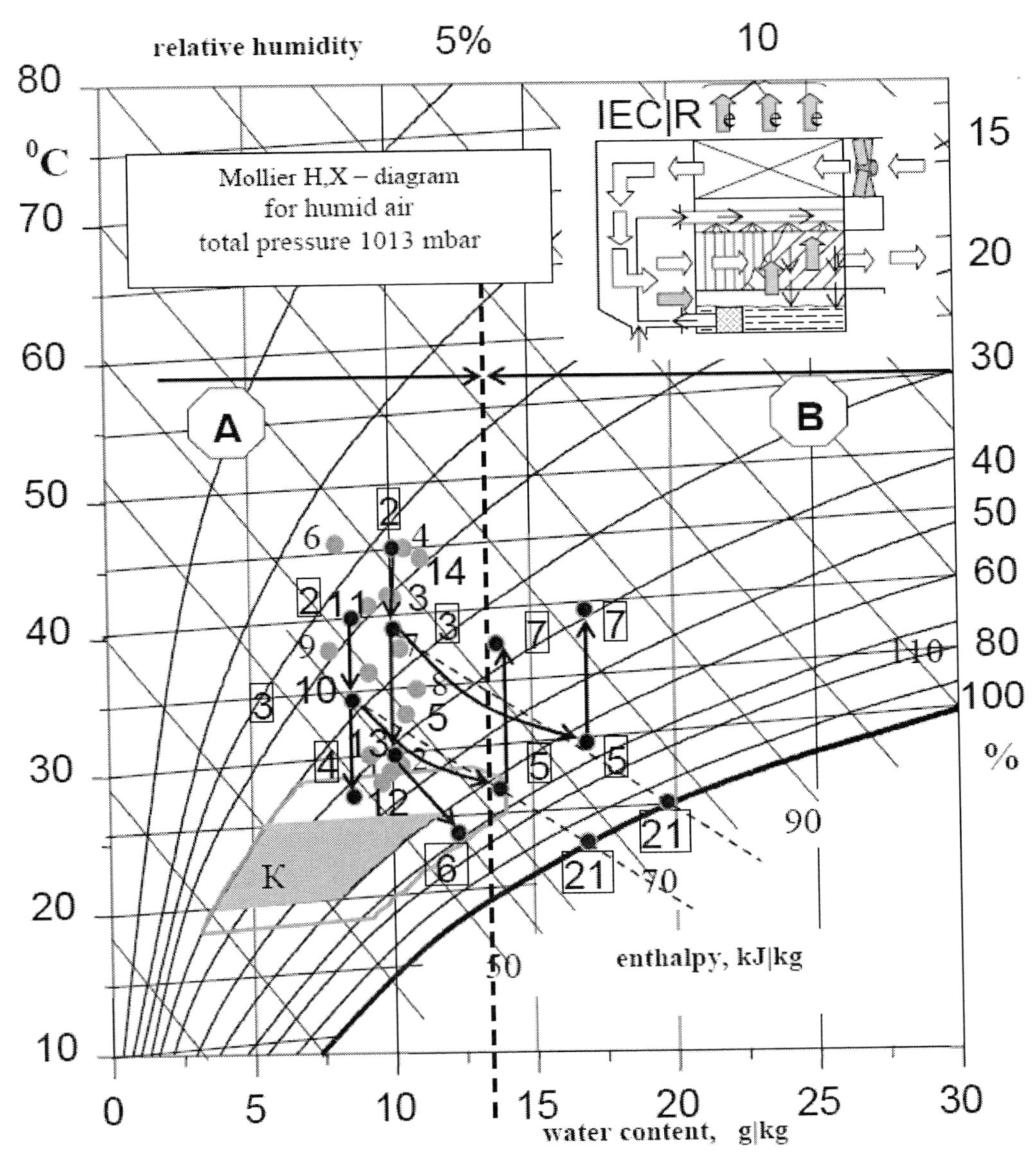

Figure 3.17. Representation of working processes in the combined evaporative cooler CEC on the basis IEC/R (the scheme is the same as in figure 3.13) on the H,X-diagram of wet air for different cities of the world (for climate conditions x_g^1 < 13 g/kg – zone A). Notations of cities are the same as in table 3.4. Progressing of the processes in IEC/R is shown on the example of two characteristic points with calculation parameters t_g = 40 °C, x_g = 8 g/kg and t_g = 45 °C, x_g = 10 g/kg, where: 2–3 and 5–7 are processes occurring in the heat exchanger HEX 13; 3–4 and 3–5 are the processes of changing the state of the main and auxiliary air flows in IEC; 21 –the state (temperature) of the water recirculating through IEC (is shown conventionally); 4–6 is the additional air cooling in DEC for the point with the calculation parameters if air t_g = 45 °C, x_g = 10 g/kg.

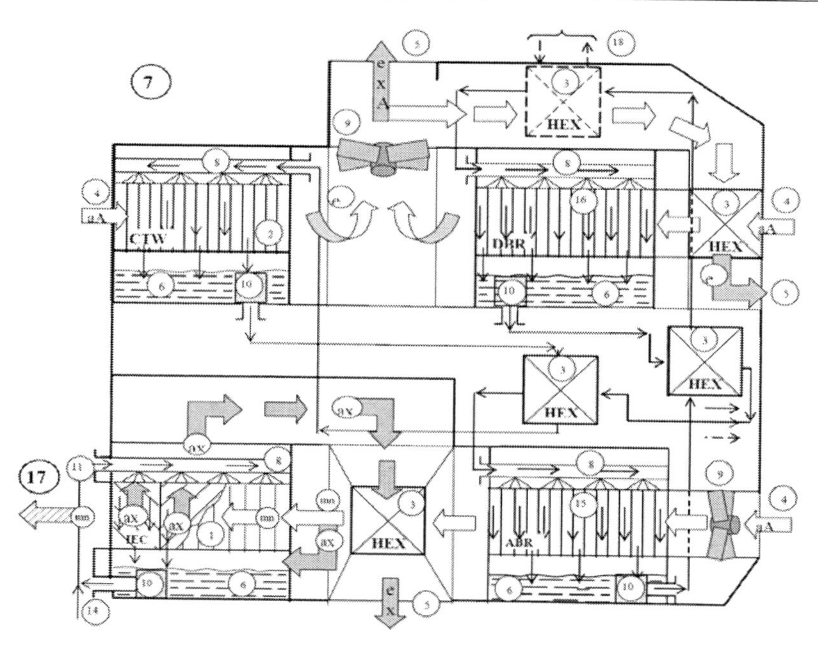

Figure 3.18. Schematic description (layout diagram) of the alternative air-conditioning system AACS on the basis of the open absorption cycle. Nomenclature: 1 – indirect evaporative cooler IEC; 2 – cooling tower CTW; 3 – heat exchangers HEX; 4 – entering (outdoor) air; 5 – air flow discharged into the atmosphere; 6 – tank for the liquid; 7 – AACS; 8 – liquid distributors; 9 – fan; 10 – liquid pumps; 11 – recirculation water loop in IEC; 12, 13 – chilled water; 14 – replenishment the system with water; 15 – absorber; 16 – desorber; 17 – air flow to the room being conditioned; 18 – heat carrier (water) from the heliosystem. In addition to the above: absorbent; water (loop of cooling); water (loop of heating).

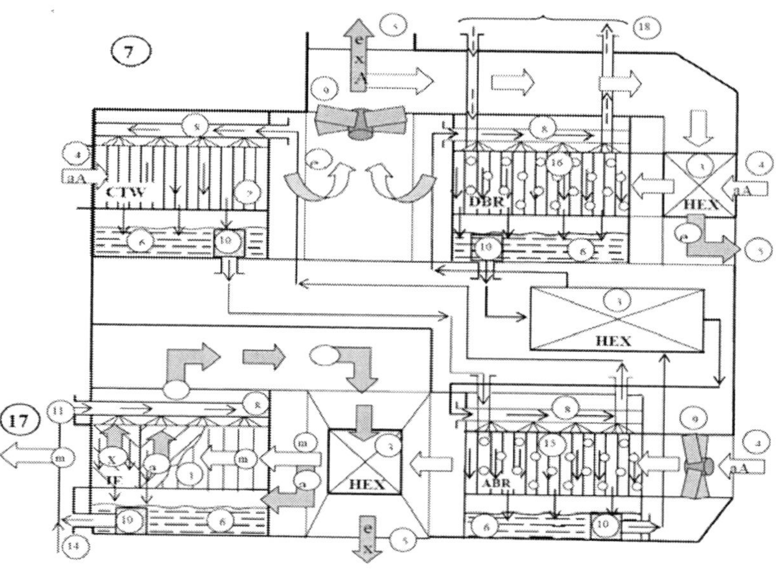

Figure 3.19. Schematic description (layout diagram) of the alternative air conditioning system AACS on the basis of open absorption cycle. Heat exchangers of the heating loop (desorber) and the cooling one (absorber) are joint with the desorber and the absorber respectively. Nomenclature is the same as in figure 3.18.

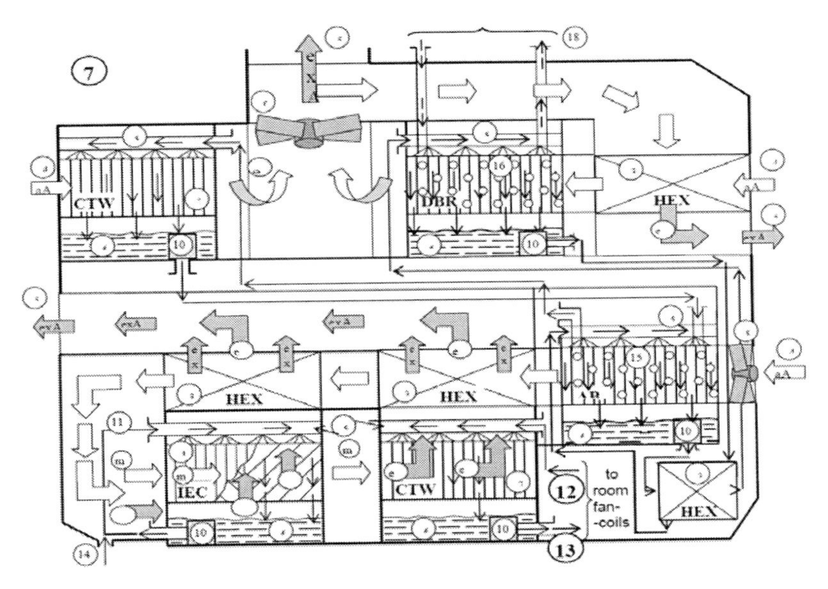

Figure 3.20. Schematic description (layout diagram) of the alternative refrigeration system ASRF on the basis of the heliosystem as a heating source (solar regeneration of the absorbent) and the supply of chilled water to heat-exchangers-coolers (room fan-coils). Nomenclature is the same as in figure 3.18.

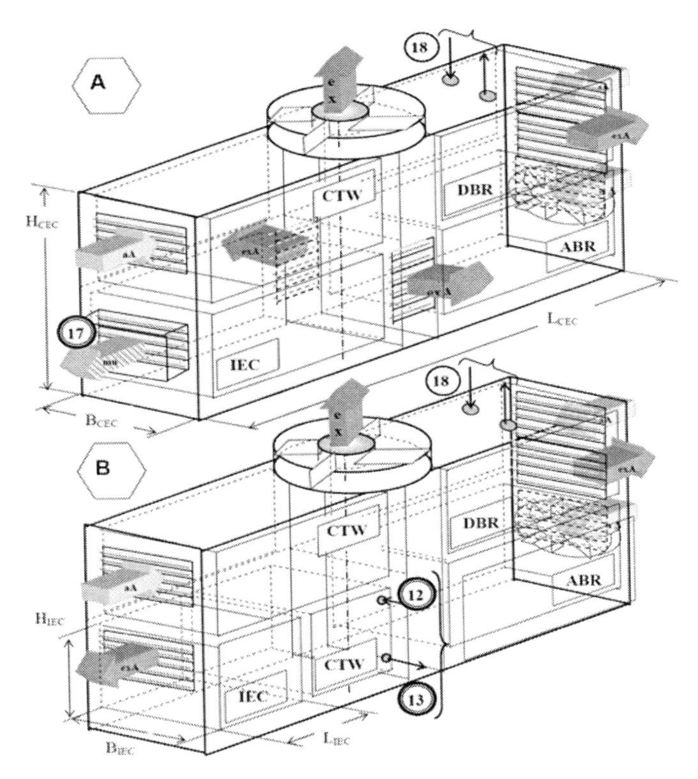

Figure 3.21. Layout diagrams of alternative air-conditioning systems AACS (A) and those of cooling ASRF (B) on the basis of the open absorption cycle. Nomenclature: IEC – indirect evaporative cooler; CTW – cooling tower; ABR – absorber; DBR – desorber; 12-13 – chilled water; 17 – air flow to the room being conditioned; 18 – heat carrier (water) from the heliosystem.

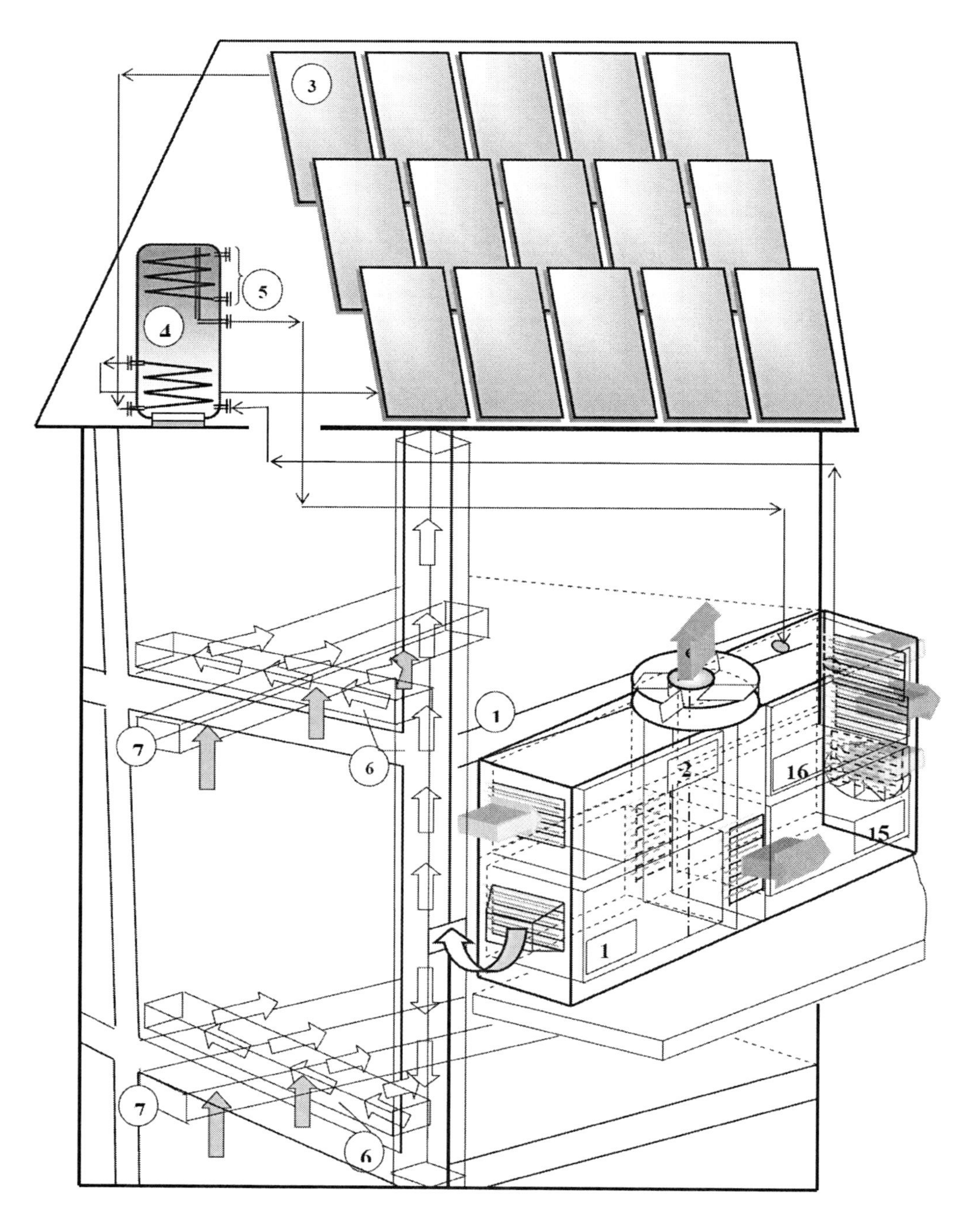

Figure 3.22. General view of AACS located outside the building with the system of air ducts in the room. Nomenclature: 1 – AACS; 3 – heliosystem; 4 – tank-heat accumulator; 5 – duplicating heater; 6 – air ducts for supplying air to rooms; 7 – exhaust ventilation the rest nomenclature is the same as in figures 3.18 and 3.21.

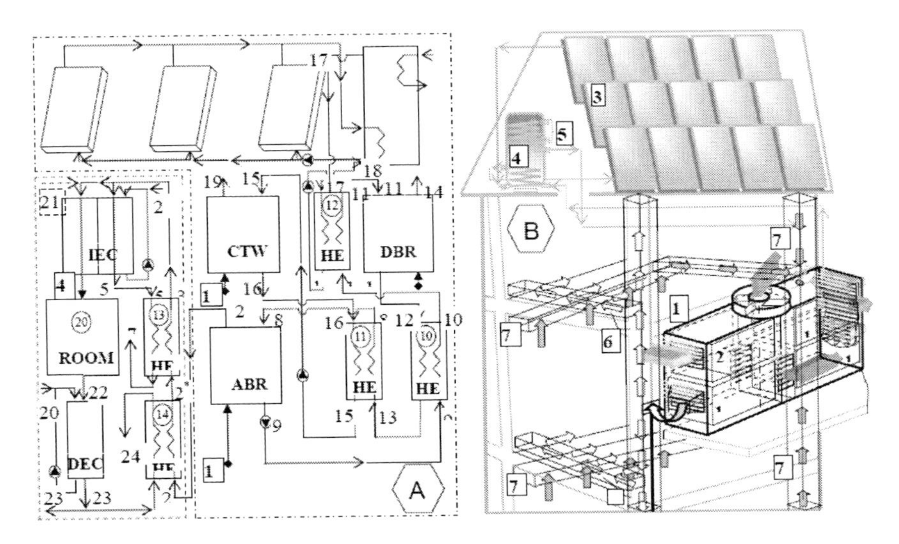

Figure 3.23. Design scheme (A) of the alternative recirculation system (recirculation mode, RM) of air conditioning. Nomenclature: 1–24 – flow parameters (numbers of design points). General view of AACS (B). The nomenclature is the same as in figures 3.18 and 3.21.

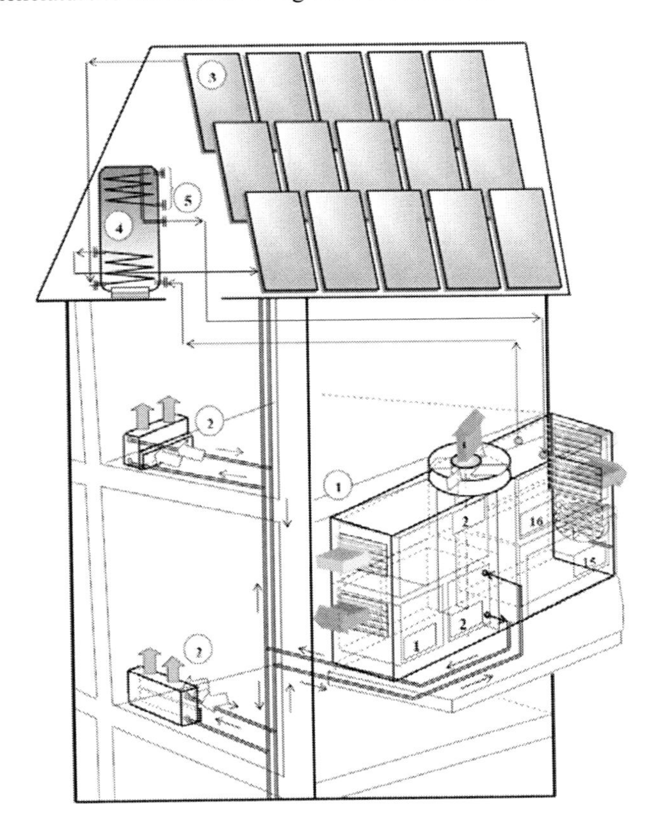

Figure 3.24. General view of AACS located outside the building with the system of water-air heat-exchangers in the rooms. Nomenclature: 1 – AACS; 2 – heat exchangers; 3 – heliosystem; 4 – tank-heat accumulator; 5 – duplicating heater; the rest nomenclature is the same as in figures 3.18 and 3.21.

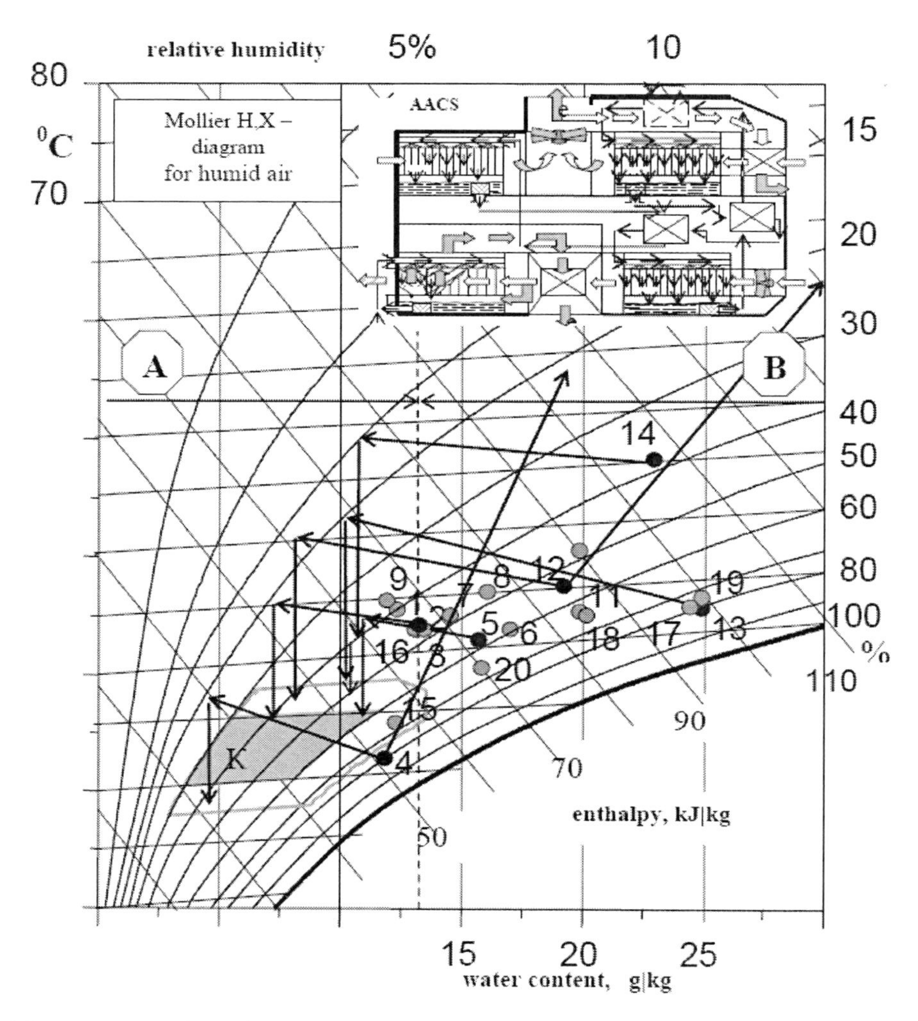

Figure 3.25. Representation of working processes in AACS (the scheme on figure 3.18) on the H,X-diagram of wet air for different cities of the world (for climate conditions of $x_g^1 > 13$ g/kg – zone B). Notations of cities are the same as in Table 3.6. Progressing of the processes in AACS is shown on the example of characteristic points with calculation parameters: $t_g = 21$ °C, $x_g = 11.99$ g/kg (Bogota), $t_g = 45.6$ °C, $x_g = 22.7$ g/kg (Port-Said), $t_g = 32.2$ °C, $x_g = 24.96$ g/kg (Douala); $t_g = 35$ °C, $x_g = 18.47$ g/kg (Tel Aviv). Shown are only processes of dehumidifying the air in the absorber and those of evaporative cooling in IEC.

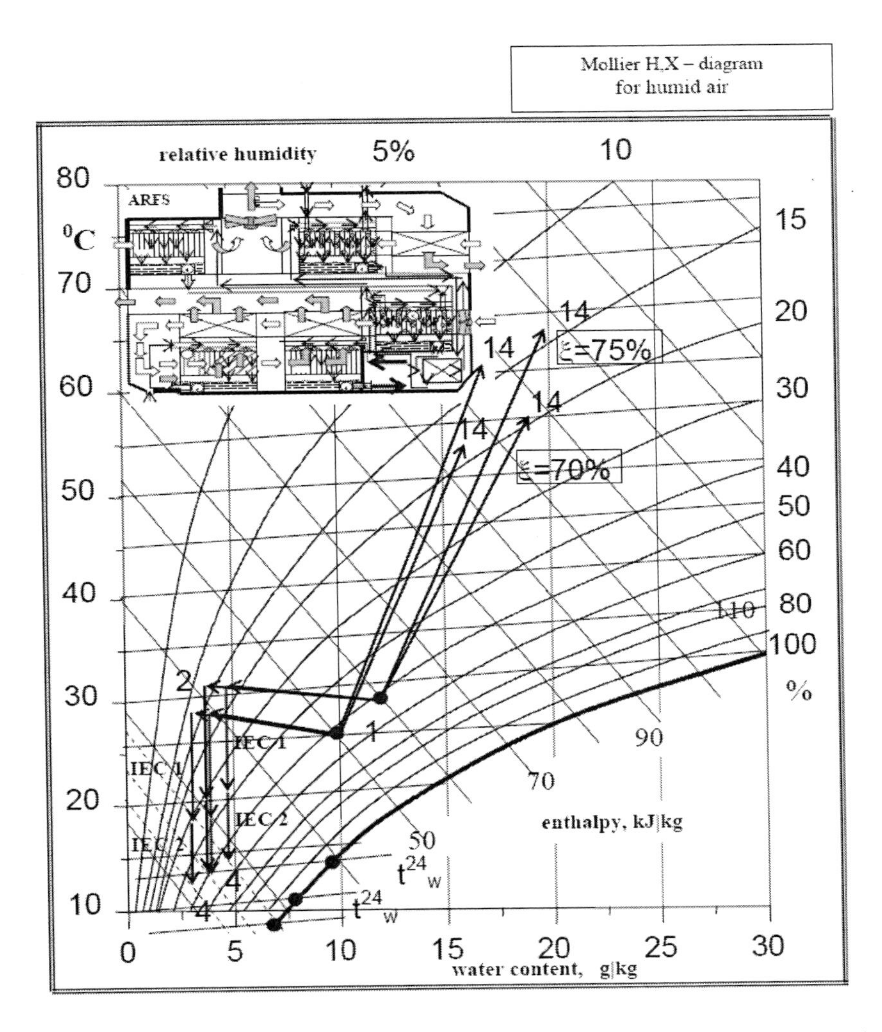

Figure 3.26. The picture of working processes in the alternative refrigerating system (the scheme is the same as in figure 3.20) on H, X diagram of wet air. The course of processes in the system is shown on the example of typical points with calculation parameters t_g = 25 °C, x_g = 10 g/kg and t_g = 28 °C, x_g = 12 g/kg (shown are the processes of dehumidifying the air in the absorber, double-stage evaporative cooling in the IEC and cooling the water in the cooling tower; the temperature of water cooling is shown conventionally in the form of a point on the line φ = 100%).

BASIC CONCLUSION

The report presented includes the thermodynamic analysis of developed schematic approaches to alternative refrigerating and air-conditioning systems, the selection of working substances (absorbents), the developed mathematical models (software) of operating processes in all the main components of alternative systems (refrigerating and air-conditioning systems), providing the interrelated calculation of such systems and the following design. A pilot plant AACS is developed on the basis of the research carried out and obtained results.

- It is preferable to use liquid sorbents (absorbents) in alternative solar systems of air-conditioning AACS and of cooling ARFS. The analysis of characteristics of working bodies used as absorbents for solar alternative systems being developed showed low suitability of $CaCl_2+H_2O$ because of high partial pressures of water vapours at absorption temperatures, the prospectiveness of solutions $LiBr+H_2O$ (LiBr), $LiBr+LiNO_2+H_2O$ (LiBr+) and $LiBr+ZnCl_2+CaBr_2+H_2O$ (LiBr++), in this case LiBr+ has better absorption capacity, and LiBr++ has less corrosive activity and allows the decrease of the required temperature of the outside heating source during the desorption, which is important in principle while using low-grade sources of heat;

The calculation analysis has been done for working characteristics in a wide range of initial parameters (temperature and moisture content of outdoor air, absorbent type and concentration, cooling water temperature, relation of gas and liquid flow rates in apparatuses of alternative systems) and of construction features of heat- and- mass transfer apparatuses which made it possible to reveal qualitative (predictable) characteristics of the system and to obtain quantitative information necessary for their further engineering calculation and designing.

- The general analysis of principal capabilities of solar air-conditioning systems AACS showed:
  - when $x_g < 13$ g/kg (zone A – dry and hot climate), providing comfort air parameters is quite possible when only evaporative methods of cooling are used, without pre-dehumidifying the air flow, at $x_g > 13$g/k evaporative cooling alone cannot provide required comfort air parameters. The need for an additional stage of cooling DEC, appears only, at high initial air temperatures ($t_g > 40$ °C).
  - when $x_g > 13$ g/kg (zone B) the alternative solar air-conditioning system is quite capable of providing the comfort air parameters for any climatic conditions of the world. And the use of a single-stage regenerative indirect-evaporative cooler IEC/R as an evaporative cooler is optimal, the need for an additional stage of cooling DEC or a double-stage system IEC/IEC appears only at high initial air temperatures ($t_g > 40$ °C);
  - the required temperature of the heating source for providing solar regeneration of the absorbent amounts to a range of values $t_{17} = 55$-125 °C. The possibility of using the simplest and cheapest flat solar collectors is limited for these systems by the range of values of low moisture contents and temperatures of outdoor air. To solve the problem of solar air conditioning in a wide range of climatic conditions it is necessary to have solar collections with selective coating or evacuated glass pipe collectors.

- The general analysis of principal capabilities of solar cooling systems ARFS showed that it is possible to achieve the cooling water temperature of 8.5–12 °C, and the required temperature of the heating source is in the range of t_{17}=95-125 °C. The possibility of functioning such a cooling system is limited by the range of rather low values of moisture content in outdoor air x_g<10-15 g/kg. The use of the double-stage cooler IE/IE as a component of ARFS greatly increases total energy expenditures for the organization of the process, since only ¼ of air dehumidified in the absorber goes out of the second evaporative stage. Further development of solar refrigerating systems ARFS requires the use of highly efficient evacuated collectors and the improvement of the cooling part of the plant.
- The pilot plant. For the calculation condition of Odessa (summer period – July, t_g^1 = 30 °C, x_g^1 = 13.3 g/kg, flow rate of the air being dehumidified G = 10000 kg/h, flow rate of the air fed to the conditioned room is 5000 kg/h ventilation scheme of air-conditioning – ventilation mode, VM), to provide a continuous cycle (absorbent regeneration) the solar collectors area of 150 m^2 (LiBr++) is necessary, in this case the calculation temperature of regeneration is 65–70°C. As compared to traditional vapour – compression systems of air – conditioning the alternative system AACS provides considerable redaction of energy consumption (30–60%), which is confirmed by not numerous data of using similar plants [24, 101, 102].

NOMENCLATURE

t, T – temperature, °C, K;
p – pressure (air), Pa;
h – entalphy, J|kg;
x – water content of air (kg water / kg dry air);
c_p – specific heat, J|kg K;
ρ – density, kg|m^3;
r – latent heat of phase transition (evaporation), J|kg;
D – diffusion coefficient, m^2|s;
g – gravitational acceleration, m|s;
φ – relative humidity of air (%);
σ – surface tension, N|m;
λ – thermal conductivity, W|m K;
ν – kinematics viscosity, m^2 |s;
μ – dynamic viscosity, kg|m s;
ξ – absorbent concentration in solution (kg absorbent / kg solution);
J – solar radiation intensity, kJ|m^2 s;
τ – time, s.

δ – liquid film thickness, m ;
δ_0 – maximum thickness of liquid film in cavities of corrugated surfaces, m;
A – transfer aria, m^2;
f (F)– surface cross-section area; surface aria, m^2;
H, L, B– height, length, width, m;
d – diameter, m;
E, P– the height and the pitch of the main corrugation of the packing bed sheets, m;
k_{opt} – optimal values of roughness parameters;
e, p – the height and the pitch of roughness riffles, m;
n – number of sheets in packing bed of heat-mass-transfer apparatus; number of tubes in SC set;
W – distance between tubes in SC, m.
G – mass flowrate, kg|s;
Q – amount of heat, kW;
w – velocity, m|s;
VR – air to water flow ratio (DEC, CTW), air to solution flow ratio (ABR, DBR), main air flow to auxiliary air flow ratio (IEC);
U – coefficient of heat losses, W|m^2 K;
E – efficiency;
COP– coefficient of performance;
θ – optical coefficient of performance;
α – heat transfer coefficient, W|m^2 K ;
k – overall heat transfer coefficient, W|m^2 K;
β – mass transfer coefficient (air side or solution side), kg|m^2 s;
Nu – Nusselt number;
Sh – Sherwood number
Pr – Prandtl number;
Re – Reynolds number;
We – Weber number;
Le – Lewis number;
Fr – Frud number;
Eu – Euler number.

Subscripts

g, L – gas, liquid;
w – water;
s – solution;
wk, str – weak and strong solution;
mn, ax – main and auxiliary air flow;
wb, db – wet bulb and dew point temperature of air;
aA, exA – ambient and exhaust air;

f – film;
H, M – heat-and-mass transfer;
e – equivalent diameter of packing elements, m;
* – interface, critical value;
in, out – inlet, outlet parameters;
id – ideal;
p – fin of absorber plate SC;
wl – wall;
1, 2 – lifting and lowering pipelines of a heliosystem;
i, k – nodal points on x and z coordinates;
min – minimum value;
max – maximum value;
☚ – difference.

Abbreviations

AACS – alternative air conditioning systems;
ARFS – alternative refrigeration systems;
CEC – combined evaporative cooler;
DEC – direct evaporative cooler;
IEC – indirect evaporative cooler;
R – regenerative scheme;
CTW – cooling tower;
ABR – absorber;
DBR – desorber;
HMA – heat-mass-transfer apparatus;
HEX – heat exchanger;
FNC – fan-coil;
SC – solar collector;
TA – tank-accumulator;
RR – regular roughness;
RP – regular packing bed;
GG – greenhouse gases;
VM – ventilation mode;
RM – recirculation mode.

Other notation are given in the text of the report in places of their use.

REFERENCES

[1] Antonopoulos, K.A., Rogdakis, E.D., 1996, Perfomans of solar-driven ammonia-lithium nitrate and ammonia-sodium thiocyanate absobtion systems operating as coolers or heat pumps in Athens, *Appl. Therm. Eng.*, vol. 16, n. 2, p. 127-147.

[2] ARI Air-Conditioning and Refrigeration Institute, *Catalog of Publications*.

[3] Beckman, Y.A., Klane, S.A., Daffi, J.A., 1982, The design of solar heat supply systems, Moscow, *Energoizdat*, pp. 72.

[4] Best, R., Pilatowsky, I., 1998, Solar assisted cooling with sorption systems: status of the research in Mexico and Latin America, *Int. J. Refrig.*, vol. 21, no. 2: pp. 100 -115.

[5] Biel, S., 1998, Sorptive luftentfeuchtung und verdunstungskühlung, *KI Luft- und Kältetechnick*, no. 7., pp. 332-336.

[6] Case, V.M., London, A.L., 1962, Compact heat exchangers, Moscow-Leningrad, *Gosenergoisdat*, p. 160.

[7] Catalogue for plate heat exchangers, Edit. P.I.Bazhan, Moscow, *Mashinostroenie*, p. 59.

[8] Chant, E.E., Jeter, S.M., 1994, A steady-state simulation of an desiccant-enhanced cooling and dehumidification system, *ASHRAE Trans.,* US, vol. 100, n. 2, p. 339-347.

[9] Chau, C.K., Worek, W.M., 1995, Interactive simulation tools for open-cycled desiccant cooling systems, *ASHRAE Trans.,* US, vol. 101, n. 1, pp. 725-734.

[10] Kholpanov, L., Doroshenko, A., 1986, Peculiarities of liquid film flowing along the elements of the packing bed with regular roughness of the surface, *Izvestiya vuzov, series – Khimiya i khimicheskaya Tekhnologiya (Chemistry and Chemical Technology)*, Vol. 29, issue 10, pp. 117-120.

[11] Claassen, N.S., 1995, Developments in multi-stage evaporative cooling, *Afr. Refrig. Air Cond., ZA,* vol. 11, n. 2, p. 55-67.

[12] Collares Pereira, M., 1995, CPC-type collectors and their potential for solar energy cooling applications, *Proceedings of 2nd Munich Discussion Meeting 'solar Assisted Cooling with Sorption System', München*, Paper No. 5.

[13] Colvin T.D., 1995, Office tower reduce operating costs with two stage eveporative cooling system, *ASHRAE*, vol. 37, n. 3, p. 23-24.

[14] Commission of the European DGXVII, 1994, The European renewable energy study, *Office for Official Publications of the European Comminities*, Luxembourg, vol. 1, p. 38.

[15] Costa, A. Et al., 1996, Potential of solar assisted cooling in Southern Europe, *Proceedings of EuroSun 96, Freiburg*, pp. 1201-1206.

[16] Czederna, A., Tillman, N.N., Herd, G.C., 1995, Polimers as advanced materials for desiccant applications. 3. Alkalisalts of PSSA and poliAMPSASS, *ASHRAE Trans.,* US, vol. 101, n. 1, p. 697-712.

[17] Daffi, J.A., Beckman W.A., 1974, Solar Energy Thermal Processes, *John Wiley & Sons*, N.Y. – London – Sydney – Toronto; Daffi, J.A., Beckman, Y.A, 1977, Heat processes with the use of solar energy, Moscow, *Mir,* pp. 566.

[18] Das Potential der sonnenergie in der EU, Eurosolar, Bonn, 1994.

[19] Deng, S.M., Ma, W.B., 1999, Experimental studies on the characteristics of an absorber using LiBr/H_2O solution working fluid, *Int. J. Refrig.*, vol. 22, P. 293-301.

[20] Dhar, P.L., Kaushik, S.V., 1995, Analysis of a fild-installed hybryi solar desiccant cooling system, desiccantaugmented evaporativecooling cycles for Indian conditions, *ASHRAE Trans.*, vol. 101, n. 1, p. 735-749.

[21] Donets, Ya.I, 1991, Mathematical modelling of flat solar collectors with liquid heat carriers. Heat pipes and heat pumps, *Minsk,* p. 131-137.

[22] Doroshenko, A., 1988, Indirect evaporative cooler for air-conditioning, *Cholodilnaya Technika,* no. 10: p. 28-33 (Russian).

[23] Doroshenko, A., 1996, Alternative Air-Conditioning, *International Conference of Research, Design and Conditioning Equipment in Eastern European Contries, September 10-13, Bucharest, Romania,* IIF/IIR: p. 102-108.

[24] Doroshenko, A., Compact heat-and-mass transfer equipment for refrigerating engineering (theory, calculation, engineering practice), Doctor dissertation, Odessa State Academy of Refrigeration.

[25] Doroshenko, A., et al., 1985, Indirect-evaporative cooling as a perspective trend in air-conditioning, *Izvestiya Akademii Nauk Turkmenskoi SSR, series – Physico-technical, chemical and geological sciences*, No 3, pp. 30-35.

[26] Doroshenko, A., 1996, New Developments of Air-Conditioning, *International Conference of Applications for Natural Refrigerants' 96, September 3-6, Aarhus, Denmark,* IIF/IIR: p. 339-345.

[27] Doroshenko, A., Karev, V., Kirillov, V., Kontsov, M., 1998, Heat and Mass Transfer in Regenerative Indirect Evaporative Colling, *Intern. Conference IIR/IIF of Advanes in the Refrigeration Systems, Food Technologies and Cold Chain- Sofia'98, September 23-26, Sofia, Bulgaria.*

[28] Doroshenko, A., Kirillov, V., 1988, Special features of liquid film flowing along surfaces roughness, *Eng.-Phys.-Journal,* vol. 54, no. 5: p. 739-745 (Russian).

[29] Doroshenko, A., Kirillov, V., Kontsov, M., 1998, Alternative Refrigerating Systems on the basis of Open Absorption Cycle Using Solar Energy as a Heat Source, *Intern. Conference IIR/IIF of Advanes in the Refrigeration Systems, Food Technologies and Cold Chain- Sofia'98, September 23-26, Sofia, Bulgaria.*

[30] Doroshenko, A., Kontsov, M., 1993, Mathematical modeling and optimization of a Solar Hot-Water System, *8-th Intern. Conference on Thermal Eugng and Thermogrammetry,* Budapest, Hungary; 1995, Theoretical and Experimental investigations of solar Hot Water Systems; The optimization of contraction of flat-plat Solar Collectors in Hot Water supply Systems, *9-th Intern. Conference on Thermal Engn and Thermogrammetry*, June 14–16 Budapest, Hungary; Doroshenko, A., Glikson., 1997, Non-Convenctional Energy Sources in To-day's Heat Supply systems, *10-th Intern. Conference on Thermal Engn and Thermogrammetry,* June 14–16 Budapest, Hungary.

[31] Doroshenko, A.V., Omelchenko, Yu.M., 1999, Alternative energetics: experience of application and real perspectives. Untraditional and renewable sources of energy, *Ministry of transport of Ukraine, State Department of marine and river transport*, Odessa, No.2, pp. 12-13.

[32] Doroshenko, A., Yarmolovich, Y., 1987, Indirect Evaporative Cooling, *Cholodilnaya Technika,* no. 12: p. 23-27 (Russian).

[33] El-Chalban, A.R., 1995, Simulation and performance analysis of a compound heat transformer – refrigeration system, *Proceedings of 2nd Munich Discussion Meeting 'solar Assisted Cooling with Sorption System', München*, Paper No. 2.
[34] Engelhorn, H.R., 1998, Solar-cooling plant Jiangxi, Fachhochschule Giessen-Friedberg.
[35] Erhard, A., Spindler, K., Hahne, E., 1995, Test and simulation of solar powered dry absorption cooling machine, *Proceedings of 2nd Munich Discussion Meeting 'solar Assisted Cooling with Sorption System', München*, Paper No. 4.
[36] EUREC Agency, 1996, The future renewable energy – Prospects and directions, *James & James Ltd*, London.
[37] Foster, R.E., Dijkastra, E., 1996, Evaporative Air-Conditioning Fundamentals: Environmental and Economic Benefits World Wide, *International Conference of Applications for Natural Refrigerants' 96, September 3-6, Aarhus, Denmark,* IIF/IIR: p. 101-109.
[38] Franke, U., 1998, Beispielsammlung für die Anwendung des SECO-Sorptionsgenerators, Fachbericht, ILK Dresden.
[39] Gandhidasan, P., 1994, Performance analysis of an open liquid desiccant cooling system using solar energy for regeneration, *Int. J. Refrig.*, vol. 17, no. 7: p. 475 - 480.
[40] Grandov, A., Doroshenko, A., 1995, Cooling Tower with Fluidized Beds for Contaminated Environment, *Int. J. Refrig.*, vol. 18, no. 8: p. 512-517.
[41] Grossman, G., Devault, R.C., Creswick, F.A., 1995, Simulation and perfomans analysis of an ammonia-water absorption heat pump based on the generator-absorber heat exchange (GAX) cycle, *ASHRAE Trans.*, vol. 1, n. 1, p. 1313-1323.
[42] Grossman, G., Zaltash, A., Devault, R.G., 1995, Simulation and perfomance analysis of a four-effect water-lithium bromide absorbtion chiller, *ASHRAE Trans.,* US, vol. 101, n. 1, p. 1302-1312; Haim, .I., Grossman, G., Shavit, A, 1992, Simulation and analysis of open cycle absorption system for solar cooling, *Solar Energy*, Vol. 49, n., 6, pp. 515-534.
[43] Guillemnot, J.J.G., Chalfen, J.B., Poyelle, F.G.R., 1995, Mass and heat transfer in consolidated adsorbing composites. Effect on the performance of a solid-adsorption heat pump, *19-th int. Congr. Refrig., The Hauge,* NL, pp. 261-268.
[44] Gusel E., Gurgor, A., 1995, Air conditioning using absorption and solar energy in an open system, *Proc. ULIBTK'95, Ankara,* TR., pp. 303-313.
[45] Hahne, E., 1996, Solar heating and cooling, *Proceedings of EuroSun 96, Freiburg,* pp. 3-19.
[46] Hamad, M., 1995, Experimental study of the performance of a solar collector cooled by heat pipes, *Renewable Energy,* GB., vol. 6, n. 1, pp. 11-15.
[47] Hammand, M.A., Audi, M.S., 1992, Performance of solar LiBr-water absorption refrigeration system, *Renewable Energy*, 2(3), pp. 275-282.
[48] Hanna, W.T., Saunders, J.H., Wilkson, W.H., Philips, D.B., *ASHRAE Trans.,* US, vol. 101, n. 1, p. 1189-1198.
[49] Hara, T., Oka, M., Nikai, I., 1996, Adsorption/desorption characteristics of adsorbent dehumidifier for air conditioners, *International Conference of Applications for Natural Refrigerants' 96, September 3-6, Aarhus, Denmark,* IIF/IIR: p. 111-120.
[50] Hartmann, K., 1998, Lithiumbromid / Wasser-Absorptions-kältemaschinen, *DIE KÄLTE & Klimatechnick*, no. 10, pp. 780-790.

[51] Heat exchange apparatuses of refrigerating plants, 1986, Edit. G.G. Danilova, Leningrad, *Mashinostroenie*, p. 34.

[52] Hellman, H.M., Grossman G., 1995, Simultation and analysis of an open-cycle dehumidifier-evaporator (DER) absorption chiller for low-grade heat utilization, *Int. J. Refrig.*, vol. 18, no. 3: p. 177-189.

[53] Hellman, H.M., Pohl, J.P., Grossman, G., 1996, *IIF/Proc. Aarhus Meat,* IIR, FR., p. 121-132.

[54] Henning, H. (Co-ordinator), Solar desiccant cooling system for an office building in Portugal, Contract EU Thermie No. REB/78/95/DE/PO.

[55] Henning, H., Erpenbeck, T., 1996, Integration of solar assisted open cooling cycles into building climatisation systems, *Proceedings of EuroSun 96, Freiburg*, pp. 1248-1253.

[56] Henning, H.-M., Häberle, A., Gerber, A., 1995, Building climatization with solar assisted open cycle solid sorption cooling systems – design rules and operation strategies, *Proceedings of 2nd Munich Discussion Meeting 'solar Assisted Cooling with Sorption System', München*, Paper No. 9.

[57] Hilali, I., Okuyan, C., Aktacir, M.A., 1995, Investigation of absorption cooling systems using solar energy in Sanliurfa, *Proc. ULIBTK'95, Ankara,* TR., p. 323-332.

[58] Höper, F., 1999, Optimierte anlagenschaltung zur solaren Kühlung mit absorptionskältetechnick, *KI Luft- und Kältetechnick*, no. 8., pp. 397-400.

[59] Idelchik, I.E., 1976, Deference-book on hydraulic resistances, Moscow, *Machinebuilding*, pp. 560.

[60] Intensification of heat exchange in evaporators of refrigerating machines, Editor A.A. Gogolin, Moscow, *Light and food industry (Legkaya and pishchevaya promishlennost),* 1982, pp. 224.

[61] Intensification of heat exchange in evaporators of refrigerating machines, 1982,Edit A.A. Gogolin, Moscow, *Light and food industries*, p. 224.

[62] International Institute of Refrigeratinf, Publications 1998.

[63] Ismail, I.M., Mahmoud, K.G., 1994, Comparative study of different air-conditioning systems incorporative air washers, *Int. J. Refrig.*, vol. 17, no. 6: p. 364-370.

[64] Iyoki, S., Uemura, T., 1988, Physical and thermal properties of the water-lithium bromide-zinc chloride-calcium bromide system, *Int. J. Refrig.,* vol. 12, Sept.: p. 272-277.

[65] Iyoki, S., Yamanaka, R., 1993, Physical and thermal properties of the water-lithium bromide-lithium nitrate system, *Int. J. Refrig.*, vol. 16, no. 3: p. 191-200.

[66] Jain, S., Dhar, P.L., Kaushik, S.C., 1995, Evaluation of solid-desiccant-based evaporative cooling cycle for typical hot humid climates, *Int. J. Refrig.*, vol. 18, no. 5: p. 287 - 296.

[67] Jin-Soon Kim, Huen Lee, Sun Il Yu, 1999, Absorption of water vapour into lithium bromide-based solutions with additives using a simple standart pool absorber, *Int. J. Refrig.*, vol. 22, pp. 188-193.

[68] John Browne, 19/5/97, Addressing climate change, a speech by the Chief Executive of BP delivered to Stanford University, California.

[69] Kang, Y.T., Christensen, R.N., 1995, Transient analysis and design model of a LiBr-H_2O absorber with rotating drums, *ASHRAE Trans.,* US, vol. 101, n. 1, pp. 1163-1174.

[70] Kassler, P., 1994, Energy for development, *Shell Selected Paper*, London.

[71] Kauffeld, M., Aarhus, Indirekte Kälteanlagen, DK, KK/9.

[72] Kaushik, S.C., Kaudinya, J.V., 1989, Open cycle absorption cooling – a review, Energy Conversion Management, 29(2), pp. 89-109.

[73] Kessling, W., Laevemann, E., Peltzer, M., 1996, Efficient energy storage for desiccant cooling systems, *Proceedings of EuroSun 96, Freiburg*, pp. 1254-1260.

[74] Kessling, W., Laevemann, E., Peltzer, M., 1998, Energy storage in open cycle liguid desiccant cooling systems, *Int. J. Refrig.*, vol. 21, no. 2, pp. 150-156.

[75] Khairullah, A., Singh, R.P., 1991, Optimization of fixed and fluidized bed freezing processes, *Int. J. Refrig.*, vol. 141, May, pp. 176-181.

[76] Khan, A.Y., 1996, Parametric analysis of heat and mass transfer performance of a packed-type liquid desiccant absorber at part-load conditions, *ASHRAE Trans.*, US, vol. 102, n. 1, p. 349-357.

[77] Kholpanov, L., Ismailov, B., Vlasak, P., 2005, Modelling of multiphase flow containing bubbles, drops and solid particles, *Engineering Mechanics*, vol.12, n.6, p.1-11.

[78] Kholpanov, L., 2008, Mathmodeling of nonlinear dynamics and mass transfer, *Czasopismo techniczne. Mechanika,* .z. 2-M, p. 95-112.

[79] Kholpanov, L., Zaporozhets V., Zibert G., Kaschickii Yu., 1998, Mathematical modeling nonlinear termohydrogasdynamic processes, Moscow, *Nauka,* pp. 320.

[80] Khrustalev, D., Faghri, A., 1996, Fluid flow effects in evaporative from liquid-vapour meniscus, *Heat Transfer,* US, vol. 118, n. 3, p. 725-730.

[81] Kirillov, V., 1994, Hydrodynamics and heat-and-mass exchange in 2-phase flows of film apparatuses for refrigerating engineering. Doctor dissertation, Odessa, Odessa State Academy of Refrigeration, (Russian).

[82] Kirillov, V., Doroshenko, A., 1988, Peculiarities film flowing along the surface with regular roughness, *Eng.-Phys.-Journal,* vol. 54, no. 5: p. 139-145 (Russian).

[83] Kirillov, V., Doroshenko, A., 1996, Maximum permissible velocity of gas in heat-and-mass transfer equipment, *Eng.-Phys.-Journal,* vol. 69, no. 2: p. 269-285 (Russian).

[84] Koepel, E.A., Klein, S.A., Mitchell, J.W., 1995, Commercial absorption chiller models for evoluation of control strategies, *ASHRAE Trans.,* US, vol. 101, n. 1., p. 1175-1184.

[85] Koo, K.K., Lee, H. R., Jeohg, S. Y., atal, 1998, Solubility and vapor pressure characteristics of H_2O/(LiBr + LiJ + $LiNO_3$ + LiCl) system for air- cooled absorbtion chillers. *International Conference " Naturals working fluids*, Oslo, Norway, IIF / IIR: p. 531-536.

[86] Kvurt, Yu., Doroshenko, A., Shestopalov K., 2008, Research and modeling evaporating cooling equipment of an indirect type, *21-th International conference on Mathematical Methods in Engineering and Technology - MMET-21, 29 May - 1 June, Saratov,* vol. 3, p. 71-72 (Russia).

[87] Kvurt, Yu., Kholpanov, L., 2008, Hydrodynamics of film current in channels with screw roughness, *3-th International conference Heat and mass transfer and hydrodynamics in swirling flow, October 21-23, Moscow*, p. 37-40.

[88] Kvurt, Yu., Kholpanov, L., 2000, Method for measuring the consumption of solid inclusions in slurries, *Proceedings of 13th Advance in Filtration and Separation Technology, March 14-17, Myrtle Beach, SC,* vol., 14, p. 789-794.

[89] Lamp, P. (Co-ordinator), Solar assisted absorption cooling machine with optimized utilization of solar energy (SACMO). Contract EU JOULE No. JOR 3CT 9500 2O.

[90] Lamp, P., Ziegler, F., 1998, Review Paper, European research on solar-assisted air conditioning, *Int. J. Refrig.*, vol. 21, no. 2, pp. 89-99.

[91] Lauritano, A., Marano, D., 1995, Open-cycle solar absorption cooling and its applicability to residential buildings in Italy, *Cond. Aria,* JT., vol. 39, p. 495-502.

[92] Lävermann, E., 1995, Desiccant cooling, Included in Potential of Solar Assisted Cooling in Southern Europe, Final Report, EU Contract RENA-CT94-0017.

[93] Lävermann, E., Keθling, W., 1995, Energy storage in open cycle liquid disiccant cooling systems, *Proceedings of 2nd Munich Discussion Meeting 'solar Assisted Cooling with Sorption System', München,* Paper No. 10.

[94] Lavrenchenko, G., Doroshenko, A., 1988, Development of Indirect Evaporative Air-Coolers for Air-Conditioning Systems, *Cholodilnaya Technika,* no. 10: p. 28-33 (Russian).

[95] Lazzarin, R., Longo, G.A., Gasparello, A., 1996, Theoretical analysis of an open-cycle heating and cooling systems, *Int. J. Refrig.*, 19(4), pp. 239-246.

[96] Lazzarin, R.M., Gasparella, A., Longo, G.A., 1999, Chemical dehumidification by liquid disiccants: theory and experiment, *Int. J. Refrig.*, vol. 22, pp. 334-347.

[97] Lowenstein, A, Novosel, D., 1995, The seasonal performance of a liquid-desiccant air conditioner, *ASHRAE Trans.,* US, vol. 101, n. 1, pp. 679-685.

[98] Lowenstein, A., 1993, Liquid desiccant air-conditioners: An attractive alternative to vapor-compression systems. *Oak-Ridge nat. Lab/Proc. Non-fluorocarbon Refrig. Air-Cond. Technol. Workshop. Breckenridge, CO, US, 06.23-25*, p. 133-150.

[99] Lowenstein, A.I., Dean, M.N., 1992, The effect of regenerator performance on a liquid-desiccant air conditioner, *ASHRAE Trans.,* US, vol. 98, n. 1, p. 704-711.

[100] Lowenstein, A.J.,Gabruk, R.S., 1992, The effect of absorber dising on the performance of a liquid-desiccant air conditionaer, *ASHRAE Trans.,* US, vol. 98, n. 1, p. 712-720.

[101] Lu, S.-M., Yan, W.-J., 1995, Development and experimental validation of a full-scale solar desiccant enhanced radiative cooling system, *Renewable Energy*, vol. 6, n. 7, pp. 821-827; Yan, W., Wu, W., Shyu, R., 1994, Study on solar liquid desiccant cooling system, *The 5th ASEAN Conference on Energy Technology*, April 25-27, The Tara-Impala Hotel, Bangkok, Thailand.

[102] Lu, S.-M., Shyu, R.-J., Yan, W.-J., Chung, T.-W., 1995, Development and experimental validation of two novel solar desiccant-dehumidification-regeneration systems, *Energy*, vol. 20, n. 8, pp. 751-757.

[103] Ma, W.B., Deng, S.M., 1996, Theoretical analysis of low-temperature hot source driven two-stage LiBr/H_2O absorption refrigeration system, *Int. J. Refrig.*, vol. 19, no. 2, pp. 141-146.

[104] Matsuda, A., Munafata, T., Yoshimura, T., Fuchi, H., 1980, Measurement of vapor pressure of lithium-bromide-water solution, *Kagaku Kagaku Ronbunshu*, no. 5: p. 119 – 122.

[105] Mendes, L.F., Collares Pereira, M., Ziegler, F., 1998, Supply of cooling and heating with solar assisted absorption heat pumps: an energetic approach, *Int. J. Refrig.*, vol. 21, no. 2, pp. 116-125.

[106] Merkle, Th., Hahne, E., 1996, Solar thermal energy concept for water heating, space heating and cooling of an industrial buildings, *Proceedings of EuroSun 96, Freiburg*, pp. 256-365.

[107] Mikheev, M., 1968, Fundamentals of heat Transfer, *MIR Publishers*, Moscow, 376 p.

[108] Mostafavi, M., Agnew, B., The effect of ambient temperature on the surface area of components of an air-cooled lithium bromide-water absorption unit, *Appl. Eng.,* GB, vol. 16, n. 4, p. 313-319.

[109] Nahredorf, F., Blank, U., Iliev, N., Saumweber, M., Stojanoff, C.G., 1998, Development of a fixed bed absorption refrigerator with high power density for the use of low grade heat sources, *Int. J. Refrig.*, vol. 21, no. 2, pp. 126-132.

[110] Ney, A., 1995, Room air conditioning by means of adiabatic evaporative cooling, *Tech. Bau,* DE, n. 1, p. 35-40.

[111] Nguen, M.M., Riffat, S.B., Whitman, D., 1996, *Appl. Therm. Eng.,* GB, vol. 16, n. 4, p. 347-356.

[112] O'gorman, T., 1995, Cooling from the sun, *Air Cond.*, GB., vol. 98, n. 1167, p. 25-26.

[113] Oertel, K., 1995, Design of stand-alone solar-hybrid cooling system for decentralized storage of food, *Proceedings of 2nd Munich Discussion Meeting 'solar Assisted Cooling with Sorption System', München*, Paper No. 12.

[114] Ono, T., 1992, High efficiency chemical refrigeration system using solar heat, Japan's Sunshine Project, 1991, Annual Summary of Solar Energy R&D Program. New Energy Development Organization.

[115] Orlov, A., Kholpanov, L., 1997, On new aspects of using solar energetics in obtaining hydrogen, *Preprint of ROAN, Institute of general physics*, Moscow, (Russian).

[116] Oppermann, G., 1996, Using heat from a depth of 1500 metres, *IEA Heat Pump Cent. Newsl.,* US, GB, vol. 16, n. 4, p. 347-356.

[117] Otternein, R.T., 1995, A theory for heat exchangers with liquid-desiccant-wetted surfaces, *ASHRAE Trans.,* US, vol. 101, p. 317-325.

[118] Palatnik, I.L., 1998, Development and investigation of heliosystems with flat solar collectors for heat and cold supply. Dissertation… Sci (Eng.), Odessa.

[119] Rats, I.I., 1962, Construction, investigation and design of heat-exchange apparatuses, Moscow, p. 230.

[120] Reference-book on heat exchangers, 1997, Edit. O. G. Martynenko et al., Moscow, Energoatomizdat, Vol. 2, p. 352.

[121] Sönke Biel, 1998, Soptive Luftentfeuchtung Und Verdunstungs-kühlung, *Ki Luft-und Kältetechnik*, no. 7: p. 332-336.

[122] Saman, N,F., Jostone, H.W., 1996, Maintaining temperature and humidity in non-humidity-generating spaces, *ASHRAE Trans.*, US., vol. 102, n. 2, p. 73-79.

[123] Schweigler, C., Demmel, S., Riesch, P., Alefeld, G., 1996, A new absorption shiller to establish combined cold, heat and power generation utilizing low temperature heat, *ASHRAE Transaction 102(1) and ASHRAE Technical Data Bulletin 12(1), Absorption/Sorption Heat pumps and Refrigeration Systems*, pp. 81-90.

[124] Sherwood, T.K., Pigford, C.R., 1982, Mass Transfer, *Chimiya*, Moscow, 697 p.

[125] Solar Energy for Refrigeration and Air Conditioning, *Proceedings of Meetings of Commissions E1 and E2,* Jerusalem, 14-15 March 1982.Int. Inst. Refrigeration, Paris.

[126] Steimle, F., Development in Air-Conditioning, *International Conference of Research, Design and Conditioning Equipment in Eastern European Contries, September 10-13, Bucharest, Romania,* IIF/IIR: p. 13-29.

[127] Stoitchkov, N. J., Dimirov, G.J., 1998, Effectiveness of Crossflow Plate Heat Exchanger for Indirect Evaporative Cooling, *Int. J. Refrig.*, vol. 21, no. 6: 463-471.

[128] Stojanoff, C., Nahredorf, F., Blank, U., Iliev, N., Saumweber, M., 1995, Development of solar-driven absorption cooling system, *Proceedings of 2nd Munich Discussion Meeting 'Solar-Assisted Cooling with Sorption System', München*, Paper No. 3.

[129] UNEP. Montreal Protocol on Substances That Deplete The Ozone Layer. Final Act: date – 11 September 1987, 6 p.

[130] United Nations on Climate Change. General Convention Kyoto, 1997.

[131] Wenjyh Yan., Weiyih Wu., Rueyjong Shyu., 1994, Study on a solar liquid desiccant cooling system, *The 5th ASEAN Conference on Energy Technology, April 25-27, Bankok, Thailand,* vol. I, pp. 304-311.

[132] West, M.K., Lyer, S.V., 1995, Analysis of a field-installed hybrid solar desiccant cooling system, *ASHRAE Trans.,* US, vol. 101, n. 1, pp. 686-696.

[133] Westerlund, L., Dahl, J., 1994, Absorbers in the open absorption system, *Applied Energy*, n. 48, pp. 33-49.

[134] Westerlund, L., Dahl, J., 1994, Use of an open absorption heat-pump for energy conservation in a public swimming-pool, *Applied Energy*, n. 49, pp. 275-300.

[135] Yang, R., Wang, P.L., 1995, Experimental study of a glased solar collector/regenerator operated under a humid climate, *INT. J. Sol. Energy,* CH, vol. 16, n. 3, pp. 185-201.

[136] Zayitsev, I.D., Aseev, G.G., Physical and chemical properties of binary and multicomponent solutions of inorganic substances Reference edit, *Chemistry (Khimiya),* Moscow, p. 416 (Russian),

[137] Zech, S., 1995, Solar cooling system based on water/zeolite absorption, *Luft Kaltetechn,* DE., vol. 31, n. 12, pp. 559-561.

In: Air Conditioning Systems
Editors: T. Hästesko, O. Kiljunen, pp. 149-194
ISBN: 978-1-60741-555-8

Chapter 2

PRINCIPLE OF LOW ENERGY BUILDING DESIGN: HEATING, VENTILATION AND AIR CONDITIONING

Abdeen Mustafa Omer
17 Juniper Court, Forest Road West, Nottingham NG7 4EU, UK

ABSTRACT

The move towards a de-carbonised world, driven partly by climate science and partly by the business opportunities it offers, will need the promotion of environmentally friendly alternatives, if an acceptable stabilisation level of atmospheric carbon dioxide is to be achieved. This requires the harnessing and use of natural resources that produce no air pollution or greenhouse gases and provides comfortable coexistence of human, livestock, and plants. This study reviews the energy-using technologies based on natural resources, which are available to and applicable in the farming industry. Integral concept for buildings with both excellent indoor environment control and sustainable environmental impact are reported in the present communication. Techniques considered are hybrid (controlled natural and mechanical) ventilation including night ventilation, thermo-active building mass systems with free cooling in a cooling tower, and air intake via ground heat exchangers. Special emphasis is put on ventilation concepts utilising ambient energy from air ground and other renewable energy sources, and on the interaction with heating and cooling. It has been observed that for both residential and office buildings, the electricity demand of ventilation systems is related to the overall demand of the building and the potential of photovoltaic systems and advanced co-generation units. The focus of the world's attention on environmental issues in recent years has stimulated response in many countries, which have led to a closer examination of energy conservation strategies for conventional fossil fuels. One way of reducing building energy consumption is to design buildings, which are more economical in their use of energy for heating, lighting, cooling, ventilation and hot water supply. Passive measures, particularly natural or hybrid ventilation rather than air-conditioning, can dramatically reduce primary energy consumption. However, exploitation of renewable energy in buildings and agricultural greenhouses can, also, significantly contribute towards reducing dependency on fossil fuels. This article describes various designs of low energy buildings. It also, outlines the effect of dense urban building nature on energy consumption, and its contribution to climate change. Measures, which would help to save energy in buildings, are also presented.

Keywords: Built environment; energy efficient comfort; ventilation; sustainable environmental impact.

NOMENCLATURE

A = Heat transfer area, m^2
A(s) = temperature area over 25°C, degree-hour
A_{max} = Maximum temperature area, degree-hour
Cp = Specific heat at constant pressure, kJ/kg K
F = Shape factor, dimensionless
F(t) = Temperature at time t truncated over 25°C
Gb = Beam solar radiation, W/m^2
Gd = Diffuse solar radiation, W/m^2
GT = Total solar radiation (Gb+Gd), W/m^2
g = Acceleration of gravity, m/s^2
h = Heat transfer coefficient, W/m^2 K
I(s) = Normalised temperature index for scenario S, dimensionless
k = Thermal conductivity, W/m K
L = Representative length, m
M = Mass, kg
Na = Nusselt number, dimensionless
Q = Thermal gains, W
Ra = Rayleigh number ($g \beta L^3 \Delta T/L\nu$), dimensionless
Rb = Beam radiation geometric projection factor, dimensionless
T = Temperature, K
t = Time, s
U = Global heat transfer coefficient, W/m^2 K

Indices

c = Convection
d = Diffuse
i = Internal, instantaneous
r = Radiation
s = Building materials
j = Windowpane

Greek Symbols

α = Absorptance, dimensionless
ε = Emittance, dimensionless
ν = Kinematic viscosity, m^2/s
ρ = Reflectance, dimensionless

σ = Stefan-Boltzmann constant, $W/m^2 K^4$

1. Introduction

Globally, buildings are responsible for approximately 40% of the total world annual energy consumption [1]. Most of this energy is for the provision of lighting, heating, cooling, and air conditioning. Increasing awareness of the environmental impact of CO_2, NO_x and CFCs emissions triggered a renewed interest in environmentally friendly cooling, and heating technologies. Under the 1997 Montreal Protocol, governments agreed to phase out chemicals used as refrigerants that have the potential to destroy stratospheric ozone. It was therefore considered desirable to reduce energy consumption and decrease the rate of depletion of world energy reserves and pollution of the environment.

One way of reducing building energy consumption is to design buildings, which are more economical in their use of energy for heating, lighting, cooling, ventilation and hot water supply. Passive measures, particularly natural or hybrid ventilation rather than air-conditioning, can dramatically reduce primary energy consumption [2]. However, exploitation of renewable energy in buildings and agricultural greenhouses can, also, significantly contribute towards reducing dependency on fossil fuels. Therefore, promoting innovative renewable applications and reinforcing the renewable energy market will contribute to preservation of the ecosystem by reducing emissions at local and global levels. This will also contribute to the amelioration of environmental conditions by replacing conventional fuels with renewable energies that produce no air pollution or greenhouse gases.

The provision of good indoor environmental quality while achieving energy and cost efficient operation of the heating, ventilating and air-conditioning (HVAC) plants in buildings represents a multi variant problem. The comfort of building occupants is dependent on many environmental parameters including air speed, temperature, relative humidity and quality in addition to lighting and noise. The overall objective is to provide a high level of building performance (BP), which can be defined as indoor environmental quality (IEQ), energy efficiency (EE) and cost efficiency (CE).

- Indoor environmental quality is the perceived condition of comfort that building occupants experience due to the physical and psychological conditions to which they are exposed by their surroundings. The main physical parameters affecting IEQ are air speed, temperature, relative humidity and quality.
- Energy efficiency is related to the provision of the desired environmental conditions while consuming the minimal quantity of energy.
- Cost efficiency is the financial expenditure on energy relative to the level of environmental comfort and productivity that the building occupants attained. The overall cost efficiency can be improved by improving the indoor environmental quality and the energy efficiency of a building.

An approach is needed to integrate renewable energies in a way to meet high building performance. However, because renewable energy sources are stochastic and geographically diffuse, their ability to match demand is determined by adoption of one of the following two approaches [2]: the utilisation of a capture area greater than that occupied by the community to be supplied, or the reduction of the community's energy demands to a level commensurate with the locally available renewable resources.

For a northern European climate, which is characterised by an average annual solar irradiance of 150 Wm^{-2}, the mean power production from a photovoltaic component of 13% conversion efficiency is approximately 20 Wm^{-2}. For an average wind speed of 5 ms^{-1}, the power produced by a micro wind turbine will be of a similar order of magnitude, though with a different profile shape. In the UK, for example, a typical office building will have a demand in the order of 300 $kWhm^{-2}yr^{-1}$. This translates into approximately 50 Wm^{-2} of façade, which is twice as much as the available renewable energies [3]. Thus, the aim is to utilise energy efficiency measures in order to reduce the overall energy consumption and adjust the demand profiles to be met by renewable energies. For instance, this approach can be applied to greenhouses, which use solar energy to provide indoor environmental quality. The greenhouse effect is one result of the differing properties of heat radiation when it is generated at different temperatures. Objects inside the greenhouse, or any other building, such as plants, re-radiate the heat or absorb it. Because the objects inside the greenhouse are at a lower temperature than the sun, the re-radiated heat is of longer wavelengths, and cannot penetrate the glass. This re-radiated heat is trapped and causes the temperature inside the greenhouse to rise. Note that the atmosphere surrounding the earth, also, behaves as a large greenhouse around the world. Changes to the gases in the atmosphere, such as increased carbon dioxide content from the burning of fossil fuels, can act like a layer of glass and reduce the quantity of heat that the planet earth would otherwise radiate back into space. This particular greenhouse effect, therefore, contributes to global warming. The application of greenhouses for plants growth can be considered one of the measures in the success of solving this problem. Maximising the efficiency gained from a greenhouse can be achieved using various approaches, employing different techniques that could be applied at the design, construction and operational stages. The development of greenhouses could be a solution to farming industry and food security.

2. Built Environment

The heating or cooling of a space to maintain thermal comfort is a highly energy intensive process accounting for as much as 60-70% of total energy use in non-industrial buildings. Of this, approximately 30-50% is lost through ventilation and air infiltration. However, estimation of energy impact of ventilation relies on detailed knowledge about air change rate and the difference in enthalpy between the incoming and outgoing air streams. In practice, this is a difficult exercise to undertake since there is much uncertainty about the value of these parameters [4]. As a result, a suitable datum from which strategic planning for improving the energy efficiency of ventilation can be developed has proved difficult to establish [4]. Efforts to overcome these difficulties are progressing in the following two ways:

- Identifying ventilation rates in a representative cross section of buildings.
- The energy impact of air change in both commercial and domestic buildings.

In addition to conditioning energy, the fan energy needed to provide mechanical ventilation can make a significant further contribution to energy demand. Much depends on the efficiency of design, both in relation to the performance of fans themselves and to the resistance to flow arising from the associated ductwork. Figure 1 illustrates the typical fan and thermal conditioning needs for a variety of ventilation rates and climate conditions.

The building sector is an important part of the energy picture. Note that the major function of buildings is to provide an acceptable indoor environment, which allows occupants to carry out various activities. Hence, the purpose behind this energy consumption is to provide a variety of building services, which include weather protection, storage, communications, thermal comfort, facilities of daily living, aesthetics, work environment, etc. However, the three main energy-related building services are space conditioning (for thermal comfort), lighting (for visual comfort), and ventilation (for indoor air quality). Pollution-free environments are a practical impossibility. Therefore, it is often useful to differentiate between unavoidable pollutants over which little source control is possible, and avoidable pollutants for which control is possible. Unavoidable pollutants are primarily those emitted by metabolism and those arising from the essential activities of occupants. 'Whole building' ventilation usually provides an effective measure to deal with the unavoidable emissions, whereas 'source control' is the preferred and sometimes only practical, method to address avoidable pollutant sources [5]. Hence, achieving optimum indoor air quality relies on an integrated approach to the removal and control of pollutants using engineering judgment based on source control, filtration, and ventilation. Regardless of the kind of building involved, good indoor air quality requires attention to both source control and ventilation. While there are sources common to many kinds of buildings, buildings focusing on renewable energy may have some unique sources and, therefore, may require special attention [5]. In smaller (i.e., house size) buildings, renewable sources are already the primary mechanism for providing ventilation. Infiltration and natural ventilation are the predominant mechanisms for providing residential ventilation for these smaller buildings.

Ventilation is the building service most associated with controlling the indoor air quality to provide a healthy and comfortable environment. In large buildings ventilation is normally supplied through mechanical systems, but in smaller ones, such as single-family homes, it is principally supplied by leakage through the building envelope, i.e., infiltration, which is a renewable resource, albeit unintendedly so. Ventilation can be defined as the process by which clean air is provided to a space. It is needed to meet the metabolic requirements of occupants and to dilute and remove pollutants emitted within a space. Usually, ventilation air must be conditioned by heating or cooling in order to maintain thermal comfort and, hence, becomes an energy liability. Indeed, ventilation energy requirements can exceed 50% of the conditioning load in some spaces [5]. Thus, excessive or uncontrolled ventilation can be a major contributor to energy costs and global pollution. Therefore, in terms of cost, energy, and pollution, efficient ventilation is essential. On the other hand, inadequate ventilation can cause comfort or health problems for the occupants.

One way of reducing building energy consumption is to design buildings, which are more economical in their use of energy for heating, lighting, cooling, ventilation and hot water supply. Passive measures, particularly natural or hybrid ventilation rather than air-conditioning, can dramatically reduce primary energy consumption. However, exploitation of renewable energy in buildings and agricultural greenhouses can, also, significantly contribute towards reducing dependency on fossil fuels.

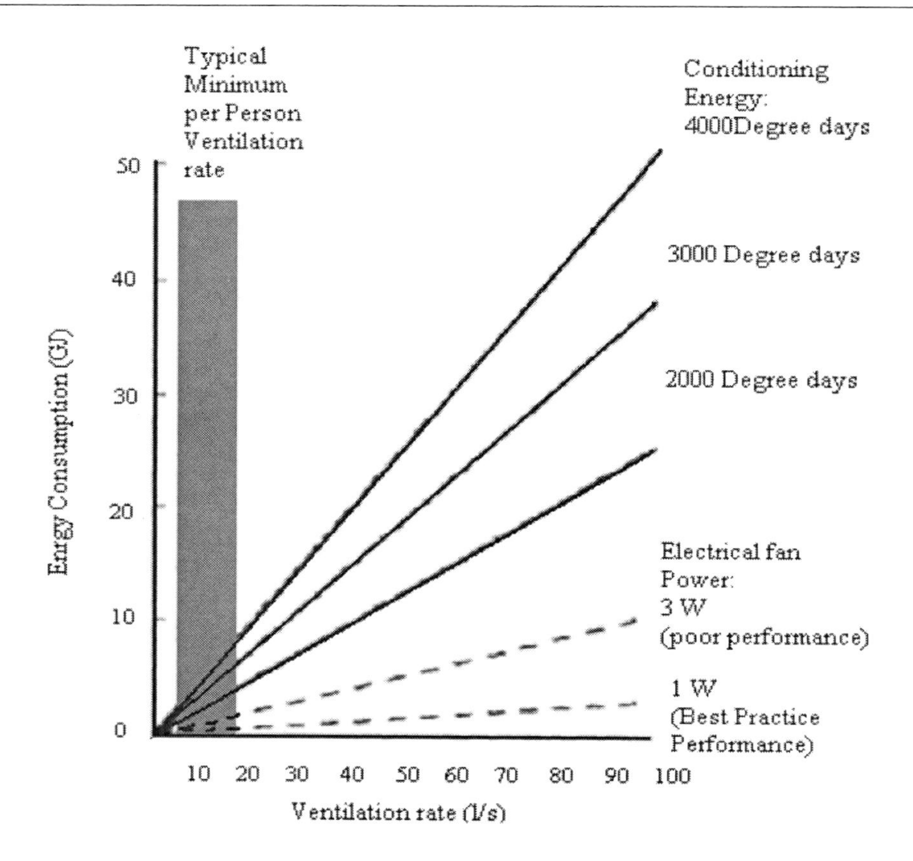

Figure 1. Energy impact of ventilation.

Good indoor air quality may be defined as air, which is free of pollutants that cause irritation, discomfort or ill health to occupants [6]. Since long time is spent inside buildings, considerable effort has focused on developing methods to achieve an optimum indoor environment. An almost limitless number of pollutants may be present in a space, of which many are at virtually immeasurably low concentrations and have largely unknown toxicological effects [6]. The task of identifying and assessing the risk of individual pollutants has become a major research activity [e.g., 5, 6, and 7]. In reality, a perfectly pollutants-free environment is unlikely to be attained. Some pollutants can be tolerated at low concentrations, while irritation and odour often provide an early warning of deteriorating conditions. Health related air quality standards are typically based on risk assessment and are either specified in terms of maximum-permitted concentrations or a maximum allowed dose. Higher concentrations of pollutants are normally permitted for short-term exposure than are allowed for long-term exposure [7].

Ventilation is essential for securing a good indoor air quality, but, as explained earlier, can have a dominating influence on energy consumption in buildings. Air quality problems are more likely to occur if air supply is restricted. Probably a ventilation rate averaging 7 l/s.p represents a minimum acceptable rate for normal odour and comfort requirements in office type buildings [8]. Diminishing returns are likely to be experienced at rates significantly above 10 l/s.p [8]. If air quality problems still persist, the cause is likely to be poor outdoor air quality (e.g., the entrainment of outdoor traffic fumes), poor air distribution or the excessive release of avoidable pollutants into space. However, the energy efficiency of ventilation can be improved by introducing exhaust air heat recovery, ground pre-heating, demand controlled ventilation, displacement ventilation and passive cooling [9]. In each case, a very careful analysis is necessary to ensure that the anticipated savings are actually achievable. Also, it is essential to differentiate between avoidable and unavoidable pollutant emissions. Achieving energy efficiency and optimum Indoor Air Quality (IAQ) depends on minimising the emission of avoidable pollutants. Pollutants inside buildings are derived from both indoor and outdoor contaminant sources, as illustrated in figure 2. Each of these tends to impose different requirements on the control strategies needed to secure good health and comfort conditions.

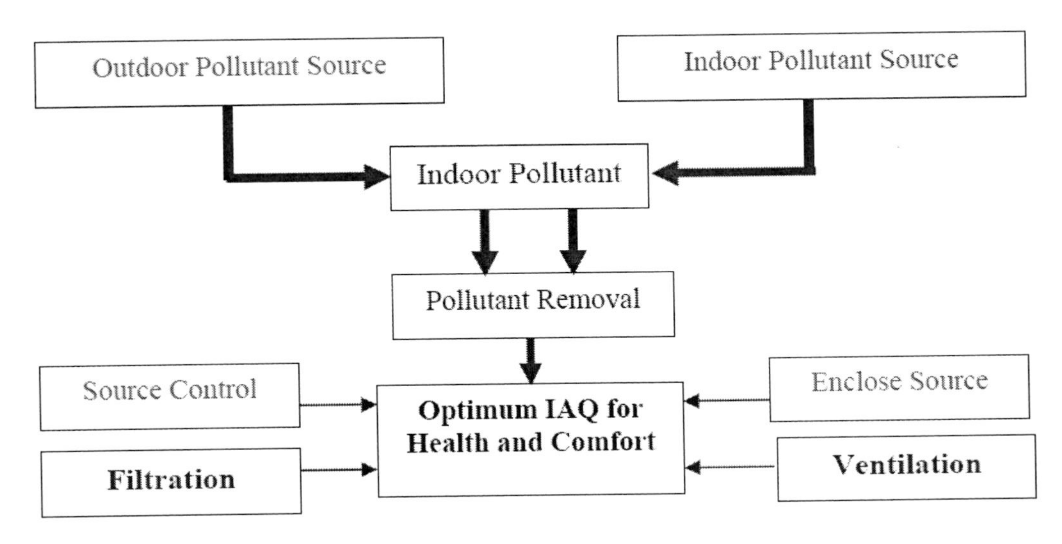

Figure 2. Strategies for controlling IAQ.

3. Achieving Energy Efficient Ventilation

In addition to conditioning energy, the fan energy needed to provide mechanical ventilation can make a significant further contribution to energy demand. A recent review of the American Society of Heating, Refrigerating, and Air-conditioning Engineers (ASHRAE) [10] concludes that the thermal insulation characteristics of buildings improve; ventilation and air movement is expected to become the dominant heating and cooling loss mechanism in buildings of the next century. Poor air quality in buildings sometimes manifests itself; refer to a range of symptoms that an occupant can experience while present in the building. Typical symptoms include lethargy, headaches, lack of concentration, runny nose, dry throat and eye and skin irritation [10]. Other examples of sick building syndrome (SBS) have been associated with the presence of specific pollutants, such as outdoor fumes entering through air intakes [11]. Improved ventilation is one way of tackling the problem. Many standards are being introduced to ensure the adequacy and efficiency of ventilation. However, to be effective, standards need to address the minimum requirements of comfort, operational performance, air tightness, provision for maintenance and durability [11]. Also, various methods have been introduced to improve the energy performance of ventilation. These include:

3.1. Thermal Recovery

Recovery of thermal energy from the exhaust air stream is possible by means of air-to-air heat recovery systems or heat pumps. In theory, such methods can recover as much as 70% of the waste heat [12]. While these methods are exceedingly popular, their full potential is, often, not achieved. This is because buildings or ductwork are often excessively leaky and, hence, additional electrical energy load is needed. To be successful, the designer must integrate such ventilation design with that of the building itself and be thoroughly aware of all the energy paths [12].

3.2. Ground Pre-Conditioning

Ventilation air can be pre-conditioned by passing the supply air ducts under-ground. This can provide a good measure of winter pre-heating and summer pre-cooling [13].

3.3. Demand Control Ventilation

Demand control ventilation, DCV, systems provide a means by which the rate of ventilation is modulated in response to varying air quality conditions. This is effective if a dominant pollutant is identifiable. In transiently occupied buildings, control by metabolic carbon dioxide concentration has become popular although must be introduced with caution. For success, it is essential to ensure that no other problem pollutants are present [14].

3.4. Displacement Ventilation

Displacement ventilation involves distributing clean air at low velocity and at a temperature of approximately $2^{o}K$ below the ambient air temperature of the space [15]. This dense air mass moves at floor level until it reaches a thermal source such as an occupant or electric equipment, causing the air to warm and gently rise. Polluted air is then collected and extracted above the breathing zone. This process reduces the mixing effect of classical dilution ventilation, thus reducing the amount of ventilation needed to achieve the same air quality in the vicinity of occupants [15]. For success, very careful temperature control is needed and there is a limit on potential cooling capacity [15].

4. Outdoors Air Pollution

Clean outdoor air is essential for good indoor air quality. Although air cleaning is possible, it is costly and not effective in the many offices and dwellings that are naturally ventilated, leaky or ventilated by mechanical extract systems. Some air quality problems are global and can only be controlled by international effort. Other pollutants are much more regional and may be associated with local industry and traffic. Nature, too, presents its own problems with large volumes of dust and gaseous emissions being associated with volcanic activity. Similarly, while naturally occurring, radon can penetrate buildings from the underlying geological strata. Even rural areas are not immune to pollution, where the presence of pollen, fungal spores and agricultural chemicals can result in poor health and cause allergic reactions. There are several pollution control strategies.

Some of them are discussed below:

4.1. Filtration

Filtration is applied primarily to remove particulates from the air. Filtration of outdoor air cannot be readily applied to the many buildings that are naturally ventilated or excessively leaky. To be effective, filtration systems must be capable of trapping the smallest of particles and of handling large volumes of airflow. Activated carbon and other absorbing filters are additionally able to remove gaseous pollutants [16].

4.2. Positioning of Air Intakes

Air intakes must be located away from pollutant sources. Particular sources include street level and car parking locations [16]. Although urban air quality is normally much improved at elevations above street level, contamination from adjacent exhaust stacks and cooling towers must, also, be avoided [16]. Determining the optimum position for air intakes may require extensive wind tunnel or fluid dynamics analysis. Further information on the positioning of air intakes is reviewed by Limb 1995 [7].

4.3. Air Quality Controlled Fresh Air Dampers

Traffic pollution in urban areas is often highly transient, with peaks occurring during the morning and evening commuting periods. At these times, it may be possible to improve indoor air quality by temporarily closing fresh air intakes and windows.

4.4. Building Air Tightness

None of the above control strategies will be effective unless the building is well sealed from the outdoor environment to prevent contaminant ingress through air infiltration. Underground parking garages must also be well sealed from occupied accommodation above. Evidence suggests that sealing is often inadequate [16].

Ventilation is the building service most associated with controlling the indoor air quality to provide a healthy and comfortable environment. In large buildings ventilation is normally supplied through mechanical systems, but in smaller ones, such as single-family homes, it is principally supplied by leakage through the building envelope, i.e., infiltration, which is a renewable resource, albeit unintendedly so.

5. Indoor Pollutants

Pollutants emitted inside buildings are derived from metabolism (odour, carbon dioxide, and bacteria), the activities of occupants (e.g., smoking, washing and cooking), emissions from materials used in construction and furnishing and emissions from machinery and processes. The preferred order of control is discussed below:

5.1. Source Control

Once a pollutant has entered a space, it can, at best, only be diluted [17]. Avoidable pollutants should, therefore, be eliminated. This means restricting or, preferably, eliminating potentially harmful pollutant emissions.

5.2. Enclosing and Ventilating at Source

Pollutants generated as part of the activity of occupants are usually highly localised. Wherever possible, source control should be applied, combined with the use of local extractors [17].

5.3. General Ventilation

General ventilation of a space is needed to dilute and remove residual pollution primarily from unavoidable contaminant sources [17].

6. Ventilation of Spaces in Humid Climate

The design of windows in modern buildings in a warm, humid climate can be influenced either by their use to provide physiological and psychological comfort via providing air and daylight to interior spaces or by using them to provide aesthetically appealing fenestration. Most spaces in modern buildings are not adequately ventilated and it is recommended that effort should be directed towards the use of windows to achieve physiological comfort. Evaluation of public housing has focused on four main aspects: economics, social and physical factors, and residents satisfaction. However, information about the physiological characteristics of spaces in a warm humid climate will aid the design of appropriate spaces with respect to the development and adequate choice of building materials and appropriate use of suitable passive energy. In this light there is a need to examine residents satisfaction with respect to these physiological issues. Proper ventilation in a space is a primary factor in determining human health, comfort and well being of the occupants. At present, getting a proper naturally ventilated space seems to be a difficult task. This is partly due to the specific environmental problems of high temperature, high humidity, low wind velocity, and variable wind direction - usually attributed to the warm humid climate, on the one hand, and the difficulty of articulating the design constraints of security, privacy and the desire of users for large spaces on the other hand. As pointed out by most researchers in the field of passive energy design, such as Givoni [18], Koenigsberger, et al. [19] Boulet [20] and Szokolay [21], the types of spaces most suited to this climate are spaces, which are cross-ventilated. This implies that these spaces must have openings at least on opposite sides of a wall, but this condition is difficult to achieve in view of the design constraints mentioned above. So in most cases, the option left to the designer is to have openings on a wall or openings on adjacent walls. The effectiveness of the above arrangement for effective ventilation of a space still depends on other parameters. Therefore, in order to optimise comfort in spaces in warm humid climate, there is a need to re-examine the factors affecting proper ventilation with respect to these design issues. In order to be thermally comfortable in interior spaces, four environmental parameters, namely air temperature, relative humidity, mean radiant temperature and air velocity, need to be present in the space in adequate proportions [21]. In a warm, humid climate, the predominance of high humidity necessitates a corresponding steady, continuous breeze of medium air speed to increase the efficiency of sweat evaporation and to avoid discomfort caused by moisture on skin and clothes. Continuous ventilation is therefore the primary requirement for comfort [18]. From the above, it is apparent that the most important of these comfort parameters in a warm, humid climate is air velocity. It should, also, be noted that indoor air velocity depends on the velocity of the air outdoors [21]. The factors affecting indoor air movement are orientation of the building with respect to wind direction, effect of the external features of the openings, the position of openings in the wall, the size of the openings and control of the openings. Cross-ventilation is the most effective method of getting appreciable air movement in interior spaces in warm, humid climates. For comfort purposes, the indoor wind velocity should be set at between 0.15 and 1.5 m/s [22]. A mathematical model based on analysis of the experimental results [23], which established the relationship between the average indoor and outdoor air velocities with the windows placed perpendicular to each other, was adapted to suit a warm, humid climate and is, usually, used to evaluate the spaces. The formula states that:

$$V_I = 0.45\,(1\text{-}\exp^{-3.84x})\,V_o \qquad (1)$$

Where:

V_I = average indoor velocity

x = ratio of window area to wall area

V_o = outdoor wind speed

6.1. Air Movement in Buildings

Natural ventilation is now considered to be one of the requirements for a low energy building designs. Until about three decades ago the majority of office buildings were naturally ventilated. With the availability of inexpensive fossil energy and the tendency to provide better indoor environmental control, there has been a vast increase in the use of air-conditioning in new and refurbished buildings. However, recent scientific evidence on the impact of refrigerants and air-conditioning systems on the environment has promoted the more conscious building designers to give serious considerations to natural ventilation in non-domestic buildings [24]. Two major difficulties that a designer has to resolve are the questions of airflow control and room air movement in the space. Because of the problem of scaling and the difficulty of representing natural ventilation in laboratory, most of the methods used for predicting the air movement in mechanically ventilated buildings are not very suitable for naturally ventilated spaces [25]. However, computational fluid dynamics (CFD) is now becoming increasingly used for the design of both mechanical and natural ventilation systems. Since a CFD solution is based on the fundamental flow and energy equations, the technique is equally applicable to a naturally ventilated space as well as a mechanically ventilated one, providing that a realistic representation of the boundary conditions are made in the solution.

6.2. Natural Ventilation

Generally, buildings should be designed with controllable natural ventilation. A very high range of natural ventilation rates is necessary so that the heat transfer rate between inside and outside can be selected to suit conditions [25]. The ventilation rates required to control summertime temperatures are very much higher than these required to control pollution or odour. Any natural ventilation system that can control summer temperatures can readily provide adequate ventilation to control levels of odour and carbon dioxide production in a building. Theoretically, it is not possible to achieve heat transfer without momentum transfer and loss of pressure. However, figures 3 and 4 show some ideas for achieving heat reclaim at low velocities. Such ideas work well for small buildings.

6.3. Mechanical Ventilation

Most of the medium and large size buildings are ventilated by mechanical systems designed to bring in outside air, filter it, supply it to the occupants and then exhaust an approximately equal amount of stale air. Ideally, these systems should be based on criteria that can be established at the design stage. To return afterwards in attempts to mitigate problems may lead to considerable expense and energy waste, and may not be entirely successful [25]. The key factors that must be included in the design of ventilation systems are: code requirement and other regulations or standards (e.g., fire), ventilation strategy and systems sizing, climate and weather variations, air distribution, diffuser location and local ventilation, ease of operation and maintenance and impact of system on occupants (e.g., acoustically). These factors differ for various building types and occupancy patterns. For example, in office buildings, pollutants tend to come from sources such as occupancy, office equipment, and automobile fumes. Occupant pollutants typically include metabolic carbon dioxide emission, odours and sometimes smoking, when occupants (and not smoking) are the prime source. Carbon dioxide acts as a surrogate and can be used to cost-effectively modulate the ventilation, forming what is known as a demand controlled ventilation system. Generally, contaminant sources are varied but, often, well-defined and limiting values are often determined by occupational standards. Ventilation can be defined as the process by which clean air is provided to a space. It is needed to meet the metabolic requirements of occupants and to dilute and remove pollutants emitted within a space.

6.4. Bioclimatic Design

Bioclimatic design cannot continue to be a side issue of a technical nature to the main architectural design. In recent years started to alter course and to become much more holistic in its approach while trying to address itself to:

- The achievement of a sustainable development.
- The depletion of non-renewable sources and materials.
- The life cycle analysis of buildings.
- The total polluting effects of buildings on the environment.
- The reduction of energy consumption and
- Human health and comfort.

Hidden dimensions of architectural creation are vital to the notion of bioclimatic design. The most fundamental ones are:

TIME, which has been called the fourth dimension of architectural space, is of importance because every object cannot exist but in time. The notion of time gives life to an object and releases it to periodic (predictable) or unperiodic repetition. Times relates to seasonal and diurnal patterns and thus to climate and the way that a building behaves or should be designed to couple with and not antagonise nature. It further releases to the dynamic nature of a building in contrast to the static image that we have created for it.

AIR, is a second invisible but important element. We create space and pretend that it is empty, oblivious of the fact that it is both surrounded by and filled with air. Air in its turn, due to air-movement, which is generated by either temperature or pressure differences, is very much there and alive. And related to the movement of air should be building shapes, sections, heights, orientations and the size and positioning of openings.

LIGHT, and in particular daylight, is a third important element. Architecture cannot exist but with light and from the time we have been able to substitute natural light with artificial lighting, many a building and a lot of architecture has become poorer so. It is not an exaggeration to say that the real form giver to architecture is not the architect himself but light and that the architect is but the forms moulder.

Vernacular architecture is beautiful to look at as well as significant to contemplate on. It is particularly interesting to realise the nature of traditional architecture where various devices to attain thermal comfort without resorting to fossil fuels can be seen. Sun shading and cross ventilation are two major concerns in house design and a south-facing façade is mandatory to harness the sun in winter as much as possible. Natural ventilation required higher ceilings to bring a cooling effect to occupants in buildings built fifty years ago, whereas modern high technology buildings have lower ceiling heights, thus making air conditioning mandatory. Admitting the human right of enjoying modern lives with a certain level of comfort and convenience, it is necessary to consider how people can live and work in an ideal environment with the least amount of energy consumption in the age of global environment problems. People in the modern age could not put up with the poor indoor environment that people in the old age used to live in. In fact, in those days people had to live with the least amount of fuels readily available and to devise various means of constructing their houses so that they would be compatible with the local climate. It is important; therefore, in designing passive and low energy architecture for the future to learn from their spirit to overcome difficulties by having their creative designs adapted to respective regional climatic conditions and to try to devise the ecotechniques in combination with a high grade of modern science. Finally, the presented theory can be used to calculate the expected effects of the reflecting wall at any particular latitude, under different weather conditions, and when the average numbers of clear days are taken into account. Thereby an assessment of the cost of a particular setup can be obtained. Under circumstances of a few clear days, it may still be worthwhile from a financial point of view to turn a classical greenhouse into one with a reflecting wall by simply covering the glass wall on the north-facing side with aluminum foil with virtually negligible expenditure.

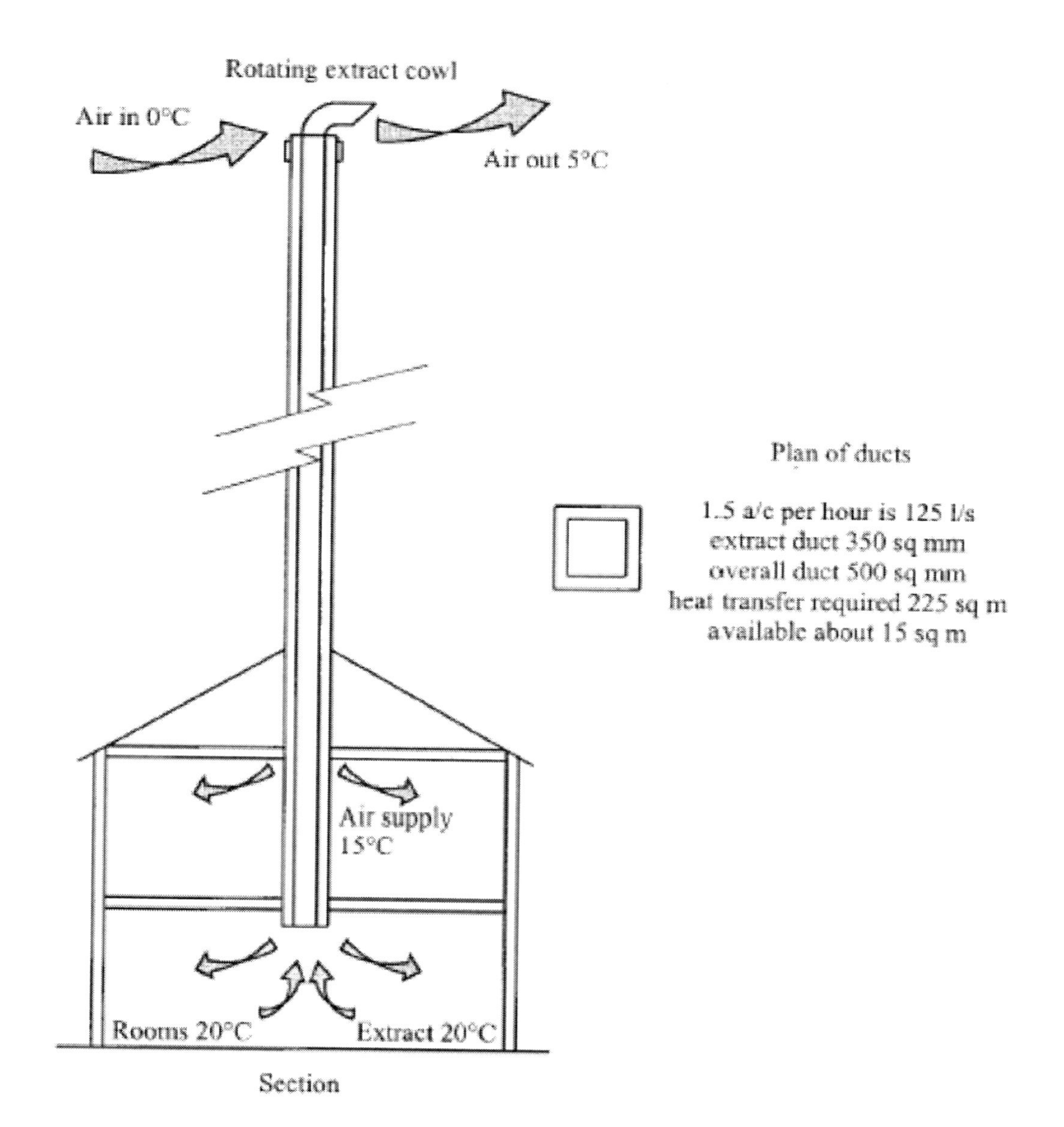

Figure 3. Small house natural ventilation with heat reclaims (A very tall chimney).

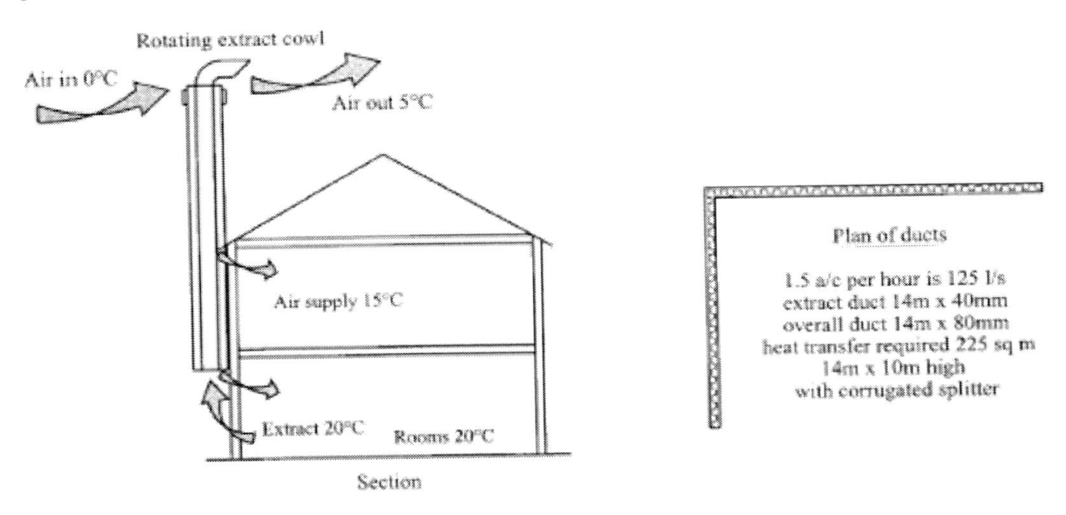

Figure 4. Ventilation duct supply and extract wraps around the building.

Heat gain in the summer is the main problem as it overheats the indoor environment of residential buildings. This forces the residents to utilise mechanical air conditioning systems to satisfy their comfort. Under today's economic crisis, energy conversation programmes and acts for respect of environment are receiving more attention. As a contribution to such efforts and in order to overcome the heat gain in houses, it is advisable to utilise passive systems, namely, producing ventilation by a solar chimney [26]. Room air is removed by ventilation produced by the metallic solar wall (MSW) as shown in figure 5. However, this is a useful effect as further increases overall energy gain. There is also an ironing out effect expressed in terms of the ratios between peak and average insolations.

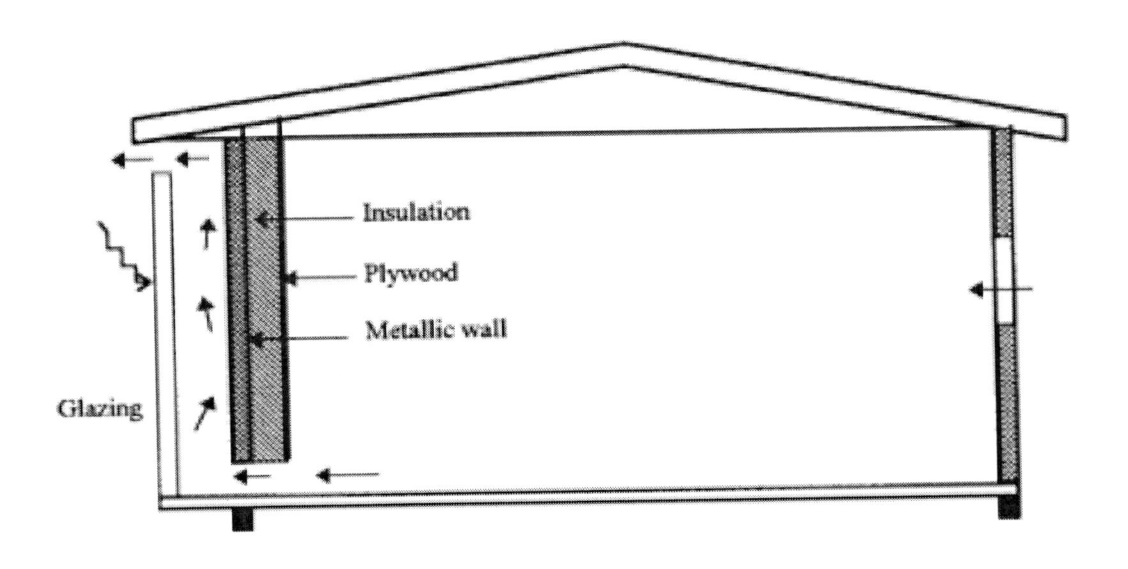

Figure 5. Schematic representations of the passive solar house and natural ventilation by metallic solar wall.

6.5. Infiltration

Infiltration is the process of air flowing in (or out) of leaks in the building envelope, thereby providing (renewable) ventilation in an uncontrolled manner [27]. All buildings are subject to infiltration, but it is more important in smaller buildings as many such buildings rely exclusively on infiltration when doors and windows are closed. In larger buildings there is less surface area to leak for a given amount of building volume, so the same leakage matters less. More importantly, the pressures in larger buildings are usually dominated by the mechanical system and the leaks in the building envelope have only a secondary impact on the ventilation rates [27]. Infiltration in larger buildings may, however, affect thermal comfort and control and systems balance. Typical minimum values of air exchange rates range from 0.5 to 1.0 h^{-1} in office buildings [27]. Buildings with higher occupant density will have higher minimum outside air exchange rates when ventilation is based on outdoor air supply per occupant, typically 7 to 10 ls^{-1} [27]. Thus, schools have minimum outdoor air ventilation rates 3 h^{-1}, while fully occupied theatres, auditoriums and meeting rooms may have minimum air exchange rates of 4 to 7 h^{-1} [27]. It is in low-rise residential buildings (most typically, single-family houses) that infiltration is a dominant force. Mechanical systems in these buildings contribute little to the ventilation rate.

6.6. Passive Ventilation Systems

Passive solar systems for space heating and cooling, as well as passive cooling techniques can significantly contribute to energy saving in the building sector when used in combination with conventional systems for heating, cooling, ventilation and lighting. The overall thermal behaviour of the building is dependent on the alternatives and interventions made on the building's shell. Passive ventilation systems share the use of renewable energy to provide ventilation with infiltration. But unlike air leakage and open windows, passive ventilation systems are designed to provide specific amounts of ventilation to minimise both energy liabilities due to excessive ventilation and periods of poor air quality due to under-ventilation [28]. However, the most common passive ventilation system is the passive stack, which is normally used to extract air from kitchen and bathrooms. In this method, prevailing wind and temperature differences are used to drive airflow through a vertical shaft. Various stack designs can be used to control or enhance the performance, based on local climate. However, careful design is required to avoid backdraughting and to insure proper mean rates. Although there is significant experience with this approach in Europe, it has been rarely used in North America [28]. Well-designed passive ventilation systems can be used to provide whole-building ventilation as well as local exhaust. Some efforts are currently underway to develop passive ventilation systems that incorporate heat recovery to minimise the need for conditioning the ventilation air [28]. These approaches aim towards a fully renewable ventilation system in that it requires no non-renewable resources for either providing the ventilation air or conditioning it. Energy efficiency and renewable energy programmes could be more sustainable and pilot studies more effective and pulse releasing if the entire policy and implementation process was considered and redesigned from the outset. New financing and implementation processes are needed which allow reallocating financial resources and thus enabling countries themselves to achieve a sustainable energy infrastructure.

6.7. Passive Cooling

In the office environment, high heat loads are commonly developed through lighting, computers and other electrical equipment. Further heat gains are developed through solar gains, occupants and high outdoor air temperature. Passive cooling methods attempt to reduce or eliminate the need for energy intensive refrigerative cooling by minimising heat gains. This involves taking advantage of thermal mass (night cooling) and introducing high levels of air change. Night ventilation techniques seem to be the most appropriate strategy for buildings. This arises as a consequence of the large diurnal temperature range during the cooling seasons and the relatively low peak air temperatures, which occur during the day [29]. Such a combination allows the thermal mass of the building to use the cool night air to discard the heat absorbed during the day. An initial examination of the weather conditions experienced during the summer months of June to September in the UK indicates that most peak conditions of external weather fall within the ventilation and thermal mass edge of the bioclimatic chart [29, and 30]. Figure 6 shows that the summer (June to September) climatic envelope is within the heating, comfort, thermal mass and ventilation effectiveness areas of the chart. The key parameters influencing the effectiveness of night cooling are summarised into the following four categories [30]:

- Internal heat gains of 10, 25 and 40 W/m^2; representing occupancy only, occupancy plus lights, and occupancy together with lights and IT load respectively.
- Envelope gains.
- Thermal response.
- Ventilation gains/losses.

Due to the complexity resulting from all the interrelated parameters affecting the effectiveness of night ventilation, it is necessary for designers to have access to a simple user friendly and yet accurate model when assessing the viability of night ventilation during the initial design stage. The followings are the key output parameters [30]:

- Maximum dry resultant temperature during the occupied period.
- Dry resultant temperature at the start of the occupied period.
- Energy savings.

The input data required are the following:

- Thermal gains related data: solar protection is assumed good, thermal gains can be varied and the user specifies the occupancy period.
- Building fabric data: glazing ratio can be any value while thermal mass can be varied at three levels.
- Ventilation data: infiltration, day ventilation and night ventilation can be specified as necessary.

- Weather data: solar data are fixed but temperature is user specified for seven days although temperature profiles need not be the same for all days. The weather data are specified in the form of maximum and minimum temperature for each day and hourly values are calculated by sinusoidal fitting.

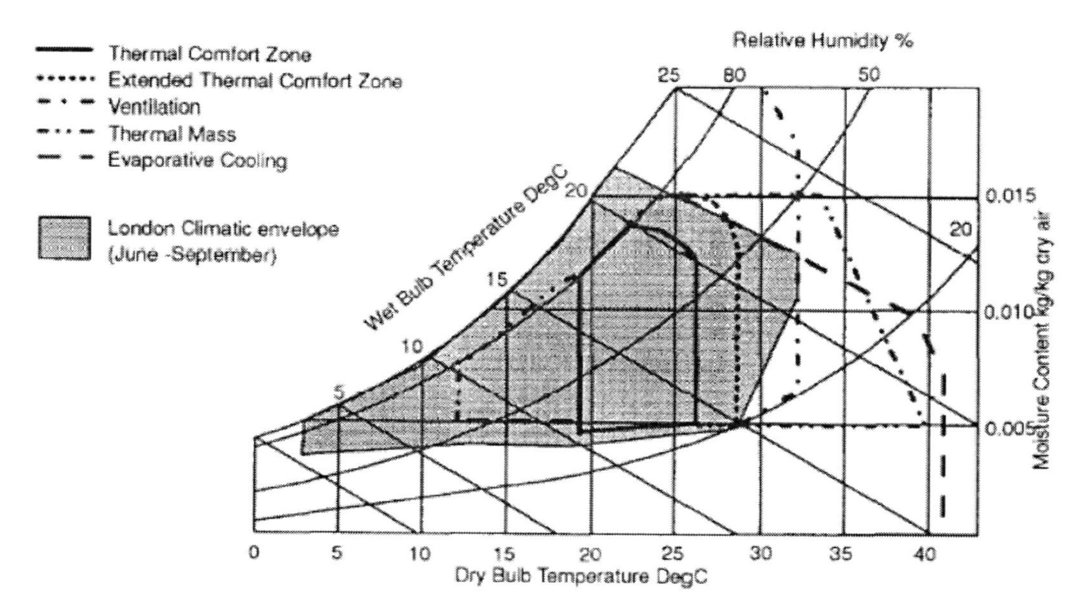

Figure 6. Bioclimatic chart with summer climatic envelope for London.

However, a primary strategy for cooling buildings without mechanical intervention in hot humid climates is to promote natural ventilation. To control the energy used for the cooling of buildings in hot-arid regions with ambient air temperatures during the hottest period between 42 to 47°C, passive cooling approaches should be implemented [30]. A solar chimney that employs convective currents to draw air out of the building could be used. By creating a hot zone with an exterior outlet, air can be drawn into the house, ventilating the structure as well as the occupants. Since solar energy in such a region is immense, the hot zone created with a black metal sheet on the glazing element can draw hotter air at a slightly higher speed [30]. Applications of solar chimneys in buildings were limited to external walls. Integrating a solar chimney with an evaporatively cooled cavity could result in a better cooling effect. However, this should be applied with care since water sources are limited [31]. Figure 7 shows the combined wall-roof solar chimney incorporated into that building. Average room and ambient air temperatures are 23 and 27°C respectively. Air velocity required to achieve thermal comfort in the room should reach a maximum of 0.3 m/s [31]. Figure 8 gives the cooling load versus air change per hour. This indicates that the inclined airflow by the combined wall-roof chimney is enough to overcome a high cooling load required to cool heavy residential buildings. This suggests that night ventilation could be improved, and incorporating a combined wall-roof solar chimney increases the cooling load. However, thermal mass and ventilation should be sufficient to cover cooling requirements in typical buildings. A high percentage of the cooling requirements can be met by night ventilation before another form of cooling is used. Finally, a simplified ventilation tool for assessing the applicability of night cooling in buildings, currently under development in terms of user inputs and typical outputs.

The encouragement of greater energy use is an essential component of development. In the short term it requires mechanisms to enable the rapid increase in energy/capita, and in the long term we should be working towards a way of life, which makes use of energy efficiency and without the impairment of the environment or of causing safety problems.

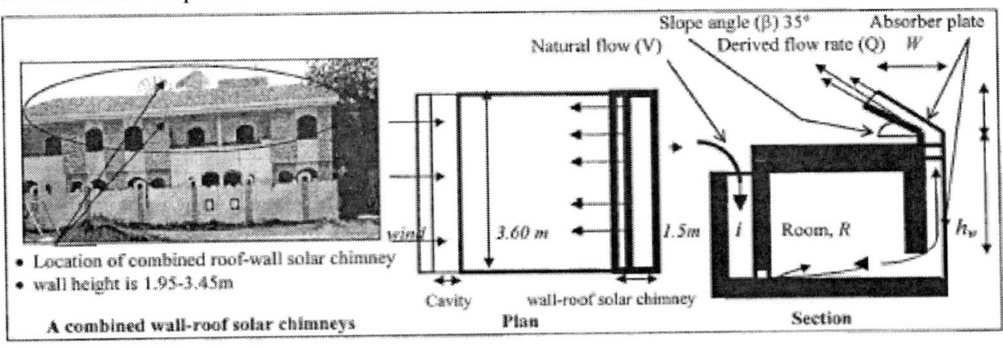

Figure 7. A combined wall-roof solar chimney incorporated into a residential building.

Night ventilation

Mean cooling load (kW)

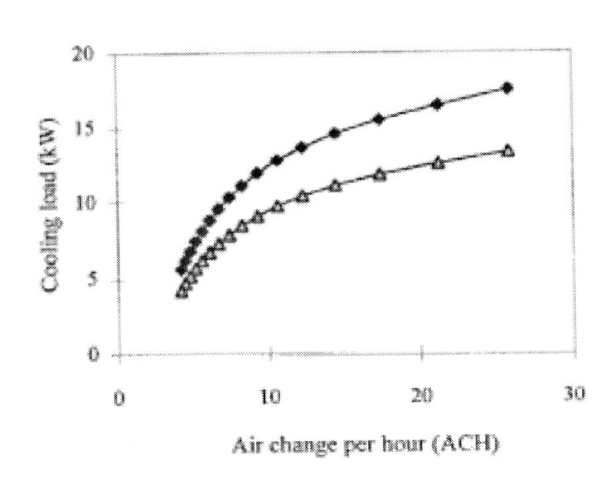

Figure 8. Cooling load by night ventilation for desired room indoor air temperatures of 23 and 25 °C.

7. Air Pollutants and Transmutation

Controlling the pollution of the present civilisation is an increasing concern. More importance is given to control global carbon dioxide, which is considered to be the main factor of green house effect. Though the complete experimental result on the fact is yet to be debated, the immense heat, temperature and turbulence of nuclear explosion oxidising the atmospheric nitrogen into nitric oxide, are considered to be similarly responsible for depletion of ozone layer [32]. At present, more importance is given for plantation to reduce the level of global carbon dioxide. The plantation over the whole earth surface may control only 50% of carbon dioxide disposed to atmosphere and its greenhouse effect. There are, also, explosions in the ozone layer time to time to add to the problem. Irrespective of the relative importance of each factor, the ozone layer protects us from harmful cosmic radiations and it is believed that the depletion of ozone layer increases the threat of outer radiations to human habitation if environmental pollution is not controlled or there is no possibility of self-sustainable stability in nature [32].

The presence of ionosphere in the outer-sphere is most probably for ionic dissociation of the gases of the outer-sphere in the presence of low pressure and cosmic radiation [32]. Moreover the ionosphere contains charged helium ions (alpha particle). Therefore, it may be concluded that the explosion in the ozone and transmission of radiations through it are the possible effects of transmutation of pollutants with exothermic reaction (emission of radiations) [32]. The existence of a black hole in the space, which is found in the photo camera of astrologist, is still unexplored. This black hole may be an effect of transmutation process with absorption of heat energy (endothermic reaction). The idea of transmutation of pollutants has been proposed for one or more of the following reasons:

- The experimental results support the transmutation of materials.
- To search the sinks of the remaining carbon dioxide not absorbed by plants or seawater.
- To find out the possible causes of explosion in the ozone layer other than the depletion of ozone layer.
- To investigate the possibilities of the self-sustaining stability of global environment.

To prove the portable of transmutation of pollutants, experimental investigations may be conducted to bombard C or CO_2 or CH_4 or other air pollutants by accelerated alpha particles in a low-pressure vacuum tube in a similar condition of ionosphere. Heating them with gamma radiation can accelerate the alpha particles. The results of such experimental investigation may prove the probable transmutation of pollutants and self-sustaining equilibrium of the global environment.

8. Greenhouses

Population growth and less availability of food material have become global concerns. The world population increases exponentially whereas food production has increased only arithmetically, meaning that the availability of food per capita has decreased. This is more pronounced in the cases of oils, vegetables, fruits and milk, whereas it is marginal, rather than minimum, in cereals. The increase in population has also resulted in the use of more urban areas for habitation, less land available for cultivation and, hence, more food requirements. The resultant need is, therefore, to increase productivity and year round cultivation. To maximise production and meet the global demand on food, vegetables, flowers and horticultural crops, it is necessary to increase the effective production span of crops. The sun is the source of energy for plants and animals. This energy is converted into food (i.e., carbohydrates) by plants through a process called photosynthesis. This process is accomplished at suitable atmospheric conditions. These conditions are provided by nature in different seasons and artificially by a greenhouse. The primary objective of greenhouses is to produce agricultural products outside the cultivation season. They offer a suitable microclimate for plants and make possible growth and fruiting, where it is not possible in open fields. This is why a greenhouse is also known as a "controlled environment greenhouse". Through a controlled environment, greenhouse production is advanced and can be continued for longer duration, and finally, production is increased [33]. The off-season production of flowers and vegetables is the unique feature of the controlled environment greenhouse. Hence, greenhouse technology has evolved to create the favourable environment, or maintaining the climate, in order to cultivate the desirable crop the year round. The use of "maintaining the climate" concept may be extended for crop drying, distillation, biogas plant heating and space conditioning. The use of greenhouses is widespread. During the last 10 years, the amount of greenhouses has increased considerably to cover up to several hundred hectares at present. Most of the production is commercialised locally or exported. In India, about 300 ha of land are under greenhouse cultivation. On the higher side, however, it is 98600 ha in Netherlands, 48000 ha in China and 40000 ha in Japan [34]. This shows that there is a large scope to extend greenhouse technology for various climates.

9. Effects of Urban Density

Compact development patterns can reduce infrastructure demands and the need to travel by car. As population density increases, transportation options multiply and dependence areas, per capita fuel consumption is much lower in densely populated areas because people drive so much less. Few roads and commercially viable public transport are the major merits. On the other hand, urban density is a major factor that determines the urban ventilation conditions, as well as the urban temperature. Under given circumstances, an urban area with a high density of buildings can experience poor ventilation and strong heat island effect. In warm-humid regions these features would lead to a high level of thermal stress of the inhabitants and increased use of energy in air-conditioned buildings.

However, it is also possible that a high-density urban area, obtained by a mixture of high and low buildings, could have better ventilation conditions than an area with lower density but with buildings of the same height. Closely spaced or high-rise buildings are also affected by the use of natural lighting, natural ventilation and solar energy. If not properly planned, energy for electric lighting and mechanical cooling/ventilation may be increased and application of solar energy systems will be greatly limited. Table 1 gives a summary of the positive and negative effects of urban density. All in all, denser city models require more careful design in order to maximise energy efficiency and satisfy other social and development requirements. Low energy design should not be considered in isolation, and in fact, it is a measure, which should work in harmony with other environmental objectives. Hence, building energy study provides opportunities not only for identifying energy and cost savings, but also for examining the indoor and outdoor environment.

Greenhouse cultivation is one of the most absorbing and rewarding forms of gardening for anyone who enjoys growing plants. The enthusiastic gardener can adapt the greenhouse climate to suit a particular group of plants, or raise flowers, fruit and vegetables out of their natural season. The greenhouse can also be used as an essential garden tool, enabling the keen amateur to expand the scope of plants grown in the garden, as well as save money by raising their own plants and vegetables.

Table 1. Effects of urban density on city's energy demand

Positive effects	Negative effects
Transport: • Promote public transport and reduce the need for, and length of, trips by private cars. Infrastructure: • Reduce street length needed to accommodate a given number of inhabitants. • Shorten the length of infrastructure facilities such as water supply and sewage lines, reducing the energy needed for pumping. Thermal performance: • Multi-story, multiunit buildings could reduce the overall area of the building's envelope and heat loss from the buildings. • Shading among buildings could reduce solar exposure of buildings during the summer period. Energy systems: • District cooling and heating system, which is usually more energy efficiency, is more feasible as density is higher. Ventilation: • A desirable flow pattern around buildings may be obtained by proper arrangement of high-rise building blocks.	Transport: • Congestion in urban areas reduces fuel efficiency of vehicles. Vertical transportation: • High-rise buildings involve lifts, thus increasing the need for electricity for the vertical transportation. Ventilation: • A concentration of high-rise and large buildings may impede the urban ventilation conditions. Urban heat island: • Heat released and trapped in the urban areas may increase the need for air conditioning. • The potential for natural lighting is generally reduced in high-density areas, increasing the need for electric lighting and the load on air conditioning to remove the heat resulting from the electric lighting. Use of solar energy: • Roof and exposed areas for collection of solar energy are limited.

9.1. Energy Efficiency and Architectural Expression

The focus of the world's attention on environmental issues in recent years has stimulated response in many countries, which have led to a closer examination of energy conservation strategies for conventional fossil fuels. Buildings are important consumers of energy and thus important contributors to emissions of greenhouse gases into the global atmosphere. The development and adoption of suitable renewable energy technology in buildings has an important role to play. A review of options indicates benefits and some problems [36]. There are two key elements to the fulfilling of renewable energy technology potential within the field of building design; first the installation of appropriate skills and attitudes in building design professionals and second the provision of the opportunity for such people to

performance of new buildings will only cut CO_2 emissions significantly in the long term. Consequently, the performance of existing buildings must be improved. For example, improving 3% of existing buildings would be more effective in cutting emissions than, say, improving the fabric standards for new non-domestic buildings and improving the efficiency of new air conditioning and ventilation systems [27]. A reduction in emissions arising from urban activities can, however, only be achieved by a combination of energy efficiency measures and a move away from fossil fuels.

9.4. Energy Efficiency

Energy efficiency is the most cost-effective way of cutting carbon dioxide emissions and improvements to households and businesses. It can also have many other additional social, economic and health benefits, such as warmer and healthier homes, lower fuel bills and company running costs and, indirectly, jobs. Britain wastes 20 per cent of its fossil fuel and electricity use. This implies that it would be cost-effective to cut £10 billion a year off the collective fuel bill and reduce CO_2 emissions by some 120 million tones. Yet, due to lack of good information and advice on energy saving, along with the capital to finance energy efficiency improvements, this huge potential for reducing energy demand is not being realised. Traditionally, energy utilities have been essentially fuel providers and the industry has pursued profits from increased volume of sales. Institutional and market arrangements have favoured energy consumption rather than conservation. However, energy is at the centre of the sustainable development paradigm as few activities affect the environment as much as the continually increasing use of energy. Most of the used energy depends on finite resources, such as coal, oil, gas and uranium. In addition, more than three quarters of the world's consumption of these fuels is used, often inefficiently, by only one quarter of the world's population. Without even addressing these inequities or the precious, finite nature of these resources, the scale of environmental damage will force the reduction of the usage of these fuels long before they run out.

Throughout the energy generation process there are impacts on the environment on local, national and international levels, from opencast mining and oil exploration to emissions of the potent greenhouse gas carbon dioxide in ever increasing concentration. Recently, the world's leading climate scientists reached an agreement that human activities, such as burning fossil fuels for energy and transport, are causing the world's temperature to rise. The Intergovernmental Panel on Climate Change has concluded that ''the balance of evidence suggests a discernible human influence on global climate''. It predicts a rate of warming greater than any one seen in the last 10,000 years, in other words, throughout human history. The exact impact of climate change is difficult to predict and will vary regionally. It could, however, include sea level rise, disrupted agriculture and food supplies and the possibility of more freak weather events such as hurricanes and droughts. Indeed, people already are waking up to the financial and social, as well as the environmental, risks of unsustainable energy generation methods that represent the costs of the impacts of climate change, acid rain and oil spills. The insurance industry, for example, concerned about the billion dollar costs of hurricanes and floods, has joined sides with environmentalists to lobby for greenhouse gas emissions reduction. Friends of the earth are campaigning for a more sustainable energy policy, guided by the principle of environmental protection and with the objectives of sound natural resource management and long-term energy security. The key priorities of such an energy policy must be to reduce fossil fuel use, move away from nuclear power, improve the efficiency with which energy is used and increase the amount of energy obtainable from sustainable, renewable sources. Efficient energy use has never been more crucial than it is today, particularly with the prospect of the imminent introduction of the climate change levy (CCL). Establishing an energy use action plan is the essential foundation to the elimination of energy waste. A logical starting point is to carry out an energy audit that enables the assessment of the energy use and determine what actions to take. The actions are best categorised by splitting measures into the following three general groups:

1. High priority/low cost:

These are normally measures, which require minimal investment and can be implemented quickly. The followings are some examples of such measures:

- Good housekeeping, monitoring energy use and targeting waste-fuel practices.
- Adjusting controls to match requirements.
- Improved greenhouse space utilisation.
- Small capital item time switches, thermostats, etc.
- Carrying out minor maintenance and repairs.
- Staff education and training.
- Ensuring that energy is being purchased through the most suitable tariff or contract arrangements.

2. Medium priority/medium cost:

Measures, which, although involve little or no design, involve greater expenditure and can take longer to implement. Examples of such measures are listed below:

- New or replacement controls.
- Greenhouse component alteration, e.g., insulation, sealing glass joints, etc.
- Alternative equipment components, e.g., energy efficient lamps in light fittings, etc.

3. Long term/high cost:

These measures require detailed study and design. They can be best represented by the followings:

- Replacing or upgrading of plant and equipment.
- Fundamental redesign of systems, e.g., combined heat and power CHP installations.

This process can often be a complex experience and therefore the most cost-effective approach is to employ an energy specialist to help.

9.5. Policy Recommendations for a Sustainable Energy Future

Sustainability is regarded as a major consideration for both urban and rural development. People have been exploiting the natural resources with no consideration to the effects, both short-term (environmental) and long-term (resources crunch). It is also felt that knowledge and technology have not been used effectively in utilising energy resources. Energy is the vital input for economic and social development of any country. Its sustainability is an important factor to be considered. The urban areas depend, to a large extent, on commercial energy sources. The rural areas use non-commercial sources like firewood and agricultural wastes. With the present day trends for improving the quality of life and sustenance of mankind, environmental issues are considered highly important. In this context, the term energy loss has no significant technical meaning. Instead, the exergy loss has to be considered, as destruction of exergy is possible. Hence, exergy loss minimisation will help in sustainability. In the process of developing, there are two options to manage energy resources: (1) End use matching/demand side management, which focuses on the utilities. The mode of obtaining this is decided based on economic terms. It is, therefore, a quantitative approach. (2) Supply side management, which focuses on the renewable energy resource and methods of utilising it. This is decided based on thermodynamic consideration having the resource-user temperature or exergy destruction as the objective criteria. It is, therefore, a qualitative approach. The two options are explained schematically in figure 9. The exergy-based energy, developed with supply side perspective is shown in figure 10.

The following policy measures had been identified:

- Clear environmental and social objectives for energy market liberalisation, including a commitment to energy efficiency and renewables.
- Economic, institutional and regulatory frameworks, which encourage the transition to total energy services.
- Economic measures to encourage utility investment in energy efficiency (e.g., levies on fuel bills).
- Incentives for demand side management, including grants for low-income households, expert advice and training, standards for appliances and buildings and tax incentives.
- Research and development funding for renewable energy technologies not yet commercially viable.
- Continued institutional support for new renewables (such as standard cost-reflective payments and obligation on utilities to buy).
- Ecological tax reform to internalise external environmental and social costs within energy prices.
- Planning for sensitive development and public acceptability for renewable energy.

Energy resources are needed for societal development. Their sustainable development requires a supply of energy resources that are sustainably available at a reasonable cost and can cause no negative societal impacts. Energy resources such as fossil fuels are finite and lack sustainability, while renewable energy sources are sustainable over a relatively longer term. Environmental concerns are also a major factor in sustainable development, as activities, which degrade the environment, are not sustainable. Hence, as much as environmental impact is associated with energy, sustainable development requires the use of energy resources, which cause as little environmental impact as possible. One way to reduce the resource depletion associated with cycling is to reduce the losses that accompany the transfer of exergy to consume resources by increasing the efficiency of exergy transfer between resources, i.e., increasing the fraction of exergy removed from one resource that is transferred to another [9].

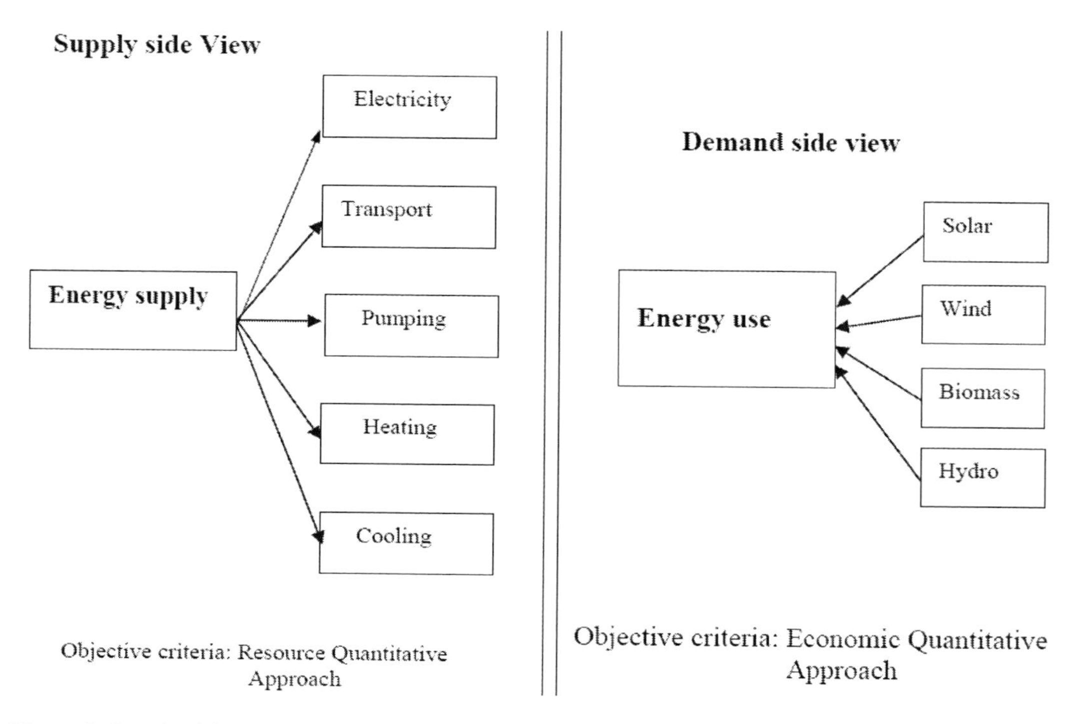

Figure 9. Supply side and demand side management approach for energy.

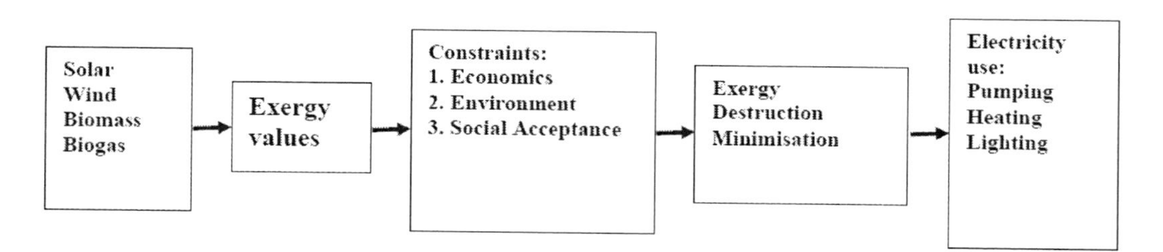

Figure 10. Exergy based optimal energy model.

As explained above, exergy efficiency may be thought of as a more accurate measure of energy efficiency that accounts for quantity and quality aspects of energy flows. Improved exergy efficiency leads to reduced exergy losses. Most efficiency improvements produce direct environmental benefits in two ways. First, operating energy input requirements are reduced per unit output, and pollutants generated are correspondingly reduced. Second, consideration of the entire life cycle for energy resources and technologies suggests that improved efficiency reduces environmental impact during most stages of the life cycle. Quite often, the main concept of sustainability, which often inspires local and national authorities to incorporate environmental consideration into setting up energy programmes have different meanings in different contexts though it usually embodies a long-term perspective. Future energy systems will largely be shaped by broad and powerful trends that have their roots in basic human needs. Combined with increasing world population, the need will become more apparent for successful implementation of sustainable development.

Heat has a lower exergy, or quality of energy, compared with work. Therefore, heat cannot be converted into work by 100% efficiency. Some examples of the difference between energy and exergy are shown in table 3.

Table 3. Qualities of various energy sources

Source	Energy (J)	Exergy (J)	CQF
Water at 80°C	100	16	0.16
Steam at 120°C	100	24	0.24
Natural gas	100	99	0.99
Electricity/work	100	100	1.00

The terms used in table 3 have the following meanings:

$$\text{Carnot Quality Factor (CQF)} = (1 - T_o/T_s) \quad (2)$$

$$\text{Exergy} = \text{Energy (transferred)} \times \text{CQF} \quad (3)$$

where T_o is the environment temperature (K) and T_s is the temperature of the stream (K).

Various parameters are essential to achieving sustainable development in a society. Some of them are as follows:

- Public awareness.
- Information.
- Environmental education and training.
- Innovative energy strategies.
- Renewable energy sources and cleaner technologies.
- Financing.
- Monitoring and evaluation tools.

The development of a renewable energy in a country depends on many factors. Those important to success are listed below:

(1) Motivation of the population

The population should be motivated towards awareness of high environmental issues, rational use of energy in order to reduce cost. Subsidy programme should be implemented as incentives to install renewable energy plants. In addition, image campaigns to raise awareness of renewable technology.

(2) Technical product development

To achieve technical development of renewable energy technologies the following should be addressed:

- Increasing the longevity and reliability of renewable technology.
- Adapting renewable technology to household technology (hot water supply).
- Integration of renewable technology in heating technology.
- Integration of renewable technology in architecture, e.g., in the roof or façade.
- Development of new applications, e.g., solar cooling.
- Cost reduction.

(3) Distribution and sales

Commercialisation of renewable energy technology requires:

- Inclusion of renewable technology in the product range of heating trades at all levels of the distribution process (wholesale, and retail).
- Building distribution nets for renewable technology.
- Training of personnel in distribution and sales.
- Training of field sales force.

(4) Consumer consultation and installation

To encourage all sectors of the population to participate in adoption of renewable energy technologies, the following has to be realised:

- Acceptance by craftspeople, marketing by them.
- Technical training of craftspeople, initial and follow-up training programmes.
- Sales training for craftspeople.
- Information material to be made available to craftspeople for consumer consultation.

(5) Projecting and planning

Successful application of renewable technologies also requires:

- Acceptance by decision makers in the building sector (architects, house technology planners, etc.).
- Integration of renewable technology in training.
- Demonstration projects/architecture competitions.
- Renewable energy project developers should prepare to participate in the carbon market by:

- Ensuring that renewable energy projects comply with Kyoto Protocol requirements.
- Quantifying the expected avoided emissions.
- Registering the project with the required offices.
- Contractually allocating the right to this revenue stream.

- Other ecological measures employed on the development include:

  - Simplified building details.
  - Reduced number of materials.
  - Materials that can be recycled or reused.
  - Materials easily maintained and repaired.
  - Materials that do not have a bad influence on the indoor climate (i.e., non-toxic).
  - Local cleaning of grey water.
  - Collecting and use of rainwater for outdoor purposes and park elements.
  - Building volumes designed to give maximum access to neighbouring park areas.
  - All apartments have visual access to both backyard and park.

(6) Energy saving measures

The following energy saving measures should also be considered:

- Building integrated solar PV system.
- Day-lighting.
- Ecological insulation materials.
- Natural/hybrid ventilation.
- Passive cooling.
- Passive solar heating.
- Solar heating of domestic hot water.
- Utilisation of rainwater for flushing.

Improving access for rural and urban low-income areas in developing countries is through energy efficiency and renewable energies. Sustainable energy is a prerequisite for development. Energy-based living standards in developing countries, however, are clearly below standards in developed countries. Low levels of access to affordable and environmentally sound energy in both rural and urban low-income areas are therefore a predominant issue in developing countries. In recent years many programmes for development aid or technical assistance have been focusing on improving access to sustainable energy, many of them with impressive results.

Apart from success stories, however, experience also shows that positive appraisals of many projects evaporate after completion and vanishing of the implementation expert team. Altogether, the diffusion of sustainable technologies such as energy efficiency and renewable energies for cooking, heating, lighting, electrical appliances and building insulation in developing countries has been slow.

Energy efficiency and renewable energy programmes could be more sustainable and pilot studies more effective and pulse releasing if the entire policy and implementation process was considered and redesigned from the outset. New financing and implementation processes are needed, which allow reallocating financial resources and thus enabling countries themselves to achieve a sustainable energy infrastructure. The links between the energy policy framework, financing and implementation of renewable energy and energy efficiency projects have to be strengthened and capacity building efforts are required.

The use of renewable energy resources could play an important role in this context, especially with regard to responsible and sustainable development. It represents an excellent opportunity to offer a higher standard of living to local people and will save local and regional resources. Energy efficiency brings health, productivity, safety, comfort and savings to homeowner, as well as local and global environmental benefits.

10. Low Carbon Building for the Future

There was a growing awareness that cutting greenhouse gases (GHGs) is a huge business opportunity. Business can both:

- Cut its energy costs and make itself more competitive, safeguarding profits and employment.
- Grow by developing and adopting products based on the new low carbon technologies.

More than 170 countries have signed up to the Kyoto Protocol. That is a huge potential market for products based on low carbon technology. Pressure on business comes from governments, reflecting the concerns of votes, directly from consumers through green purchasing and indirectly through shareholder democracy. Non-governmental organisations (NGOs) have done a great deal to help create the market conditions that allows business to go down the low carbon route. Many governments are investing considerable sums in low carbon technology. There is a huge amount of research and development (R&D) going on, which could help to make a reality of Kyoto. The CO_2 emissions can be cut by around 60% over the next few decades, using existing and emerging technologies, which depend on:

- Active and positive engagement from the business world.
- Governments setting a firm policy context, so those innovative companies can profit.
- A strong input from NGOs who have also played a key role.

The business and the public sectors play their full part in delivering GHG reductions and prepare the ways for a low carbon economy up to 2050 and beyond. A truly low carbon economy is impossible without business involvement and support business development and carbon reduction go hand in hand. The vision and challenges are needed. Tackling carbon emissions is good business and introducing low carbon technology is good for the bottom line.

The Kyoto Protocol to the United Nations Framework Convention on Climate Change (UNFCCC) was initially designed to reduce GHG emissions from industrialised countries by 5%. There were many debates over the controversial area of restricting the buying, selling and banking of emissions reductions, particularly with regard to carbon sinks (the temporary storage of carbon in forests, soils, etc.). The potential for renewables is vast, uncontroversial, yet under-appreciated. In the case of solar PVs, for example even in a cloudy, rainy country like UK, modern PV technology applied to all available UK roofs would generate more electricity than the nation currently consumes in a year.

Global warming is in the process of teaching us that over security is best built by making sure our neighbours are secure as well. Its beachheads are becoming clear in proliferating climate extremes like the long-running drought. Unless we cut the burning of oil, gas and coal deeply the effects of global warming would ultimately be second only to nuclear war. A new global energy-security paradigm is urgently required. Budgets and policies should be consisted with the newly convergent imperatives of environmental and global security. New technologies were credited with offering low carbon solutions with fuel cells providing primary. Fuel cell technology has yet to prove itself and consideration should be given to the many innovative devices currently available that are grossly under-utilised. There is a need to face up to the rehabilitation of nuclear power despite any additional worries. Energy efficiency meaning improvements to the performance of power conversion and energy using devices would have a crucial role to play. The electricity supply industry talked, rather altruistically, of bringing power to the 1.6 billion people in the world that do not have access to it.

10.1. Low-Energy Device for Integrated Heating, Cooling and Humidity Control in Greenhouses

A combination of plant transpiration, wet soil, and warm temperatures leads to high humidity in greenhouse. High humidity promotes the spread of disease inside greenhouse. The growth of various types of fungi, such as downy mildew and grey mould, is greatly enhanced in a humid environment and these diseases can have a critical effect on crop quality and yield. The best way to control these fungal diseases is by humidity control. However, although ventilation is commonly used to control humidity during warm weather, the high cost of heating the inlet air during cold periods results in many greenhouse operations using fungicides rather than humidity control during cold weather. A cost effective, non-chemical method is required to replace fungicides for several reasons:

- There is growing consumer demand for pesticide-free products.
- Fungicide registrations are being cancelled.
- Botrytis cinerea and other fungi are developing resistance to fungicides.

Resistance to major groups of fungicides in botrytis has been documented in the United Kingdom, Italy, Canada and the USA. However, there are several important crops for which no fungicides effective against botrytis are registered. Thus, for several reasons, non-chemical methods of botrytis control must be exploited wherever possible. Humidity reduction is the most important, non-chemical method would offer a further advantage in controlling – calcium related disorders, which are humidity dependent and result in loss of both quality and yield.

The greenhouse effect is one result of the differing properties of heat radiation when it is generated by at different temperatures. The high temperature sun emits radiation of short wavelength, which can pass through the atmosphere and through glass. Inside the greenhouse or other building this heat is absorbed by objects, such as plants, which then re-radiate the heat. Because the objects inside the greenhouse are at a lower temperature than the sun the radiated heat is of longer wavelengths, which cannot penetrate glass. This re-radiated heat is therefore trapped and causes the temperature inside the greenhouse to rise. The atmosphere surrounding the earth also behaves as a large greenhouse around the world. Changes to the gases in the atmosphere, such as increased carbon dioxide content from the burning of fossil fuels, can act like a layer of glass and reduce the quantity of heat that the planet earth would otherwise radiate back into space. This particular greenhouse effect therefore contributes to global warming. Meeting the target of a 60% reduction in carbon dioxide (CO_2) emissions on environmental pollution is both technologically feasible and financially viable.

10.2. Bioclimatic Approach

The question of thermal comfort is increasingly generating an intensive debate. The subject is not new but the exact solution is illusive to the fact that considerable part of the thermal sensation can only be evaluated by subjective means. Consequently the thermal comfort ranges do differ and seem to be complex functions of culture, physiology and geographical location. The human body is not exempt from the effects of the second law of thermodynamics. When the body heat cannot be dissipated to the surrounding environment; a condition that occurs when the ambient temperature is higher than the body temperature, then thermal discomfort starts. The three common body index temperatures are 36.6°C (oral), 37°C (anal), and 35°C (skin temperature). While 37°C is the temperature of the internal organs, the skin temperature is the reference datum for the thermal comfort sensation.

Buildings in the tropical area of the world are constantly exposed to solar radiation almost everyday. As a result, building design should aim at minimising heat gain indoors and maximising adequate thermal cooling so that user of these spaces can have adequate thermal comfort. To achieve this objective, buildings in this part of the world should have shapes and frames which should (1) be responsive to this objective (2) be properly oriented, and (3) the fabric of the buildings should be specified to prevent and have minimal use of active energy for economic viability. In order to meet the above requirements, it implies that buildings should be bioclimatic responsive. Observation of most buildings (both traditional and contemporary) in the built environment and also of building design approaches- past and present- reveals that most of the above criteria have not been strictly adhered to. Traditional buildings have laid too much emphasis on social-cultural and economic factors. Also, contemporary buildings, especially housing, have depended on imported building materials. Various problems have emanated from the present design approaches and philosophy. First, most buildings seem to be replicas of buildings in European countries in shape and form despite marten differences in climatic conditions. Secondly, despite observed climatic differences in various cities, forms and shapes of buildings tend to look alike. Thirdly, windows of buildings have not been properly oriented to maximise air movement for space cooling indoors. Window sizes and openings have not responded to physiological comfort. Finally, material specification for buildings in the housing sector has followed the same pattern despite the difference in climate. The tropical areas of the world are generally referred to as the overheated regions. For building design purposes, overheated regions of the world are classified into three categories: (1) hot/warm, (2) arid/semi arid regions, and (3) temperate, both arid and humid regions. In order to properly access the effect of climate in any particular location and to determine appropriate climatic responsive design solution for buildings, it is necessary to evaluate the characteristics for the combined effect of thermal comfort parameters. These parameters should be carefully analysed for various areas. Two methods of climate analysis must be used so that a proper design solution for buildings can be made. The methods are: (1) the general atmospheric circulation model, and (2) the control potential technique. The general atmospheric circulation model (GACM) is based on the principle that the climate of a particular point in space and time is a result of three forces created by the earth's revolution, rotation and vertical heat transport. In other words, more characteristics of the climate of region will be considered. The vertical height is limited to the biosphere. The climatic data relevant to building design were then collected and analysed. The maximum and minimum monthly temperatures, the relative humidity both for morning and afternoon, solar radiation, and precipitation levels were collected from past records from meteorological stations.

It is suggested that the climate of a given location should be analysed in its own terms and that this analysis should directly lead to a certain architectural response type, i.e., of the appropriate control potential strategies. For a comfortable indoor environment to be achieved, the microclimate of the locality in which design is taking place should be carried out. The analysis of climate data closely related to the design environment will lead to more adequate and precise design decisions in terms of adequate orientation, spatial organisation, prevention of heat gain into spaces and better choice of building materials.

10.3. Applications of Solar Energy

These design strategies would provide the most effective combination for heating, natural ventilation and day lighting. Passive solar systems for space heating and cooling, as well as passive cooling techniques when used in combination with conventional systems for heating, cooling, ventilation and lighting, can significantly contribute to the energy saving in the buildings sector [36-40]. The available environmentally sensitive design strategies for improving the energy performance of spaces without appropriate orientation are shown in table 4.

Design strategies that could be cost-effective and easily constructed, is the most important factor for the selection. A described by Baker, et al., [41] a skylight is an opening located on a horizontal or tilted roof. It allows the zenithal entry of daylight increase the limit level of the lower space under the skylight. It can be opened to admit ventilation. In winter, a reflector can be used to enhance solar gain, since the amount of solar energy transmitted on a horizontal surface is considerably poor. In summer, exterior shading is required to avoid excessive solar gain [42].

Table 4. Passive solar strategies for spaces without north-facing façade

Passive solar strategies / for	Natural illumination	Heating	Cooling	Ventilation
Skylight	X	X		X
Clerestory	X	X		X
Monitor	X	X		X
Saw tooth facing south	X			X
Light shelves	X			
Light pipe systems	X			
Thermal storage mess on roofs (roof pond)		X		
Thermosiphon (roof-floor)		X		
Roof greenhouse		X		
Black attic		X		
Wind tower			X	X
Solar chimney			X	X

The clerestory roof windows are vertical or tilted openings projecting up from the roof plane. They permit zenithal penetration of daylight, redirecting it towards the spaces below. They also allow natural ventilation. They are particularly effective for heating by direct sunlight entering a space and onto an interior thermal storage wall. Roof monitors raised section of roof with north and south openings. They permit the zenithal entry of daylight towards the lower zone increasing luminic level and allowing ventilation through the apertures. They have similar attributes to the clerestory in that sunlight is directed on to an internal thermal storage wall. Spaces without northerly orientations have an impact on the energy behaviour of a building.

In this article considered the envelope area, shape and tilt of the glassed area and the arrangement of the elements across the roof area. Further studies are required to measure the influence of diffuse glazing in the distribution approaches. Different approaches could be appropriate for different situations and environmental conditions depending on heating, day-lighting, ventilation requirements, or construction facilities. The more important aspects to be considered are the glazing tilt and orientation of the element. The arrangement of the zenithal strategies has little influence in the thermal performance; however, regions with heavy snowfalls in winter will have an impact on the design strategy. Snow accumulation can cause problems for large elements in terms of roof structure and insulation, and smaller elements will be a structure and insulation, and smaller elements will be a better solution. In addition, these arrangements are more expensive and have more building, and structural complications. The element and roof join is usually a weak point in the roof structure so that the use of many elements will cause problems with roof insulation water leakage. The roof monitors with their north and south-facing openings allow a better quality of illuminance giving the whole spectrum of colour. It is also important to note that roof monitors had the best-combined results in terms of lighting and thermal effects. Coupled with the possibilities of ventilation by having the two openings (north and south windows) operable, this will also allow the removal of the exhaust air in summer.

11. THEORETICAL FOUNDATIONS

The analysis of thermal exchange among building components can be performed for any number of elements. However, the basic limitation to its calculation remains the choice of pertinent heat transfer coefficients. In this paper three basic elements are considered in a lumped-parameter approach: the building materials (structure and fittings, including finishing, of floor, walls and roofs), a single window, and the inside air. The value of this last variable is taken to represent the pertinence of design.

A limiting condition to lumped parameter analyses is the definition of adducted heat transfer coefficients among the system, components, and among these and the environment. These coefficients are needed to solve the quasi-steady-state equations that describe the heat balance in each of the relevant system components. In the absence of ventilation, the windowpane heat balance equation can be written as:

$$M_v\, Cp_v\, dT_v/dt = Q_v\, A_s\, U_{sv}\, (T_v-T_s) - A_v\, U_{vi}\, (T_v-T_i) - A_v\, U_{va}\, (T_v-T_a) \tag{4}$$

In the preceding equation, the windowpane has an exposed area Av, thermal mass Mv, specific heat cp_v and its temperature T_v varies with time t as a result of solar heat Q_v received on its external face. T_s, T_a and T_i are respectively, the building, ambient and internal air temperature.

Solar gains Q_v are calculated by the following relationship:

$$Q_v = A_v\, (\alpha_v\, G_b\, R_b\, (1-\rho_i) + G_d\, (1-\rho_d)\, \alpha_v) \tag{5}$$

where the first inside the parenthesis describes direct solar gains, and the second, indirect or diffuse solar gains, as described in the nomenclature. Similarly, for the building materials,

$$M_s\, Cp_s\, dT_s/dt = Q_s - A_v\, U_{sv}\, (T_s-T_v) - A_s\, U_{si}\, (T_i-T_s) - A_s\, U_{sa}\, (T_s-T_a) \tag{6}$$

In the prior equation, the building inside temperature Ts is associated with a total thermal mass M_s, specific heat Cp_s and varies with time t as a result of solar heat Q_s received through the window. Total heat can be approximated by:

$$Q_s = A_v\, (G_b\, R_b\, (1-\alpha_b-\rho_i) + G_d\, (1-\alpha_v\, -\rho_d)) \tag{7}$$

where, as in the preceding case, the first term inside the parenthesis describes direct solar gains, and the second, indirect or diffuse solar gains. And finally, for the air inside the enclosure, neglecting air exchange with the ambient,

$$Mi\, Cp_i\, dT_i/dt = A_s\, U_{si}\, (T_s-T_i) - A_v\, U_{vi} - (T_i-T_v) \tag{8}$$

In this case, M_i, Cp_i and T_i refer to inside air mass, specific heat and temperature, respectively.

The universally accepted nomenclature by Duffie and Beckman [43] is retained in equations (4) through (8). Indices v, s and i refer in all cases to, respectively, the windowpane, the building mass, the inside air, and the ambient temperature. The equations are solved simultaneously by means of the variable, non-linear internal heat transfer coefficients U_{si}, U_{vi} and U_{sv}. Coefficient that refers to thermal exchange with the external environment, U_{va} and U_{sa}, are calculated according to Watmuff, et al., [44]. In general, any U_{jk} value that depicts heat transfer between element j and element k can be approximated by:

$$U_{jk} = h_{rjk} + h_{cjk} \quad (9)$$

In eq. (9), the term with sub-index r refers to radiation and the one with c, to convection. Both are very sensitive to temperature. The radiation term is usually approximated by Duffie and Beckman [43]:

$$h_{rjk} = \sigma\varepsilon\alpha\, F\, (T_j^2 + T_k^2)\,(T_j + T_k) \quad (10)$$

where σ stands for Stefan-Boltzman constant, α is the thermal absorptance, ε is thermal emittance and F is the geometric shape factor, in such a way that radiative transfer per unit area is calculated as:

$$q_{rjk} = h_{rjk}\,(T_j - T_k) \quad (11)$$

No generally acceptable coefficients are available for this application. However, it is possible to employ adapted expressions for heat transfer by natural convection in closed cavities, such as can be found in Thomas [45], making use of the adequate aspect-ratio relationships. For the convection term,

$$h_{cjk} = N_u\, k/L\ CR\ \alpha_L^{m}\,(H/L)^n \quad (12)$$

In this case, the convective heat transfer coefficient is sensitive to the characteristic distance L, conductivity k of air at the mean temperature inside the envelope, Ray-Leigh number R_a, aspect-ratio H/L, constant C and powers m and n are adjusted to experimental results below:

For U_{vi} and U_{si}, h_r =0, and h_c values are calculated with eq. (12) using C=0.162, m=0.29, n=0, H/L=1 and L=1.2 for U_{vi} and L=1.7 for U_{si}.

Finally, the required temperature dependent air transport properties were evaluated by the following expression, which are valid between 2°C and 77°C with temperature expressed in k:

Thermal diffusivity, $\alpha = 1.534 \times 10^{-3}\ T - 0.2386\ (10^{-4}\ m^2s^{-1})$
Kinematics viscosity, $v = 0.1016\ T - 14.8\ (10^{-6}\ m^2s^{-1})$
Thermal conductivity, $k = 7.58 \times 10^{-5}\ T + 3.5 \times 10^{-3}\ (Wm^{-1}K^{-1})$, and
Thermal expansion coefficient, $\beta = T^{-1}\ (K^{-1})$

In order to depict the relative contribution of each of these techniques to inside temperature, a dimensionless index is defined as follows. When interior temperature exceeds 25°C, it will be considered as a temperature discomfort condition. This reference temperature is widely elements. Then the following expression:

$$F(t) = \max\,(T_t - 25.25) \quad (13)$$

I_s a time function of truncated temperature and it will be able to estimate the overall discomfort by means of the integration along the day for each different scenarios S:

$$A(S) = \int_S F(t)\,dt \quad (14)$$

Then, for each passive technique, let:

$$A_{max} = \max\ [A(S)\text{: for all scenarios S}] \quad (15)$$

Finally, the normalised temperature index (figures 11-12) for each scenario S is:

$$I(S) = A(S)/A_{max} \quad (16)$$

Naturally, it would be preferred, for comfort reasons that this index would be small, preferably nil. It may be seen that the variable is directly related to temperature discomfort: the larger the value of the index, the farthest will inside conditions be from expected wellbeing. Also, the use of electricity operated air conditioning systems will be more expensive the higher this variable is. Hence, energy expenditure to offset discomfort will be higher when comparing two index values; the ratio of them is proportional to the expected energy savings. When the external shade blocks the windowpane completely, the excessive heat gains belong to the lowest values in the set, and the dimensionless index will be constant with orientation. For the climate conditions of the locality, it can be seen that a naked window can produce undesirable heat gains if the orientation is especially unfavourable, when the index can have an increase of up to 0.3 with respect to the totally shaded window.

The most favourable orientation, which is due north, results in diminished excessive solar gains through the windows. However, most buildings cannot be oriented at will. If the only possible orientation is due south, and no external shade is used, the index reveals extra heat gains of some 0.26 over the value of totally shaded window. Application of the model results from exploring the relative importance of the thermal inertia of walls, floor and ceiling. Heat stored in building materials, as proven in old, massive buildings, can be compensated during high insolation hours with thermal losses at night and early morning hours, when ambient temperatures are below 25°C. Temperature variation will be lower for higher thermal capacities of building materials. However, it is known while thermal capacity increases the relative importance of individual heat flows change. For example, for lower wall temperatures, the contribution of radiative heat transfer will be reduced, and the relative importance of convective processes will increase, and thus the difficulty to calculate accurately the overall heat flows. The relevance of certain passive techniques is variable with prevailing weather. Where ambient temperature is mostly stable, thermal mass is no advantage, as Lee, et al., [47] have shown for very light housing in Korea. Vernacular architecture, where massive buildings are common, suggests the use of some passive techniques. These are provided with thick walls and small windows and when properly shaded and ventilated, can result in very acceptable temperature levels without the need of active systems in the extreme varying weather [47]. Bioclimatic design of buildings is one strategy for sustainable development as it contributes to reducing energy consumption and therefore, ultimately, air pollution and greenhouse gas emissions (GHG) from conventional energy generation.

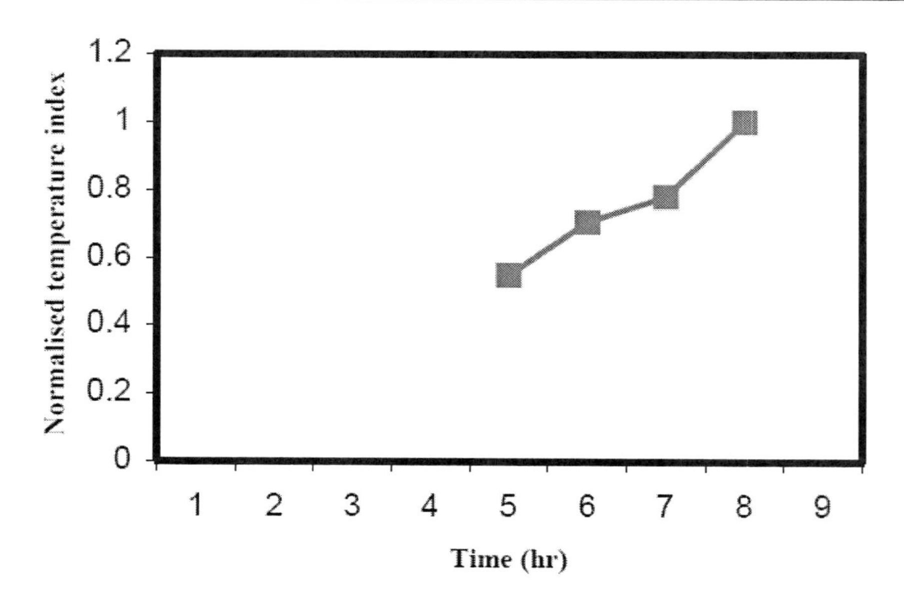

Figure 11. Effect of roof extension on normalised inside temperature.

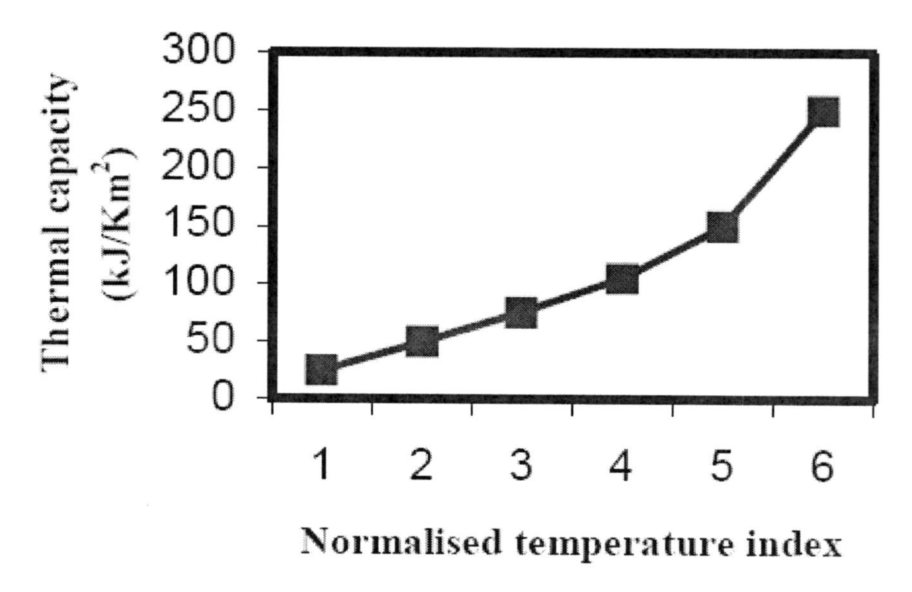

Figure 12. Effect of thermal capacity on normalised inside temperature.

12. Climate Change

The scientific consensus is clear-climate change is occurring. Existing renewable energy technologies could play a significant mitigating role, but the economic and political climate will have to change first. Climate change is real, it is happening now, and greenhouse gases produced by human activities are significantly contributing to it. The predicted global temperature changes of between 1.5 and 4.5 degrees C could lead to potentially catastrophic environmental impacts-including sea level rise, increased frequency of extreme weather events, floods, droughts, disease migration from various places and possible stalling of the Gulf stream. This is why scientists argue that climate change issues are not ones that politicians can afford to ignore. And policy makers tend to agree, but reaching international agreements on climate change policies is no trivial task [48].

Renewable energy is the term to describe a wide range of naturally occurring, replenishing energy sources. The use of renewable energy sources and the rational use of energy are the fundamental inputs for a responsible energy policy. The energy sector is encountering difficulties because increased production and consumption levels entail higher levels of pollution and eventually climate changes, with possibly disastrous consequences. Moreover, it is important to secure energy at acceptable cost to avoid negative impacts on economic growth. On the technological side, renewables have an obvious role to play. In general, there is no problem in terms of the technical potential of renewables to deliver energy and there are very good opportunities for renewable energy technologies to play an important role in reducing emissions of greenhouse gases into the atmosphere-certainly far more than have been exploited so far [48]. But there are still technical issues to be addressed to cope with the intermittency of some renewables, particularly wind and solar. However, the biggest problem with replying on renewables to deliver the necessary cuts in greenhouse gas emissions is more to do with politics and policy issues than with technical ones. The single most important step governments could take to promote and increase the use of renewables would be to improve access for renewables to the energy market. That access to the market would need to be under favourable conditions and possibly under favourable economic rates. One move that could help-or at least justify-better market access would be to acknowledge that there are environmental costs associated with other energy supply options, and that these costs are not currently internalised within the market price of electricity or fuels. It could make significant difference, particularly if, appropriate subsidies were applied to renewable energy in recognition of environmental benefits it offers. Cutting energy consumption through end-use efficiency is absolutely essential. And this suggests that issues of end-use consumption of energy will have to come onto the table in the foreseeable future.

13. Conclusions

Thermal comfort is an important aspect of human life. Buildings where people work require more light than buildings where people live. In buildings where people live the energy is used for maintaining both the temperature and lighting. Hence, natural ventilation is rapidly becoming a significant part in the design strategy for non-domestic buildings because of its potential to reduce the environmental impact of building operation, due to lower energy demand for cooling. A traditional, naturally ventilated building can readily provide a high ventilation rate. On the other hand, the mechanical ventilation systems are very expensive. However, a comprehensive ecological concept can be developed to achieve a reduction of electrical and heating energy consumption, optimise natural air condition and ventilation, improve the use of daylight and choose environmentally adequate building materials. Energy efficiency brings health, productivity, safety, comfort and savings to homeowner, as well as local and global environmental benefits. The use of renewable energy resources could play an important role in this context, especially with regard to responsible and sustainable development. It represents an excellent opportunity to offer a higher standard of living to local people and will save local and regional resources. Implementation of greenhouses offers a chance for maintenance and repair services. It is expected that the pace of implementation will increase and the quality of work to improve in addition to building the capacity of the private and district staff in contracting procedures. The financial accountability is important and more transparent. Various passive techniques have been put in perspective, and energy saving passive strategies can be seen to reduce interior temperature and increase thermal comfort, reducing air conditioning loads. The scheme can also be employed to analyse the marginal contribution of each specific passive measure working under realistic conditions in combination with the other housing elements. In regions where heating is important during winter months, the use of top-light solar passive strategies for spaces without an equator-facing façade can efficiently reduce energy consumption for heating, lighting and ventilation. The use of renewable energy resources could play an important role in this context, especially with regard to responsible and sustainable development. It represents an excellent opportunity to offer a higher standard of living to local people and will save local and regional resources. Implementation of greenhouses offers a chance for maintenance and repair services. Various passive techniques have been put in perspective, and energy saving passive strategies can be seen to reduce interior temperature and increase thermal comfort, reducing air conditioning loads.

REFERENCES

[1] Jeremy, L. The energy crisis, global warming and the role of renewables. *Renewable Energy World.* 2005; 8 (2).

[2] Omer, A. Low energy building materials: an overview. In: Proceedings of the Environment 2010: Situation and Perspectives for the European Union. p. 16-21. Porto: Portugal. 6-10 May 2003.

[3] UNEP. Handbook for the international treaties for the protection of the ozone layer. *United Nations Environment Programme.* Nairobi: Kenya. 2003.

[4] Viktor, D. Ventilation concepts for sustainable buildings. In: Proceedings of the World Renewable Energy Congress VII, p. 551, Cologne: Germany. 29 June – 5 July 2002.

[5] Lam, J.C. Shading effects due to nearby buildings and energy implications. *Energy Conservation and Management.* 2000; 47 (7): 647-59.

[6] Raja, J., Nichol, F., and McCartney, K. Natural ventilated buildings use of controls for changing indoor climate. In: Proceedings of the 5th World Renewable Energy Congress V., p. 391-394. Florence: Italy. 20-25 September 1998.

[7] Limb, M.J. Air intake positioning to avoid contamination of ventilation. AIVC. 1995.

[8] Miller, G. Resource conservation and management. Wadsworth Publishers. California: USA, p.51-62. 1990.

[9] Erlich, P. Forward facing up to climate change, in global climate change and life on Earth. R.C. Wyman (Ed), Chapman and Hall, London. 1991.

[10] ASHRAE. Energy efficient design of new building except new low-rise residential buildings. BSRIASHRAE proposed standards 90-2P-1993, alternative GA. American Society of Heating, Refrigerating, and Air Conditioning Engineers Inc., USA. 1993.

[11] Molla, M. Air pollutants and its probable transmutation in the ionosphere. *Renewable Energy*. 10 (2/3): 327-329. 1997.

[12] Bahadori, M. A passive cooling/heating system for hot arid regions. In: *Proceedings of the American Solar Energy Society Conference*. Cambridge. Massachusetts, p.364-367. 1988.

[13] Dieng, A., and Wang, R. Literature review on solar absorption technologies for ice making and air conditioning purposes and recent development in solar technology. *Renewable and Sustainable Energy Review*. 2001; 5 (4): 313-42.

[14] Lobo, C. Defining a sustainable building. In: Proceedings of the 23rd National Passive Conference. *American Solar Energy Society*. (ASES'98). Albuquerque: USA. 1998.

[15] Crisp, V., Cooper, I., and McKennan, G. Daylighting as a passive solar energy option: an assessment of its potential in non-domestic buildings. Report BR129-BRE. Garston. UK. 1988.

[16] Horning, M., and Skeffington, R. Critical loads: concept and applications. Institute of Terrestrial Ecology. HMSO Publishers Ltd. London: UK, p. 23-27. 1993.

[17] Humphrey's, M. Outdoor temperatures and comfort indoor. *Building Research and Practice*. 6 (2). 1978.

[18] Givoni, B. Man climate and architecture. Applied Science Publishers Ltd, p.289-306. London: UK. 1976.

[19] Koenigsberger, O., Ingersoll, T., Mayhew, A., and Szokolay, S. Manual of tropical housing and building. Part 1: Climate design. Longmas, p.119-130. London: UK. 1973.

[20] Boulet, T. Controlling air movement: a manual for architects and builders. McGraw-Hill, p.85-138, New York: USA. 1987.

[21] Szokolay, S. Design and research issues: passive control in the tropic. *Proceedings First World Renewable Energy Congress,* p.2337-2344, Reading: UK. 1990.

[22] Borda-Daiz, N., Mosconi, P., and Vazquez, J. Passive cooling strategies for a building prototype design in a warm-humid tropical climate. *Solar and Wind Technology*. 6, p.389-400. 1989.

[23] Givoni, B. Laboratory study of the effect of window sizes and location on indoor air motion. *Architectural Science Review*. 8: 42-46. 1965.

[24] Fanger, P. Thermal comfort: analysis and applications in environmental engineering. Danish Technical Press. 1970.

[25] Fordham, M. Natural ventilation. *Renewable Energy*. 19: 17-37. 2000.

[26] Awbi, H. Ventilation of buildings. Spon Publisher. London: UK. p. 9-13. 1991.

[27] Givoni B. Man climate and architecture. *Applied Science Publisher Ltd,* p.289-306. London: UK. 1976.

[28] BS 5454. Storage and exhibition archive documents. British Standard Institute. London. 1989.

[29] Lazzarin, R. D'Ascanio, A., and Gaspaella, A. Utilisation of a green roof in reducing the cooling load of a new industrial building. In: *Proceedings of the 1st International Conference on Sustainable Energy Technologies (SET),* p. 32-37, Porto: Portugal. 12-14 June 2002.

[30] David, E. Sustainable energy: choices, problems and opportunities. *The Royal Society of Chemistry*. 2003; 19: 19-47.
[31] Zuatori, A. An overview on the national strategy for improving the efficiency of energy use. *Jordanian Energy Abstracts*. 2005; 9 (1): 31-32.
[32] Anne, G., and Michael, S. Building and land management. 5th edition. Oxford: UK. 2005.
[33] Randal, G., and Goyal, R. Greenhouse technology. New Delhi: Narosa Publishing House. 1998.
[34] Yadav, I., Chauadhari, M. Progressive floriculture. Bangalore: The house of Sarpan, p.1-5, 1997.
[35] EIBI (Energy in Building and Industry). Constructive thoughts on efficiency, building regulations, inside committee limited. Inside Energy: Magazine for Energy Professional. UK: KOPASS, p.13-14. 1999.
[36] Jermey, L. The positive solution: Solar Century. *Energy and Environmental Management,* 2002. p.4-5.
[37] WEC Commission, editor. Energy for tomorrow's world. London: UK. St. Martin's Press, 1993.
[38] EUREC Agency, editor. The future for renewable energy. London: James and James, 1996.
[39] Plaz, W. Renewable energy in Europe: statistics and their problems, London: James and James, 1995.
[40] Hohmeyer, O. The social costs of electricity-renewables versus fossil and nuclear energy. *Int. J. Solar Energy.* 1992; 11: 231-50.
[41] Baker N., Fanchiotti, A., and Steemers, K. Day-lighting in architecture. European Commission Directorate-General XII for Science, Research and Development, 1993.
[42] Mazria, E. Direct gain systems: clerestories and sky-lights. The passive solar energy book, chap. 10. Emmaus, PA: Rodale Press, 1979.
[43] Duffie, J. A., and Beckman, W. A. Solar engineering of thermal processes. John Wiley and Sons, 1991.
[44] Watmuff, J. H., Charters, W. W. S., and Proctor D. Solar and wind induced external coefficients for solar collectors. *Internationale Heliotechnique.* 19977; 2:56.
[45] Thomas L. C. Heat transfer. Prentice Hall, 1992.
[46] Lee, K., Han, D., and Lim, H. Passive design principles and techniques for folk houses in Cheju Island and Ullung Island of Korea. *Energy and Buildings*. 1996; 23: 207-16.
[47] JETFAN. Greenhouse ventilation. The thermal engineering contributes to better plant health and lower heating costs. *Devon.* 1996.
[48] John, A.A., and James, S.D. The power of place: bringing together geographical and sociological imaginations. 1989.

In: Air Conditioning Systems
Editors: T. Hästesko, O. Kiljunen, pp. 195-223
ISBN: 978-1-60741-555-8

Chapter 3

DESIGN AND IMPLEMENTATION OF A SOLAR-POWERED AIR-CONDITIONING SYSTEM

***J. N. Lygouras*[*1] *and V. Kodogiannis*[†2]**
[1] Laboratory of Electronics, School of Electrical and Computer Engineering, Democritus University of Thrace, Vas. Sofias 12, 67100 Xanthi, Greece
[2] Centre for Systems Analysis, School of Computer Science, University of Westminster, London, HA1 3TP, UK

ABSTRACT

A result of the projected world energy shortage is an important effort in the last years towards the usage of solar energy for environmental control. This effort is receiving much attention in the engineering sciences literature. The design and implementation of an exclusively solar-powered air-conditioning system is described in this chapter. Its operation principle, the controller design and experimental results are presented. A variable structure fuzzy logic controller for this system and its advantages are investigated. Two DC motors are used to drive the generator pump and the feed pump of the solar air-conditioner. The first affects the temperature in the generator while the second the pressure in the power loop. Two different control schemes for the DC motors rotational speed adjustment are implemented and tested for the system considered initially as Single-Input/Single-Output (SISO): the first, is a pure fuzzy controller, its output being the control signal for the DC motor driver. The second scheme is a two-level controller. The lower level is a conventional PID controller, and the higher level is a fuzzy controller acting over the parameters of the low level controller. Step response of the two control loops are presented as experimental results. Next, the design and implementation of a Two-Input/Two-Output (TITO) variable structure fuzzy logic controller is considered. The difficulty of Multi-Input/Multi-Output (MIMO) systems control is how to overcome the coupling effects among each degree of freedom. According to the characteristics of the system's dynamics coupling, an appropriate coupling fuzzy controller (CFC) is incorporated into a traditional fuzzy controller (TFC) to compensate for the dynamic coupling among each degree of freedom. This control

* Tel. +30 25410 79578; fax: +30 25410 79570; E-mail address: ilygour@ee.duth.gr
† Email: kodogiv@wmin.ac.uk

strategy can not only simplify the implementation problem of fuzzy control, but can also improve control performance. This mixed fuzzy controller (MFC) can effectively improve the coupling effects of the systems, and this control strategy is easy to design and implement. Simulation results from the implemented system are presented.

Keywords: Solar air-conditioner, PID controller, Fuzzy logic controller, Mixed fuzzy controller, MIMO system, Microcontroller

1. Introduction

The use of solar power instead of electric power in air conditioning systems is gaining increasing interest during the last years. The trend towards reusable energy sources is a significant reason for this interest and especially for hybrid (solar-electric power) systems. Energy consumption in commercial and residential buildings is steadily increasing every year and represents approximately 40% of Europe's energy budget. The final energy consumption for 2002 in the building sector amounted to 435 Mtoe or 40.3% of the total Europe's final energy use [1]. The higher living and working standards, the adverse outdoor conditions in urban environments and reduced prices of air-conditioning units, have caused a significant increase in demand for air conditioning in buildings, even where there was hardly any before. The number of installed air-conditioning systems in Europe with cooling capacity over 12 kW has increased by a factor of 5 in the last 20 years [2]. Total air-conditioned floor space has grown from 30 million m^2 in 1980 to over 150 million m^2 in 2000. Annual energy use of room air conditioners was 6 TJ in 1990, 40 TJ in 1996 and is estimated to reach 160 TJ in 2010. As a result of the projected world energy shortage, the use of solar energy for environmental control is receiving much attention in the engineering sciences literature. With suitable technology, solar cooling can help alleviate the problem. The fact that peak cooling demand in summer is associated with high solar radiation offers an excellent opportunity to exploit solar thermal technologies that can match heat-driven cooling technologies. Of particular interest are urban areas where adverse outdoor conditions, as a result of higher outdoor pollution and the urban heat island effect, encourage the use of mechanical air-conditioning with a direct impact on peak electrical energy use [3]. Commercial application of solar energy for air conditioning purposes is relatively new. Lamp and Ziegler [4] give an overview of the European research on solar-assisted air conditioning up to 1996. Tsoutsos *et al.* [5] present a study of the economic feasibility of solar cooling technologies.

Different heat-driven cooling technologies are available on the market, particularly for systems of above 40 kW, which can be used in combination with solar thermal collectors. The main obstacles for large-scale application, beside the high first cost, are the lack of practical experience and acquaintance among architects, builders and planners with the design, control and operation of these systems. For smaller scale systems, there is no market available technology. Therefore, the development of low power cooling and air conditioning systems is of particular interest [6].

However, operating this type of system presents certain particular issues that must be addressed by the control strategy. Firstly, the primary energy source (the sun) cannot be manipulated, as in the case of any other conventional thermal process. Secondly, great disturbances exist in the process, mainly due to changes in environmental conditions. There are also dead-times related to fluid transportation which are variable depending on the operating conditions. Finally, the cooling demand is variable since it depends on the occupancy rate and the kind of activity that is being carried out in the cooled space. Several control strategies have been tested at solar power plants to address these problems [7, 8, 9]. Model Predictive Controllers have also received a lot of attention in the last few decades, both within the research control community and in industry [10]. The basic idea is to calculate a sequence of future control signals in such a way that it minimizes a multistage cost function defined over a control horizon.

In the recent years, the development of sophisticated control systems which are often including standard processors: microprocessors (μP), microcontrollers (μC), Digital Signal Processors (DSP) [11-13] and specific hardware components: Application Specific Signal Processors (ASSP), Field Programmable Gate Arrays (FPGA), Application Specific Integrated Circuits (ASIC) [14-15], has gained increasing interest. A new trend is to use configurable logic circuits (FPGA, Configurable Programmable Logic Devices-CPLD, etc.) with processor units on the same board. This allows rapid and efficient adaptation of the used board to new applications [16]. An interesting area of application of the above type of controllers is the implementation of DC motor controller, due to the wide range of applications of this type of motor [17-19]. With microcontrollers being available in different configurations, low cost, small die sizes, large on-chip memory, power management features and high clock rates, it is reasonable to apply fuzzy logic in an inexpensive microcontroller to implement the fuzzy control loop [20]. Here, the implementation of the fuzzy logic controller for a solar-powered air conditioning system is presented, based on the use of inexpensive microcontrollers to carry out all the appropriate control operations is presented.

This chapter is organized as follows: In section 2 the description of the solar air conditioning system is presented. In section 3 some general aspects about the fuzzy logic control of single-input/single-output and multi-input/multi-output systems are given. In section 4 the combination of the fuzzy logic controller with conventional design techniques is outlined while in section 5 the two-input/two-output air conditioning system controller is described. In section 6 some ideas about the knowledge-based design are outlined, experimental results are presented in section 7 and finally the conclusions are given in section 8.

2. Description of the Solar Air Conditioner

Several techniques have been proposed in the literature for the direct conversion of the solar energy to heating, cooling or both. The most common way for cooling is to exert mechanical pressure to the cooling liquid, usually using an electric compressor and to pass it through a valve to a lower pressure chamber. The evaporation of the liquid at this point is accompanied by heat absorption and thus the "*cooling*" effect is produced. In the case of solar air condition a different principle is adopted. The cooling cycle is based on the operation of a special ejector, as shown in figure 1. There are two loops in this configuration. The first loop (1-2-3-4-1) includes the *Ejector*, the *Condenser*, the *Feed Pump* and the *Generator*. The second loop (1-2-5-6-1) includes the *Ejector*, the *Condenser*, the *Throttle valve* and the *Evaporator*. The first loop is called *Power loop* while the second *refrigeration loop*. The purpose of the heating loop is the energy transfer in the form of heat from the falling solar radiation in the oil reservoir. The generator pump is circulating the oil in the loop which is heated passing through the solar panel. The achieved oil temperature is depended on the intention of the solar radiation as well as the time the oil remains inside the panels. The circulating oil transfers the heating energy to the generator. The oil temperature in the output of the solar panels must be about 125-135^{0}C. The pump must adjust the oil circulation in the solar-panel loop in order to achieve the desired temperature. If the solar radiation is low then the generator pump must lower its speed and the oil to remain for longer time in the solar panel. On the other hand, if the solar radiation is high, then the oil temperature is increasing rapidly. In this case the rotational speed of the pump must be high for the oil to remain inside the panels for less time. For the implementation of the above logic, the control system must observe the oil temperature in the solar panel outlet $T_{oilpout}$. The purpose of the above control system is to retain the oil temperature in the desired limits.

The oil is going out at lower temperature T_{oilpin} from the generator and is driven again in the solar panels. The transfer of energy in the generator unit increases the temperature of the cooling liquid and it is converted to vapour. This in turn is followed by increment of the pressure P_G of the vapour. Thus, the rotational speed of the generator pump is directly related to the heat energy transfer from the solar panels to the refrigeration loop and the rotational speed of the feed pump affects the pressure of vapour in the power loop and consequently the quantity of heat absorption from the environment. Next table 1 indicates the upper and lower limits for some critical parameters of the solar air condition system.

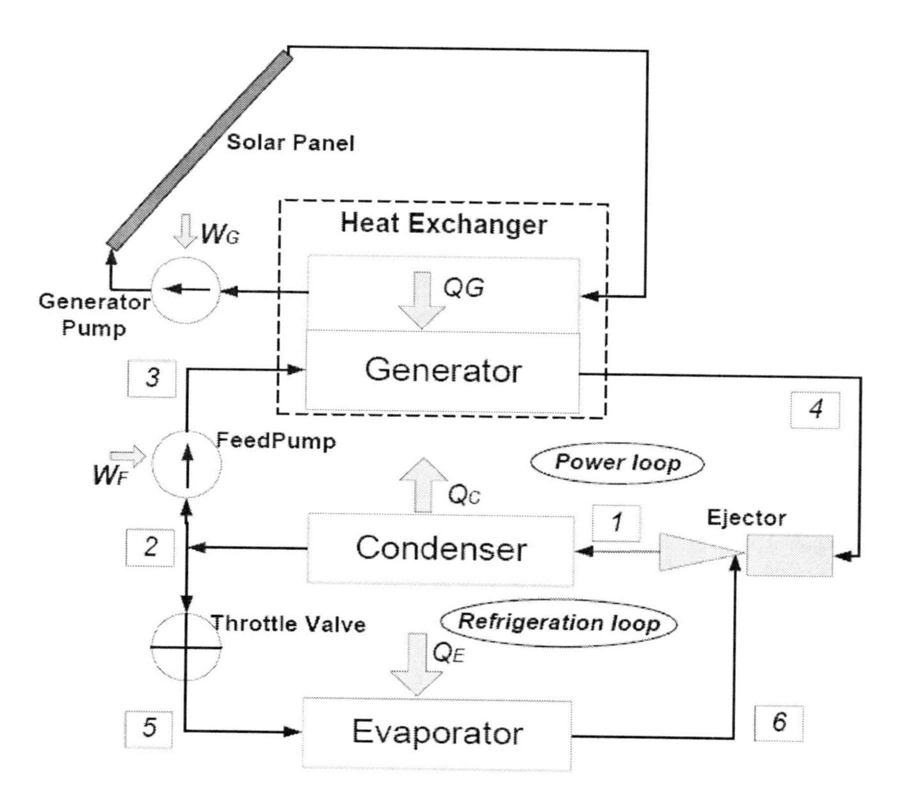

Figure 1. Block diagram of the solar air-conditioner.

Table 1. Limit values for some critical parameters of the solar air conditioning system

	Pressure (P_G)	Temperature (T_G)
Lower operation limit	10.5bar	102^0C
Normal operation point	14.5bar	116^0C
Upper operation limit	16.5bar	125^0C

The two parameters T_G and P_G are affected by the generator and feed pump speed respectively and they are playing an important role in the solar air-conditioning system operation.

The heating energy Q_G derived from the solar panels is given to the power loop, in the generator, through the *Heat Exchanger* and by the aid of the *Generator Pump*. At this point the evaporation of the cooling liquid is achieved through heating, when it is passing from point 3 to 4. In point 4 there is vapour of low speed and high pressure. This high pressure vapour is passing through the small orifice of the ejector to the condenser and consequently it results in high speed. In the condenser the vapour is converted again to liquid. This phase change is accompanied from heat dissipation Q_C which is rejected to the environment.

The cooling liquid coming out from the condenser is passing through the pump, thus closing the power cycle. A part of the cooling liquid, passing through the throttle valve to the point 5, is fed at low pressure to the evaporator. Here, the evaporation of the cooling liquid in point 6 is achieved due to the operation of the ejector. When the high pressure vapour coming from point 4 is injected in the ejector the resultant low-pressure in the orifice coming from point 6, is forcing the conversion to vapour phase of the liquid in the evaporator. Thus, vapour coming from the evaporator is mixed with vapour coming from the generator and is converted to liquid in the condenser. The evaporation is accompanied from heat absorption Q_E from the environment, thus producing the cooling effect in the evaporator. In the air-conditioning system the evaporator section is the indoor unit, while the condenser section corresponds to the outdoor unit.

In the application described in this chapter, a DC motor is used to drive the generator pump in the heating cycle as it is described bellow and a second DC motor is used to drive the feed pump in the cooling cycle of a solar-powered air-conditioner. In the first cycle (*Power loop*), the parameter which affects the rotational speed of the DC motor is the temperature of the heated liquid (measured in the generator). It must be retained constant, as possible. It is supposed that the solar panels are capable of providing the appropriate energy Q_G to do so. In the second cycle (*refrigeration loop*), the parameter which affects the rotational speed of the DC motor is the pressure of the cooling liquid (measured in the generator's output). It must be also retained constant, as possible. Since the two control loops are identical just one of them is described in the following.

For simplicity reasons the following assumption is initially adopted: The rotational speed of the generator pump in the above described system is considered to affect the quantity of energy supplied to the power loop and the rotational speed of the feed pump to affect the quantity of heat absorption from the environment. This assumption is true if the two processes are considered as completely independed and thus the Two-Input/Two-Output (TITO) system is considered uncoupled. In practice, the situation is not exactly so simple, since there is a considerable dependence between the working parameters of the system. In the next, this dependence is considered and a more complex TITO model is studied. In this system the two inputs are the control voltages applied to the two motors, while the two outputs are the rotational speeds of the two motors related directly with the temperature of the heated liquid in the first loop and the pressure of the liquid in the second loop. In figure 2, the block diagram for each DC motor controller is shown. A tacho-generator is used in the feedback for the actual speed reading. There is a linear relation between the voltage of the tacho-generator output waveform and the rotational speed of the motor shaft. Pulse Width Modulation technique (PWM) is used to drive the DC motor at the desired speed. The microcontroller computes the duration *d* of the PWM waveform accordingly, and the appropriate signals are applied to the driving circuit (DC chopper).

A fuzzy logic controller (FLC), is based on fuzzy logic and constitutes a way of converting a linguistic control strategy into an automatic one, by generating a rule base which controls the behavior of the system. Several controllers of this type have been reported, with direct usage in industrial applications and robotics [17-22].

In other relevant works as for example in [23] in order to control a MIMO system, a mixed fuzzy controller is developed, to improve performance and to solve the problem of computational burden and dynamic uncertainty associated with MIMO systems. In this work, a traditional fuzzy controller (TFC) is first designed from the viewpoint of a single-input/single-output (SISO) system for controlling each degree of freedom of the MIMO system. Then, an appropriate coupling fuzzy controller is also designed according to the characteristics of the system's dynamics coupling and incorporated into a TFC, to compensate for coupling effects between the degrees of freedom. In [24] a self-learning fuzzy logic system is proposed for control of unknown multiple-input multiple-output plants. To verify the applicability of the proposed controller a two link robotic manipulator with a complex dynamic model is studied, while in [25] this method is applied to a dynamic absorber with two-level mass-spring damper structure.

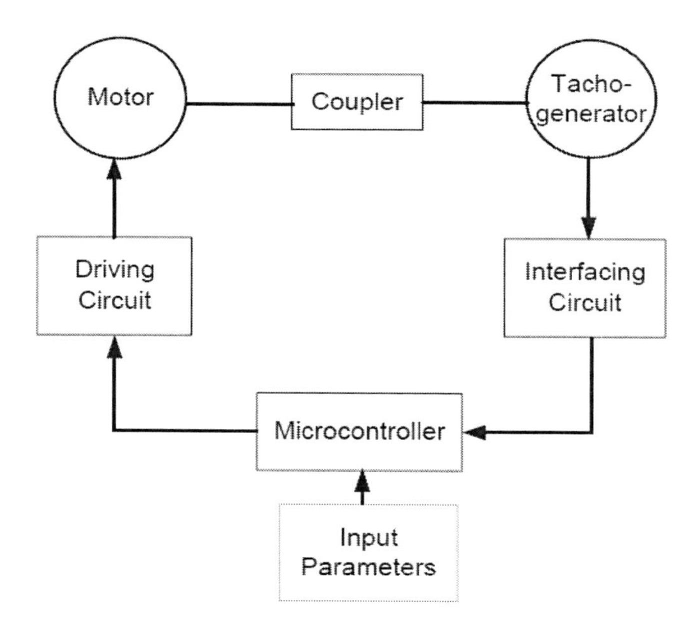

Figure 2. Block diagram of the configuration for DC motor rotational speed control.

In [26] analytical structures of TITO Mamdani fuzzy PI/PD controllers are investigated with respect to conventional PI/PD control and variable gain control and the characteristics of the fuzzy controller are analysed in relation to those of the corresponding TITO linear PI/PD control. The application of a self-organizing fuzzy proportional-integral-derivative (SOF-PID) controller to a MIMO nonlinear revolute-joint robot arm is studied in [27], where the SOF controller is a learning supervisory controller, making small changes to the values of the PID gains while the system is in operation. The software (S/W), as well as the hardware (H/W) implementation of a TITO fuzzy logic controller for each DC motor rotational speed control of a solar air conditioner is presented in [28].

3. THE STRUCTURE AND THE DESIGN OF FUZZY LOGIC CONTROLLER

Initially, two control schemes are presented and discussed considering the two loops of the system as independed: the first makes use of a pure fuzzy controller and the second of a two level controller, which uses the fuzzy controller as a supervisor in the operation of the system. A fuzzy matrix for the two control schemes is also defined. The structure of the fuzzy logic controller consists of the following three major components, as shown in figure 3:

i. the fuzzifier; this component carries out the following functions: a) measurement of the input b) definition of the fuzzy sets that will be applied c) transformation of the input values into membership degrees,
ii. the fuzzy control; which provides the system with the necessary decision-making logic based on the rule base that determines the control policy and
iii. the defuzzifier; which combines the actions that have been decided and produces a single non-fuzzy output that is the control signal of the system.

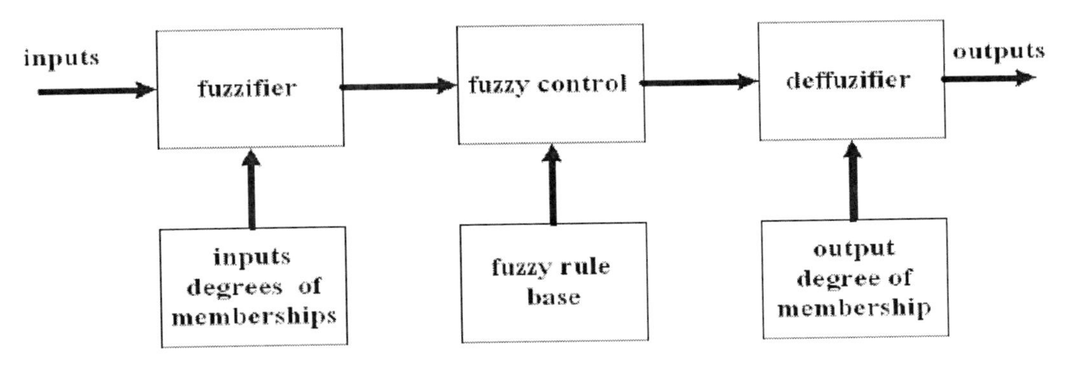

Figure 3. The structure of the fuzzy logic controller.

3.1. The Fuzzification Process

The input variables for the fuzzy controller are the *error* defined as the difference between the desired value of the rotational speed ω_{ref} and the actual value ω_{act} and its derivative. The determination of the control variables for the microcontroller for the above inputs is as follows:

The error:

$$Error(k) = e(k) = \omega_{ref}(k) - \omega_{act}(k) \quad (1)$$

The derivative of the error:

$$\Delta[Error(k)] = \Delta e(k) = e(k) - e(k-1) \quad (2)$$

The following seven fuzzy sets: positive large (PL), positive medium (PM), positive small (PS), zero (ZE), negative small (NS), negative medium (NM), negative large (NL), represented as triangular functions of the membership degrees, are defined for each of the input variables. The number of the fuzzy sets which will be used must be decided taking into account the desired sensitivity of the controller and the unavoidable increase of the rule base. Actually, the production of the fuzzy sets is achieved by defining only the zero fuzzy set. The rest of them can easily be produced by appropriately shifting this set:

$$\mu_a(x) = \mu_{ZE}(x+q), \quad (3)$$

$$\mu_b(x) = \mu_{ZE}(x-q),$$

where a and b can be any negative or positive defined sets, respectively, x is the input variable and q is the shifting parameter.

As it results from the previous paragraph the size of the generated rule base is 7x7=49 rules. The general form of each rule is:

$$R_i: \text{IF } e_i \text{ is } A_i \text{ AND } de_i \text{ is } B_i \text{ THEN output is } C_i \quad (4)$$

where A, B and C can be any of the fuzzy sets defined. We also define thirteen fuzzy sets for the output variable (NL, NM, ...,ZE, ..., PM, PL), the reason being the need of a very sensitive control signal. We can consider this as another method to influence the refinement of the control action. The fuzzy control rules expressed in this form provide a way of simulating the human thinking and decision making. The new constructed fuzzy matrix shown in table 2:

Table 2. The fuzzy matrix of the controller

de \ e	NL	NM	NS	ZE	PS	PM	PL
PL	ZE	PS	PS	PM	PL	PL	PL
PM	NS	ZE	PS	PS	PM	PL	PL
PS	NS	NS	ZE	PS	PS	PM	PL
ZE	NM	NS	NS	ZE	PS	PS	PM
NS	NL	NM	NS	NS	ZE	PS	PS
NM	NL	NL	NM	NS	NS	ZE	PS
NL	NL	NL	NL	NM	NS	NS	ZE

The behavior of the system is also influenced by the percentage of the overlapping of the fuzzy sets, i.e. the limits of the defined functions. Every time that a new error, *e*, (and its derivative, *de*) is computed, the software produces a vector with seven elements, each one representing the membership degree μ of the input variables to the fuzzy sets. In order to improve the performance of the FLC, the rules and membership functions are adjusted using a Variable Structure Control (VSC) technique, as shown in figure 4. The VSC strategy consists in switching to a different control structure in each side of a given switching surface according to a predefined switching function. An interesting characteristic of the VSC is that under certain conditions the system responds with a sliding mode on the switching surface and in this mode the system is insensitive to parameter variations and disturbances. If the desired dynamics of the system are specified in terms of the switching surface, as shown in figure 4, the error trajectory approaches the switching surface gradually with the slopes $-m_f$ and $-m_s$ corresponding to fast and slow dynamics, respectively.

With reference to figure 4, the slopes of the error trajectory m_f and m_s can be adjusted according to the desired response time and sensitivity in the steady state.

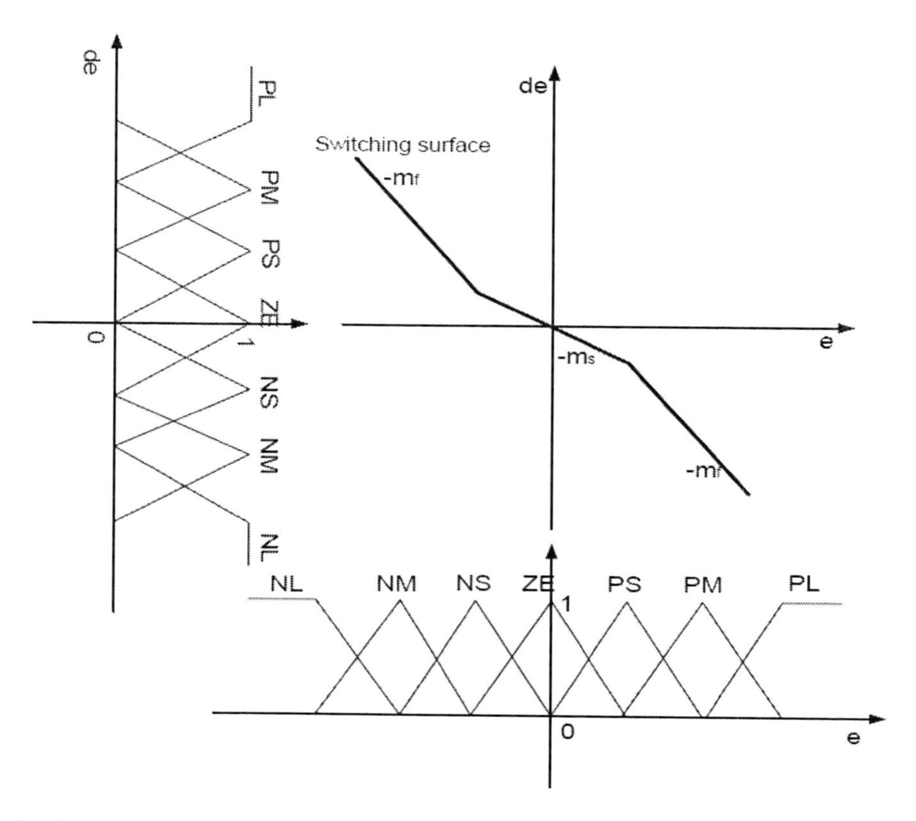

Figure 4. The defined fuzzy sets for membership functions adjustments, with slope $-m_s$ and $-m_f$, represented as triangular functions.

The design of the controller (i.e. the decision that the fuzzy sets will overlap) indicates that four rules of the form of eqn. (4) of the fuzzy matrix will fire, since every input variable will have two non-zero elements of its membership degree vector. The output of each of these active rules is calculated using Zadeh's definition concerning the logical operation AND between two fuzzy-logic variables [29]:

$$\text{action} = \min(\mu_A(e), \mu_B(de)) = \mu_c(\text{output}) \quad (5)$$

The above definition provides a way of interpreting the "weights" of each rule or, in other words, the degree to which each rule fires.

3.2. The Defuzzification Process

At the stage of the defuzzification, all the actions that have been activated are combined and converted into a single non-fuzzy output signal, which is the control signal of the system. An important factor here is the definition of the fuzzy sets for the control signal (the output of the controller). As it was mentioned in subsection 3.1, the output levels are thirteen and their representation will be also triangular, their position depending on the non-linearities existing to the system. To achieve this, we develop the control curve of the system representing the I/O relation of the system and, based on that information, we define the output degree of the membership function with the aim to minimize the effect of the non-linearity.

The method that was adopted here for the estimation of the control signal is the centroid method. The method results in an output action that is associated with the center of gravity of the active rule outputs. For a vertically symmetric degree of the membership function, as in our case, the centroid of a fuzzy set always corresponds to a single output value, regardless of the value of the input degree of membership. Figure 5 shows the output fuzzy set consisting of the 0(=ZE), +1(=PS), +2(=PM) and +3(=PL) fuzzy sets for different degrees of membership of the output variable.

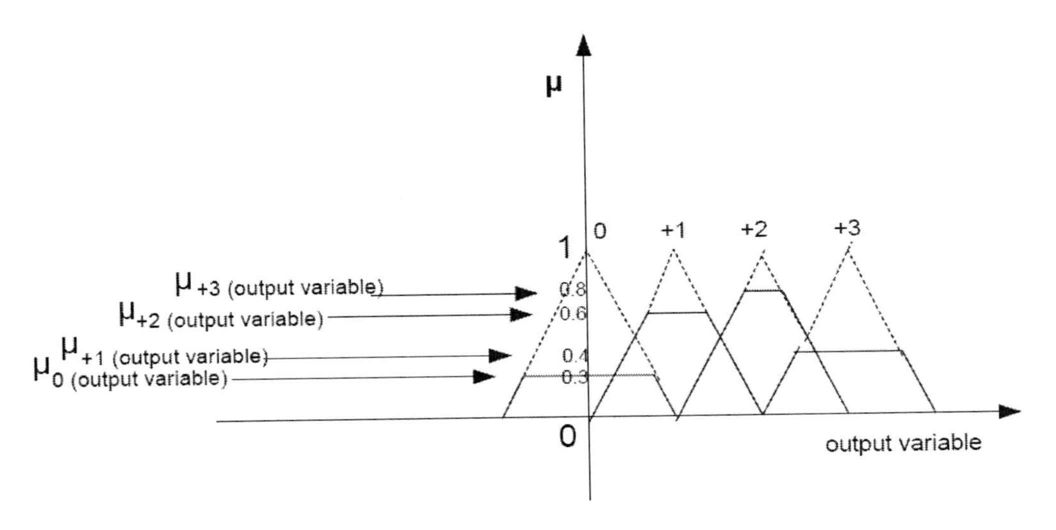

Figure 5. The output fuzzy set defined by the centroid method.

The control signal is computed using the following relation taking into account all the fuzzy rules that fire, as well as the weights of each rule:

$$output\_signal = R = \frac{\sum_{j=1}^{n} action_j x centroid_j}{\sum_{j=1}^{n} action_i} xK \qquad (6)$$

The variable K is a design parameter and is used to scale the output signal and to improve the performance of the controller. The above control variables, in proportion to the values they are taking each time the loop is implemented, will determine the one and only output of the system, which is the Duty Cycle of the PWM.

4. The Combination of the Fuzzy Logic Controller with Conventional Design Techniques

The second control scheme that was implemented in this work is described as a two level controller. The lower level consists of a conventional PID controller, whereas the higher level consists of the fuzzy controller that acts over the parameters of the lower level controller the purpose being to improve its performance. Several techniques are well known for tuning the parameters of a PID controller. The Ziegler-Nichols is a quite simple and easy to execute technique. Firstly, we ignore the integral and the derivative terms of the controller and, secondly, we increase the analog gain until the system constantly oscillates. The period of the oscillations is T and the analog gain is K_G. Then the parameters of the PID controller are calculated as follows:

analog gain, K_p=0.6 K_G
integral time constant, T_i=0.5 T, (7)
derivative time constant, T_d=0.25 T

The fuzzy matrix used is similar to that shown in table 2, but the fuzzification process converts the input variables to fuzzy singletons, so that each input variable is assigned to only one fuzzy set. Each fuzzy output, sets a single output value, defined by the matrix, corresponding to the control parameter. The following relations describe the PID controller equation, which is affected by the fuzzy controller:

P_member= (K_p + *FLC_output* x *u*) x (*error[1]-error [2]*)
I_member=(K_i + *FLC_outpur* x *v*) x *error[1]*
D_member= (K_d + *FLC_output* x *w*) x(*error[1]-2* x *error[2]* + *error[3]*) (8)
PID_output= *PID_output*+(*P_member*+*I_member*+ *D_member*)

where K_p, K_i-=K_p (T_s/T_i) and K_d=K_p (T_d/T_s) are the analog, integral and derivative gains of the PID controller, respectively, and T_s is the sampling period of the system. The variable FLC_output is an element of the fuzzy matrix (the control parameter) and its choice depends on the fuzzy sets assigned to the error and its derivative. The parameters u, v and w are adjusting parameters used for the optimal tuning of the controller.

The fuzzy controller affects the PID controller. In order to improve the response of the system independently of the input signal, the following points must be taken into account:

- Before the system reaches the set value, the proportional term (P_member) has to be increased to reduce the rise time, and then it should immediately be reduced to avoid oscillations.
- The integral term (I_member) has to be increased during the rise time for the same reason as before, but it should be reduced later to avoid large overshoot (integral wind-up phenomenon).
- The derivative term (D_member) has to be increased. Derivative action compensates for a changing measurement. Thus, derivative takes action to inhibit more rapid

changes of the measurement than proportional action. When a load or set-point change occurs, the derivative action causes the controller gain to move the "wrong" way when the measurement gets near the set-point. Therefore, an increase of the derivative parameter will decrease overshoot, and damp out any transient oscillations. The larger the derivative term, the more rapidly the controller responds to changes in the process's output (improvement of system's steady state).

The design of the system has been followed by the computer simulation using the MATLAB/SIMULINK software to simulate the closed-loop control system. The simulation is performed in the time domain and the sampling period of A/D converter is set at $T=4ms$. The hardware implementation of the system (for both the control schemes) has been experimentally verified in the solar air conditioning system. The control algorithm described has been used to drive the main two DC motors (the generator pump and the feed pump) of the solar air-conditioner. The control algorithms are implemented using a 16-bit microcontroller including an 8-channels 12-bit A/D converter and in order to avoid the use of an external D/A converter, the PWM technique is used (since the microcontroller provides such capability) to drive directly the power circuits. The block diagrams for the two control algorithms implementation are shown in figures 6 and 7, respectively: The command signal for the first DC motor driving the *Feed Pump* is the pressure measured in the Generator's output while the command signal for the DC motor driving the *Generator Pump*, is the oil temperature measured in the solar panel output. In order to achieve the desired correspondence between the input and the output variables in the two controlled loops (pressure-rotational speed and temperature-rotational speed) a look up table is possible to be used which is stored in external memory in the microcontroller. The look-up table operation can be executed after each A/D conversion. In figure 8 the flow chart of the implemented in software fuzzy controller is shown.

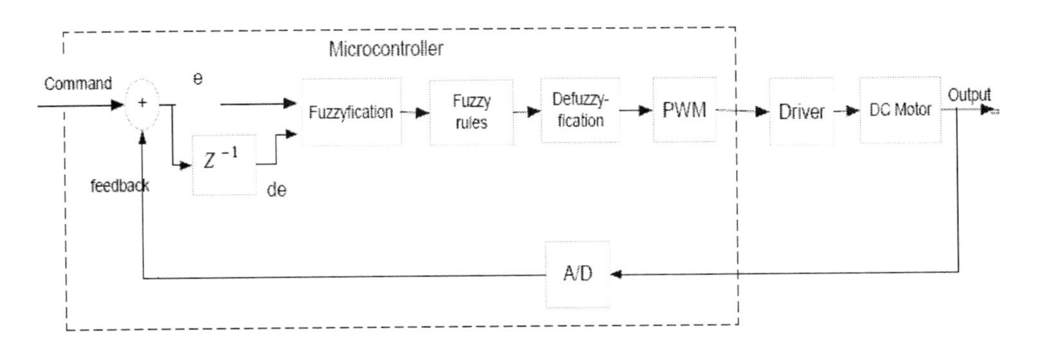

Figure 6. Block diagram for the system under Fuzzy Logic Control.

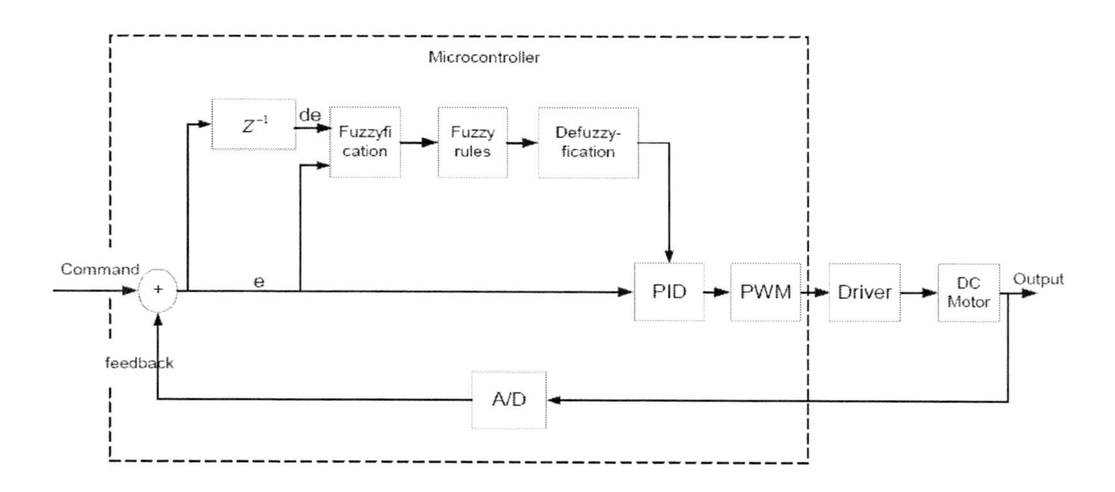

Figure 7. Block diagram for the system under Fuzzy Logic / PID control.

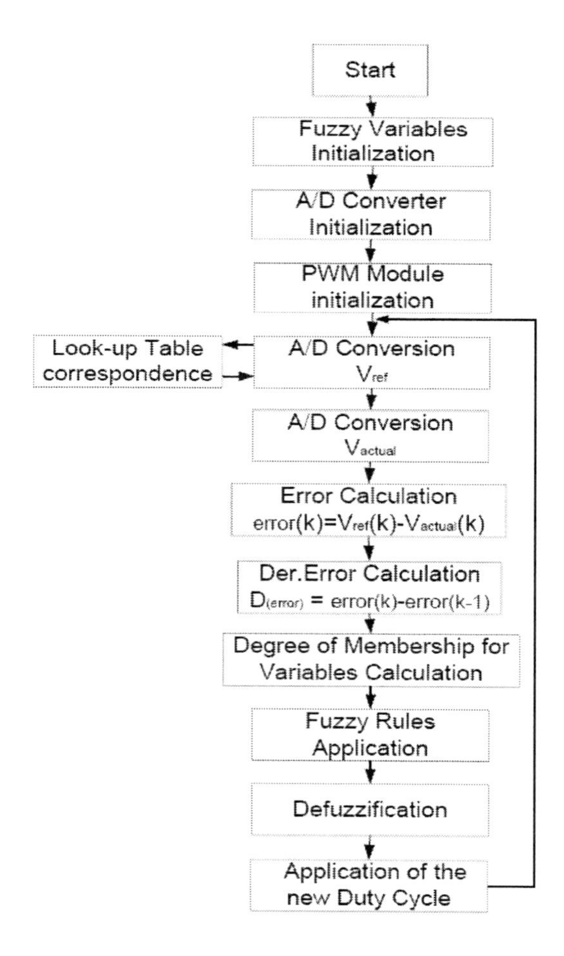

Figure 8. Flow chart of the implemented in software SISO Fuzzy controller.

5. Fuzzy Logic Controller for MIMO Systems

Initially, and for simplicity reasons the two loops in the solar air conditioning system were considered as completely independent and thus two identical SISO control systems have been designed, based on this assumption. In practice, this is not true since the change in each input (rotational speed of the generator and feed pump) affects the measured parameters in both loops. Thus, there is a considerable dependence/coupling between the working parameters of the system. In the following this dependence is considered and a more complex TITO model is studied. The above system is a two input (generator pump and feed pump speed control voltage) /two output (temperature T_G and pressure P_G) closed loop control system. The fuzzy inference logic applied the Max-Min product composition to operate the fuzzy control rules. Finally, this work employs the area method to defuzzify the output variables to attain the accurate objectives for controlling this system.

In order to demonstrate this idea of mixed fuzzy controller, a practical application case is proposed in the following sections, which employs numerical simulation with experimentation to verify this idea. The general structure of a MIMO fuzzy logic controller (FLC) for a closed loop MIMO system is shown in figure 9. The controlled system is supposed to have Q inputs and J outputs. The FLC calculates the Q inputs to the plant. The closed loop system has totally J inputs and J outputs, thus J errors are produced. The errors $e_j(t)$ and their changes $ce_j(t)$, $j=1, 2, \ldots, J$, are treated as linguistic FLC input variables. Although fuzzy control theory has been successfully employed in many control engineering fields its control strategies were mostly used to control SISO systems or the fuzzy controller was designed based on the SISO system point of view, in spite of the dynamic coupling of a MIMO control system. The difference between MIMO systems control and SISO systems control is how to estimate and compensate the interaction among each degree of freedom.

The matrix of fuzzy control rules for SISO systems has a dimension of F^2, where F is the number of used fuzzy sets. The direct fuzzy controller design for MIMO systems arises the following problems (direct design means one controller for the overall system):

- The quantity of fuzzy rules increases exponentially as the system's inputs increase.
- The cross coupling exists between inputs and outputs of a physical system. In ideal case, it *can* be assumed that each input manipulates only one output. However, it is a special case to find in practice a decoupled system which requires an equal number of inputs and outputs.

Generally, the maximum possible (upper bound) number of rules N_R is given by:

$$N_R = r \times F^{(m,n)} \tag{9}$$

where:

r : the dimension of input vector u,
m : the dimension of output vector y,
F : the number of fuzzy sets,
n : the number of input variables in the fuzzy rule (e.g. error, change of error, e.t.c)

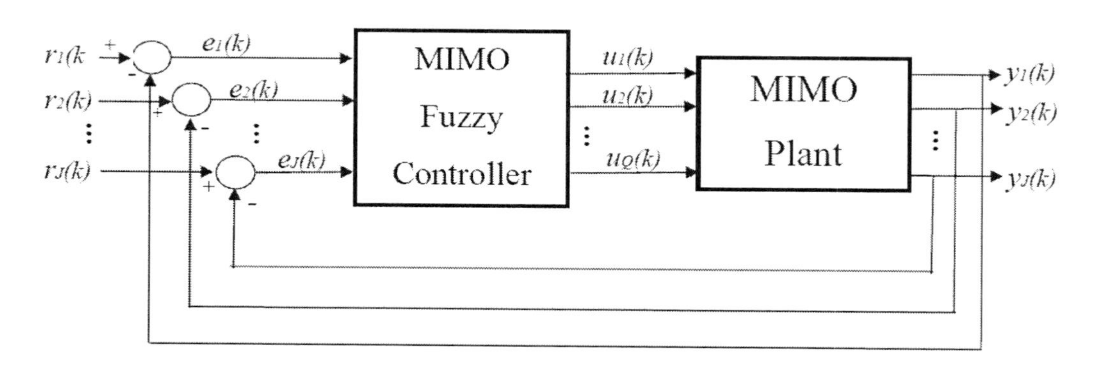

Figure 9. Structure of a MIMO fuzzy logic controller for a closed loop MIMO system.

From equation (9) N_R has a linear relation for increasing r while m is constant. But it has an exponential relation for increasing m while r is constant. Therefore, if N_R is very large, this leads to speed down the inference mechanism of fuzzy controller and much storage memory is required. This is a significant problem especially in case where a microcontroller is used to implement the control functions, due to the limited internal memory space included in those devices. In case the internal memory is not enough, external memory mast be added. The above equation affirms that the direct synthesis of fuzzy controller for a multivariable system is not practical. Thus it is preferred to develop an alternative design methodology to overcome this problem.

A traditional fuzzy logic controller operating with system output error $e_i(t)$ and error change $ce_i(t)$ was chosen as the main controller to control each degree of freedom of the MIMO system. Let the vector $Y(t)$ represent the collection of J plant outputs, and let the vector $R(t)$ be the collection of J desired plant outputs. Also, let the collection of Q plant inputs be denoted by $U(t)$. The vector of errors is given by:

$$\boldsymbol{E}(\mathrm{k})=\begin{bmatrix} e_1(k) \\ e_2(k) \\ \cdot \\ \cdot \\ \cdot \\ e_J(k) \end{bmatrix}=\boldsymbol{R}(\mathrm{k})-\boldsymbol{Y}(\mathrm{k})=\begin{bmatrix} r_1(k)-y_1(k) \\ r_2(k)-y_2(k) \\ \cdot \\ \cdot \\ \cdot \\ r_J(k)-y_J(k) \end{bmatrix} \quad (10)$$

$$\boldsymbol{C_E}(\mathrm{k})=\begin{bmatrix} e_1(k)-e_1(k-1) \\ e_2(k)-e_2(k-1) \\ \cdot \\ \cdot \\ \cdot \\ e_J(k)-e_J(k-1) \end{bmatrix} \quad (11)$$

Several types of fuzzy control rules have been employed in the literature as for example the method of Takagi and Sugeno, by which the consequent of a control rule is a linear combination of the components of the crisp vector $\boldsymbol{H(t)}$. A control rule can be given by:

Rule i: IF (x_1 is $s_1(h_1(t))$ and x_2 is $s_2(h_2(t))$ and … and x_M is $s_M(h_M(t))$

$$\text{THEN } \sigma_k(t)=\alpha_{k,0}+\alpha_{k,1}h_1(t)+\alpha_{k,2}h_2(t)+\ldots+\alpha_{k,M}h_M(t) \quad (12)$$

The difficulty of employing traditional fuzzy control theory for MIMO systems control is how to overcome the coupling effect among each degree of freedom. Because the dynamic model of MIMO systems possesses the usual characteristics of nonlinear and complicated dynamics coupling, it is difficult to establish the accurate dynamic model and to decouple it for controller design and thus model-free intelligent control strategies are employed for designing the controller of MIMO systems. Hence the traditional model based SISO system control scheme is not easy to implement on complicated MIMO systems due to the large computation burden. Therefore, the model-free intelligent control strategy is gradually drawing attention. In addition, the control rules and controller computation are growing exponentially with respect to the number of considered variables. It is obvious that in MIMO systems an appropriate coupling fuzzy controller must be incorporated into a TFC to control the MIMO systems to compensate for the dynamics coupling among each degree of freedom. This mixed fuzzy controller (MFC) can effectively improve the coupling effects of the systems, and this control strategy is easy to design and implement.

The concept of adding an appropriate coupling controller to compensate for the coupling effect is proposed to enhance the control performance of the TITO control system of the solar air conditioning system. The new control approach is developed here by combining a traditional fuzzy controller and an appropriate coupling fuzzy controller. Its control block diagram is shown in figure 10. The design procedure of the mixed fuzzy control strategy is to design a TFC first, to control each degree of freedom of this MIMO system individually. This design has been described previously. Then, an appropriate coupling fuzzy controller is designed to compensate for the coupling effects of system dynamics among each degree of freedom. Thus, a good control performance is obtained and also the total number of the rules can be kept small. In this diagram, the two inputs of the closed loop system are the desired temperature T_G and the pressure P_G of the coolant in the generator. The two outputs of the controller are the speed of generator and feed pump. In order to simplify the controller design, the two inputs/two outputs system, is simplified considering it as two subsystems having two inputs/one output.

Generally, the membership function of a fuzzy logic controller is difficult to decide for an unknown control system. In the solar air conditioning system, triangle-type membership functions [20] have been used to convert these input and output variables ($e_i(t)$, $ce_i(t)$ and $\Delta u_i(t)$) into linguistic control variables (NL, NM, … , PM, PL), where the subscript $i = 1, 2, 3, \ldots, m$ is used for indicating the temperature or pressure error, the error change and the control output in each degree of freedom. The fuzzy control rules employed to control a MIMO solar air conditioning system are listed in table 3.

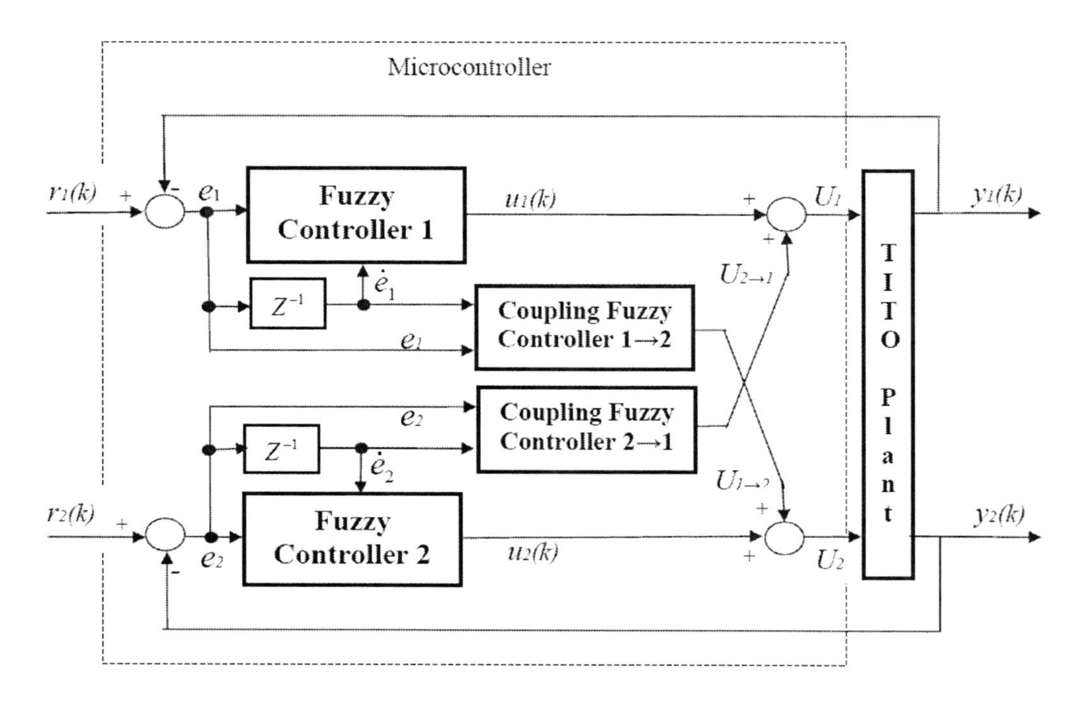

Figure 10. Block diagram of the TITO mixed fuzzy control strategy considered for the air-conditioner. In dashed lines the section implemented using a 12-bit microcontroller.

Table 3. Fuzzy control rules employed to control a TFC for the solar air conditioning system

<table>
<tr><th>ce_i \ e_i</th><th>NL</th><th>NM</th><th>NS</th><th>ZE</th><th>PS</th><th>PM</th><th>PL</th></tr>
<tr><td>PL</td><td>ZE</td><td>PS</td><td>PM</td><td colspan="2">PL</td><td colspan="2" rowspan="3">PL</td></tr>
<tr><td>PM</td><td>NS</td><td>ZE</td><td>PS</td><td>PM</td><td>PL</td></tr>
<tr><td>PS</td><td>NM</td><td>NS</td><td>ZE</td><td>PS</td><td>PM</td></tr>
<tr><td>ZE</td><td>NL</td><td>NM</td><td>NS</td><td>ZE</td><td>PS</td><td>PM</td><td>PL</td></tr>
<tr><td>NS</td><td colspan="2" rowspan="3">NL</td><td>NM</td><td>NS</td><td>ZE</td><td>PS</td><td>PM</td></tr>
<tr><td>NM</td><td>NL</td><td>NM</td><td>NS</td><td>ZE</td><td>PS</td></tr>
<tr><td>NL</td><td colspan="2">NL</td><td>NM</td><td>NS</td><td>ZE</td></tr>
</table>

By way of the above design process, the actual control input voltage of the actuator for this TFC can be written as:

$$u_i(k) = u_i(k - 1) + \Delta u_i(k) \tag{13}$$

where $\Delta u_i(k)$ indicates the voltage increment of the *ith* degree on the *k* step sampling interval, $u_i(k)$ and $u_i(k - 1)$ express the system control input voltage of the *ith* degree on the *k* step and $k-1$ step sampling intervals, respectively.

In a real MIMO system, the control output is influenced by more than one variable. According to the system characteristics, these system variables are obviously interactive.

$$Y_i = P_i(u_1, u_2, \ldots, u_m, Y_1, Y_2, \ldots, Y_n) \tag{14}$$

or

$$Y_l = P_l(u_l, Y_1, Y_2, \ldots, Y_n) \tag{15}$$

where $P_i(.)$ and $P_l(.)$ are complicated coupling functions that are difficult to define and derive.

According to the analysis of the dynamic equation (14) or (15), it is well known that u_i is the main effect and u_l is the secondary effect for the output Y_i. Similarly, for the output Y_l, u_l is the main effect and u_i is the secondary effect. The main effect on the system is controlled using a TFC. The secondary effect on the system is controlled by designing an appropriate coupling fuzzy controller. It is clear that the control structure of the fuzzy control system is very complicated when the input variable is multi-degree and the output variable is one degree, or more than one degree. These parameters of a fuzzy control system are not easily decided because the fuzzy control rules will be generated as a geometric series, and much computing time will be required such that a general CPU is not acceptable. Hence, the design procedure of this coupling fuzzy controller should be simplified to reduce the computing burden during implementation. The input variable of the coupling fuzzy controller is chosen mainly as two coupling effected factors such that the structure of the fuzzy rules can be built conventionally. However, both the input variable and the output variable generally exceed two control degrees for MIMO systems, such that the design of the coupling fuzzy controller becomes very complicated, and is difficult to implement. According to the coupling situation among each degree of freedom for MIMO systems which by way of dynamics coupling characteristics analysis, a few of the mainly affected coupling factors are first considered to design an appropriate coupling fuzzy controller to compensate for dynamic coupling effects and enhance control performances of the systems. For example, output responses of nearer two degrees should be heavily affected for a MIMO system which is chosen as an input variable of a coupling fuzzy controller, and neglected output responses of other degrees will be weakly affected. This design method for the coupling fuzzy controller will be able to effectively decrease the difficulties of the coupling fuzzy controller design.

The fuzzy inference and defuzzification methods are similar to those of the TFC. The output of the coupling fuzzy controller is chosen directly as the coupling control input voltage. The main reason is that there is a different coupling effect for each sampling interval and it does not have an accumulating feature. The coupling effect is incorporated into the TFC for each step to improve system performance and robustness. Therefore, the total control input voltage of this MFC is represented as:

$$U_i(k) = u_i(k) + U(k)_{l \to i}, \; i \neq l \tag{16}$$

where $u_i(k)$ expresses the system control input voltage of the *ith* degree of a TFC. $U(k)_{l \to i}$, represents the coupling effect control of the *lth* degree relative to the *ith* degree of the coupling fuzzy controller and since we are dealing with a TITO system, $i = \{1, 2\}$ and $l = \{1, 2\}$.

The fuzzy sets that have been chosen for each variable for the coupling fuzzy controller are three: NL (Negative Large), ZE (Zero) and PL (Positive Large). The membership functions for each fuzzy set have been determined based on the rules that have been previously described, i.e. the variation of the temperature will affect "more" the speed of feed pump, while the pressure will affect "more" the generator's pump. Therefore, the membership functions that have been set for the control of the generator's pump are slightly different from those that have been set for the control of the feed pump.

B. The Defuzzification Process

The defuzzification process is in general the same as it was described earlier. All the actions that have been activated are combined and converted into a single non-fuzzy output signal, which is the control signal of the system. The method that was adopted here for the estimation of the control signal is also the centroid method.

7. Experimental Results

The hardware implementation of the system has been experimentally verified in the solar air conditioning system. The control algorithms described above have been used to drive the main two DC motors (the generator pump and the feed pump) of the solar air-conditioner. A tacho-generator is used in the feedback of each motor for the actual rotational speed reading. There is a linear relation between the voltage of the tacho-generator output waveform and the rotational speed of the motor shaft. Pulse Width Modulation technique (PWM) is used to drive the DC motor at the desired speed. The microcontroller computes the duration d of the PWM waveform accordingly, and the appropriate signals are applied to the driving circuit (DC chopper).

The control algorithms are implemented using a 16-bit microcontroller including an 8-channels 12-bit A/D converter and in order to avoid the use of an external D/A converter, the PWM technique is used (since the microcontroller provides such capability) to drive directly the power circuits. The command signal for the first DC motor driving the *Feed Pump* is the pressure measured in the Generator's output while the command signal for the DC motor driving the *Generator Pump*, is the oil temperature measured in the solar panel output. Step function responses of the SISO system have been derived initially by simulation. The experimental results show minimization of the response time of the system when FLC is used to adjust the parameters of a PID controller (settling to steady state after 400 sampling periods) compared with simple fuzzy logic control (settling to steady state after 530 sampling periods), as well as lower overshoot/undershoot of the responses obtained. Furthermore the simulation results, carried out before the system implementation, using MATLAB/SIMULINK are in very good agreement with those obtained experimentally. The DC motors that have been used in the generator and the feed pump are two 24V/4A DC motors, on the steady state and with constant load conditions. The power circuit includes Insulated Gate Bipolar Transistors (IGBT) with a maximum collector's current I_C=30A. Also, for driving the IGBT, the IR2121 interface chip is used. Figures 11(a) and 11(b) show the experimental responses of the first and the second control scheme, respectively, the input being a step function. Comparison of the two control schemes shows the superiority of the two level controller, since 2.30% overshoot is presented in the first case while 1.00% overshoot is presented in the case of combined Fuzzy/PID control as well as shorter response time.

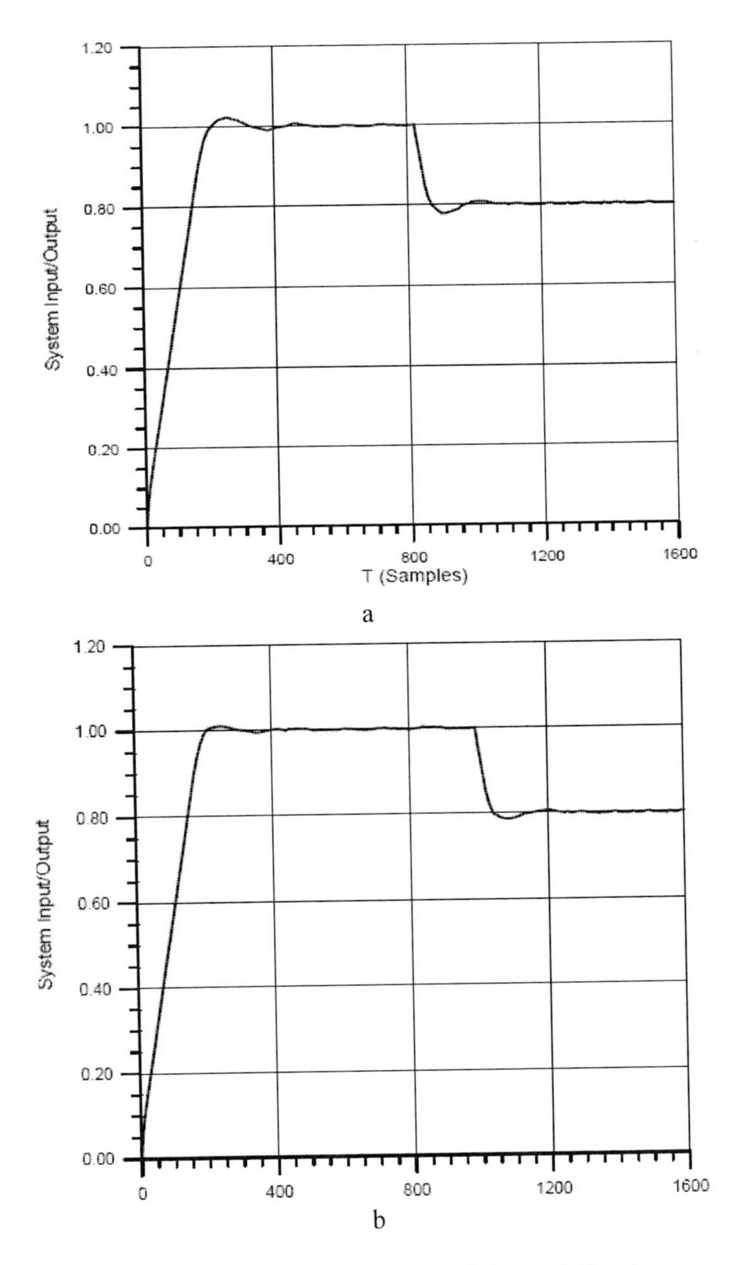

Figure 11. (a) Step response for the system under Fuzzy Logic Control, (b) Step response for the system under Fuzzy Logic / PID control.

In order to achieve the desired correspondence between the input and the output variables in the two controlled loops (pressure-rotational speed and temperature-rotational speed) a look up table is used which is stored in external memory in the microcontroller. The look-up table operation is executed after each A/D conversion.

In figure 12 the flow chart of the implemented in software MFC TITO fuzzy logic controller is shown. Step function responses of the MFC TITO Fuzzy controller system have been derived initially by simulation also. In figure 13, step responses of the system with SISO PI tuning and MFC TITO Fuzzy controller are shown. It is obvious that the system under MFC TITO controller behaves better compared with the SISO PI tuning especially at the transients. Finally, in figure 14 the cooling capacity vs the hot oil inlet temperature of the constructed solar-powered air conditioning system is shown.

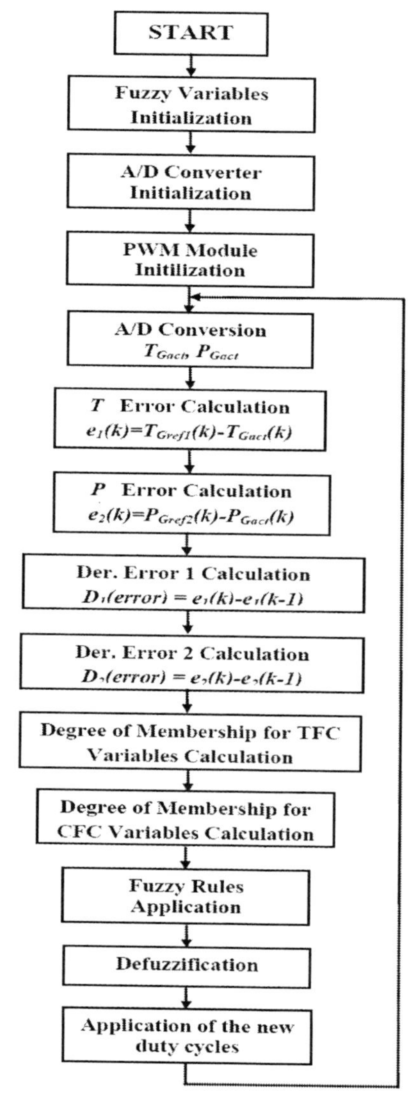

Figure 12. Flow chart of the implemented in software TITO Fuzzy controller.

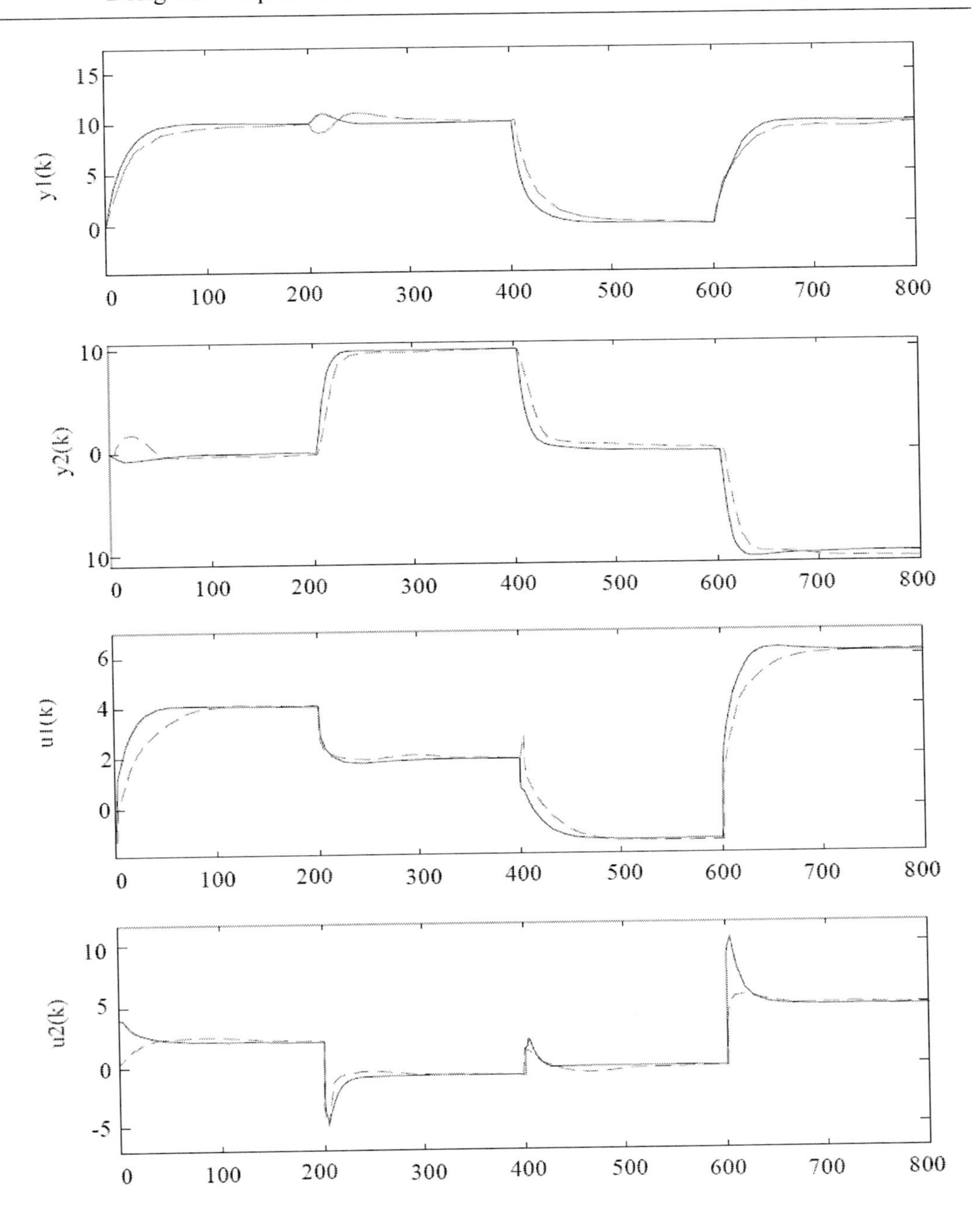

Figure 13. Step responses of the system with SISO PI tuning (---) and MFC TITO Fuzzy controller (___).

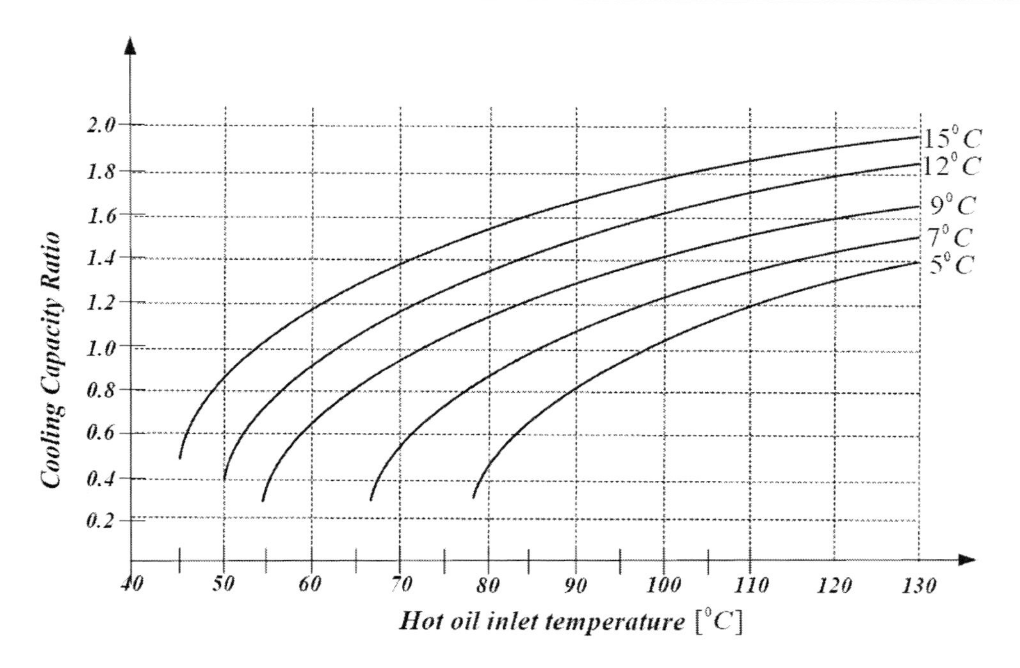

Figure 14. Cooling capacity vs the hot oil inlet temperature of the constructed solar-powered air conditioning system.

8. Conclusions

An intelligent fuzzy logic controller to adjust the rotational speed of two DC motors of a solar powered air conditioner has been described in this chapter combining the classical control schemes with fuzzy logic theory. Comparison of the two control schemes implemented for the SISO system presented here, shows that the two level controller behaves better in all situations. Fuzzy logic controller is implemented here using conventional control methods, and a general purpose low-cost microcontroller for the speed control of DC motors. A disadvantage of the two control schemes is the intensive computation required and the size of memory needed especially during the stages of fuzzification and defuzzification. On the other hand, by using the microcontroller, the controller can be implemented by using only a small amount of components and easily modified to be an adaptive fuzzy controller. Comparison of conventional and fuzzy logic control techniques shows that fuzzy logic can reduce the effects of nonlinearities in a DC motor and improve the performance of the controller. Finally, FLC algorithms can be easily transferred to different motor sizes and allows a strong tolerance for electric motor parameter oscillations.

The aim of this work was to investigate whether fuzzy control can be used to control a TITO process. The original idea is to design a fuzzy controller for the process without any modeling phase. In a second stage the interactions between each degree of freedom are considered based on information available from the real system data. This information is introduced in the system controller via the fuzzy control rules employed for the coupling fuzzy controller. A mixed fuzzy controller was then developed for controlling the TITO system. Following the TFC design for each level only to control a SISO system, a coupling fuzzy controller was introduced into the traditional fuzzy controller to improve control performances of the TITO solar air-conditioning system. The system has been experimentally demonstrated and its performance was verified. It can be observed from the experimental step input response that the control performance of this mixed fuzzy control strategy is obviously improved. Therefore, this mixed intelligent control strategy could be employed to control MIMO systems and improve the control performance.

Another contribution of this design is that in the control system, the fuzzy logic is implemented through software in a common, inexpensive, 16-bit microcontroller, which does not has special abilities for fuzzy control. All these properties makes FLC suitable for this application.

ACKNOWLEDGMENTS

The work described in this article was fully supported by the European Union in the context of the project: "Solar Air Conditioning System Using Very Low Cost Variable Plastic Ejector with Hybrid Potential for Different Markets (SACPEH).

REFERENCES

[1] European Union Energy & Transport in Figures (2004 edition). Part 2: Energy. European Commission, Directorate General for Energy and Transport, Brussels. p. 138.

[2] Adnot J., Giraud D., Colomines F., Riviere P., Becirspahic S., Benke G., et al., (2002). Central (commercial) air-conditioning systems in Europe. *Proceedings of second international conference on improving electricity efficiency in commercial buildings (IEECB)*, Nice, May 27–29, pp. 143–149.

[3] Papadopoulos A.M., Oxizidis S., Kyriakis N., (2003). Perspectives of solar cooling in view of the developments in the air-conditioning sector. *Renew. Sustain Energy Rev.* Vol. 7, pp. 419–38.

[4] Lamp P., Ziegler F., (1998). European research on solar-assisted air conditioning. *Int. J. Refrig.* Vol. 21, pp. 89–99.

[5] Tsoutsos T., Anagnostou J., Pritchard C., Karagiorgas M., Agoris D., (2003). Solar cooling technologies in Greece-An economic viability analysis. *Appl. Thermal. Eng.* Vol. 23, pp. 1427–39.

[6] Grossman G., (2002). Solar-powered systems for cooling, dehumidification and air-conditioning. *Solar Energy.* Vol. 72, pp. 53–62.

[7] Camacho E.F., Berenguel M., Rubio F.R., (1997). Advanced Control of Solar Plants, Springer, London,.

[8] Camacho E.F., Rubio F.R., Hughes F.M., (1992). Self-tuning control of a solar power plant with a distributed collector field, *IEEE Control Systems Magazine.* Vol. 12, pp.72–78.

[9] Silva R.N., Rato L.M., Lemos J.M., (2002). Observer based non-uniform sampling predictive controller for a solar plant, *15th IFAC World Congress*, Barcelona,.

[10] Nunez-Reyes A., Normey-Rico J.E., Bordons C., Camacho E.F., (2005). A Smith predictive based MPC in a solar air conditioning plant. *Journal of Process Control.* Vol. 15, pp. 1–10.

[11] Zhang W., Feng G., Liu Y.F. and Wu B., (2004). A Digital Power Factor Correction (PFC) Control Strategy Optimized for DSP. *IEEE Trans. Power Electronics.* Vol. 19, pp. 1474-1485.

[12] He D. and Nelms R.M., (2004). Fuzzy Logic Average Current-mode Control for DC/DC Converters Using an Inexpensive 8-bit Microcontroller. *IEEE Industry Applications Conf.*, Vol. 4, pp. 2615-2622.

[13] Navin Govind, (2005). Fuzzy Logic Control with the Intel 8XC196 Embedded Microcontroller. Intel Corporation, Internal Report.

[14] Peyravi H., Khoei A. and Hadidi K., (2002). Design of an Analog CMOS Fuzzy Logic Controller Chip. *Fuzzy Sets and Systems.* Vol. 132, pp. 245-260.

[15] Takahashi T. and Goetz J., (2004). Implementation of Complete AC Servo Control in a Low –cost FPGA and Subsequent ASSP Conversion. *19th Ann. IEEE Applied Power Electronics Conf. and Exposition*, Vol. 1, pp. 565-570.

[16] Saoud S. B., Gerstlauer A. and Gajski D., (2005). Codesign Methodology of Real-time Embedded Controllers for Electromechanical Systems. *American Journal of Applied Sciences.* Vol. 9, pp. 1331-1336.

[17] Tipsuwan Y. and Chow M. Y., (1999). Fuzzy Logic Microcontroller Implementation for DC Motor Speed Control. *IEEE IECON'99*, San Jose, CA, pp. 1271-1276.

[18] Mrozek B. and Mrozek Z., (2000). Modelling and Fuzzy Control of DC Drive. *14th European Simulation Multiconf. ESM 2000*, May 23-26, Ghent, pp. 186-190.

[19] Dimitriadis C. M., Lygouras J. N. and Tsalides Ph. G., (1994). An Intelligent Fuzzy Logic Controller for One-Dimension Motion Control. *IEEE Computer Society Press.* pp. 133-139.

[20] Lygouras J.N., Botsaris P.N., Vourvoulakis J. and Kodogiannis V.S., (2007). Fuzzy logic controller implementation for solar air-conditioning system, *Applied Energy. (Elsevier)*, Vol. 84 (12), pp. 1305-1318.

[21] Denai M. A. and Attia S. A., (2002). Fuzzy and Neural Control of an Induction Motor. *Int. J. Appl. Math. Comput. Sci.* Vol. 12, pp. 221-233.

[22] Muškinja N. and Tovornik B., (2000). Controlling of Real Inverted Pendulum by Fuzzy Logic. *Controlo'2000, Conference Proceedings,* Guimaraes, pp. 354-358.

[23] Ruey-Jing L, Bai-Fu L., (2005). Design of a mixed fuzzy controller for multiple-input multiple-output systems. *Mechatronics.* Vol. 15, pp. 1225-1252.

[24] Chunshien L, Priemer R., (1999). Fuzzy control of unknown multiple-input multiple-output plants. *Fuzzy Sets and Systems.* Vol. 104, pp. 245-267.

[25] Ruey-Jing L, Shiuh-Jer H., (2001). A mixed fuzzy controller for MIMO systems. *Fuzzy Sets and Systems.* Vol. 120, pp. 73-93.

[26] Ying H., (2000). TITO Mamdani Fuzzy PI/PD Controllers as Nonlinear, Variable Gain PI/PD controllers. *International Journal of Fuzzy Systems.* Vol. 2, pp. 191-196.

[27] Kazemian H.B., (2002). The SOF-PID Controller for the Control of a MIMO Robot Arm. *IEEE Trans. on Fuzzy Systems.* Vol. 10, pp. 523-532.

[28] Lygouras J.N., Kodogiannis V.S., Pachidis T.P., Tarchanidis K.N. and Koukourlis C.S., (2008). Variable Structure TITO Fuzzy Logic Controller Implementation for a Solar Air-conditioning System, *Applied Energy.* Vol. 85 (4), pp. 190-203.

[29] Zadeh L.A., (1973). Outline of a New Approach to the Analysis of Complex Systems and Decision Process. *IEEE Transactions on Systems, Man and Cybernetics.* Vol. 3, pp. 354-359.

In: Air Conditioning Systems
Editors: T. Hästesko, O. Kiljunen, pp. 225-242

ISBN: 978-1-60741-555-8

Chapter 4

ROOM AIR DISTRIBUTION, INDOOR ENVIRONMENTAL QUALITY AND ENERGY PERFORMANCE

Zhang Lin and Wai Sun Shum
Building Energy & Environmental Technology Research Unit,
Division of Building Science & Technology, City University of Hong Kong

ABSTRACT

The application and performance on thermal comfort, indoor air quality and energy efficiency of a variety of room air distribution modes, namely, mixing ventilation, displacement ventilation, task/personalized ventilation, impinging jet ventilation, confluent jet ventilation and stratum ventilation, are discussed. There is a worldwide plea to reduce carbon dioxide emission. In response to such a call, there is a trend to adopt a higher indoor temperature in summer, particularly in East Asia. To promote such an idea, the public need to be convinced that such practice would not sacrifice indoor environmental quality, especially thermal comfort of the occupants. To implement such a measure, existing room air distribution modes are screened according to a set of criteria. Suitable room air distribution mode(s) that work under the unconventional conditions are identified. Design principles are discussed. Experimental and computational results show that with properly designed supply air velocity and volume, locations of supply and exhaust, the proposed stratum mode has the potential to maintain better thermal comfort with a smaller temperature difference between head and foot level, lower energy consumption, and better indoor air quality (IAQ) in the breathing zone. In addition, efficient energy performance of heating, ventilating and air-conditioning (HVAC) systems due to active and passive control of energy consumption are addressed.

1. INTRODUCTION

In 1998, commercial sector accounted for 59% and residential sector accounted for 26% of electricity consumption in Hong Kong, where 40-50% of the total energy consumption comes from HVAC systems. Due to the increasing demand of energy for the booming economy that causes environmental degradation, Hong Kong Energy Efficiency Office (EEO) presented codes of practice for building energy efficiency in 1998 (Chan and Yeung, 2005). These set of codes represented maximum allowable energy consumption by various installations. In 1992 alone, it was estimated that power stations in Hong Kong generated 25 million tones of carbon dioxide, 350,000 tonnes of sulphur dioxide and 120,000 tonnes of nitrogen oxide and nitrogen dioxide. To reduce energy usage that based on Building Energy Code (BEC), high efficacy electronic lamps are replacing conventional fluorescent lamps; variable speed drive is recommended for fan system in air-conditioning and water pump system to save energy. These implementations not only save energy but also lower emission of greenhouse gases.

However, there is a trade-off between maintaining a good indoor air quality and at the same time to reduce energy consumption. Different legislation or norms arose in different countries in order to ensure good indoor air quality, where indoor air quality depends on outdoor air quality, interior installation, ventilation system employed as well as system management and maintenance. There are ASHRAE standard norm in America, prENV 1752 norm in Europe and RITE norm in Spain. Rey and Velasco (2000) used photoacoustic spectroscopy and tracer gas technique to take indoor air measurement with results obtained and compiled according to different norms. From the data achieved, they concluded that air supply rate at 17 l/s per person could maintain good indoor air quality and energy saving simultaneously.

It can be generally perceived that thermal comfort and indoor air quality are inversely related to building energy consumption. Atthajariyakul and Leephakpreeda (2004) presented an optimization methodology for real-time performance of HVAC systems to account for thermal comfort, indoor air quality and building energy consumption. This methodology was to optimize the performance index which is the summation of square errors between actual parameter indices and their desired values, where the actual parameter indices included thermal comfort, indoor air quality and energy consumption indices, which employed predicted mean vote, carbon dioxide concentration and cooling/heating load respectively as indicators. The desired values were obtained from experimental results.

An international project carried out by the International Energy Agency (IEA) in 1989 was to design and build a number of low energy houses with total energy consumption of space heating, domestic hot water and electricity that was 25% of those new houses in participating countries (Thomsen et al., 2005). Energy saving technique used in those houses included insulation, high-performance windows, ventilation heat recovery systems, ground coupled heat exchangers, sunspaces, thermal storage, active solar water systems, photovoltaic systems, home automation systems, energy-efficient lights and appliances. Although target of achieving 75% reduction of energy for typical houses were not realistic, a 60% reduction of energy were recorded for those houses.

2. Issues for Building Ventilation

Regardless of the kind of ventilation mode of an indoor environment, designers are generally concerned about its performances in three aspects, namely thermal comfort, indoor air quality and energy efficiency.

2.1. Thermal Comfort Issue

Lin *et al.* (2005a) studied about the effect of air supply temperature of displacement ventilation on thermal comfort. Numerical studies on the differences between mixing ventilation and displacement ventilation on thermal comfort were performed by Lin *et al.* (2005b). Air temperature distribution, percentage of dissatisfied people (PD) due to draft and predicted percentage of people dissatisfied (PPD) on thermal comfort are the major criteria to evaluate thermal comfort. Meanwhile, two models for predicting discomfort due to skin humidity and insufficient respiratory cooling were proposed by Toftum and Fanger (1999). The skin humidity model predicted discomfort as a function of relative humidity of skin. The respiratory model predicted discomfort as a function of driving forces for heat loss from temperature and humidity of surrounding air. At present, air humidity affecting comfort was not completely known. However, air humidity did affect evaporation of moisture from human skin and respiratory tract. Thus, it had an influence on human heat loss and thermal sensation. The predominant factors for humidity discomfort were basically because of high level of skin humidity and insufficient cooling of mucous membranes in upper respiratory tract. Skin humidity was determined by human sweat and skin wetness when in contact with clothing. Insufficient respiratory cooling was caused by high air temperature or high air humidity when air was inhaled through mucous membranes in upper respiratory tracts. High air humidity caused condensation in ducts which resulted in growth of fungi or other microorganisms that would trigger respiratory infections, eye irritations and skin rashes.

2.2. Indoor Air Quality Issue

Common health symptoms for sick building syndrome include nausea, headache, eye, nose and throat irritation, coughing, fatigue and general feelings of illness. Indoor air pollutants included but were not limited to combustion products from fossil fuels, building material furnishings and outside sources such as radon. Three major concerns on indoor air quality (IAQ) are raised for building occupants, notably, volatile organic compounds such as toluene, benzene and formaldehyde, airborne pathogens and indoor carbon dioxide concentration level. Lin *et al.* (2005c) performed a study about the effect of air supply temperature of displacement ventilation on IAQ. Lin *et al.* (2005d) stated that pollutants transportation and distribution depended in general upon ventilation system, building geometry, pollutant source characteristics and thermal/fluid boundary conditions such as airflow rates, locations of supply inlets and exhaust outlets and diffuser characteristics. For certain kind of building geometry and pollutant sources, increasing ventilation rate could improve IAQ. However, this would result in higher energy consumption rate which in most cases was not economically feasible.

2.3. Energy Efficiency Issue

Although energy savings seem to be unrelated to IAQ, high energy usage is nevertheless a direct cause of greenhouse effect by increasing outdoor carbon dioxide concentration, which indirectly increases indoor carbon dioxide concentration level. China, being the second largest energy consumer after US, has 75% of energy supply come from coal consumption according to Niu and Liao (2002). This energy source entails by-products which imposes a heavy strain on natural environment. With energy consumption in direct proportion to national economic growth, China has recorded the fastest growing economy and will double its GDP based on 1997 in ten years time. Hence energy savings is an immediate measure to reduce CO_2 concentration level before any other energy substitutes are available for practical use. Two levels of measures are possible for control of energy consumption and carbon dioxide emission, the active and the passive control of energy usage. The active method involves development of renewable energy and the passive method is to reduce heating/cooling demand by insulation of building envelop, which includes roof, walls, window glazing and floor of a building. Verbeeck and Hens (2005) examined the economic viability on residence with building envelop insulation and window glazing as well as solar panel installation for generation of electricity. Considerations were particularly put on investment on retrofit installations and environmental impact by reducing carbon dioxide emission. Conclusions were although insulation of building envelop and adoption of renewable solar energy were economically viable, the payback time for these long term investments will be as long as 30 years.

3. Ventilation Modes

Various ventilation modes have been proposed for HVAC systems. Ventilation refers to the process of introducing and distributing outdoor air into a building or a room. The amount of air circulated per unit volume per unit time is termed ventilation rate. Air distribution pattern in a ventilation process is a mode of air movement within buildings. Ventilation process can involve airflow by natural wind pressure, thermal buoyancy or convective air stream generated by fans. Air movement can be controlled and affected by physical obstacles in buildings or movements of occupants. Evaluation of ventilation effectiveness and thermal comfort of various industrial ventilation schemes were carried out by real size or scale model experiments. The objective was to compare the performances of supply and return air terminals located at different heights for a range of flow rates and heat loads. Small scale model experiments were tested to predict full scale conditions by keeping the same Reynolds number, the Archimedes number, the Prandtl number and the Schmidt number. The largest ventilation effectiveness occurred for a low supply/high return configuration, followed by high supply/high return configuration and then low supply/low return configuration. Increasing the number of diffusers in occupied zone provided increasing ventilation effectiveness. For a given configuration, ventilation efficiency was generally increasing when heat load was increasing and/or flow rate was decreasing. Ventilation modes under present discussions are mixing ventilation, displacement ventilation, task/personalized ventilation, impinging jet ventilation, confluent jet ventilation and stratum ventilation.

3.1. Mixing Ventilation

Mixing ventilation is a traditional mode of ventilation with supply inlets and return outlets located at ceiling level which is still widely used today. It could take care of both heating and cooling with highly varying load patterns. Mixing is achieved by supplying ventilation air as a high velocity jet to entrain air already in the room. Mixing ventilation could provide comparably uniform air temperature distribution in occupied zone. However, it also leads to problems such as poor IAQ, air draft in occupied zone and some other thermal discomfort. In addition, it normally consumes more energy than displacement ventilation.

3.2. Displacement Ventilation

Displacement ventilation has been used quite commonly in Scandinavia during the past twenty-five years as a mean of ventilation in industrial facilities to provide better IAQ and to save energy. This ventilation has supply inlets located near floor level and return outlets at ceiling level. More recently, its use has been extended to ventilation of offices, classrooms, commercial buildings and other non-industrial premises. In contrast to mixing ventilation, buoyancy forces (induced by heat sources) govern the airflow in displacement ventilation. Because airflow is thermally driven, this mode of ventilation is only satisfactory when excess heat is needed to be removed. In a room ventilated by displacement ventilation, air quality in breathing zone is usually better than that in mixing ventilation operated with the same airflow rate. Ventilation efficiency of a displacement ventilated room was also significantly better than that of mixing ventilated room (Awbi, 2003).

3.3. Task/Personalized Ventilation

As far as IAQ of breathing zone and energy efficiency are concerned, task/personalized ventilation is the most effective. It may be used to remove excessive cooling load and maintain a comfortable indoor environment. Task/personalized ventilation systems supply air through nozzles located near occupants (e.g. at an edge of a desk). Potential draft exists because of the short distance between supply gears and occupants. A field study found that workers in a task ventilated office satisfied with thermal conditions because they could individually control local environment (Bauman *et al.* 1993). Occupants can control temperature, flow rate and direction of air from nozzles. Measurements conducted by Faulkner *et al.* (1993, 1995) showed that age of air at breathing level with task/personalized ventilation was approximately 30% younger than that with mixing ventilation. However, application of task/personalized ventilation depended much on indoor furnishings. On one hand, it was difficult or expensive to equip nozzles and connect duct in various indoor spaces. On the other hand, some occupants did not usually stay in a fixed location, for instance, customers in a retail shop. These limitations restrict the use of task/personalized ventilation. In another study, Pan *et al.* (2005) evaluated the performance of partition-type-fan-coil unit (PFCU), which is a kind of personalized cooling system, with a central air-conditioning system, which was a large ceiling-hanged water-cooled fan-coil unit, in terms of thermal comfort and cooling energy cost. By achieving the same level of thermal comfort in the experiment, PFCU system could save up to 45% of energy consumed by central system. This was because central system was required to remove additional heat from walls and near ceiling level.

3.4. Impinging Jet Ventilation

Another ventilation method known as impinging jet ventilation, developed by Karimipanah and Awbi (2002), drives a jet of air with high momentum downward to the floor. This ventilation mode was shown to have slightly better mean age of air and velocity distribution than displacement ventilation.

3.5. Confluent Jet Ventilation

Cho *et al.* (2008) investigated the characteristics of wall confluent jets ventilation both experimentally and numerically, where wall confluent jets are a line of air-jet outlets which are aligned vertically along the corner of two adjacent walls of a room, air with velocity as high as 12 m/s is injected into a room by air-jet outlets that are pointing towards the walls of a room. In their numerical simulation between wall confluent jet ventilation and displacement ventilation, they concluded that wall confluent jets system produced a greater horizontal spread over the floor than displacement jet. Karimipanah *et al.* (2007) conducted a numerical and experimental study on confluent jets ventilation as well as displacement ventilation for classrooms. Their results showed that confluent jet system was in general performed better than displacement ventilation system in terms of ventilation effectiveness, heat removal efficiency, comfort for air quality and comfort for thermal sensation. Experimental and numerical experiments performed by EL-Taher (1983) had been used to determine the behavior of a wall jet.

3.6. Stratum Ventilation

Specific limitations exist in mixing ventilation, displacement ventilation and/or task/personalized ventilation. To overcome these problems, a new ventilation mode has been developed. The new system should be able to provide good IAQ in breathing zone, to have minimum temperature difference between head and ankle levels to obtain thermal comfort in occupied zone, to equip duct and diffusers conveniently and to have high energy efficiency.

The underlying principle of displacement ventilation implies that in an air-conditioned room, the conditions of IAQ and thermal comfort beyond occupied zone and beneath breathing zone (approximately $z > 2$ m and $z < 0.8$ m, where z is measured height from the floor.) are of little interest. For conventional ventilation modes, breathing air is transported by boundary layer around the body of an occupant and air quality is a weighted average air quality in a room from the breathing level to floor level. Ventilation efficiency would be maximized if air is supplied directly into breathing zone and air form a well controlled "fresher air layer" to fill a breathing zone. The thickness of fresher air layer depends on the nature of occupancy. Meanwhile, a quasi-stagnant zone is also formed between breathing zone and floor (approximately $0 < z < 0.8$ m). If the temperature within quasi-stagnant zone is not as low as displacement ventilation, the problem of "cold ankles" could be solved. Energy is also saved by avoiding over-cooling of lower part of a room.

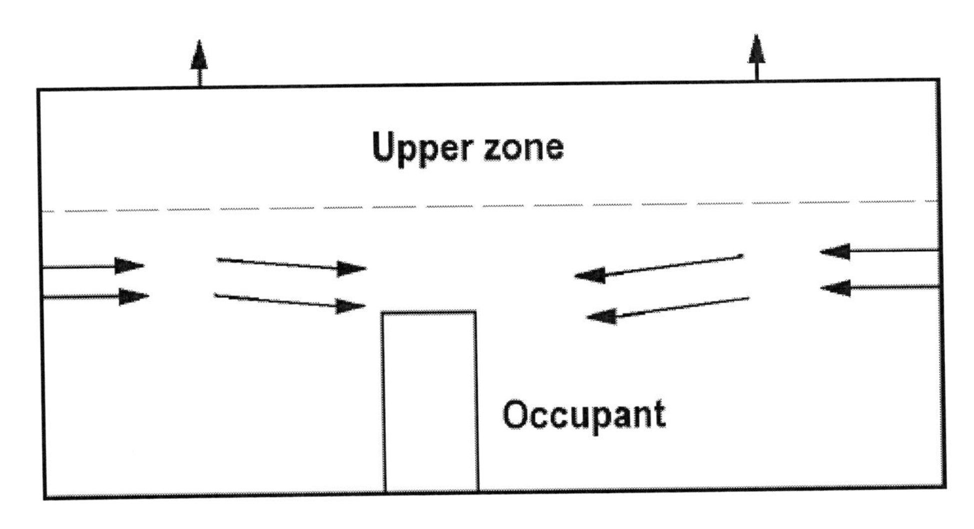

Figure 1. Conceptual diagram of stratum ventilation.

Stratum ventilation works by creating a layer of fresher air in occupants' breathing zone. This is done by placing supply inlets at the side-wall of a room slightly above the height of occupants, standing or sitting, depending on application. Fresh air enters the room and gradually loses momentum farther away from supply (figure 1). Fresh air supply is sufficient to provide adequate air of young age. The range of face velocity and locations of air supply and exhaust should be carefully optimized to break boundary layer around the body of an occupant to minimize risks of draft and cross contamination. A knowledge based system consisting of a database relating drop distance of an air jet based on the Archimedes number (*Ar*) should be developed to decide the appropriate air supply velocity (Leonard and McQuitty 1986, Zhang and Strom 1999). The thickness of fresh air layer required depends on the nature of occupancy. At the same time, a quasi-stagnant zone is also form between breathing zone and floor (say $0 < z < 0.8$ m). Temperature within quasi-stagnant zone should be reasonably controlled. In addition, the supply of air at this level increases the convection effect of heat and helps to displace contaminants into unoccupied zone. It therefore brings fresher air to breathing zone than that of conventional modes of ventilation. Indoor air should be mixed well and air temperature gradient should be low in order not to cause thermal discomfort. Because air is supplied directly to breathing zone, less fresh air bypasses occupants. Thus, there are also possibilities to reduce fresh airflow rate for energy saving. Since air is supplied directly to breathing zone, air supply temperature should be usually above 20°C. This implies condensing temperature of refrigerating plant could also be elevated accordingly, which could result in higher coefficient of performance (COP).

4. ACTIVE ENERGY CONTROL

Energy saving due directly to HVAC systems is known as active energy control. These control methods include but are not limited to application of renewable energy, design modification of air-conditioning systems and use of fluid agents other than air as the heat transfer medium for ventilation systems.

4.1. Renewable Energy

Fossil fuels included coal, oil and gas. They are formed from the organic remains of prehistoric plants and animals. Currently, world's energy generation is mostly dependent on fossil fuel. However, burning of fossil fuels produces carbon dioxide which contributes to greenhouse effect and sulphur dioxide which contributes to acid rain. An alternative is to use nuclear energy. Nevertheless, nuclear waste produced from uranium for nuclear power is very hazardous to handle. Both of these energy resources are not renewable. On the other hand, other energy resources such as solar, wind, tidal, hydroelectric, wave, geothermal and biomass energy are renewable sources of energy (http://www.darvill.clara.net/altenerg).

Bakos (2003) performed a thermal analysis on the passive-solar-heated residence with an attached sunspace. Predictions of energy saving were obtained from passive-solar-heated, unutilizability and solar/load ratio methods. Four different water heating systems were studied by Li and Yang (2009) including solar water heating systems, electric water heaters, towngas water heaters and solar assisted heat pump systems. Economic comparisons showed that solar water heating systems required the lowest payback period among the four systems studied. Li and Yang (2009) further showed that towngas boosted solar water heating system had the greatest potential to reduce greenhouse gas emission under any design conditions.

A solar chimney can cause hot air inside it to rise when sun shines on the solar chimney. A wind tower has a water tank located at the upper part of it which can cause cooler air to sink when water in the water tank is evaporated by dry air that enters the wind tower. By placing a solar chimney and a wind tower on two opposite sides of a room, warm air is exiting the room through the solar chimney while cooler denser air is entering the room through wind tower. This kind of cooling system had been studied recently by Kalantar (2009) who numerically simulated the natural convention process of a cooling tower system. Another application of wind energy is through the use of wind turbines.

4.2. Modified Air-Conditioning Systems

Yu and Chan (2007) concluded from their study that electricity saving up to 10.1% could be achieved when a chiller plant of six chillers with three different sizes instead of four equally sized chillers were installed. A low energy consumption air-conditioning system which comprised cooled ceiling (CC), microencapsulated phase change material (MPCM) slurry storage and evaporative cooling technologies was proposed by Wang *et al.* (2008). Assessment of the system was done on five representative cities in China which included Hong Kong, Shanghai, Beijing, Lanzhou and Urumqi. Results indicated that proposed air-conditioning system had the potential of energy saving up to 80% for northwestern Chinese climate in Urumqi and 10% for southeastern Chinese climate in Hong Kong. For the other three cities, energy saving applying the new system is in between 10% to 80%.

To reduce the moisture of indoor air by air-conditioning system, silica gel desiccant has been applied for this purpose. Hirunlabh *et al.* (2007) conducted an experimental study on a room with split type system (1.5 ton refrigeration capacity) and central air conditioning system (100 ton refrigeration capacity) with desiccant beds. Optimum percentage of air ratios were 15% outdoor air, 15% return air and 70% indoor air with outdoor and return air mixed together at desiccant bed to produce dry air. Although split type desiccant air conditioning system could save energy as much as 24%, the cost of implementing this system would require a payback period of a little over 12 year. Nevertheless, for system with large cooling capacity such as central air system, the cost analysis indicated it was economically feasible with a payback period of only 4 years.

A heating, ventilating and air-conditioning (HVAC) system combined with chilled-ceiling with desiccant cooling was proposed by Niu *et al.* (2002) for hot and humid climate. This cooling system ran chilled water through ceiling panels with a desiccant cooling unit to remove moisture from the ceilings. Results showed that chilled-ceiling with desiccant cooling systems could save up to 44 % of primary energy in comparison to all-air systems, where all-air systems were HVAC systems which employed air for ventilation as well as heat transfer.

4.3. Non-Air Cooled Air-Conditioning Systems

Water cooled air-conditioning systems are supposed to be more energy efficient than air cooled air-conditioning systems. Fong *et al.* (2006) applied an optimization process on simulating heating, ventilating and air-conditioning (HVAC) system of a subway station. Mathematical models of water-cooled chiller as well as cooling coil were set up. Evolutionary programming (EP), which is similar to genetic algorithm (GA) in terms of mutation process where EP chooses to use real-values instead of binary codes in GA, was employed to minimize the objective function of total energy usage by the HVAC systems. This objective function had a direct connection to water-cooled chiller and cooling coil and the simulation result showed a 7% saving in energy compared to HVAC systems with existing operational settings.

In order to meet the energy savings requirement set by the Hong Kong government as a precursor to reduce greenhouse gas emissions, Yik *et al.* (2001) proposed three different energy efficient water-cooled air conditioning schemes as alternatives to air-cooled air conditioning system. These three water-cooled systems were (i) Centralized piped seawater supply for condenser cooling (CPSSCC) which composed of a centralized seawater pumping station, supply piping that delivered seawater to connected buildings for cooling and return piping back to CPSSCC for discharging seawater, (ii) Centralized piped seawater supply for cooling towers (CPSSCT) which is similar to CPSSCC but with slower seawater flow rate with no return piping, where seawater could be disposed of through sewage systems at local buildings and (iii) District cooling systems (DCS) which delivered chilled water at 7°C to a group of buildings with returned chilled water at 12°C back to DCS. Among these three more energy efficient water-cooled schemes than the air-cooled scheme, CPSSCT was less cost effective than CPSSCC and DCS was the least expensive scheme.

Chang (2007) solved an optimal chiller loading model by Evolution Strategy (ES). The optimal chiller loading model which predicts power consumption of a chiller is a function of cooling water return temperature, chilled water supply temperature and chiller load. Chang also pointed out that ES has the advantage over Lagrangian method and genetic algorithm (GA) to solve the same model for the reason that Lagrangian method is only useful when the model composes of convex functions and GA requires complicated evolution process.

Split type air conditioner with air cooled (AAC) and water cooled (WAC) options were experimentally studied by Chen *et al.* (2008). Coefficient of power (COP) for WAC was recorded to be 17.4% on the average higher than AAC, which obviously meant WAC was more energy efficient than AAC. Measured data from experiment were used to validate the mathematical models for predicting AAC and WAC, where a value of 94% correlation between experimental data and the AAC model was recorded. It was estimated that WAC could achieve an annual energy saving of 16.2% which confirmed that wide use of WAC was beneficial.

Bakos et al. (2006) proposed a low energy air-conditioning heating system to reduce energy consumption. The system involved the use of a liquid agent for heat transfer instead of water for conventional central heating system. Energy usage of two air-conditioning heating systems, one with water and the other one with liquid agent as heat transfer fluids, were measured by installing the two systems in two different chambers. Results showed that energy saving was up to 30% by using the liquid agent rather than water as the heat transfer fluid.

5. Passive Energy Control

Since modern city like Hong Kong has 90% of its population sheltering in high-rise buildings, a passive strategy to conserve building energy could be through the design of building envelope. Cheung *et al.* (2005) examined six passive strategies, notably, insulation, thermal mass, glazing type, window size, color of external wall and external shading devices. A standard residential design adopted by the Housing Authority of Hong Kong was simulated for energy consumption by the computer program TRNSYS. By applying thermal mass for insulation, white wall finishes and overhangs for shading, energy savings were at 20%, 12.6% and 5% respectively. Together with using low-e window glazing an annual energy saving of 31.4% could be achieved. Since air-conditioning heating and cooling accounts for more than 50% of total electricity usage in residential sector, Yu *et al.* (2008) used eQUEST software package to analyze energy saving based on low energy envelope building design. Low energy envelope building design included exterior wall thermal insulation, solar radiation absorptance of exterior wall, area ratio of window to wall and use of window glazing. The study showed that envelope shading and exterior wall thermal insulation decreased air-conditioning (AC) electricity consumption by 11.31% and 11.55% respectively. With optimization of the two different energy control strategies, heating and cooling electricity consumption of AC could be reduced by 21.08% and 34.77% respectively. Different strategies of passive energy control are discussed as follows.

5.1. Energy Management

Demand side management (DSM) programs are rebate incentives offered by power companies to cut power consumption by consumers. Yik and Lee (2005) pointed out that if a power company was satisfied with a steady growth in electricity demand and could satisfy that demand while electricity price was under control at a level below a marginal cost, the power company was willing to offer rebates to consumers. However, due to the lack of information on the marginal gains of power companies and consumers in Hong Kong, government intervention on implementing the DSM programs became inevitable.

Increased awareness of energy saving sparks the retail energy company to provide information on customers' past and current energy consumption, load profiles, customer segmentations and geographical location of customers. This is expected to motivate consumers to save energy. Räsänen *et al.* (2008) collected data from 8000 customers who were grouped into twelve clusters and were processed by self-organizing map and *K*-means algorithm. Energy saving guidance resulted from the study included (i) switch off TV when not in use instead of standby, (ii) use task lighting to target work and leisure activities, (iii) use energy saving light bulb, (iv) defrost fridge and freezer on a regular basis and (v) use energy efficient electrical appliances. This conservation policy save more energy than new energy sources can produce per unit of investment, which also reduce the need of investment in power plants.

Bakos *et al.* (1999) developed an energy saving management method that saved fuel for daily heating cycle by using burners and circulators that were controlled by microprocessors. This method divided the daily heating operation cycle into a number of periods that was based on outdoor temperature. The proposed management method could save up to 30% of fuel compared to conventional heating system. An optimal control strategy on heating operation could be applied to buildings to satisfy basic requirements of minimum cost and/or comfort. Bakos (2000, 2002) employed a minimum-cost control method for a house with passive solar system. The objective function based on the heat transfer rates of the house, the wall of the house and the sunspace was then optimized.

With smart meter and broadband internet technology, Wood and Newborough (2007) recommended energy saving could be motivated by installing energy consumption displays (ECDs) at home. Three major factors are considered for installing ECD, (i) the location of the ECD at home, (ii) motivation of energy saving which could be achieved by goal setting and cash reward, (iii) energy consumption data which could be displayed as a total expenditure over a long time period.

A concept of building environmental performance model (BEPM) was developed by Mui (2006) to optimize energy usage in HVAC systems without sacrificing indoor air quality and thermal comfort. This BEPM is further divided into adaptive comfort temperature (ACT) module, which involved a mathematical formula to determine indoor temperature set point by outdoor temperature, and new demand control ventilation (nDCV) module, which measure indoor carbon dioxide concentration level. With BEPM integrated into Grade A private office buildings in Hong Kong, an annual total value worth $122 million Hong Kong dollars of energy saving was calculated while indoor air quality and thermal comfort were maintained.

5.2. Thermostat Offset

Air-conditioning system for some commercial and institutional buildings with part-day occupancy could be running on a continuous basis including weekends. Maheshwari *et al.* (2001) experimentally studied the amount of energy saved by using a temperature offset control of a programmable thermostat. Measurements were performed on a kindergarten, a polyclinic and a mosque. The thermostat automatically turned off the air-conditioning system as well as the air circulation fan during the non-occupancy period. Experimental data showed that energy savings of 46, 25 and 37% were accomplished for kindergarten, polyclinic and mosque respectively.

5.3. Solar Glazing Window

High intensity solar radiation shining through window in summer leads to excessive cooling load. There are a number of glazing systems for buildings that can preserve cooling and heating energy. Vacuum window, low-e window and airflow window are good for cold climate, while reflective window and venetian blind window work well in warm climate (Tanimoto and Kimura, 1997, McEvoy *et al.*, 2003). Chow *et al.* (2006) proposed a numerical model for predicting energy saving on using solar screen glazing window, which was applicable to both summer and winter climate. Results were compared with previously established experimental data. This solar screen glazing window system composes of two glass panes, a clear glass pane for weatherproof seal and a shaded absorptive glass pane with top and bottom vent openings. The system is equipped with a 180 degree switching mechanism. The absorptive glass pane is facing outward of a room and inward to a room during summer and winter times respectively.

In order to lower heating/cooling loads and artificial lighting, Chow *et al.* (2007) suggested the use of photovoltaic (PV) ventilated window for building architecture. This PV ventilated window consists of two glazing assemblies, the clear glazing inner pane and the PV glazing outer pane with semi-transparent a-Si solar cells on its surface and vent openings at its top and bottom ends. They further established energy performance model for a PV ventilated window system to show that a solar cell transmittance between 0.45-0.55 would give the best performance.

Bojić and Yik (2007) examined the energy saving of using single low-e, single low-e-reverse and double glazing with comparison to single clear glazing windows of public housing estates in Hong Kong. The differences of low-e, low-e-reverse and clear glazing lie in their solar transmittance and reflectance, where low-e and low-e-reverse glazing windows had a lower transmittance and higher reflectance of solar than clear glazing. Five different window glazing types were simulated using EnergyPlus. Results showed that low-e + clear double glazing window would conserve most energy for the buildings simulated. However, due to the higher costs of advanced glazing than clear glazing, the installment of advanced glazing was not economically viable.

5.4. Overhangs

One consequence of applying shading devices is it may reduce natural daylighting and hence increases artificial lighting which causes more energy for cooling load. Shading maybe classified into three categories, they are natural shading from trees, external horizontal shading devices and internal Venetian blinds shading devices. Ossen *et al.* (2005) investigated the impact of horizontal shading devices to reduce solar heat gain and the amount of natural daylight penetrating a building office. Energy saving was achieved with various sizes of horizontal overhangs.

5.5. Daylighting

Increasing daylighting not only decrease artificial light consumption but also decrease the heat dissipated by electric lamps. It had been reported that daylighting could improve student performance (Li and Lam, 2003). A field measurement on the performance of daylighting by integrating with photosensors which provided a closed loop control dimming system was conducted by Li and Lam. A substantial energy saving was recorded between 11:00 to 17:00. This amounted to a 65% energy reduction yearly round. Bodart and De Herde (2002) discussed about the integration of daylighting and thermal aspects of cooling or heating systems to maximize energy saving. They pointed out that window positions, window transmittance and wall reflection coefficient all have influences on daylight distribution in a room.

5.6. Anidolic Integrated Ceiling

The contrast of a bright window in the front part of a room and the darker rear part of the room is likely to create visual discomfort. Wittkopf *et al.* (2006) applied anidolic integrated ceiling (AIC) for electrical lighting. This device collects daylight from an anidolic external collector as an overhang that transmits daylight to a reflective ceiling plenum. At the exit apertures of the plenum located at the rear of a room, daylight is then discharged to the deep gloomy zones of the room which could save as much as 20% of energy for using electricity to light the rear part of the room.

5.7. Material Cold Storage

A material called phase change material (PCM) can collect and store 'cold' during the night time and cool the interior of buildings during the warmer daytime. This concept is known as free cooling which can reduce the cooling load for HVAC systems. Stritih and Butala (2007) conducted experimental and numerical studies on a system using PCM cold storage with both results in good agreement with each other.

5.8. Vacuum Insulation

Vacuum insulation panels (VIP) are thermal insulation for building envelops which have five to eight times higher thermal resistivity than conventional insulation materials. Brunner and Simmler (2008) monitored data from a terrace with VIP for 3 years with results compared with laboratory data at constant conditions. Because of the agreement between the results, thermal performance for 25 years of VIP installation was predicted.

5.9. Green Roofs

Niachou *et al.* (2001) analyzed a passive cooling technique of green roofs protection from solar radiation. Green roofs could contribute to cooling and warming of spaces beneath roofs during summer and winter periods respectively. This is because biological functions of photosynthesis, respiration, transpiration and evaporation absorbed a significant portion of solar radiation and green roofs acted as a heat insulation layer. Both experimental and mathematical studies on the performance of green roofs were carried out. Conclusions were (i) lower temperature were measured when covered by thick dark green vegetation and higher temperature were measured when covered by sparse red vegetation or soil only, (ii) estimated heating and cooling loads were lower in buildings with green roof regardless of the kind of roof insulation, (iii) during a whole year round, energy savings for non-insulated, moderately insulated and insulated buildings with green roofs were 37% - 48%, 4% - 7% and 2% respectively.

5.10. Others

Residential electricity usage is one of the major contributors to world energy consumption. There exists a great potential for energy saving up to 70% in space heating alone which would indirectly reduce carbon dioxide emission. Effort of improving energy efficiency has been mainly placed on thermal insulation of building envelop, improve efficiency of boiler, utilizing renewable energy and control of distribution system. Besides imposing energy control on HVAC systems discussed in this presentation, energy saving could be achieved through other residential channels. Liao and Dexter (2004) introduced an ideal PI control scheme to improve boiler heating control system. The investigation was performed through surveys, simulation and experimentation. Results showed that the improved boiler control could save up to 20% of energy as well as an improvement in thermal comfort. Although it is easy to understand that electricity is required to heat up water and bring water from reservoirs to residential areas by water pumps, the public are in general not aware of the direct linkage of water usage to energy consumption. In addition, water treatment is necessary from reservoirs to buildings and sewage treatment is necessary from buildings to dump sites, these urban municipal facilities consume a significant amount of energy as well. Cheng (2002) had made a quantitative analysis on electricity consumption due to house pumps, residential water heaters, water treatment plants and sewage treatment plants to show that a considerable amount of energy was consumed because of water usage.

REFERENCES

Atthajariyakul S. and Leephakpreeda T., "Real-time determination of optimal indoor-air condition for thermal comfort, air quality and efficient energy usage", *Energy and Buildings*, 36, 720-733 (2004).

Awbi H.B., 2003. Ventilation of buildings. London: Spon Press.

Bakos G., "Energy management method for auxiliary energy saving in a passive-solar-heated residence using low-cost off-peak electricity", *Energy and Buildings*, 31, 237-241 (2000).

Bakos G.C., "Improved energy management method for auxiliary electrical energy saving in a passive-solar-heated residence", *Energy and Buildings*, 34, 699-703 (2002).

Bakos G.C., "Electrical energy saving in a passive-solar-heated residence using a direct gain attached sunspace", *Energy and Buildings*, 35, 147-151 (2003).

Bakos G.C., Spirou A. and Tsagas N.F., "Energy management method for fuel saving in central heating installations", *Energy and Buildings*, 29, 135-139 (1999).

Bakos G.C., Tsioliaridou E. and Tsagas N.F., "Experimental investigation of a low energy consumption air conditioning system based on conventional central heating installation", *Energy and Buildings*, 38, 45-52 (2006).

Bauman FS, Zhang H, Arens EA and Benton CC. 1993. Localized comfort control with a desktop task conditioning system: laboratory and field measurements. *ASHRAE Transactions*. 99(2): 733–749.

Bodart M. and De Herde A., "Global energy savings in offices buildings by the use of daylighting", *Energy and Buildings*, 34, 421-429 (2002).

Bojić M. and Yik F., "Application of advanced glazing to high-rise residential buildings in Hong Kong", *Building and Environment*, 42, 820-828 (2007).

Brunner S. and Simmler H., "In situ performance assessment of vacuum insulation panels in a flat roof construction", *Vacuum*, 82, 700-707 (2008).

Chan A.T. and Yeung V.C.H., "Implementing building energy codes in Hong Kong: energy savings, environmental impacts and cost", *Energy and Buildings*, 37, 631-642 (2005).

Chang Y.C., "Optimal chiller loading by evolution strategy for saving energy", *Energy and Buildings*, 39, 437-444 (2007).

Chen H., Lee W.L. and Yik F.W.H., "Applying water cooled air conditioners in residential buildings in Hong Kong", *Energy Conversion and Management*, 49, 1416-1423 (2008).

Cheng C.L., "Study of the inter-relationship between water use and energy conservation for a building", *Energy and Buildings*, 34, 261-266 (2002).

Cheung C.K., Fuller R.J. and Luther M.B., "Energy-efficient envelope design for high-rise apartments", *Energy and Buildings*, 37, 37-48 (2005).

Cho Y., Awbi H.B. and Karimipanah T. (2008) "Theoretical and experimental investigation of wall confluent jets ventilation and comparison with wall displacement ventilation", *Building and Environment*, 43, 1091-1100.

Chow T.T., Fong K.F., He W., Lin Z. and Chan A.L.S., "Performance evaluation of a PV ventilated window applying to office building of Hong Kong", *Energy and Buildings*, 39, 643-650 (2007).

Chow T.T., Lin Z., He W., Chan A.L.S. and Fong K.F., "Use of ventilated solar screen window in warm climate", *Applied Thermal Engineering*, 26, 1910-1918 (2006).

Department of Energy (1979) The report of the alcohol fuels policy review. DOE/PE-0012, Washington, D.C., June.

EL-Taher R.M. (1983) "Experimental investigation of curvature effects on ventilated wall jets", *AIAA Journal*, 21,1505-1511.Faulkner C, Fisk WJ and Sullivan D. 1993. Indoor airflow and pollutant removal in a room with desktop ventilation. *AHARAE Transactions* 99 (2): 750–758.

Faulkner C, Fisk WJ and Sullivan D. 1995. Indoor airflow and pollutant removal in a room with floor-based task. *Building and Environment* 30 (3): 323–332.

Fong K.F., Hanby V.I. and Chow TT., "HVAC system optimization for energy management by evolutionary programming", *Energy and Buildings*, 38, 220-231 (2006).

Hirunlabh J., Charoenwat R., Khedari J. and Teekasap S., "Feasibility study of desiccant air-conditioning system in Thailand", *Building and Environment*, 42, 572-577 (2007).

Kalantar V., "Numerical simulation of cooling performance of wind tower (Baud-Geer) in hot and arid region", *Renewable Energy*, 34, 246-254 (2009).

Karimipanah T. and Awbi H.B. (2002) "Theoretical and experimental investigation of impinging jet ventilation and comparison with wall displacement ventilation", *Building and Environment*, 37, 1329-1342.

Karimipanah T., Awbi H.B., Sandberg M., and Blomqvist C. (2007) "Investigation of air quality, comfort parameters and effectiveness for two floor-level air supply systems in classrooms", *Building and Environment*, 42, 647-655.

Leonard JJ and McQuitty JB. Archimedes number criteria for the control of cold ventilation air jets. *Canadian Agricultural Engineering* 28(2): 117–123 (1986).

Li D.H.W. and Lam J.C., "An investigation of daylighting performance and energy saving in a daylit corridor", *Energy and Buildings*, 35, 365-373 (2003).

Li H. and Yang H., "Potential application of solar thermal systems for hot water production in Hong Kong", *Applied Energy*, 86, 175-180 (2009).

Liao Z. and Dexter A.L., "The potential for energy saving in heating systems through improving boiler controls", *Energy and Buildings*, 36, 261-271 (2004).

Lin Z, Chow TT, Tsang CF, Chan LS and Fong KF, 2005a. Effect of air supply temperature on performance of displacement ventilation (Part I) — thermal comfort. *Indoor and Built Environment,* 14(2): 103-116.

Lin Z, Chow, TT, Fong, KF, Wang Q and Li Y, 2005b. Comparison of performances of displacement and mixing ventilations (Part I) — thermal comfort. *International journal of Refrigeration*, 28: 276-287.

Lin Z, Chow TT, Tsang CF, Chan LS and Fong KF, 2005c. Effect of air supply temperature on performance of displacement ventilation (Part II) — indoor air quality. *Indoor and Built Environment,* 14(2); 117-132.

Lin Z, Chow TT, Fong KF, Tsang CF and Wang Q, 2005d. Comparison of performances of displacement and mixing ventilations (Part II) — indoor air quality. *International journal of Refrigeration*, 28: 288-305.

Lin Z., Chow T.T., and Tsang C.F. (2006) "Validation of CFD model for research into stratum ventilation", *International Journal of Ventilation*, 5, 345-363.

Maheshwari Gopal P., Al-Taqi Hanay, Al_Murad Raba'a and Suri Rajinder K., "Programmable thermostat for energy saving", *Energy and Buildings*, 33, 667-672 (2001).

McEvoy M.E., Southall R.G. and Baker P.H., "Test cell evaluation of supply air windows to characterize their optimum performance and its verification by use of modeling techniques", *Energy and Buildings*, 35, 1009-1020 (2003).

Mui K.W., "Energy policy for integrating the building environmental performance model of an air conditioned building in a subtropical climate", *Energy Conversion and Management*, 47, 2059-2069 (2006).

Niachou A., Papakonstantinou K., Santamouris M., Tsangrassoulis A. and Mihalakakou G., "Analysis of the green roof thermal properties and investigation of its energy performance", *Energy and Buildings*, 33, 719-729 (2001).

Niu J.L. and Liao Z., "Forecasting Residential Energy Demand in China: An approach to technology impacts", *Journal of Asian Architecture and Building engineering*, 1, 95-103 (2002).

Niu J.L., Zhang L.Z. and Zuo H.G., "Energy savings potential of chilled-ceiling combined with desiccant cooling in hot and humid climates", *Energy and Buildings*, 34, 487-495 (2002).

Ossen D.R., Ahmad M.H. and Madros N.H., "Optimum overhang geometry for building energy saving in tropical climates", *Journal of Asian Architecture and Building Engineering*, 4, 563-570 (2005).

Pan C.S., Chiang H.C., Yen M.C. and Wang C.C., "Thermal comfort and energy saving of a personalized PFCU air-conditioning system", *Energy and Buildings*, 37, 443-449 (2005).

Räsänen T., Ruuskanen J. and Kolehmainen M., "Reducing energy consumption by using self-organizing maps to create more personalized electricity use information", *Applied Energy*, 85, 830-840 (2008).

Rey F.J. and Velasco E., "Experimental study of indoor air quality, energy saving and analysis of ventilation norms in climatised areas", *Energy and Buildings*, 33, 57-67 (2000).

Stritih U. and Butala V., "Energy saving in building with PCM cold storage", *International Journal of Energy Research*, 31, 1532-1544 (2007).

Tanimoto J. and Kimura K.I., "Simulation study on an air flow window system with an integrated roll screen", *Energy and Buildings*, 26, 317-325 (1997).

Thomsen K.E., Schultz J.M. and Poel B., "Measured performance of 12 demonstration projects–IEA Task 13 "advanced solar low energy buildings"", *Energy and Buildings*, 37, 111-119 (2005).

Toftum J. and Fanger P.O. 1999. Air humidity requirements for human comfort. *ASHRAE Trans* 105, 641–647.

Verbeeck G. and Hens H., "Energy savings in retrofitted dwellings: economically viable?", *Energy and Buildings*, 37, 747-754 (2005).

Wang X., Niu J. and van Paassen A.H.C., "Raising evaporative cooling potentials using combined cooled ceiling and MPCM slurry storage", *Energy and Buildings*, 40, 1691-1698 (2008).

Wittkopf S.K., Yuniarti E. and Soon L.K., "Prediction of energy savings with anidolic integrated ceiling across different daylight climates", *Energy and Buildings*, 38, 1120-1129 (2006).

Wood G. and Newborough M., "Energy-use information transfer for intelligent homes: Enabling energy conservation with central and local displays", *Energy and Buildings*, 39, 495-503 (2007).

Yik F. W. H., Burnett J. and Prescott I., "A study on the energy performance of three schemes for widening application of water-cooled air-conditioning systems in Hong Kong", *Energy and Buildings*, 33, 167-182 (2001).

Yik F.W.H. and Lee W.L., "Rebate as an economic instrument for promoting building energy efficiency in Hong Kong", *Building and Environment*, 40, 1207-1216 (2005).

Yu F.W. and Chan K.T., "Strategy for designing more energy efficient chiller plants serving air-conditioned buildings", *Building and Environment*, 42, 3737-3746 (2007).

Yu J., Yang C. and Tian L., "Low-energy envelope design of residential building in hot summer and cold winter zone in China", *Energy and Buildings*, 40, 1536-1546 (2008).

Zhang G and Strom JS. Jet drop models for control of non-isothermal free jets in a side-wall multi-inlet ventilation system. *Transactions of the ASAE* 42(4): 1121–1126 (1999).

In: Air Conditioning Systems
Editors: T. Hästesko, O. Kiljunen, pp. 243-265
ISBN: 978-1-60741-555-8

Chapter 5

SOLAR THERMAL ENERGY TECHNOLOGY: EXPERIMENTAL STUDY ON THE DESIGN AND PERFORMANCE OF AN INDIRECT SOLAR DRYER FOR AGRICULTURAL PRODUCTS

Sabah A. Abdul-Wahab
Mechanical & Industrial Engineering Department, Sultan Qaboos University,
P.O. Box 33, Al Khoud, P.C. 123, Muscat, Sultanate of Oman

ABSTRACT

This chapter investigated the use of solar energy for drying. The objective of the study was to study a simple batch dryer for agriculture products in order to improve the design and to determine the most effective variables influencing its performance. For this purpose, an indirect conventional solar dryer unit consisting of a solar air heater (i.e., inclined solar collector with corrugations) and a drying chamber (i.e., basically a batch dryer) was designed and built at the College of Engineering, Sultan Qaboos University (SQU) in Muscat (Oman). This unit can be used for drying various agricultural products like fruits and vegetables. In this chapter, the solar dryer unit was used to dry radish crops as the test samples. The dryer unit was investigated experimentally and its performance was evaluated under the climatic conditions of Oman (21° 00 N, 57° 00 E). To this effect, a series of experiments were performed on the developed solar dryer during summer conditions in Oman (between May 4 and June 8, 2008). The performance of the solar dryer was computed and expressed in terms of the moisture evaporation (crop mass during drying). Several influencing design parameters, involved in the dehydration process, that were expected to affect on this performance and on the drying time were tested and discussed. The design parameters studied were the number of glazing cover, the type of the absorber of the solar collector, the loading capacity, and the drying airflow rate. In terms of the number of glazing cover sheet, experiments were performed with a single glazing and a double glazing. With regard to the type of the absorber plate, vertical and horizontal corrugated absorber plates were used. The thermal performance of the solar dryer was also evaluated under no absorber plate conditions. In addition, the thermal performance of the solar dryer was evaluated under various load conditions and the unit was also tested under various drying air velocity. The temperature and relative humidity data during the solar drying were also recorded for each experiment. The comparison revealed that the performance of the solar dryer was improved remarkably with the use of the vertical corrugated duct collector and the drying time was consequently reduced. The results also affirmed that the performance increased with reducing the loading capacity. The other parameters were found to have a negligible effect on the performance.

Keywords: Indirect solar dryer; design parameters; moisture content; drying time; corrugated absorber; Oman.

1. Introduction

Drying is an excellent process to preserve vegetables and other food crops. It is used extensively in industry and in agriculture to dry different materials and agricultural products. Due to the environmental considerations and the increasing cost of conventional energy sources, greater attention is given in the use of solar energy as an alternative energy source in the drying process. While conceptually a simple technology and well known, solar drying technology is a complex process of simultaneous heat and mass transfer. It involves the transport of moisture to the surface of the material and subsequent evaporation of the moisture by thermal heating (Jain and Tiwari, 2004). Perfecting this technology, therefore, was targeting by several researchers to adopt and to design the right type of solar dryer.

Solar dryers are classified primarily according to their heating modes (the manner in which the solar heat is utilized) or their operation modes. They can be classified into three types; namely direct mode dryers, indirect mode dryers and mixed mode dryers. Solar dryers may also be classified according to the mode of air flow into passive (natural convection) and active dryers (forced convection). Ekechukwa and Norton (1999) reviewed many solar dryers and compared their performance and applicability in rural areas. Pangavhane and Sawhney (2002) reviewed the research and development work on various types of solar dryers, with a special attention given to the dryers used for grape drying. A review of the parameters involved in the testing and evaluation of different types of solar food dryers and results in the literature can be found in Leon et al. (2002). They presented a detailed classification of available solar food dryers based on the design of system components and the mode of utilization of solar energy.

Considerable research efforts in the literature were made to design and develop solar dryers for drying of agricultural products. Several theories were presented to explain the kinetics of drying and therefore, the improvement of the behavior of solar dryers passed through many theoretical studies to develop different models for simulation. It was shown that the complexity of the mechanisms of the drying and the variable character of the products prevented a single model to represent all the situations. Therefore, several experimental studies in this field were performed on the development and testing of various types of solar dryers to demonstrate their characteristics and performances (Lutz, 1987; Das and Kumar, 1989; Mahapatra and Imre, 1990; Lawrence et al., 1992; Tiwari et al., 1994; Ayensu, 1997; Ekechukwu and Norton, 1997; Nijmeh et al., 1998; El-Sebaii et al., 2002; Pangavhane et al., 2002; Torres-Reyes et al., 2002; karim and Hawlader, 2004; Shanmugam and natarajan, 2006; Gbaha et al., 2007; Sreekumar et al., 2008). Various previous studies in the literature showed that the use of solar dryers were financially attractive in comparison with the substitution of commercial fuels (purohit and Kandpal, 2005). Moreover, the issues related to the comparison of the benefits and costs of substituting open-sun drying with solar drying was also addressed (Purohit et al., 2006). In their study, Purohit et al. (2006) attempted to make a simple framework to facilitate a comparison of the financial feasibility of solar drying as against open sun drying. Further, studies on the drying process and the cost economic analysis under open sun and solar biomass (hybrid) drying can be found in Prasad et al. (2006).

Research efforts were also focused on both theoretical and experimental investigations of solar dryers to compare the predicted results with the experimental data. Examples of such studies can be found in Sodha et al. (1985); Hachemi et al. (1998); Queiroz and Nebra (2001); Yaldiz et al. (2001); Jain and Tiwari (2004); Karim and Hawlader (2005); Ait Mohamed et al. (2005). The feasibility of the solar dryer depends largely upon the crop to be dried and the climatic conditions. In general, there are many factors affecting the drying rate of the agricultural products (Bolin et al., 1975; Eissen et al., 1985; bains et al., 1989; Ayensu, 1997; Leon et al., 2002; Bennamoun and Belhamri, 2003; Karim and Hawlader, 2004).

In this chapter, the detail design and performance of an indirect cabinet solar dryer unit was reported. The unit consisted of two parts: a collector to heat the incoming ambient air using solar radiation and; a drying chamber in which commodities to be dried were spread on three trays. The solar collector consisted of a corrugated sheet as absorber. The unit was designed on the principles of convective heat flow and it was constructed and tested at the SQU in Oman during the months of May and June 2008. It was oriented south to allow for the capture of the maximum amount of solar radiation. The dryer was used to dry radish samples. The weight of the radish samples and the drying parameters (relative humidity, temperature and airflow rate) were measured at different hours of the day, until constant weights were observed. The experimental tests were conducted under varying various design parameters. A total of eight tests were undertaken. Details of the experiments conducted and the results drawn were discussed.

2. EXPERIMENTAL

2.1. Experimental Set-Up

Figure 1 presents the design of the solar cabinet dryer used in this chapter. The solar dryer, as shown in figure 1, consisted mainly of two main parts: an inclined solar collector (i.e., solar air heater) and a drying chamber (i.e., dryer). The photograph of actually fabricated solar dryer is shown in figure 2. The inclined solar collector part (figure 3) was a single pass air-heater and was made of steel of dimensions 100 x 165 x 22 cm. It was used for supplying hot air to the drying chamber. It comprised of a base plate, suspended matt black painted sheet absorber, and glazing cover sheet. The suspended corrugated black absorber plate was made of 2 mm thick steel of dimensions 124 cm x 165 cm. It was used to collect the energy from the sun to supply the required heat for drying. This absorber was kept suspended to allow flow of air both below and above it for good heat transfer. Hence, the drying air through the absorber was heated as it flowed over and under the absorber plate providing two channels through which air flowed in the same direction on both sides of the absorber plate. This arrangement provided twice as much surface area for heat transfer to the air compared to the single-pass air heater. The design of the solar collector was such that only the absorber-plate was changed when it was needed to vary the collector configuration from horizontal absorber to vertical absorber or to conduct the experiments without absorber. The thickness of the glazing cover sheet was 4 mm. The effect of one and two cover glass sheets were investigated and compared. To maintain this comparison, the solar collector was assembled to be dismantled in such away no need to use two collectors. In this setup of the solar collector, only the second cover glass sheet of the solar collector was removed when it was needed to work with one cover glass sheet. The whole assembly of the solar collector was positioned due south and it was inclined at an angle of 23.37° with horizontal optimized for 23.37 N latitude of Muscat to maximize the capture of solar radiation. Polystyrene of thickness of 3 cm was used as insulation to cover the base plate from bottom and the solar collector from both its sides. The thermal conductivity of this polystyrene material was very law (0.03 W/m °C).

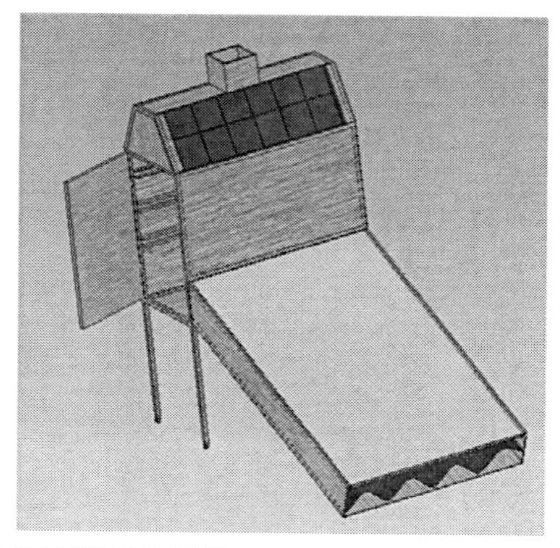

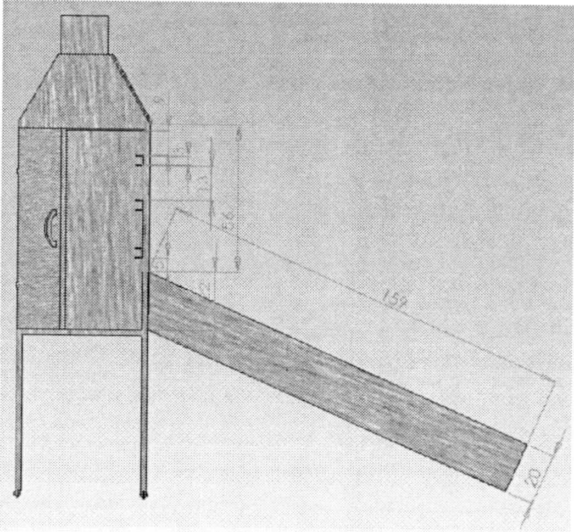

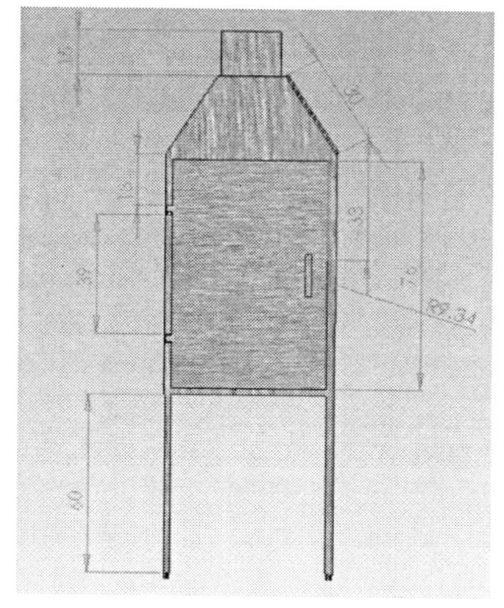

Figure 1. Design of the solar cabinet dryer (dimensions are given in cm).

Figure 2. Photographs of fabricated solar dryer in the field.

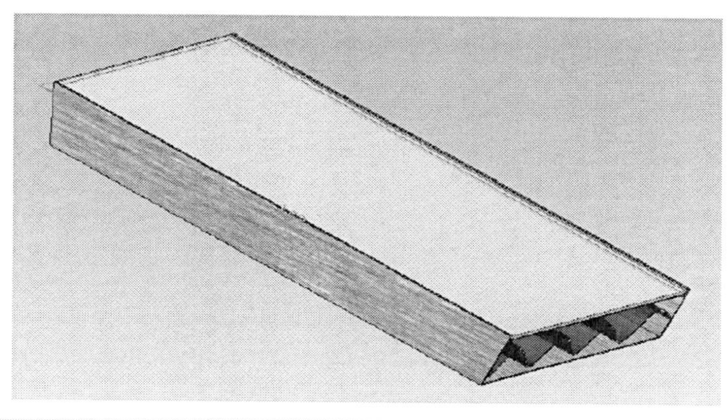

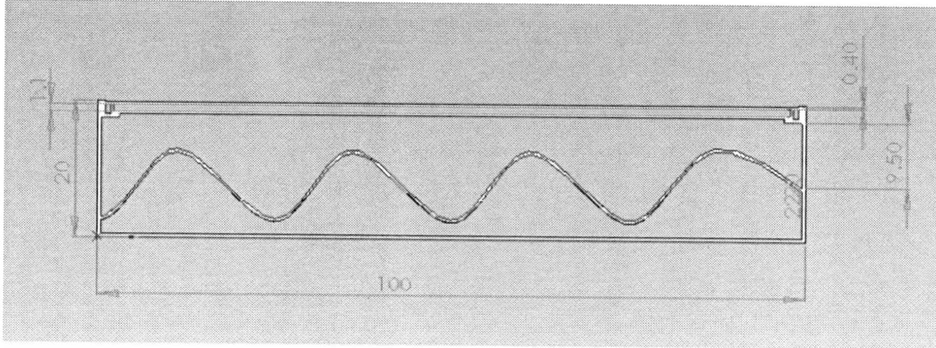

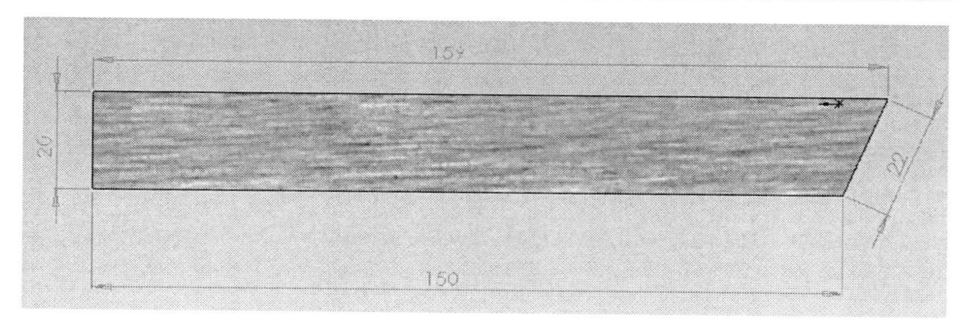

Figure 3. The inclined solar collector part.

The drying chamber part (figure 4) was basically a batch cabinet dryer, made of 4 mm metal steel plate. It was a drying cabinet in which the product to be dried was placed. It had a perforated bottom through which air flows. It comprised of shelves, trays (i.e., product beds where the product was placed) and door. A thick rubber gasket was used to make it airtight. In order to prevent losses of heat to the surrounding, styrofoam insulation, 3 cm thick, was used between the inner and outer surfaces of the drying chamber. In addition, all sides of the drying chamber were insulated. The drying trays, which were the drying area, were made of metal. They were made from wire mesh base with a fairly open structure to allow drying air to pass through the product and to prevent the product items from falling into the drying chamber by giving the support to the trays. There were 3 trays inside the dryer which provided a total drying area of around 1.32 m^2. Each drying tray can slide on bar rails welded onto the inside of the drying chamber at 95 cm apart. The trays can be removed from the dryer for the purpose of cleaning and the product loading and unloading. The door was used to provide an access to the inside of the dryer. It was placed for loading and unloading of the trays of the product inside the dryer. The exhaust vent of the drying chamber had a cross section area of 20 cm x 20 cm, allowed the moist air to escape into the surrounding. A fan was fixed at the outlet port of the drying chamber for the purpose of forcing the inlet ambient air into the drying chamber and expelling the moist air out of it. The power energy of this fan was provided by using solar cells. Moreover, a manually adjustable resistance was attached behind the fan to adjust the air velocity at the drying chamber. Four wheels were assembled under the frame of the drying chamber part, two in the front and two in the back. These wheels were used to allow for mobility of the dryer. The front wheels allowed movement only in one direction, whereas the back wheels helped the system to rotate freely. Also, the back wheels made the system to be stable by providing the break to the system.

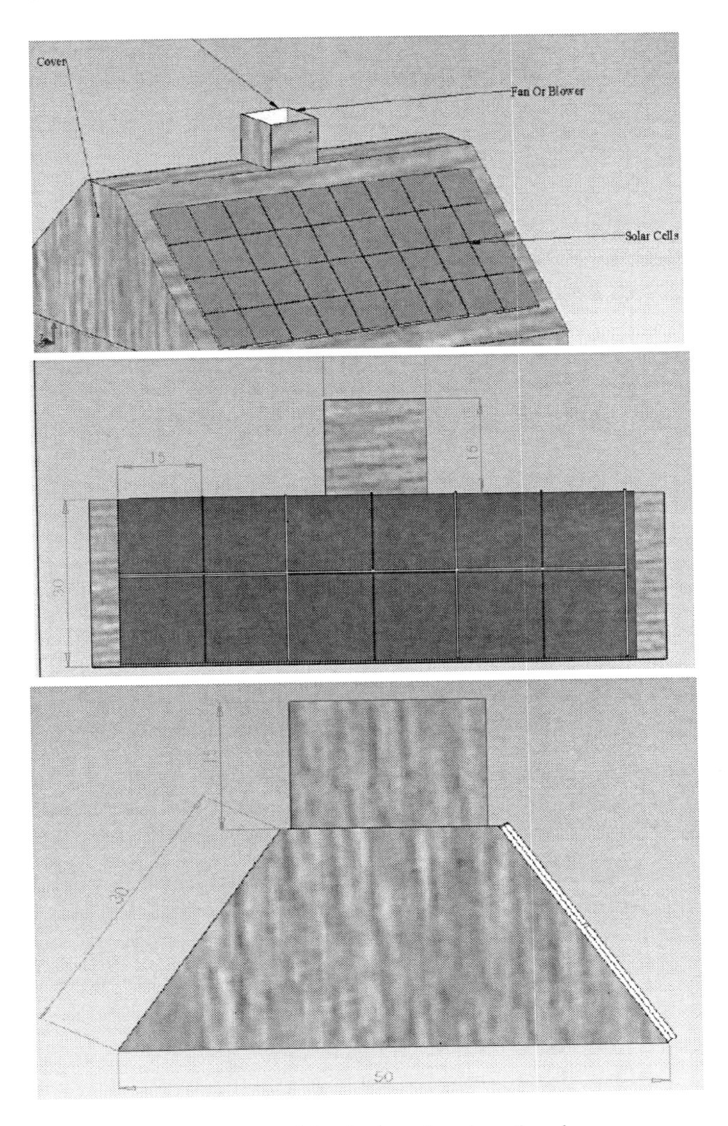

Figure 4. Systematic drawing of the cover of the drying chamber chamber.

2.2. Instrumentation

The mass loss of the product during the drying process was carried out using a digital display electronic balance of 0.001 g precision. Three thermocouples (K type) were used to measure temperatures at different points of the solar dryer. Two of them were fixed to measure the dry bulk temperatures of air at the entrance of the solar collector and inside the drying chamber (upper exit end) whereas the third one was fixed to measure the temperature at the outer surface of the absorber plate. All thermocouples were calibrated using standard thermometers. These thermocouples were further connected to a multi point data recorder (i.e., thermocouples reader) to record temperatures every 30 min. A hygrometer with a minimum count of 1 % was used for measurement of relative humidity of air at the entrance of the solar collector and inside the drying chamber.

2.3. Experimental Procedure

The solar dryer was tested under meteorological conditions of Oman, Muscat. Drying experiments were performed during the period May-June 2008 (i.e., good sunshine during the harvest season). The dryer was designed to operate with only solar radiation as the main source of energy. In the drying experiments, radishes were used as the test samples in the dryer unit. It is worth to mention that chemical pretreatment of the drying materials is usually required to increase the diffusion of moisture from the interior of the product and thus, increase the rate of evaporation from the surface. However, radish does not have hard cover; hence the chemical pretreatment of radish was not required. As a result, the moisture was easily evaporated from the surface of the radish without any chemical pretreatment. During the drying experiments, the weather was sunny and no rain appeared. There was no shading due to trees, buildings or other structures around the solar collector.

The product to be dried was uniformly distributed evenly on the metallic mesh of the drying tray that was then placed on the first shelf of the drying chamber (i.e., top shelf). The amounts of radish used were 200, 400 and 600 g (200 g per tray). When the solar dryer was operating, solar radiation was passed through the glazing transparent cover sheet of the solar air collector, and was then absorbed by the suspended black absorber. Air entered the bottom of the solar collector and was heated by the black absorber. Due to the trapped energy, the air around the absorber was heated and rise into the drying chamber. A negative pressure was created by the rising air which resulted in drawing additional fresh air up through the inlet lower side of the solar collector. This air was then heated and the process continued. As a result, a continuous flow of heated air taken place over/through the radish placed on the perforated trays. This heated air entered the drying chamber below the perforated trays and flowed upwards through the samples. It passed to the dryer part, through the third bed, then the second and finally the first bed (top). The hot air dehydrated the radish, picked up its moisture, and got out through the vent at the top of the drying chamber. The door of the dryer was properly closed and sealed to ensure no leakage of hot air.

The product was weighed to determine the amount of water evaporated by the product (humidity lost). Therefore, during each drying experiment, the reduction in moisture content of the product was determined by weighing the product at every half an hour. The weight of the product on the tray was measured by removing it from the drying chamber for approximately 10-15 s. Weighing a large product by removing it from the drying chamber was found most impractical (Nijmeh et al., 1998). To overcome this problem and to avoid any errors that may occur from cooling, a small sample of radishes was used (200-600 g).

The weight measurements were continued to be undertaken every 30 min at regular intervals between 8:30 a.m to 4:00 p.m. The dry bulk temperature of the air and its relative humidity levels, at the entrance of the solar collector and inside the drying chamber, were also measured simultaneously every 30 min, to determine its absolute humidity. All these measurements were continued to be undertaken until constant weights of the products were observed.

3. Analysis

The performance of the solar dryer was evaluated by conducting a series of runs. Several parameters that were expected to affect the performance were studied. To predict the interaction of these parameters, only one parameter at a time was allowed to vary, while the other variables were held constant at their intermediate levels (reference). The parameters that were considered are shown in table 1. They include the number of glazing cover sheet, the type of sheet absorber of the solar collector, the loading capacity, and the drying air velocity. Hence, a total of eight experimental tests were undertaken. An overview of these experimental tests is presented in table 2.

Table 1. Parameters considered in computing the performance of the designed solar lar dryer

Parameter	Values	Reference (intermediate) position
Number of glasse	Single glass Double glasses	Double glasses
Type of absorber	Vertical Horizontal None	Vertical
Number of shelves	One shelf Two shelves Three shelves	Two shelves
Flow rate of the fan	Low Medium High	Medium
Total number of runs	8	

Table 2. Overview of experimental tests

Seq	Day of experiment	Variable parameter	Fixed parameters	Remarks
1	4 May 2008	Double glasses	Vertical correlated Two shelves Medium flow rate	Reference
2	6 May 2008	High speed of fan	Double glasses Two shelves Vertical correlated	
3	19 May 2008	Without black plate	Double glasses Medium flow rate Two shelves	
4	2 June 2008	Single glass	Medium flow rate Two shelves Vertical correlated	
5	3 June 2008	Medium speed of fan	Double glasses Two shelves Vertical correlated	Reference
6	4 June 2008	Three shelves	Double glasses Medium flow rate Vertical correlated	
7	7 June 2008	One self	Double glasses Medium flow rate Vertical correlated	
8	8 June 2008	Horizontal plate (new)	Double glasses Medium flow rate Two shelves	

The performance of the solar dryer was evaluated by calculating the moisture content of the product. The moisture content on dry basis (db) is the weight of moisture present in the product per unit weight of dry matter in the product and was computed as

Initial moisture content $$M_i = (W_i - W_d)/W_d \quad (1)$$

where

M_i is the initial moisture content at t = 0 (%db)
W_i is the weight of the sample at t = 0 (g)
W_d is the weight of dry matter (g)

The instantaneous moisture content (on dry base) of the sample at any given time (t) during the drying process was calculated from Eq. (2) or Eq. (3).

$$M_t = ([(M_i + 1)(W_t/W_i)] - 1)\ 100\% \quad (2)$$

$$M_t = (W_t - W_d) / (W_d) \quad 100\% \quad (3)$$

where

M_t is the instantaneous moisture content of the product to be dried at any time (%db)

M_i is the initial moisture content of the product to be dried (%db)

W_i is the initial weight of the product to be dried (g)

W_t is the weight of product to be dried at any given time (g). It was calculated by weighing the drying tray with its load of sample at any time during the drying process.

The hourly average values of the solar radiation according to the months of the year were computed and are presented in table 3. These values are important as they can be used to find out how much solar energy is available at any time of the year in the location of the study.

4. Results and Discussion

The diurnal variation of the temperature data over the entire test series are presented in table 4. The experimental tests commenced at about 8:0 a.m. local time of Oman. It can be seen from table 4 that the mean drying air temperatures inside the drying chamber recorded over the entire test series were between 55.2 °C and 68.9 °C (with peak values of 62.3 °C-78.8 °C) compared with mean ambient temperatures of approximately 40.1 °C-44.3 °C (i.e., at the entrance of the solar collector). The absorber plate temperatures were consistently above the drying air temperatures with mean of the order of 72.8 °C-76.0 °C, with peak values between 81.1 °C and 83.7 °C (i.e., the plate is heated up 40 °C above the ambient temperatures). Hence while operating, the designed dryer produced temperatures of 55.2 °C - 78.8 °C, which is a desirable range for most food drying.

During the days of the experiment, the diurnal variations of the ambient air temperature (entrance of the solar collector) and dryer air temperature were plotted and are shown in figures 5-7. From these figures it is observed that the rise in air temperature due to the generated air flow rate in the collector were sufficient for the purpose of radishes drying. It was found that temperatures increased with the time of day until reaching their maximum values at noon. Then the temperatures observed to be decreased. Such diurnal variations in temperatures were expected as they closely followed the insolation pattern. This agreement was true of all the experimental tests. It is evident from figure 5 that the solar dryer with vertical plate tended to function at temperatures considerably higher than that of the horizontal plate.

The comparison between the temperature profiles at different loads for the solar dryer is shown in figure 6. It can be seen from this figure that the maximum temperature occurred at the same time for all loads tested due to the dependency of the product drying temperature on the temperature of the air entering the drying chamber, which in turn depended on the solar radiation incident at the time of measurement. It should be noted that higher loads tended to have a lower temperature profile, due to the existence of cooling through vaporization of the water in the product. However, in this chapter, this agreement was not seen for the case of the lowest load (200 g). This might be attributed to the position of this tray at the upper end of the solar dryer; whereas the other two trays that positioned at the lower end (inlet air) of the dryer were unloaded.

Table 3. Hourly average solar radiation

Hr.	Dec	Jan	Feb	Mar	Apr	May	Jun	Jul	Aug	Sep	Oct	Nov	Dec
0:0	0	0	0	0	0	0	0	0	0	0	0	0	0
1:0	0	0	0	0	0	0	0	0	0	0	0	0	0
2:0	0	0	0	0	6	7	8	0	0	0	0	0	0
3:0	0	3	33	46	67	83	94	91	63	33	20	10	0
4:0	61	102	189	207	253	270	281	280	233	185	162	126	61
5:0	270	314	**452**	**471**	**494**	**508**	**510**	**548**	**502**	**403**	364	210	270
6:0	**455**	**510**	**616**	**620**	**667**	**692**	**710**	**690**	**556**	**517**	**449**	**406**	**455**
7:0	**587**	**644**	**784**	**787**	**814**	**837**	**843**	**896**	**801**	**699**	**583**	**533**	**587**
8:0	**690**	**782**	**894**	**901**	**915**	**936**	**936**	**990**	**867**	**764**	**666**	**592**	**690**
9:0	**698**	**827**	**920**	**928**	**946**	**959**	**963**	**1003**	**903**	**832**	**745**	**604**	**698**
10:0	**646**	**783**	**873**	**883**	**889**	**900**	**907**	**949**	**864**	**845**	**763**	**627**	**646**
11:0	**543**	**664**	**713**	**736**	**764**	**777**	**782**	**808**	**771**	**646**	**548**	**505**	**543**
12:0	360	**494**	**556**	**600**	**612**	**611**	**610**	**624**	**498**	**456**	352	304	360
13:0	180	254	320	353	377	396	**416**	370	307	293	200	181	180
14:0	24	63	107	135	166	181	187	151	120	94	72	52	24
15:0	0	0	50	55	56	58	40	16	3	0	0	0	0
16:0	0	0	0	0	0	0	0	0	0	0	0	0	0
17:0	0	0	0	0	0	0	0	0	0	0	0	0	0
18:0	0	0	0	0	0	0	0	0	0	0	0	0	0
19:0	0	0	0	0	0	0	0	0	0	0	0	0	0
20:0	0	0	0	0	0	0	0	0	0	0	0	0	0
21:0	0	0	0	0	0	0	0	0	0	0	0	0	0
22:0	0	0	0	0	0	0	0	0	0	0	0	0	0
23:0	0	0	0	0	0	0	0	0	0	0	0	0	0

Table 4. Diurnal variation of the temperature data obtained during solar drying of radish under various experimental conditions

	Reference			High speed fan			Without plate		One glass			Low speed fan			Three shelves			One shelf			Horizontal plate		
Time Hour	T_{in} °C	T_{out} °C	T_P °C	T_{in} °C	T_{out} °C	T_P °C	T_{in} °C	T_{out} °C	T_{in} °C	T_{out} °C	T_P °C	T_{in} °C	T_{out} °C	T_P °C	T_{in} °C	T_{out} °C	T_P °C	T_{in} °C	T_{out} °C	T_P °C	T_{in} °C	T_{out} °C	T_P °C
8	40.7	53.2	62.3	36.5	47.6	61.1	38	43.4	40.7	53.2	62.3	37.8	49.8	60.2	38.4	51.3	60.3	34.4	44.5	59.6	38.6	50.3	58.5
8.3	41.7	63.9	66.8	37.5	52.5	65.6	39	54.1	41.7	63.9	66.8	39.5	60.5	64.7	39.4	62	64.8	36	44.7	64.5	39.6	61	63
9	44.7	67.5	72.4	40.5	56.1	71.2	42	57.7	44.7	67.5	72.4	42.5	64.1	70.3	42.4	65.6	70.4	38.8	46.5	70.5	42.6	64.6	68.6
9.3	45.1	73.5	79.2	40.9	62.1	78	42.4	63.7	45.1	73.5	79.2	42.9	70.1	77.1	42.8	71.6	77.2	44.1	51.8	76.6	43.8	67.7	76.4
10	45.2	74.4	79.5	41	63	78.3	42.5	64.6	45.2	74.4	79.5	43	71	77.4	42.9	71	77.5	44.2	54.7	78	43.9	68.6	76.7
10.3	46.3	74.6	80.1	42.1	63.2	78.9	43.6	64.8	46.3	74.6	80.1	44.1	71.2	78	44	71.2	78.1	45.3	55.9	78.7	45	68.8	77.3
11	46.9	77.4	81.6	42.7	66	80.4	44.2	67.6	46.9	77.4	81.6	44.7	74	79.5	44.6	74	79.6	45.8	56.6	79.3	45.6	71.3	78.8
11.3	47.3	77.6	82.3	43.1	66.2	81.1	44.6	67.8	47.3	77.6	82.3	45.1	74.2	80.2	45	74.2	80.3	46.2	59.8	79.9	45.9	71.5	79.5
12	47.5	78	83.1	43.3	66.6	81.9	44.8	68.2	47.5	78	83.1	45.3	74.6	81	45.2	74.6	81.1	46.8	60.7	80.4	46.2	71.9	80.8
12.3	46.5	78.3	83.4	42.3	66.9	82.2	43.8	68.5	46.5	78.3	83.4	44.3	74.9	81.3	44.2	74.9	81.4	45.4	62	82.7	45.2	72.2	81.1
13	46.6	78.8	83.7	42.4	67.4	82.5	43.9	69	46.6	78.8	83.7	44.4	75.4	81.6	44.3	75.4	81.7	45.8	62.3	78.6	45.7	72.7	81.4
13.3	43.6	74	82.7	39.4	62.6	81.5	40.9	64.2	43.6	74	82.7	41.4	70.6	80.6	41.3	70.6	80.7	42.3	58	78.1	42.7	67.9	79.2
14	42.8	71.6	78.8	38.6	60.2	77.6	40.1	61.8	42.8	71.6	78.8	40.6	68.2	76.7	40.5	68.2	76.8	41.8	57.2	77.1	41.9	65.5	75.3
14.3	42.3	69.3	75.2	38.1	57.9	74	39.6	59.5	42.3	69.3	75.2	40.1	65.9	73.1	40	65.9	73.2	40.5	57.9	74.6	41.4	62.2	71.7
15	42.4	57.5	69.7	38.2	46.1	68.5	39.7	47.7	42.4	57.5	69.7	40.2	54.1	67.6	40.1	54.1	67.7	39.9	56.9	69.1	41.5	54.6	66.2
15.3	41.7	51.3	66.9	37.5	39.9	65.7	39	41.5	41.7	51.3	66.9	39.5	47.9	64.8	39.4	47.9	64.9	39.6	56.1	65.8	40.8	47.2	63.3
16	41.1	50.9	64.3	36.9	39.5	63.1	38.4	41.1	41.1	50.9	64.3	38.9	47.5	62.2	38.8	47.5	62.3	39.5	53.4	61.4	39.8	44.8	60.6
Ave.	44.3	68.9	76.0	40.1	57.9	74.8	41.6	59.1	44.3	68.9	76.0	42.0	65.5	73.9	42.0	65.9	74.0	42.1	55.2	73.8	43.0	63.7	72.8

T_p: Temperature of the plate

Reference (Double glasses, two shelves, medium flow rates, vertical absorber plate).

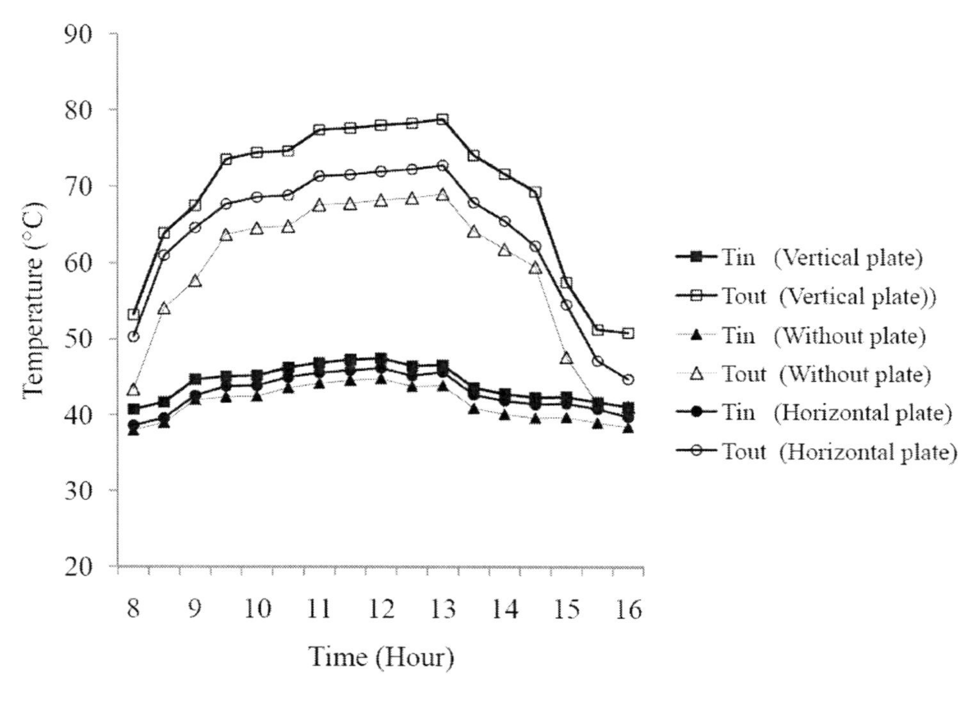

Figure 5. Diurnal variation (temperature profiles) for various absorber plates of the solar dryer (vertical plate, horizontal plate and without plate).

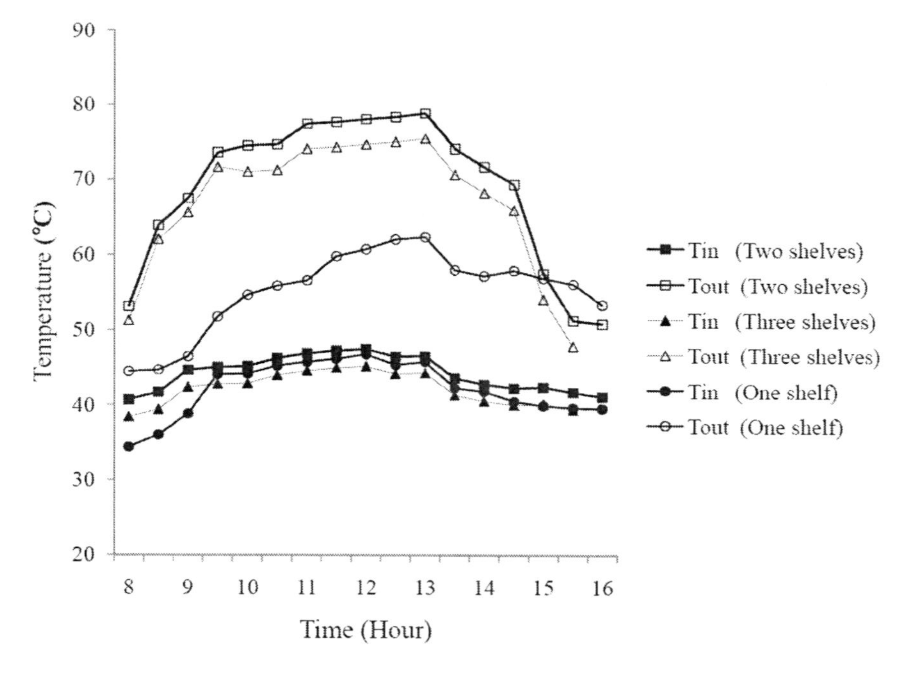

Figure 6. Diurnal variation (temperature profiles) of the solar dryer for various loading capacity (One shelf, two shelves and three shelves).

The comparison between the temperature profiles at different air speed for the dryer is shown in figure 7, where the profiles showed clearly that there was no time lag of the higher speeds compared to that of the lower. Such diurnal variations in the drying conditions were expected as they closely followed the insolation pattern

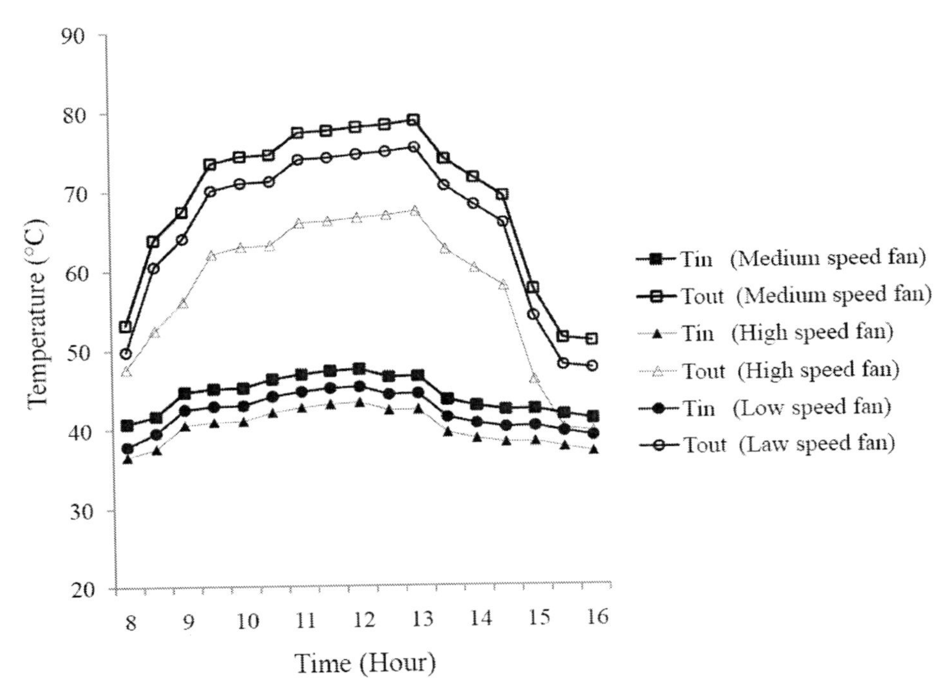

Figure 7. Diurnal variation (temperature profiles) of the 400 g radish for various flow rates (law speed fan, medium speed fan and high speed fan).

Using eq. (2) or Eq. (3), calculations were made for the change in product moisture content with time. The results are illustrated in figures 8-11. It is clear from these figures that the moisture content of the product decreased exponentially with drying time. The evaporated mass of water decreased as the drying time increased because at the initial stages of drying, the moisture removal was rapid, following the constant drying rate period. The effect of absorber plate type on the product moisture content is given in figure 8. As seen from the figure, the lowest product moisture reduction was seen with the case of using no absorber plate. It was observed after 2 hours of drying (at 10:00 a.m.) that the moisture content was reduced from 312.4%db to 114.4 %db (198%db) for the case of using no absorber plate, from 471.4%db to 171.4%db (300%db) for the case of using horizontal plate, from 543%db to 168.3%db (365.7%db) for the case of using vertical absorber plate. In the light of this, the drying achieved with the vertical plate absorber was the best.

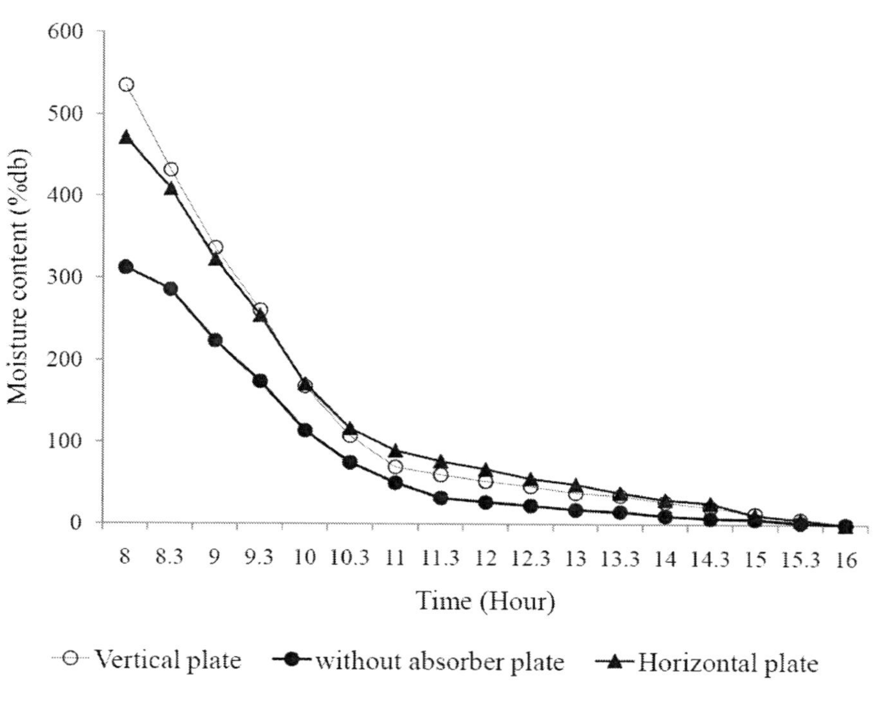

Figure 8. Variation of moisture content with drying time for various absorber plates (vertical plate, horizontal plate and without plate).

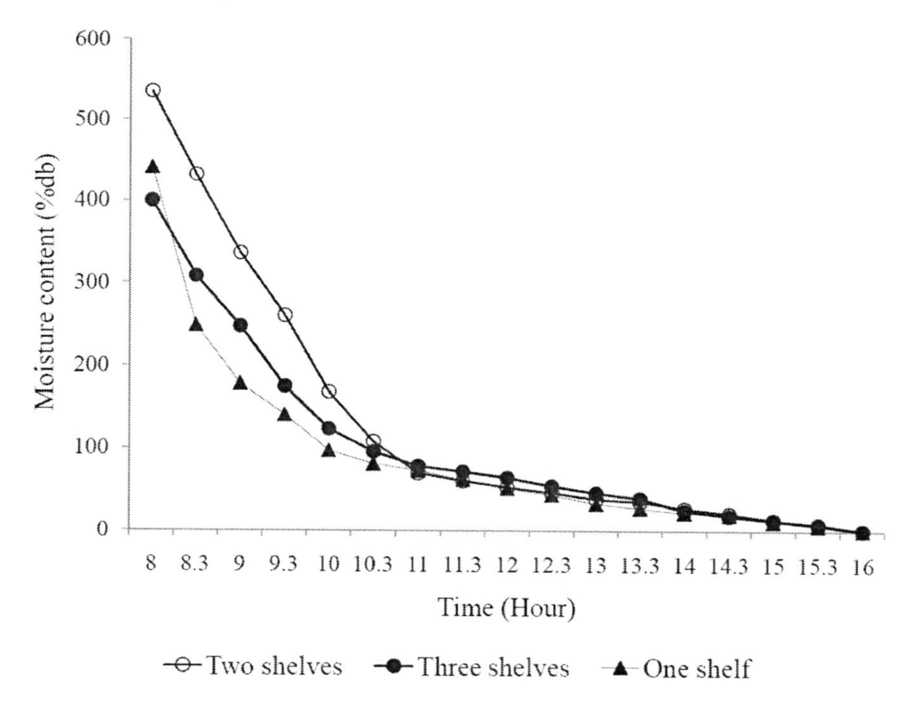

Figure 9. Variation of moisture content with drying time for various loading capacity (One shelf, two shelves, and three shelves).

The effect of the loading capacity on the product moisture content is given in figure 9. As seen from the figure, the lowest product moisture reduction was seen with the case of using three shelves. It was observed after 2 hours of drying (at 10:00 a.m.) that the moisture content was reduced from 440.5%db to 140.5 %db (300%db) for the case of using one shelf, from 555.7%db to 268.9%db (286.8%db) for the case of using two shelves, from 400%db to 175%db (225%db) for the case of using three shelves. Hence it is evident from figure 9 that the lesser the loading capacity, the higher is the moisture content reduction.

Variations of the product moisture content with one glass and two glasses of the solar collector when drying 400 g of radishes are shown in figure 10. It can be seen that no clear difference of the moisture content of the dried material was observed with changing the number of glass. As a result, the drying curves deviate insignificantly with the number of the glass cover.

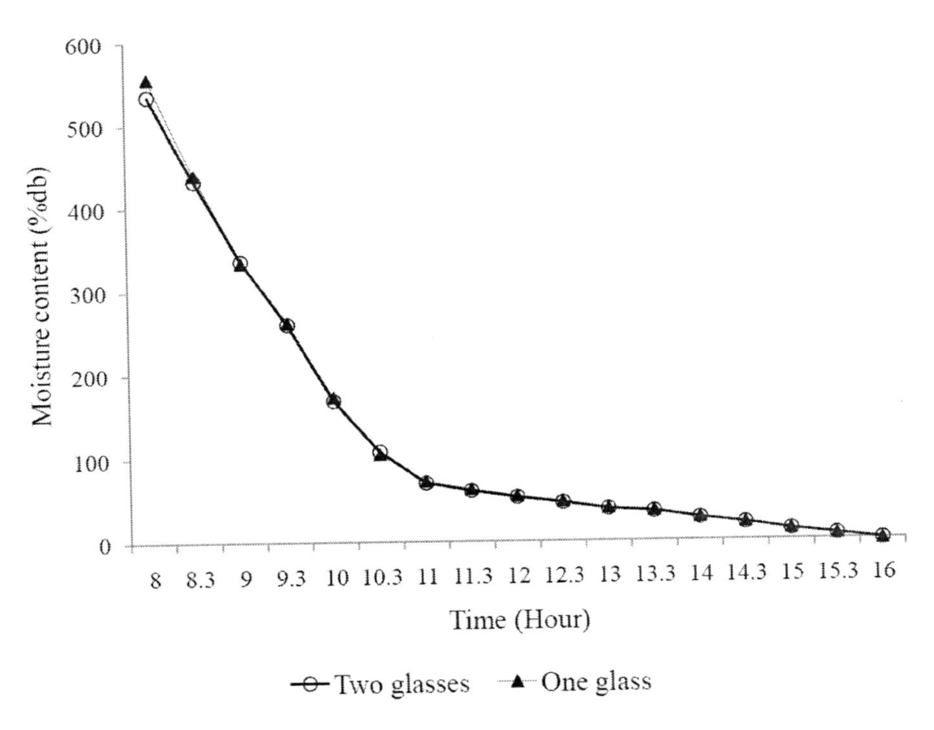

Figure 10. Variation of moisture content with drying time for one glass and two glasses cover sheet.

Hourly variation of the product moisture content for various airflow rates is presented in figure 11. Airflow rate is an important parameter that influences the drying process. In this work, a relatively high product moisture reduction occurred when low airflow rate was used, compared to the medium and high speed fans. It was observed after 2 hours of drying (at 10:00 a.m.) that the moisture content was reduced from 589.7%db to 175.9 %db (413.8%db) for the case of using low speed fan, while it was reduced only from 534.9%db to 168.9%db (366.6%db) for the case of using medium speed fan, and from 555.7%db to 173.8%db (381.9%db) for the case of using high speed fan.

Table 5. Diurnal variation of the mass and the moisture content of radish dried under various experimental conditions

	Reference		High speed fan		Without plate		One glass		Low speed fan		Three shelves		One shelf		Horizontal plate	
Time Hour	MoP g	MC %db	MoP g	MC %db	MoP g	MC %db	MoP g	MC %db	MoP g	MC %db	MoP g	MC %db	MoP g	MC %db	MoP g	MC %db
8.0	400	534.9	400	555.7	400	312.4	400	555.7	400	589.7	600	400.0	200	440.5	400	471.4
8.3	335	431.7	333	445.9	374	285.6	329	439.3	326	462.1	490	308.3	129	248.6	356	408.6
9.0	275	336.5	273	347.5	314	223.7	265	334.4	266	358.6	417	247.5	103	178.4	296	322.9
9.3	227	260.3	225	268.9	266	174.2	221	262.3	218	275.9	330	175.0	89	140.5	248	254.3
10.0	169	168.3	167	173.8	208	114.4	166	172.1	160	175.9	268	123.3	73	97.3	190	171.4
10.3	131	107.9	129	111.5	170	75.3	125	104.9	122	110.3	235	95.8	67	81.1	152	117.1
11.0	107	69.8	105	72.1	146	50.5	105	72.1	98	69.0	214	78.3	64	73.0	133	90.0
11.3	101	60.3	99	62.3	128	32.0	99	62.3	92	58.6	206	71.7	60	62.2	124	77.1
12.0	96	52.4	94	54.1	123	26.8	94	54.1	87	50.0	197	64.2	56	51.4	117	67.1
12.3	92	46.0	90	47.5	119	22.7	90	47.5	83	43.1	185	54.2	53	43.2	109	55.7
13.0	87	38.1	85	39.3	114	17.5	85	39.3	78	34.5	175	45.8	49	32.4	104	48.6
13.3	85	34.9	83	36.1	112	15.5	83	36.1	76	31.0	167	39.2	47	27.0	97	38.6
14.0	80	27.0	78	27.9	107	10.3	78	27.9	71	22.4	149	24.2	45	21.6	91	30.0
14.3	76	20.6	74	21.3	104	7.2	74	21.3	67	15.5	141	17.5	44	18.9	88	25.7
15.0	71	12.7	69	13.1	103	6.2	69	13.1	62	6.9	135	12.5	41	10.8	78	11.4
15.3	67	6.3	65	6.6	99	2.1	65	6.6	59	1.7	129	7.5	39	5.4	74	5.7
16.0	63	0.0	61	0.0	97	0.0	61	0.0	58	0.0	120	0.0	37	0.0	70	0.0

MoP: Mass of product

MC: Moisture content

Reference (Double glasses, two shelves, medium flow rates, vertical absorber plate).

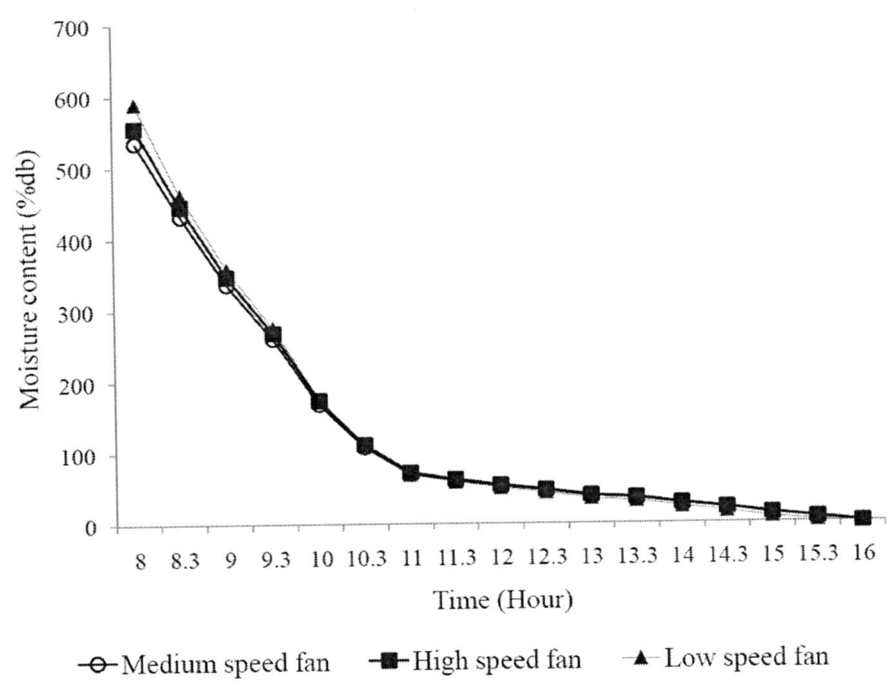

Figure 11. Variation of moisture content with drying time for different speed fans (low speed fan, medium speed fan and high speed fan).

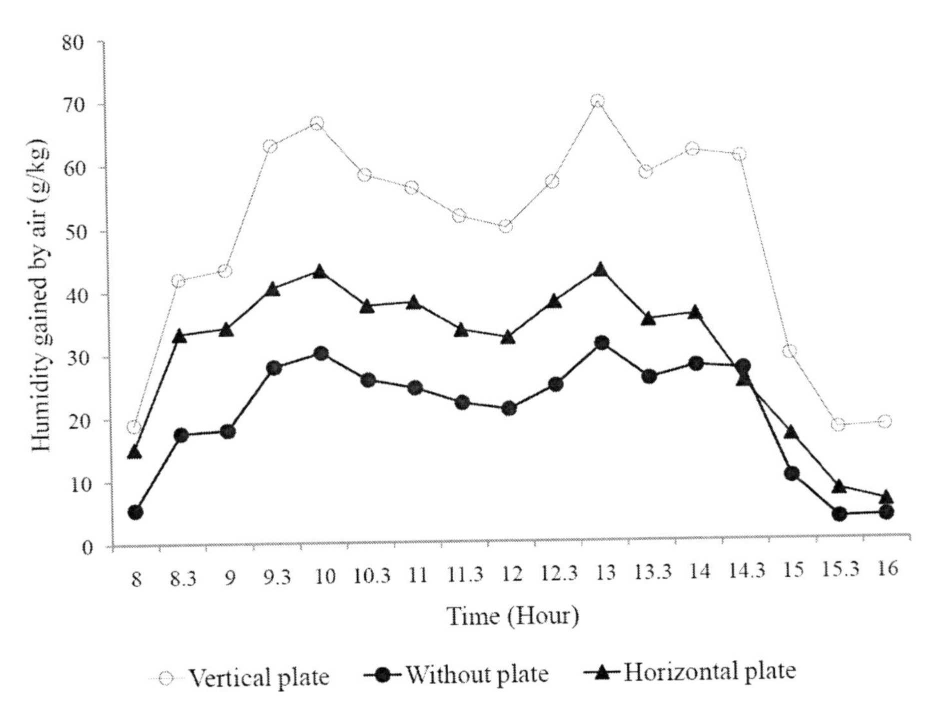

Figure 12. Diurnal variation of humidity absorbed by the drying air for various absorber plates (vertical plate, horizontal plate and without plate).

Hourly variations of the instantaneous absolute humidity absorbed by the drying air (evaporated moisture from the product) are depicted in figures 12-14. It can be seen from the figures that the drying air absorbed the highest absolute humidity with vertical absorber plate, low capacity loading and low airflow rates. Detailed information about the diurnal variation of the mass and the moisture content of radish for all the experimental tests were presented in table 5.

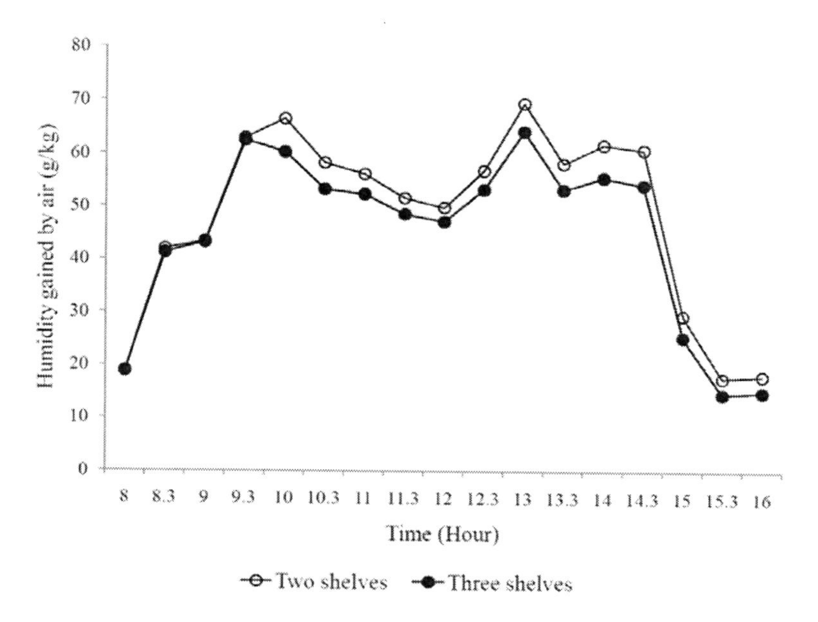

Figure 13. Diurnal variation of humidity absorbed by the drying air for various loading capacity (two shelves, and three shelves).

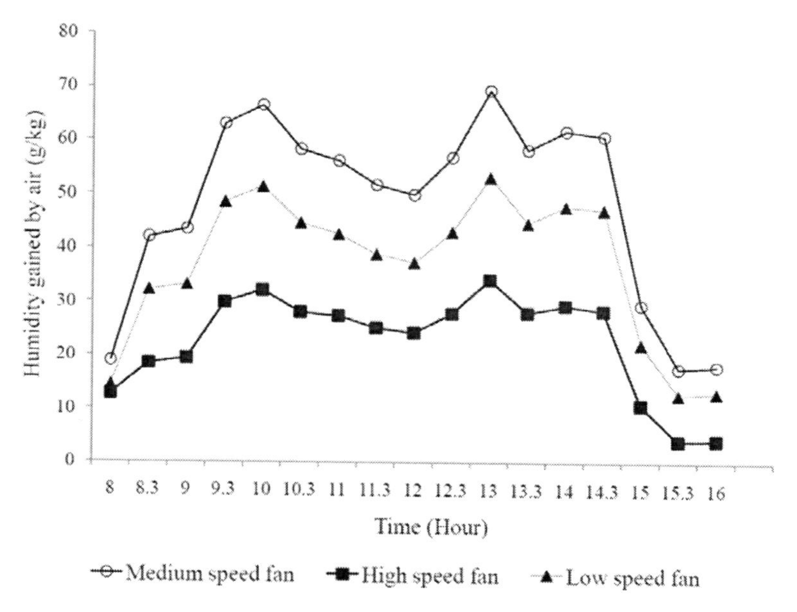

Figure 14. Diurnal variation of humidity absorbed by the drying air for various airflow rates (low speed fan, medium speed fan and high speed fan).

5. Conclusion

This chapter investigated the potential of using indirect cabinet solar dryer unit fabricated from locally materials under Oman climatic conditions for drying radish crop. The chapter was an attempt to explore some of the design variables that can affect on the dryer's performance. The drying experiments were performed for the summer months of Muscat, in Oman during the period May-June 2008. Results of various experiments carried out on the dryer indicated its superior performance. The dryer was capable of producing the temperatures which are optimum for drying of the most of the fruits and vegetables. Therefore, the designed solar dryer can also be used for drying other agricultural products like fruits and vegetables. Further, due to its indirect mode of heating inside the cabinet drying chamber, the designed unit is very much useful for drying of herbal products, which are sensitive to direct sunlight

The best performance was attained with the vertical absorber type. The corrugations of this absorber and its alignment to the air flow direction enhanced the heat transfer. Accordingly, the moisture content reduction was more with the vertical absorber plate in comparison to the horizontal plate type. The vertical plate was also found to produce the highest temperature data.

Also it was found that double glazing was not necessary for increasing the performance of the drying as it only produced slightly higher temperatures. On the other hand, the double glazing design can increase the cost of the dryer significantly. Hence, the cost of the solar dryer unit and the time to construct it can be reduced by eliminating the second layer of glazing.

References

Ait Mohamed, L., Kouhila, M., Jamali, A., Lahsasni, S., Kechaou, N., & Mahrouz, M. (2005). Single layer solar drying behavior of Citrus aurantium leaves under forced convection. *Energy Conversion and Management*, 46, 1473-1483.

Ayensu, A. (1997). Thermodynamic Dehydration of food crops using a solar dryer with convective heat flow. *Solar Energy*, 59(4-6), 121-126.

Bennamoun, L., & Belhamri, A. (2003). Design and simulation of a solar dryer for agriculture products. *Journal of Food Engineering*, 59, 259-266.

Bains, M.S., Ramaswamy, H.S., & Lo, K.V. (1989). Tray drying of apple puree. *Journal of Food Engineering*, 9(3), 195-201.

Bolin, H.R., Petruccia, V. & Fuller, G. (1975). Characteristics of mechanically harvested raisins produced by dehydration and by field drying. *Journal Food Science*, 40, 1036-1038.

Das, S.K., & Kumar, Y. (1989). Design and performance of a solar dryer with vertical collector chimney suitable for rural application. *Energy Conversion and management*, 29(2), 129-135.

Eissen, W., Muhlbauer, W., & Kutzbach, H.D. (1985). Solar drying of grapes. *Energy Drying Technol*, 3(1), 63-74.

Ekechukwu, O.V., & Norton, B. (1997). Experimental studies of integral-type natural-circulatio solar-energy tropical crop dryers. *Energy Conversion and management*, 38(14), 1483-1500.

Ekechukwu, O.V., & Norton, B. (1999). Review of solar-energy drying systems II: An overview of solar drying technology. *Energy Conversion and management*, 40, 615-655.

El-Sebaii, A.A., Aboul-Enein, S., Ramadan, M.R.I., & El-Gohary, H.G. (2002). Experimental investigation of an indirect type natural convection solar dryer. *Energy Conversion and Management*, 43, 2251-2266.

Gbaha, P., Andoh, H.Y., Saraka, J.K., Koua, B.K., & Toure, S. (2007). Experimental investigation of a solar dryer with natural convective heat flow. Renewable *Energy*, 32, 1817-1829.

Hachemi, A., Abed, B., & Asnoun, A. (1998). Theoretical and experimental study of solar dryer. *Renewable Energy*, 13(4), 439-451.

Jain, D., & Tiwari, G.N. (2004). Effect of greenhouse on crop drying under natural and forced convection II: Thermal modeling and experimental validation. *Energy Conversion and management*, 45, 2777-2793.

Karim, M.A., & Hawlader, M.N.A. (2005). Mathematical modeling and experimental investigation of tropical fruits drying. *Heat and Mass Transfer*, 48, 4914-4925.

Karim, M.A., & Hawlader, M.N.A. (2004). Development of solar air collectors for drying applications. *Energy Conversion and management*, 45, 329-344.

Lawrence, S.A., Pole, A., & Tiwari, G.N. (1992). Performance of a solar crop dryer under PNG climatic conditions. *Energy Conversion and Management*, 30, 333-342.

Leon, M.A., Kumar, S., & Bhattacharya, S.C. (2002). A comprehensive procedure for performance evaluation of solar food dryers. *Renewable and Sustainable Energy Reviews*, 6, 367-393.

Lutz, K., Muhlbauer, W., Muller, J., & Reisinger, G. (1987). Development of a multi-purpose solar crop dryer for arid zones. *Solar Wind Technology*, 4, 417-424.

Mahapatra, A.K., & Imre, L. (1990). Role of solar-agricultural-drying in developing countries. *International Journal of Ambient Energy*, 11(4), 205-210.

Nijmeh, M.N., Ragab, A.S., Emeish, M.S., & Jubran, B.A. (1998). Design and testing of solar dryers for processing food wastes. *Applied Thermal Engineering*, 18, 1337-1346.

Pangavhane, D.R., Sawhney, R.L., & Sarsavadia, P.N. (2002). Design, development and performance testing of a new natural convection solar dryer. *Energy*, 27, 579-590.

Pangavhane, D.R., & Sawhney, R.L. (2002). Review of research and development work on solar dryers for grape drying. *Energy Conversion and Management*, 43, 45-61.

Prasad, J., Prasad, A., & Vijay, V.K. (2006). Studies on the drying characteristics of zingiber officinale under open sun and solar biomass (hybrid) drying. *International Journal of Green Energy*, 3, 79-89.

Purohit, P., & Kandpal, T.C. (2005). Solar crop dryer for saving commercial fuels: A techno-economic evaluation. *International Journal of Ambient Energy*, 26, 3-12.

Purohit, P., Kumar, A., & Kandpal, T.C. (2006). Solar drying vs. open sun drying: A framework for financial evaluation. *Solar Energy*, 80, 1568-1579.

Queiroz, M.R., & Nebra, S.A. (2001). Theoretical and experimental analysis of the drying kinetics of bananas. *Journal Food Engineering*, 47, 127-132.

Shanmugam, V., & Natarajan, E. (2006). Experimental investigation of forcd convection and desiccant integrated solar dryer. *Renewable Energy*, 31, 1239-1251.

Sodha, M.S., Dang, A., Bansal, P.K., & Sharma, S.B. (1985). An analytical and experimental study of open sun drying and a cabinet type dryer. *Energy Conversion and Management*, 25(3), 263-271.

Sreekumar, A., Manikantan, P.E., & Vijayakumar, K.P. (2008). Performance of indirect solar cabinet dryer. *Energy Conversion and management*, 49, 1388-1395.

Tiwari, G.N., Singh, A.K., & Bhatia, P.S. (1994). Experimental simulation of a grain drying system. *Energy Conversion and management*, 35(5), 453-458.

Torres-Reyes, E., Navarrete-Gonzalez, J.J., & Ibarra-Salazar, B.A. (2002). Thermodynamic method for designing dryers operated by flat-plate solar collectors. *Renewable Energy*, 26, 649-660.

Yaldiz, O., Ertekin, C., & Uzun, H.I. (2001). Mathematical modeling of thin layer solar drying of sultana grapes. *Energy*, 26, 457-465.

In: Air Conditioning Systems
Editors: T. Hästesko, O. Kiljunen, pp. 267-291
ISBN: 978-1-60741-555-8

Chapter 6

PASSIVE AND AVAILABLE HOUSE AIR-CONDITIONING BY EXPLOITING RENEWABLE ENERGY SOURCES

Majdi Hazami*[*1], *Sami Kooli*[1], *Meriem Lazaar*[1], *Abdel Hamid Farhat*[1] and *Ali Belghith*[2]
[1] Laboratoire de Maîtrise des Technologies de l'Energie;
Technopôle de Borj. Cedria Hammam-Lif BP 95
[2] Faculté des Sciences de Tunis ; Campus, El belvédère Tunis-1060

ABSTRACT

In Tunisia, the buildings' space heating sector represents a major part of the total energy consumption budget. These issues have been increasingly prominent concerns since the energy crisis. Hence, interest has been growing to adopt renewable energies as viable sources of energy that offer a wide range of exceptional benefits for present and future applications, with an important degree of promise especially in the buildings sector. However, the management of the renewable energy sources for space air heating/cooling is usually not economically feasible compared with the traditional carriers. In this chapter, we present a passive energy system, called an air-conditioning cupboard which exploits renewable energies [hot water supplied from a solar collector (40–50°C) and cold groundwater (19°C)] as thermal sources, conceived and tested in our laboratory (LMTE, Tunisia).

To evaluate the air-conditioning cupboard efficiency, indoor experiments were carried out under varied Tunisian environmental conditions for several days. Results show that the air-heating system has a good thermal effectiveness (80 %). It permits us to maintain the temperature inside the experimented room at the range of [24–27°C] during the cold months and [20–23°C] during hot months. A theoretical model is employed for the sizing of the air-conditioning cupboard to obtain the required temperature values. This model allows also the determination of the air-cupboard conditioning thermal performances.

Keywords: Solar energy, solar storage collector, air-conditioning cupboard.

---

* Hazamdi321@yahoo.fr

NOMENCLATURE

A_c Surface area of the collector (m^2)
A_d Surface area of doors (m^2)
A_g Surface area of glasses of windows (m^2)
A_w Surface area of walls (m^2)
D Diffusive intensity of radiation
F Lighting coefficient (F=1.2 for florescent lamps and 1 for others)
$C_{p,c}$ Specific heat of air at constant pressure (kJ/kg K)
C_p Specific heat of water at constant pressure (kJ/kg K)
$c_{p,a}$ Specific heat of air at constant pressure
G Solar radiation (W/m^2)
h_i Heat transfer coefficient (W/m^2 .°C)
h_{iR} Heat transfer coefficient of the air under the roof (W/m^2 .°C)
P Power consumption for lighting system
$\dot{m}$ Water mass flow rate (kg/s)
$\dot{m}_a$ Air mass flow rate (kg/s)
M_c Concrete mass (kg)
Δt Time period (s)
T Temperature (°C)
$\dot{V}$ Volumetric flow rate of air
U_c Heat loss coefficient (W/m^2 .°C)

Greek Letters

α Absorption coefficient
α_1 Input-output test coefficient (m^2)
α_2, α_3 Input-output test coefficient (MJ/K)
λ Latent heat if vaporisation
ρ Density of air
τ_G Transmission coefficient of glass for direct solar radiation
τ_D Transmission coefficient of glass for diffuse solar radiation
η_0 Optical yield (dimensionless)
$\overline{\eta_j}$ Thermal efficiency (dimensionless)

Subscripts

a Ambient air temperature (°C),
Abs Absorber
av Average
c Concrete

e Environment
f Final
g Glass
i Inlet
o Outlet
Ri Internal roof surface
w Water
wi Interior surface of the wall

1. Introduction

1.1. Background

The demand in Tunisia for energy during hot months (for cooling) and cold months (for heating) is increasing very fast. Indeed, in Tunisia six months (from May to October) are the hot months (32°C) of the year, and June, July and August are the months of harsh heat conditions (40°C). The other six months (from November to April) are characterized by fairly humid and cold climates [15–20°C]. Hence, it is clearly necessary to create comfortable conditions inside buildings. Figure 1 shows the typical daily temperature changes during one year obtained from the meteorological station installed in our laboratory (LMTE, Tunisia).

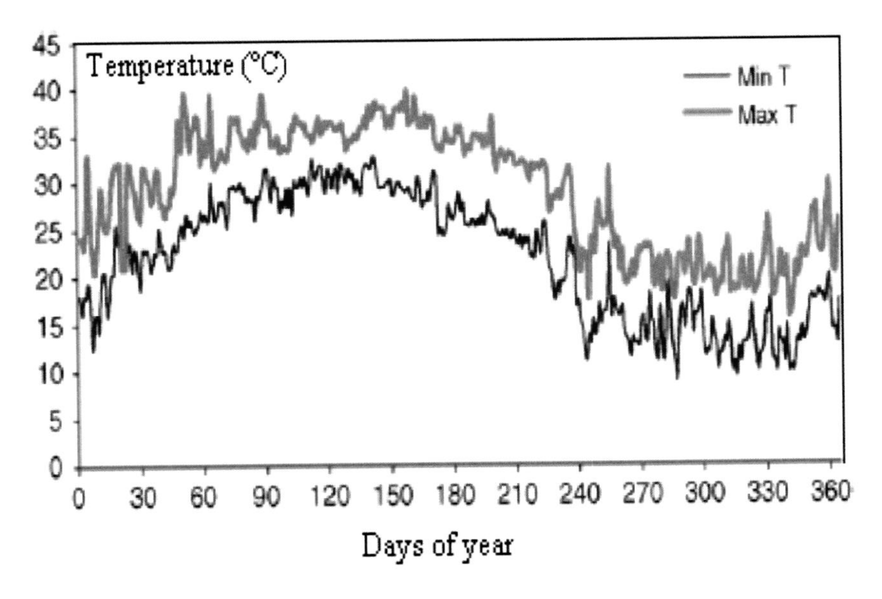

Figure 1. Temperature changes during one year.

Unfortunately, in Tunisia the investment costs for the commercialized heating systems, ventilating and air conditioning systems, which consume sizeable quantities of primary energy, are still relatively high. In fact, the highest amounts, approximately 50 %, of the total electric energy and fossil fuel—which becomes scarce and expensive—is spent on heating/cooling purposes in the buildings sector.

In an environmental context, wide efforts have been undertaken to alleviate global warming of the earth, caused by the emission of carbon dioxide into the atmosphere generated by intensive burning of fossil fuels, which is leading to environmental degradation in heavily populated areas [1, 2, 3]. There seems to exist, therefore, a pressing need to study alternative means of air conditioning and to identify and implement strategies to save energy while maintaining a comfortable environment in buildings. Accordingly, interest has been strongly growing to adopt solar energy and cold groundwater as viable sources of energy that offer a wide range of exceptional benefits for present and future applications with an important degree of promise. These resources represent a marketable massive energy potential, which greatly exceeds fossil fuel resources [4, 5, 6, 7, 8, 9]. Indeed, Tunisia is blessed with abundant sunshine thus making it appropriate to select solar energy for different domestic purposes. In fact, the maximum intensity of solar energy at sea level is approximately 2.59 MJ/h. m^2 and the amount of solar energy that falls on one square meter ranges from about 2.8–9.4 GJ/year depending on the location. The ambient weather conditions are also suitable for introducing solar energy technologies.

Deep groundwater used for buildings' air-cooling during the hot months of the year involves using naturally cold water as a heat sink in a heat exchange system, eliminating the need for conventional air-conditioning. The expected economic and environmental benefits were realized, but barriers to large-scale adoption of the technology were apparent. This technology requires that a client with a large cooling need be near a deep, cold body of water.

As per the environmental point of view, the use of solar energy and cold groundwater, respectively, for buildings air-heating and air-cooling can play an important role to save the environment and to improve our quality of life, while reducing fossil energy consumption. However, the management of the renewable energy sources is usually not economically feasible compared with the traditional carriers. On the other hand, the problems of renewable energy are associated with distribution, access, and security of supply. To overcome the present economic, regulatory and institutional barriers, it is necessary to develop an efficient system capable of exploiting renewable energy sources with satisfactory costs.

1.2. Objective of the Work

The originality in this work is due to the fact that it deals with an innovative system that can be included in management programs aiming to reduce at least the cost of air-conditioning by the application of renewable energy sources [10, 11]. Thus, a trial prototype for a passive air-conditioning system is presented in this chapter. It consists of an air-conditioning cupboard designed and manufactured in our laboratory (LMTE, Tunis) (figure 2). The passive air-conditioning system leads us to exploit hot water [40–50°C] supplied from a solar collector for air-heating during cold months and cold groundwater at a constant temperature (19 °C) for air-cooling. This system is expected to satisfy the demands of household heating and cooling with lower costs than standard air-conditioning systems.

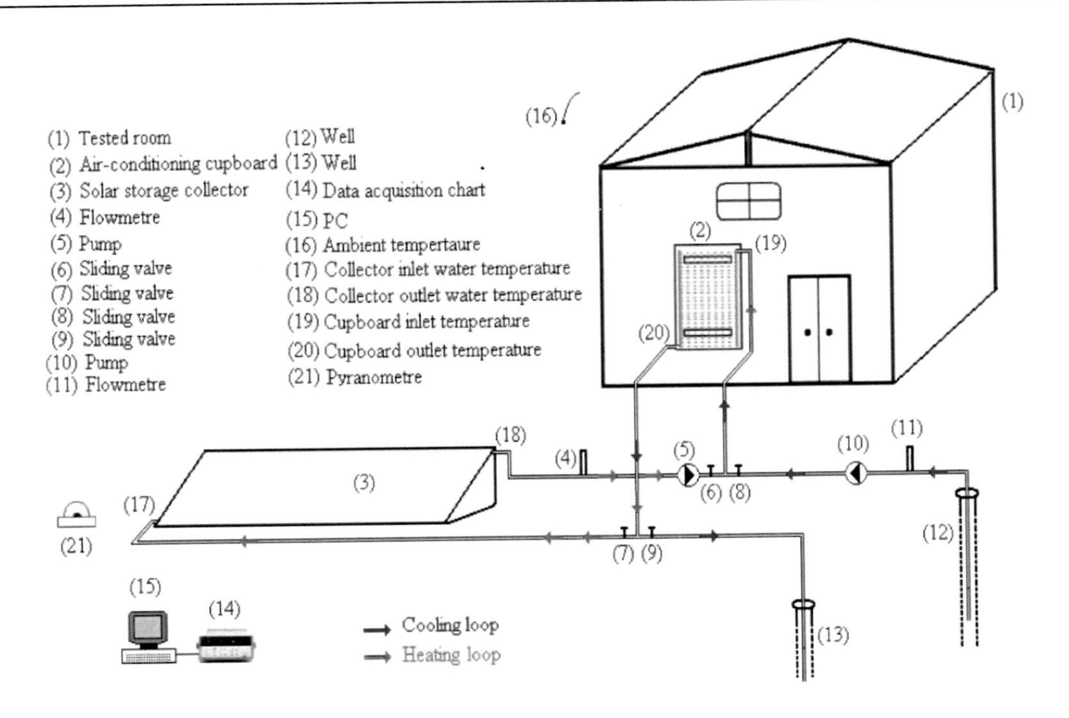

Figure 2. Diagram of the air-conditioning system.

2. Experimental Investigation

A schematic diagram for the air-conditioning system consisting of an air-conditioning cupboard (2), a solar storage collector (3) and two wells (12, 13) bored next to the laboratory is shown in figure 2.

2.1. Air-Conditioning Cupboard Description

The air-conditioning cupboard, made of Plexiglas (2 m x 1.5 m x 0.1 m), is equipped with two traps placed at the top and the bottom (figures 3, 4) [12]. On its two lateral sides a network of low-cost polypropylene capillary heat exchangers are fixed. An air flow circulates naturally between the two traps permitting the exchange of the thermal energy through the air-conditioning cupboard placed inside an experimental room (the width is 3 m, the length is 4 m, and the height is 3 m). The experimental room has one window and one door. Room walls consists of masonry of a thickness of 0.2 m, thermal conductivity of 1.6 $W.m^{-1}.K^{-1}$, density of 990 kg/m^3, specific heat of 1260 $J.kg^{-1}K^{-1}$, and coefficient of the surface radiation emission of 0.56. In table 1 is represented the room's technical specifications.

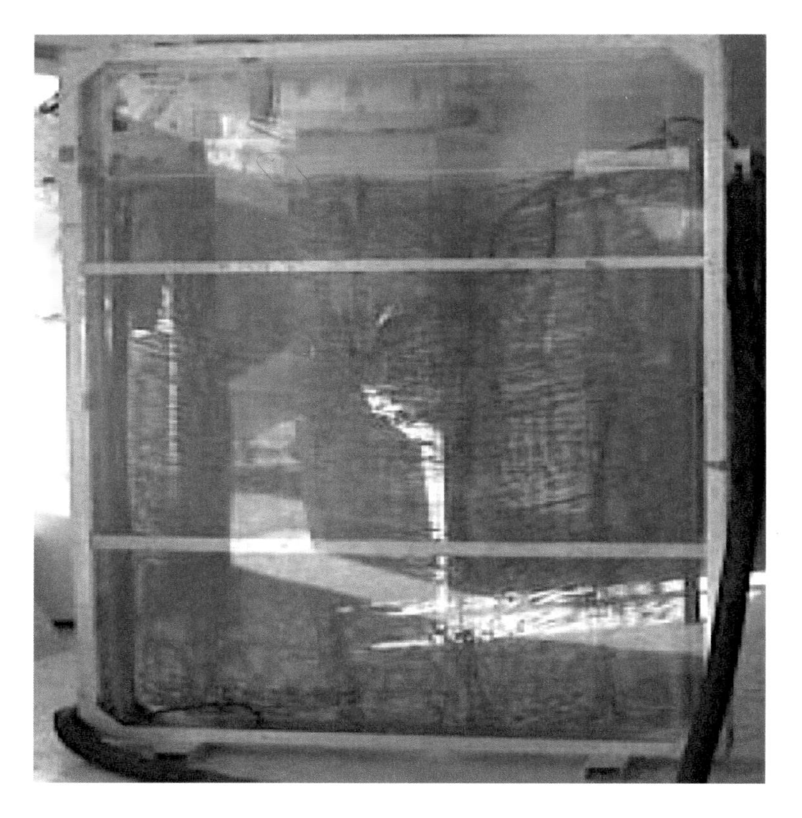

Figure 3. Photo of the air-conditioning cupboard.

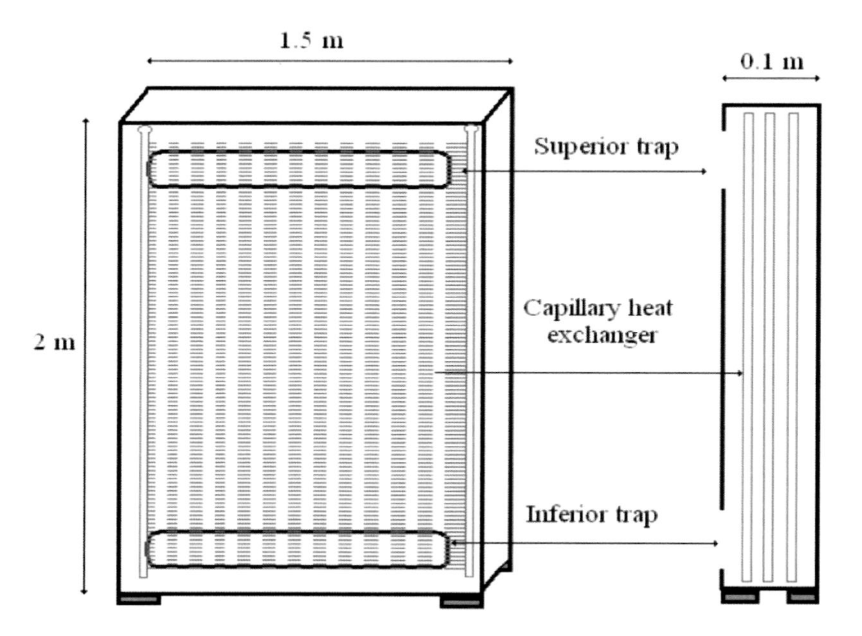

Figure 4. Descriptive diagram of the air-conditioning Cupboard.

Table 1. The room technical specifications

Parameter	Description
Wall	Bricks, thickness =20 cm (k = 1.6 $W.m^{-1}.°C^{-1}$; ρ = 990 $kg.m^{-3}$; solar heat absorption=0.56)
Roof	Concrete blocks, thickness = 36cm (k=1.6 $W.m^{-1}.°C^{-1}$; solar heat absorption = 0.98)
Door	4 cm-thick woods
Window	Iron frame, 1.5m x 1.5m
Glass	Reflex glass, 4 mm thick
1800 m^2	Built over area
Lighting	Lighting heat load = 15 $W.m^{-2}$

2.2. Description of the Solar Storage Collector

A low-cost solar storage collector of total aperture area of 5 m^2, that does not need a storage tank, is conceived and realized in (LMTE, Tunisia) (3) [13]. The absorber matrix is in concrete painted black and having a 50 mm thickness. The concrete matrix performs the function of a storage tank. The thermal conductivity and specific heat of the concrete absorber are equal to 0.81 $W.m^{-1}.K^{-1}$ and 880 $J.kg^{-1}.K^{-1}$ respectively. A polypropylene capillary heat exchanger with an aperture area of 5 m^2 is embedded inside the concrete absorber. The capillary heat exchanger permits us to carry the heat transfer water inside the solar storage collector. A transparent double-layer PVC cover with a thickness of 3 mm (transmission and emissivity coefficients are respectively, 81 and 94 %) is placed over the solar storage collector. An air space gap with a thickness of 20 mm is maintained between the cover and absorber (front-pass). A 30 mm-thick polyurethane foam with a thermal conductivity and specific heat respectively equal to 0.32 $W.m^{-1}.K^{-1}$ and 2090 $J.kg^{-1}.K^{-1}$ was placed on the back of the absorber plate and on the casing sides of the solar collector to minimize heat losses. In figure 5 is represented some stages of the solar storage collector construction.

Figure 5. Continued on next page.

Figure 5. Different stages of the solar storage collector construction.

2.3. Description of the Frigorific Source

For the purpose of air cooling, we have bored two wells next to the laboratory (50 m). The first well, bored at 20-m depth (12), supplies cold water at a constant temperature (about 19 °C). The groundwater extracted from this well is used as a frigorific source. The second well (13), bored at 4-m depth, is used to evacuate the water at the outlet of the air-conditioning cupboard (figure 2).

2.4. The Air-Conditioning System's Operation

2.4.1. Heating Loop

Throughout the day, a great part of the absorbed solar energy is stored inside the concrete matrix. When the solar radiation deceases, the concrete matrix supplies the stored energy to the water within the heat exchanger embedded inside the absorber. This heat is used to provide house heating during the cold season (November to February). When the air-room temperature decreases seriously, a pump installed inside the heating loop (5) starts to operate and circulate the hot water supplied from the solar collector, at a temperature in the range of [40–50°C], to the air-conditioning cupboard placed inside the tested room. The pump operates only when the temperature in the house is lower than the required values (26 ± 2 °C). The hot water supplied by the solar collector replenishes through the capillary heat exchanger placed inside the air-conditioning cupboard. This will continue until the room's air temperature reaches the required value.

It is useful to consider the solar storage collector as having three basic modes of operation, depending on the conditions that exist in the system at a particular time:

i. If solar energy is available and heat is not needed in the room, energy gained from the collector is added to the storage inside the concrete matrix. To preserve the stored energy inside the concrete absorber during the night-time, the solar collector is covered by a double layered panel of 4 cm thick of polystyrene,

ii. If solar radiation is available and heat is needed in the room, energy gained from the collector is used to supply the room's needs,

iii. If solar radiation is not available (after 17:00 h, in the winter season), the stored energy inside the concrete matrix is used to supply the room's needs.

2.4.2. Cooling Loop

During hot months of the year (May to October), the air-conditioning cupboard is connected to the first well (12) by using the sliding valve (8). When the air-room temperature increases seriously, the second pump installed inside the cooling loop (10) circulates the cold water (at 19 °C) extracted from the first well to the air-conditioning cupboard. The return of the water from the air-conditioning cupboard is refuelled in the second well (13) by using a valve (9). This operation will continue until the air temperature in the tested room reaches the required value (23°C).

2.4.3. Data Measurement

Solar radiation is measured by a solar cell placed on a horizontal plane 1 m away from the solar storage collector and calibrated by an Eppley pyranometer (21) with the accuracy of ± 5%. All temperatures are measured by type T and K thermocouples, including ambient temperature (16), inlet and outlet water temperatures (19, 20) and the temperature field in the room. The thermocouples were calibrated individually under different temperatures to obtain the calibration curves to be used for data processing with errors less than 0.1 °C. The four measuring points in tested room were set vertically apart with the distance of 500 mm. The operation of the water pump was initially controlled by means of a timing device, so that water circulated between 09: 00 and 23: 00. The rate of flow was controlled by means of valve mounted at the outlet of the two pumps (5, 10) and the two flow-metres (4, 11). The signals, then, are all automatically collected and recorded by a data acquisition system based on a HP-Micrologger and PC (14, 15). All data were measured at 5 s intervals, from which 15 minute averages were calculated and recorded.

3. Energy Analysis

The numerical model is employed for the study of the air-conditioning system performances and to estimate the optimal air-conditioning cupboard sizing to reach the required indoor climate. This model allows: (i) the evaluation of the useful heat delivered by the solar storage collector and the stored heat inside the concrete absorber matrix, (ii) the overall heat exchange coefficient of the air conditioning cupboard, (iii) the determination of the thermal efficiency of the air-conditioning system, (iv) the determination of the heating/cooling needs and (v) the estimation of the polypropylene capillary heat exchanger surface to integrate inside the air-conditioning cupboard to obtain an optimal function.

3.1. Solar Storage Collector Characterization

According to the INPUT/OUTPUT standard [14, 15], the useful heat delivered by the solar storage collector can be represented by the following empirical equation:

$$Q_u(W) = \alpha_1\ G + \alpha_2\ (T_{a,av} - T_{i,av}) + \alpha_3 \qquad (1)$$

The overnight heat loss coefficient (U_C) of the hot water storage system is determined by measuring the temperature loss of the water during a 12 h nocturnal period [16]. The formula used is:

$$U_C = \frac{M_c C_{p,c}}{A_c\ \Delta t} \ln \frac{T_i - T_{a,av}}{T_f - T_{Abs,av}} \qquad (2)$$

The daily solar collector's thermal efficiency $\overline{\eta}_j$ is given by the expression:

$$\overline{\eta}_j = \eta_o - U_c \frac{T_{Abs,av} - T_{a,av}}{G_{av}} - M_c\ C_{p,c} (T_{Abs,t_2} - T_{Abs,t_1}) \qquad (3)$$

where; t_1 is the initial instant and t_2 the final instant of the test period.

3.2. Thermal Behavior of the Air-Conditioning Cupboard

The role of the air-conditioning cupboard is to maintain an accurate temperature in the tested room. In order to understand the thermal exchanges better inside the room, a numerical model has been developed. This model permits us to evaluate the energy needs of the room. We represent in figure 6 a schematic view of the different types of heat exchanged inside the room. The total heating/cooling load can be calculated by taking into account the effect of various parameters according to the following [17, 18]:

$$Q_P = Q_{Transmission} + Q_{Solar} + Q_{Internal} + Q_{Airflow} \qquad (3)$$

Q_p represents the total heat transferred to the room air when water is flowing through the polypropylene heat exchanger integrated inside the air-cupboard. It was calculated from the temperature difference between the entrance and the exit of the capillary heat exchanger [19]:

$$Q_p = \dot{m} C_p (T_{w,o} - T_{w,i}) \qquad (4)$$

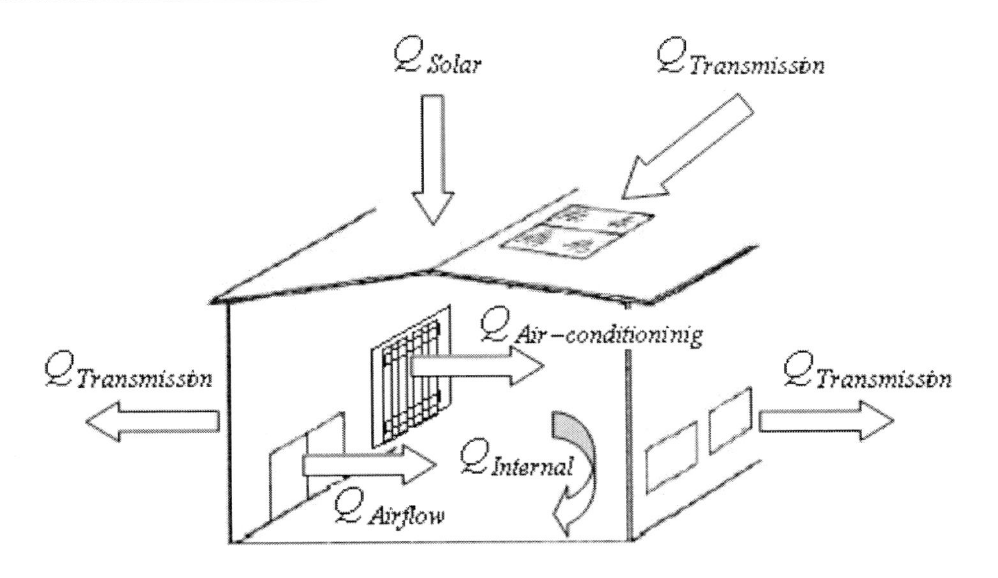

Figure 6. Schematic view different types of exchanges that occurs inside the tested room.

Due to convection between the air surrounding and the water inside the polypropylene heat exchanger, the transferred heat can be also written as:

$$Q_p = S\,U \frac{T_{w,o} - T_{w,i}}{Ln \dfrac{T_{w,o} - T_a}{T_{w,i} - T_a}} \tag{5}$$

The determination of the overall heat transfer coefficient, U, is obtained by using the electric analogy. The overall heat transfer coefficient, U, is given by [20, 21]:

$$U = \frac{1}{\dfrac{1}{h_{tube-air}} + \dfrac{1}{h_{tube}} + \dfrac{1}{h_{water}}} \tag{6}$$

where: $h_{tube\text{-}air}$, h_{tube} and h_{water} were calculated by using the relationsship given from appropriate tables [22, 23].

Eliminating Q_p from (4) and (5) gives the exponential relation for the outlet temperature of the tube as a function of the ambient air and inlet temperature:

$$T_{w,o} = T_a + (T_{w,i} - T_a)\ \exp(-(\frac{US}{\dot{m}\,C_p})) \tag{7}$$

The indoor average temperature is given by the relation:

$$T_m = \frac{1}{\exp(NUT) - 1}[T_{w,i}.\exp\ (NTU) - T_{w,o}] \quad (8)$$

$NTU = \dfrac{U\,S}{\dot{m}\,C_p}$ represents the non-dimensional group called the Number of Transfer Units.

The effectiveness of the air-conditioning cupboard is given by:

$$E = \frac{T_{w,o} - T_{w,i}}{T_a - T_{w,i}} \quad (9)$$

Using (7) the effectiveness becomes:

$$E = 1 - \exp(-(\frac{U\,S}{\dot{m}\,C_p})) \quad (10)$$

$Q_{\text{Transmission}}$ represents the heat transfer from exterior walls, windows, door and envelopes [24]:

$$Q_{Transmission} = h_i \sum_{j=1}^{Nw} A_{wj}(T_{wi} - T_a)_j + A_{gj}(T_g - T_a)_j + A_{dj}(T_d - T_a)_j + h_{iR} \sum_{j=1}^{NR} A_{Rj}(T_{Ri} - T_a)_j \quad (11)$$

Wall temperature must be calculated by mathematical modelling of the unsteady state heat transfer in the wall. It depends on radiation intensity, variations of external air temperature and accumulation of heat inside the wall. Rate of heat absorption due to radiation on black bodies can be calculated by:

$$Q_{Rad} = \alpha(G + D) \quad (12)$$

Q_{Solar} represents the heat due to sun radiation transmitted from windows. Thermal heat due to this phenomenon is calculated by using:

$$Q_{Solar} = \tau_G\, G + \tau_D\, D \quad (13)$$

$Q_{Internal}$ represents internal heat generated by lighting system, persons in the building and home appliances can be calculated by using [25]:

$$Q_{Internal} = Q_{Lihting} + Q_{Persons} + Q_{Appliances} \quad (14)$$

Heat generated by lighting system can be calculated by using:

$$Q_{Lighting} = P.F \quad (15)$$

Heat generated by persons and appliances can be found from appropriate tables.

$Q_{Airflow}$ is the heat due to airflow into the building (sensible and latent heat). Heat load due to outside air flow infiltration can be calculated by [26]:

$$Q_{Airflow} = Q_s + Q_L \quad (16)$$

where:

$$Q_s = \dot{V}\,\rho\, c_{p,a}\,(T_{a,0} - T_{a,i}) = \dot{m}_a\; c_{p,a}\,(T_{a,0} - T_{a,i})$$

(Sensible heat by entering air)

$$Q_L = \dot{V}\,\rho\,(W_i - W_0)\,\lambda$$

(Latent heat by entering air)

The optimal value of the exchange surface necessary to maintain the indoor climate to the required conditions is given by the relation:

$$S = \frac{\dot{m}\,Cp}{U}\; Ln\left[\frac{T_{w,i} - T_a}{T_{w,o} - T_a}\right] \quad (17)$$

4. Results and Discussions

4.1. Solar Storage Collector Characterisation

The solar storage collector constitutes the basic element for the air-heating loop. Therefore, outdoor experiments were carried out under winter weather conditions according to Input-Output test method. The solar collector's efficiency for a selected winter day of December is shown in figure 7. It can be seen that the efficiency starts to increase after dawn until it reaches its maximum value at noon, and then decreases again. In fact, the efficiency changed between 22 % and 40 %.

The solar storage collector enjoys a daily efficiency ($\bar{\eta}_j$) of 30 % achieved at an average solar radiation level of 620 W.m^{-2} and an ambient temperature of 20 °C. To appraise the solar storage collector heat loss coefficient, U_c, we have covered the solar storage collector by a 30 mm-thick polyurethane foam while the circulation of water inside the heat exchanger is stopped. Then we followed the concrete absorber temperature decrease during about 24 h. The overnight heat loss coefficient, U_c, is determined by equation 2. It is assumed to be equal to 14 W. °C^{-1}. The solar radiation and the solar storage outlet water temperature increases are represented in figure 8 (a, b). As expected, the outlet water temperature of the solar storage collector depends on solar radiation. During a typical sunny day of December the outlet water temperature increases to a maximum value of 50 °C at 12:30 pm in the noon and remains almost constant during 3 h before it starts to decrease later in the afternoon.

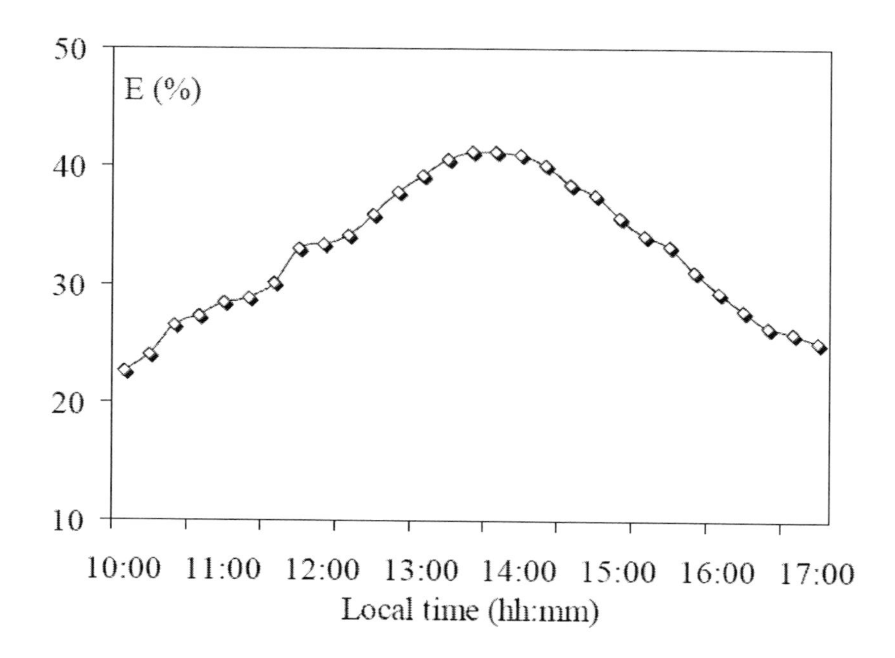

Figure 7. The variation of the solar storage collector efficiencies versus daytime.

During a typical winter cloudy day of December, characterized by a harsh solar radiation oscillation, we established that the outlet water temperature increases to a maximum value of 43 °C at 12:30 pm and remains almost constant during 3–2 h before it starts to decrease later in the afternoon. We noted also that the temperature of the water at solar collector outlet was not affected by solar radiation oscillation. In fact, the absorbed solar heat stored inside the concrete absorber is delivered to the capillary heat exchanger whilst the insulation starts to decrease (during night) or swing ruthlessly (in winter days). This stored energy is used for the air-room heating during the cold season until the night. A variation of stored energy for different times during a typical winter day of December is presented in figure 9. A maximum peak of stored energy, 3.5 kW, is obtained at 13:00 h and remains almost constant during 3 h then the stored energy decreases. To improve the stored thermal energy conservation we opt for covering the collector surface with an insulation blanket before the sharp decreasing of solar radiation (after 17:00 h).

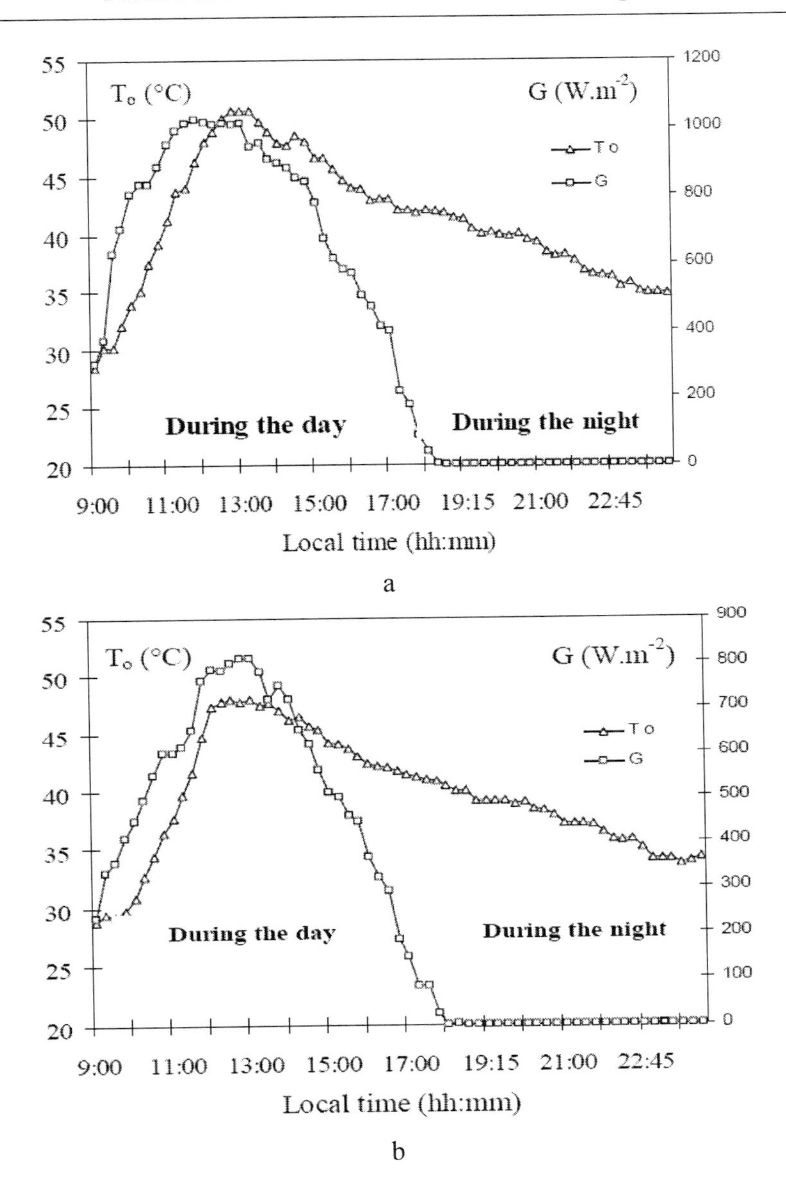

Figure 8. Outlet water temperature increase versus day times for mass flow rates and equal to 0.0416 $kg.s^{-1}$ for: (a) Sunny day of December (20 °C), (b) Cloudy day of December (16 °C).

Nightly, the solar storage collector supplies about 400 liters of hot water at a temperature of 45 °C (above the human's body temperature, 37 °C) until during winter. This quantity of heated water is circulated thought a tubing system to the air-conditioning cupboard. Compared to the conventional solar collector, the concrete solar collector, that does not need a storage tank, presents satisfactory performances and a low cost compared with a conventional solar collector (table.2).

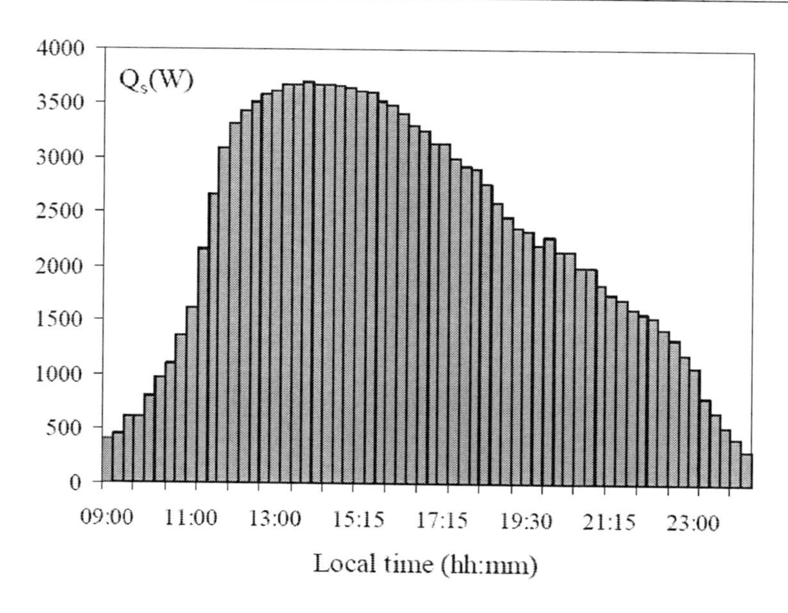

Figure 9. Stored energy within concrete variation versus time of day.

Table 2. Comparison between a conventional solar collector and solar storage collector

	Conventional solar collector	Solar storage collector
Thermal efficiency	38%	30%
System cost	800 $	250 $
1kWh selling price	0.05 $	0.02 $
Payback time	8 years	3 years

4.2. Thermal Behavior of the Air-Conditioning Cupboard

4.2.1. The Air-Conditioning Cupboard Sizing

The model allows calculating the optimal exchange surface of capillary heat exchanger to place inside the air-conditioning cupboard to reach the required air-room temperature. The estimation of the heat exchanger surface to place inside the air-conditioning cupboard should take into account many parameters; (i) the heat exchanger performances (overall heat exchange coefficient, efficiency and the optimal function parameters (heat exchanger length and water masse flow), (ii) the air-conditioning cupboard inlet water temperature (For heating and cooling load), (iii) the room climatic needs and (iv) the difference between indoor and outdoor temperature.

An analytical method is developed for the accurate calculation of the overall heat transfer coefficient and the thermal efficiency of the air-conditioning cupboard. It consists of measuring the amount of the temperatures of the inlet and the outlet of the heat exchanger and the air-room average temperatures. Experiments were carried out, varying the flow rate, the heat exchanger length (The width is fixed to 1 m) and the inlet temperature. Results show that the maximal value of overall heat exchange coefficient (28 $W.m^{-2}.°C^{-1}$) is obtained for a network of heat exchanger with 4 m length crossed by water at a masse flow equal to 200 $l.h^{-1}$ (figure 10).

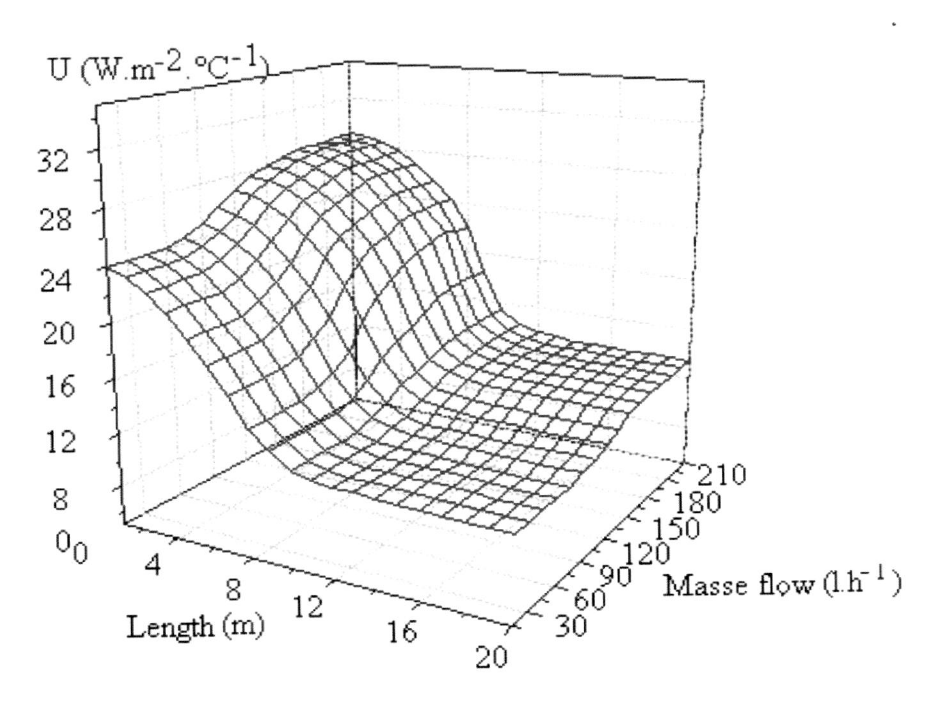

Figure 10. Numerical variation of the overall heat exchange coefficient according to the length and masse flow rates and a temperature equal to [20 °C – 70 °C].

The numerical and experimental efficiency profiles, for three network of capillary heat exchanger with a fixed width (1 m) and different lengths (3m, 4m and 5m) and for different masse flow rates are plotted in figure 11. We noted that the net energy efficiency increase with masse flow rates and heat exchanger lengths. The efficiency reaches the higher value of 80 % achieved for a heat exchanger length and masse flow rates respectively equals to 4m and 200 $l.h^{-1}$. We established that using a network of heat exchanger with 1 m width and a length superior of the optimal value of 4m dose not increase the capillary heat exchanger efficiency. We established also that for this optimal length (4m) corresponds an optimal value of masse flow rate equal to 200 $l.h^{-1}$. It is obvious from the figure 11 that using a masse flow and a length lower than 200 $l.h^{-1}$ and 4 m decrease seriously the cupboard efficiency.

The sizing of the air-conditioning cupboard depends also on the difference between indoor and outdoor temperature and the tested room annual heating/cooling load needs. The numerical model described in this chapter (§ 3.2) permit the estimation of the tested room energy needs to maintain the required indoor climate. Numerical results are shown in figure 12 (heating need is shown as positive and cooling need is shown as negative). It is seen that for heating and cooling cases the maximal needs are 4.2 kW and 4.6 kW, respectively.

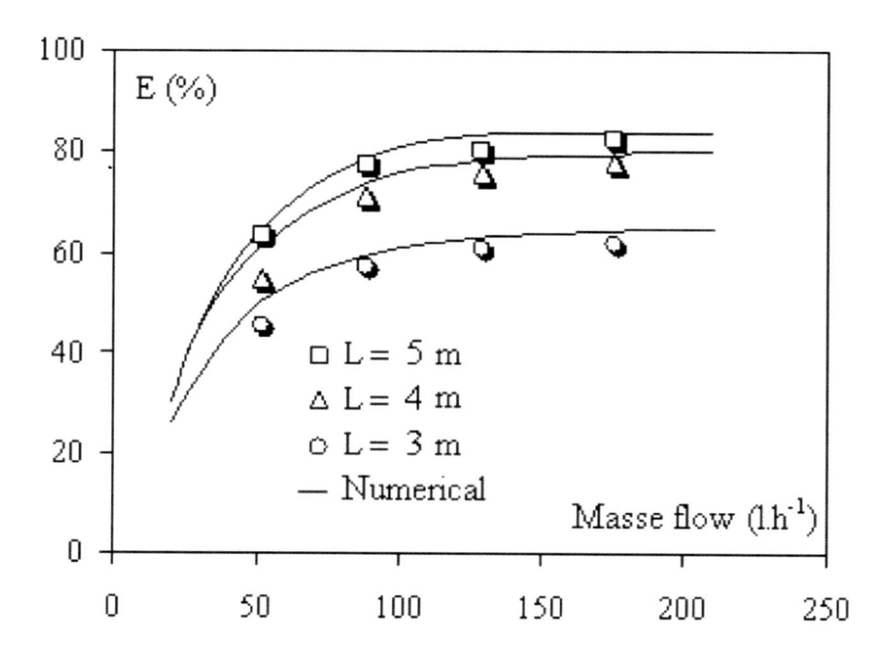

Figure 11. The variation of the solar storage collector efficiencies for different heat exchanger length and versus masse flow rates.

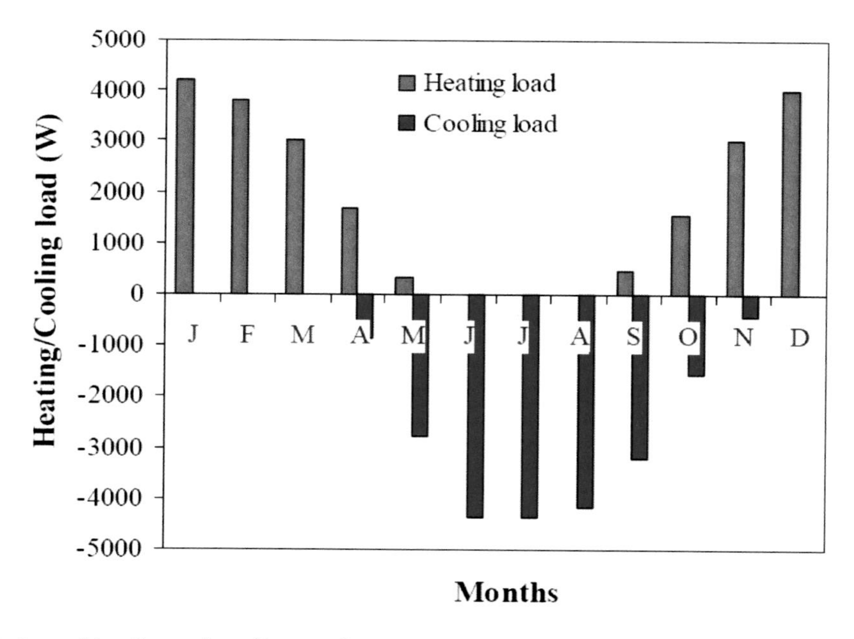

Figure 12. Annual heating and cooling needs.

For the Heating Load

To maintain the temperature of the tested room to the value of 26 °C it is necessary to place inside the air-conditioning cupboard 9m^2 (Almost 2 networks of heat exchanger with the size; 1m x 4m) of the capillary heat exchanger crossed by hot water with a constant masse flow and temperature respectively equal to 200 l.h^{-1} and [40-50°C]. In figure 13, is represented the calculated variation of the thermal heating load as well as the thermal heating needs to reach required temperature value within the tested room. Numerical results show that the air-heating loop provides a maximal value of thermal energy about 3.5 kW to increase the air room temperature of about [7-10 °C]. It is also seen that the heat load provided by the solar storage collector represents about 83 % of the annual thermal heat needed to maintain the air-room temperature at about 26 ± 3°C.

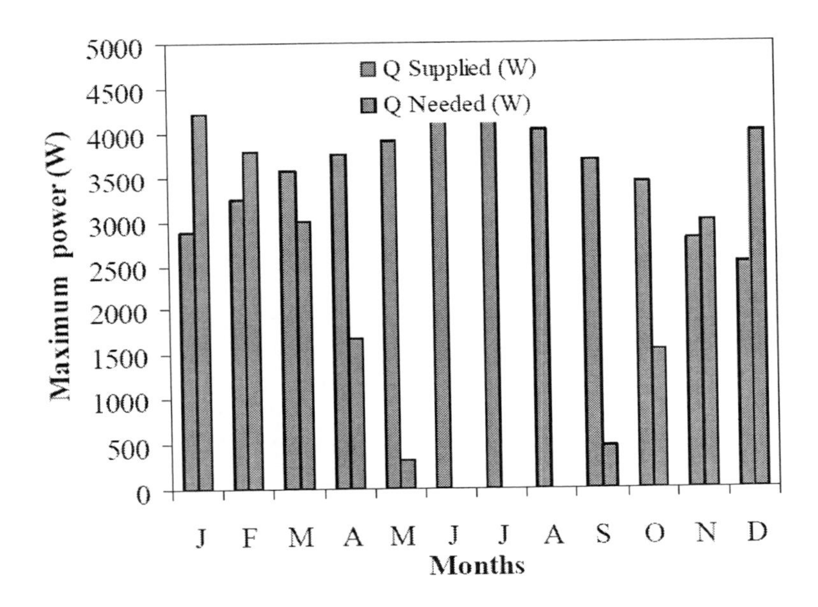

Figure 13. Maximum power in excess estimated and contribution heating necessary to maintain the air-room temperature, 26 °C.

For the Cooling Load

The heat exchanger surface to place within the air-conditioning cupboard is assumed to be equal to 11 m² (almost 3 networks of heat exchanger with the size; 1m x 4m) crossed by cold groundwater (19°C) with a masse flow about 200 l.h^{-1}. In figure 14, is represented the calculated variation of the thermal cooling load as well as the variation the thermal energy needs which should be evacuated to maintain the required temperature in the tested room at the value of 23°C. The results show that it is necessary to provide a refrigerated energy of the order of 4.3 kW to maintain the room temperature at the required value. The heat load that should be extracted by air-conditioning system represents 95 % of the annual thermal heat needed to maintain the air-room temperature at the required value.

Table 3 summarises the needed and loaded energy rates according to the required indoor temperature. It is seen that both the heating and cooling loops cover easily the energy needs especially for cooling purposes. This is explained by the fact that the frigorific source

(Groundwater) represent a constant temperature for all over the year (19 °C) while the hot water supplied by the solar collector present a variable temperature [40-47°C]. This is due to the fact that the water at the solar storage collector outlet depends on solar insoltaion.

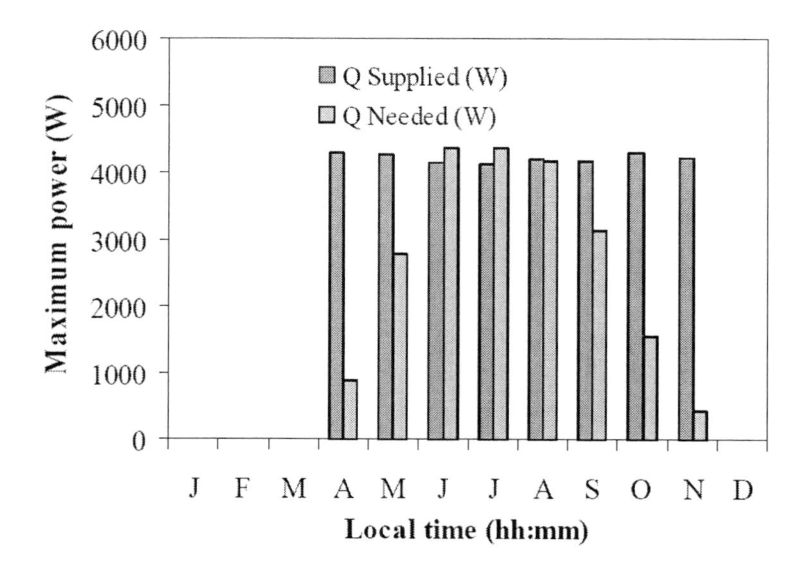

Figure 14. Maximum power in excess estimated and contribution cooling necessary to maintain the air-room temperature, 26 °C.

Table 3. Summary of the loads/needs

	Maximum (kW)	Annual consumption (MWh)	(%) consumption
Heating load	3.5	70.4	83
Heating need	4.2	86.4	
Cooling load	4.1	89.2	96
Cooling need	4.3	93.5	

4.2.2. Performances of the Air Conditioning System

The air-heating loop and the air-cooling loop were respectively investigated during cold months of the year (November to April) and all through the hot months of the year (May to October). Inside the air-conditioning cupboard was fixed 12 m^2 (3 networks of polypropylene heat exchanger with the size; 1m x 4m) of a capillary heat exchanger. The hot/cold water masse flow was fixed within the air heating/cooling loops at the value 200 $l.h^{-1}$. Many thermocouples (Types T and K) were set in different places inside the tested room to indicate the evolution of the indoor climate for different applications and for the overall experimental investigation.

For the Air-Heating Load

The hot water supplied by the solar storage collector is pumped to the capillary heat exchanger positioned within the air-conditioning cupboard. The figure 16 represents the air temperature field in the tested room. The results show that the use of hot water [30 °C-45 °C] provided by the solar storage collector, permits to increase the air room temperature about 7 to 10 °C. We noted that the indoor temperature reach it is maximum value after 40 minutes of air-conditioning function. Figure 15 shows also that the more elevated temperature reach in the local (30 °C) is gotten at 2m-height. In fact; inside the tested room gets settled a temperature gradient about 2.5 °C/m.

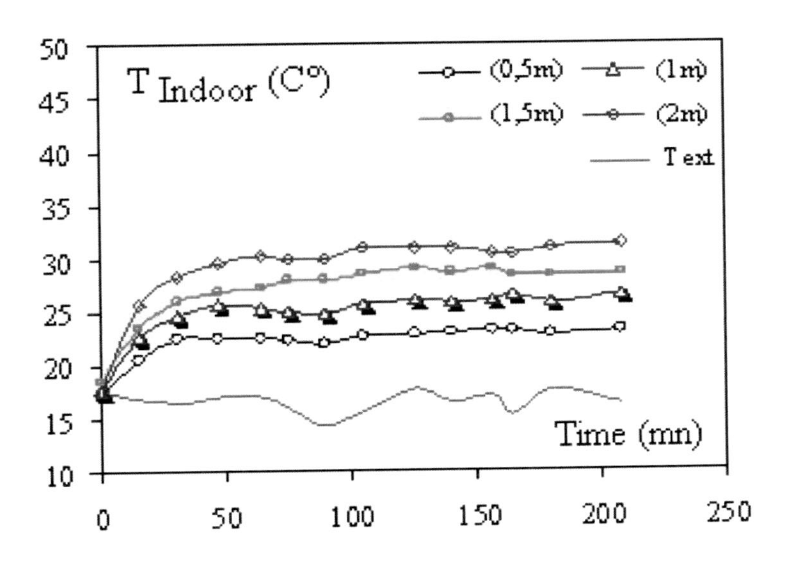

Figure 15. The air temperature vertical profile in the tested room.

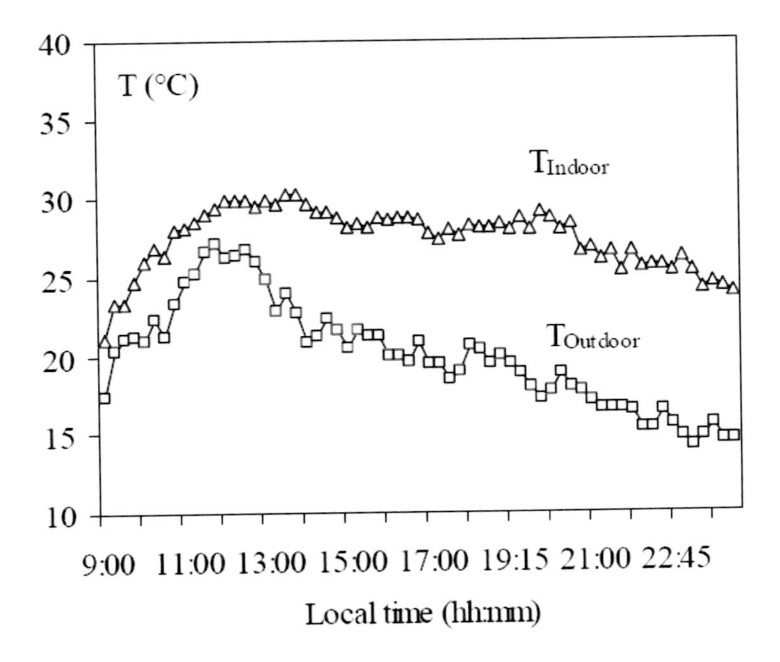

Figure 16. Variations of indoor and outdoor temperature.

In an independent test we kept the air-heating system operating continuously for about 14 h. The result of the experience is drawn in figure 16 which shows the inside and outside temperatures variation for a selected winter day (December. 15th, 2006). The figure shows that the indoor temperature is lower in the early morning and then starts to increase until it reaches its maximum value (30°C) in the afternoon beforė decreasing again. This is typical of Tunisian climate during winter due to the fact that a great period of the days is sunny.

The air heating loop demonstrates a satisfactory ability to meet a moderate indoor climate compared to the outdoor climate. In fact, the gap between the indoor and the outdoor temperature reach 4-10 °C especially during the night. The average solar storage energy inside the concrete absorber is 21.74 kW in the whole operation (14 h), and it attains 25.04 kW during the working hours from 9:00 to 17:00, which is sufficient to keep indoor thermal environment at the average temperature 26 ± 2 °C.

Figure 16 shows also that the inside temperature was kept around the desired value (26°C) for a great period of the day and night (From 10:00 to 21:00). With respect to the whole heating period, the air-heating loop was capable of meeting heating requirement in 60 days. However, during a cloudy day the fluctuation of the air temperature in the experimented room becomes more serious due to the daily insolation oscillation. In this case the stored solar energy is also affected by the critically insolation decreasing.

For the air-Cooling Load

The cooling loop usually used in summer when the average outdoor temperature and relative humidity are respectively about 37.3 °C and 50 %. Figure 17 represents the indoor temperature field. Results show that the use of the groundwater well (19 °C), for the air-conditioning cupboard, permits us to decrease air room temperature about 8 °C. On the other hand, the air room temperature vertical profile shows that the weakest temperature reached in the local (23 °C) is gotten to a height of 0.5 m from soil. Same as the heating case we found inside the tested room the same temperature gradient (about 2 °C/m). Until in the cooling case, the indoor temperatures reach their maximum values after 40 minutes of air-conditioning function.

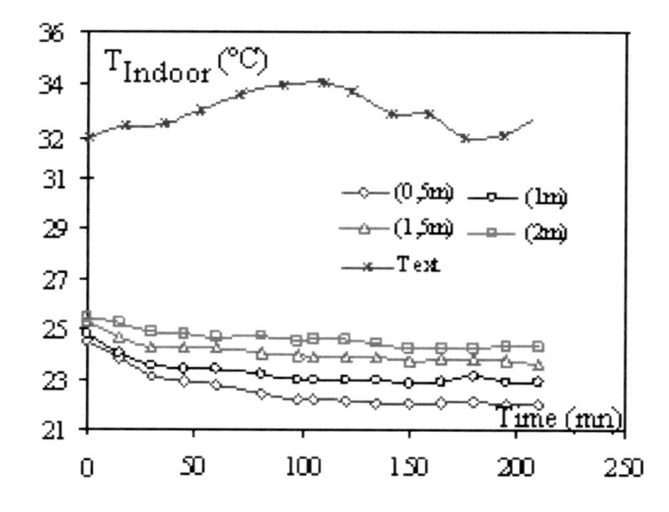

Figure 17. The air temperature field in the tested room.

Figure 18 represents the numerical and experimental variation of the air room temperature as well as the daily outside air temperature evolution. The results show that the air cooling loop permits the decreasing of the room air temperature by about [4–13°C]. We also note that although the outdoor temperature increases seriously, mainly during the included time period between 11:00 h and 16:00 h, the indoor temperature doesn't vary; it is about 23 °C.

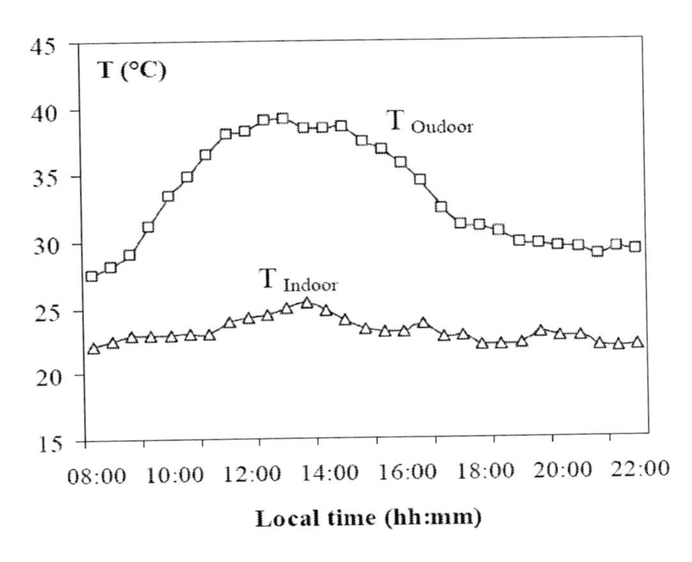

Figure 18. Variations of all-day room temperatures (heated and non heated).

CONCLUSION

Based on the results obtained from the performance analysis of the passive air-conditioning system in a demonstration room, the following results are obtained:

i. The air-conditioning system enjoys a maximum overall heat exchange coefficient (28 $W.m^{-2}.°C^{-1}$) and a total efficiency of about 80% achieved for an optimal heat exchanger length and masse flow rates respectively equal to 4m and 200 $l.h^{-1}$.
ii. The numerical model allows the estimation of the tested room heating/cooling energy needs. For heating and cooling loads the maximal needs are estimated to be respectively equal to 4.2 kW and 4.6 kW.
iii. It is also seen that to maintain the temperature of the tested room to the value of 26 °C (during heating loop function) it is necessary to place inside the air-conditioning cupboard 9 m^2 of the capillary heat exchanger crossed by hot water with a constant masse flow and temperature range respectively equal to 200 $l.h^{-1}$ and [40–50°C]. The air-heating loop provides a maximal value of thermal energy about 3.5 kW to increase the air room temperature about [7–10 °C].
iv. For the cooling load, the heat exchanger surface to place within the air-conditioning cupboard is assumed to be equal to 11 m^2 crossed by cold groundwater (19°C) with a

masse flow about 200 $l.h^{-1}$. The thermal cooling load provides 4.3 kW of frigorific energy to maintain the room temperature at the value of 23 °C.

v. For both heating/cooling loops, the heat load that should be provided/extracted by the air-conditioning system covers easily the energy needs. They represent between 83 % and 95 % of the annual thermal heat needed to maintain the air-room temperature at the required value during the entire heating/cooling periods. In fact, the air heating/cooling loops demonstrate a satisfactory ability to meet a moderate indoor climate compared to the outdoor climate by maintaining during the night a gap between the indoor and the outdoor temperature (about 4–13°C).

Acknowledgments

The authors gratefully acknowledge the support of this work through Laboratoire de Maitrise des Technologies de l'Energie, LMTE de Borj Cedria, Tunisia.

References

[1] Niu JL, Kooi Jvd, Ree Hvd. (1995). Energy saving possibilities with cooled-ceiling systems. *Energy and Buildings*. 23, 147–58.

[2] Niu JL, Kooi Jvd. (1994). Indoor climate in rooms with cool ceiling systems. *Building and Environment*. 29, 283–90.

[3] Mertz G. (1992). Chilled ceilings and ventilating systems—thermal comfort and energy saving. *Air Infiltration Review*.13, 7–10.

[4] Jager F. (1981). *Solar energy applications in houses*. Luxembourg, Commission of the European Communities.

[5] ASHRAE. (1977). *Handbook of Fundamentals*. New York: American Society of Heating and Air Conditioning Engineers.

[6] Yellott JI. (1971). Solar heating and cooling of houses. In: Sayigh AM, editor. *Solar Energy Engineering*. Academic Press, New York, (chapter 17).

[7] Kiruma K. (1983). Utilization of solar energy, the Japanese experience in solar energy application in the tropics. New York: Reidel Publishing Company.

[8] McVeigh J. (1983). Sun power, an introduction to the application of solar energy. Oxford (UK): Pergamon Press.

[9] Lof G, Karaki. (1983). System performance for the supply of solar heat. *Mechanical Engineering*, Des 33±47.

[10] Lundsager P. (1996). Integration of renewable energy into local and regional power supply. World Energy Council,117–22.

[11] Archibald J. (1999). Building integrated solar thermal roofing systems history, current status, and future promise. *Proceedings of the solar 99 conference*. Maine: American Solar Energy Society (ASES). 95–100.

[12] M. Hazami, S. Kooli, M. Lazaar, A. Farhat, A. Belghith. (2004). Heat transfer characteristics of a capillary heat exchanger based air conditioning cupboard. *Desalination*. 166, 435-442.

[13] M. Hazami, S. Kooli, M. Lazaar, A. Farhat, A. Belghith. (2005). Performance of a solar storage collector. *Desalination*. 183 167–172
[14] R. Ranjit (1990). A primer on the Taguchi method. Dearborn, MI: Society of Manufacturing Engineers.
[15] Tsui K-L. (1992). An overview of Taguchi method and newly developed statistical methods for robust design. *IIE Trans*;24(5):44–57.
[16] Bourges B., Rabl A., Leide B., Carvalho M. J. and Collares. Pereira M. (1991). Accuracy of the European solar water heater test procedure. Part 1: Measurement errors and parameter estimates. Part 2: Prediction of long-term performance. *Solar Energy*. 47, 1–25.
[17] T. Hong, Y. Jiang. (1997). A new multizone model for the simulation of building thermal performance, *Building and Environment*. 32 (2), 123–128.
[18] M.S. Hatamipour, H. Mahiyar, M. Taheri. (2007). Evaluation of exising cooling sustems for reducing cooling power consumption. Energy and Buildings. 39, 105-112.
[19] M. Hazami. (2008). Etude expérimentale et numérique du phénomène de stockage et du déstockage de l'Energie thermique en utilisant un échangeur capillaire en polypropylène. Thesis. 124-127
[20] K. S. Lee et J. Y. Yun. 1999. Investigation of heat transfer characteristics on various kinds of fin and tube heat exchangers with interrupted surfaces", *Int. J. Heat and Mass Transfer*. Vol. 42.
[21] J. Claesson. (2005). Correction of logarithmic mean temperature difference in a compact brazed plate evaporator assuming heat flux governed flow boiling heat transfer coefficient. *International Journal of Refrigeration*. 28 573–578.
[22] A. Batasani, M. Feibig, N. K. Mitra. (1992). Numerical studies of a compact fin tube heat exchanger", *Design and Operation of Heat Exchangers*, Springer-Verlag, Berlin. 154-163.
[23] M. Feibig, A.Grosse-Gorgemann, Y. Chen, N.K. Mitra. (1995). Congugate heat transfer of finned tube part A : heart transfer behavior and occurrence of heat transfer reserval", *Numerical Heat Transfer*. Part A 28, pp 133-146
[24] A. Sharian, B. Shalabi, A. Rousan, B. Tashtoush. (1998). Effects of the absorptance of external surfaces on heating and cooling loads of residential buildings in Jordan. *Energy Conversion & Management*. 39 (3/4) 273-284.
[25] Y. Jiang. (1981). State space method for analysis of the thermal behavior of rooms and calculation of air-conditioning load, *ASHARE Transactions*. 88, 122–132.
[26] E. Shaviv, A. Yezioro, I. G. Capelurto. (2001). Thermal mass and night ventilation as passive cooling design strategy. *Renewable Energy*.24, 445-452.

In: Air Conditioning Systems
Editors: T. Hästesko, O. Kiljunen, pp. 293-314
ISBN: 978-1-60741-555-8

Chapter 7

MULTI-PURPOSE OPERATION PLANNING OF THE MICRO-ENERGY SOURCE FOR AIR CONDITIONING WITH RENEWABLE ENERGY

Shin'ya Obara *

Department of Electrical and Electronic Engineering, Kitami Institute of Technology
165 Kouen-cho, Kitami, HOKKAIDO 0908507, Japan

ABSTRACT

The complex system for air conditioning heat sources composed from a fuel cell system and renewable energy has a multi-purpose operation. So, the analysis method for operational planning of the complex system, when two or more pieces of energy equipment were introduced in an individual house, was developed. The cooperative control with two or more objective functions of the complex system was simulated using a genetic algorithm. In this Section, details of the proposed analysis method are explained, and a case study is described. An active energy device, a renewable energy device, and an unutilized energy device were connected by an energy network, and a preliminary survey of the system design required for cooperative operation was conducted. A genetic algorithm, which can analyze nonlinear problems and many variables at a time, was used for the software for the operational planning of the complex system developed in this study. As for the energy network to reduce energy costs and the environmental load, successive introductions are predicted.

NOMENCLATURE

C : cost (Dollar)

D_i : energy device

E : electric power (W)

E_c : chemical equivalent (equivalent)

* E-mail: obara@indigo.plala.or.jp; Phone and FAX: +81-157-26-9262

E_{EL} : amount of electric power storage (J)
E_V : voltage (V)
ΔE : consumption of electric power (W)
F : quantity of fuel mass flow (g/s)
F_d : Faraday constant (C/g)
f_m : objective function
H : thermal energy (kW)
ΔH : consumption of thermal energy (W)
M : number of energy devices
N_{dv} : number of chromosome models
P : probability
Q : number of select switches
R : operation period of system (s)
S_E : amount of electric power storage (J)
S_{St} : amount of thermal energy storage (J)
t_k : sampling time (s)
Δt : sampling time interval (s)

Greek Symbols

α : calorific power of fuel (J/g)
ϕ : efficiency
ρ : density of thermal storage medium (g/m^3)

Subscripts

EL : water electrolyzer
FS : fuel cell stack
H : electric heater
HP : geothermal heat pump
Rad : radiator
SL : solar module
St : thermal storage tank

1. Introduction

Until now, various energy devices with individual controls have been used in buildings. The ultimate goal of this research is to connect distributed active energy, renewable energy and unutilized energy using an energy network and developing an analysis method for an operational plan to conduct a cooperative operation. The energy network is structured using an electric power system, a hot water system, and a fuel system. As for how the energy network will reduce energy costs and the environmental load, successive introductions are predicted. For the operational plan of the energy network that conducts the cooperative operation of complex energy devices, it is necessary to solve the nonlinear problem of many variables with objective functions provided beforehand. In the optimization calculations of system operational planning of a complex energy system, linear approximation calculations based on the mixed-integer plan-making method was used [1]. However, to analyze the operational planning of a complex energy system with high accuracy, it is necessary to solve the nonlinear problem with many variables. A genetic algorithm [2] is therefore introduced to analyze operational planning in this study. Previously, an analysis method of a large-scale energy system that combined a genetic algorithm and an annealing algorithm [3] was developed [4, 5, 6]. However, an analysis method that optimizes the operational pattern with the application of a genetic algorithm (GA) to the compound system built using an active energy device, a renewable energy device, and an unutilized energy device has not yet been developed. In this Section, a GA analysis method for a compound energy system is developed as a preliminary survey of the energy network that conducts cooperative operation. The analysis software using GA developed in this study is introduced in an individual house in Sapporo, Japan, which is a cold, snowy area, and the operational plan is investigated. The operational planning of the compound energy system is analyzed using the minimization of operational costs and the maximization of renewable energy use, and the operational planning of an active energy device is considered. Although operation costs and facility costs need to be considered for a feasibility study of the system, the facility costs of a solid polymer membrane fuel cell are changing greatly. Since estimating facility costs is difficult, the analysis in this study only considers operation costs. Furthermore, the device capacity for the accumulation of electricity and thermal storage is estimated.

2. The Energy Network and Combined Energy System

2.1. Network of Distributed Energy Devices

In a dispersed arrangement of small energy devices, a reduction in power transmission loss and heat dissipation loss is expected. Since the discharge of carbon dioxide is predicted, renewable energy devices and unutilized energy devices are connected along with established active energy devices in an energy network, and research on supplying energy to two or more houses is required. A network model of the fuel cell cogeneration (CGS) installed in individual houses, as assumed for the final target of this research, is shown in figure 1. The fuel cell CGS installed in each house is connected with hydrogen gas system piping, an electric system power line, and hot water piping of an exhaust heat system. The hot water system recovers heat from fuel cells and supplies thermal energy to individual houses. Hot water flows in one direction, as shown by the arrows in figure 1. The energy devices are connected to the electric power and thermal energy network, and the operational planning of a system that fulfills the energy demands of individual buildings is considered. The energy devices installed in each house were controlled by autonomous distribution. The objective of an energy network is to control the devices linked to the network cooperatively, and to obtain a better effect than conventional autonomous distribution control.

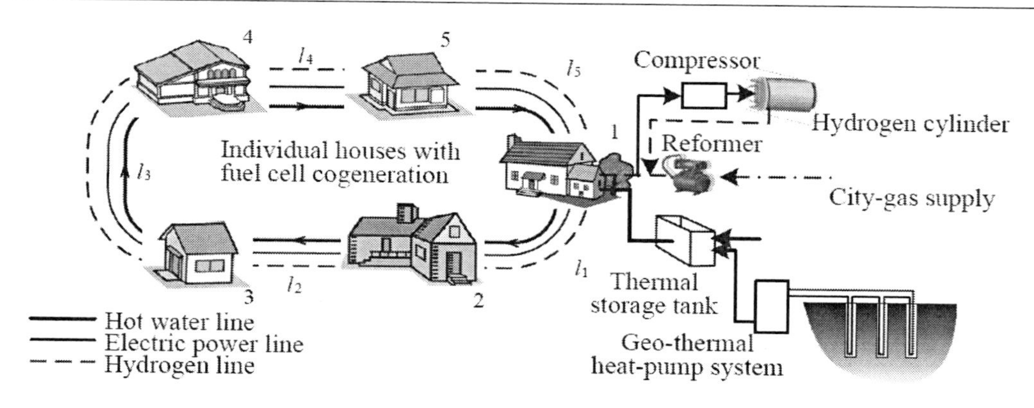

Figure 1. Fuel cell network system model.

Figure 2 shows a model of the cooperative operational control of an energy network. The control device of the energy network is composed of a computer, a communication device, and a LAN that communicates control information for each energy device. In this system, the operational state of each energy device linked to LAN, weather information and maintenance information can be communicated to the outside.

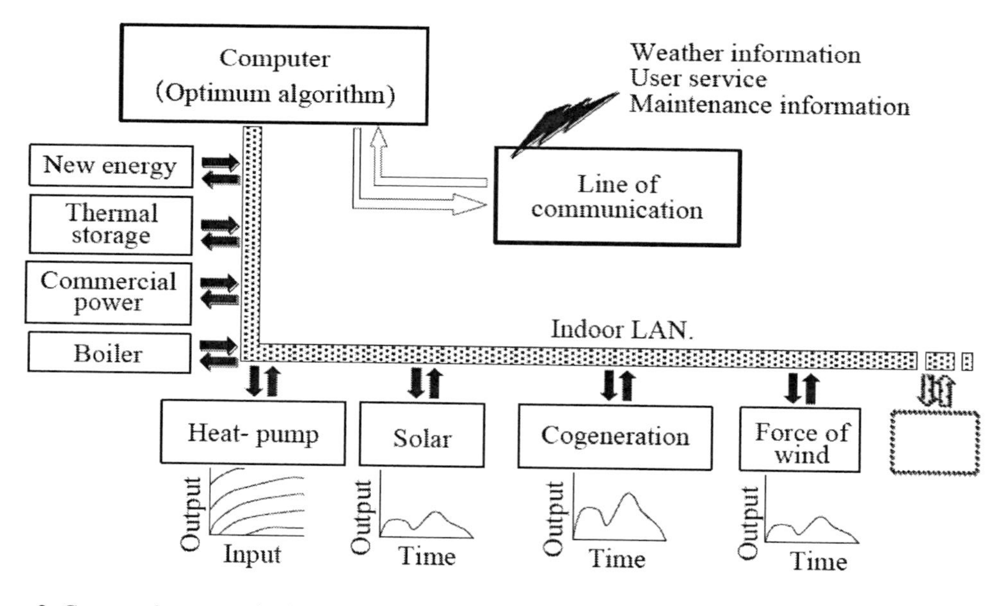

Figure 2. Cooperative control of energy equipments for local energy network.

2.2. The Combined Energy System

A feasibility study of the operational planning analysis method of the energy network with cooperative control, shown in figure 1, is the target of this study. The analysis method in the case of operating the compound energy system consisting of an active energy device, a renewable energy device, and an unutilized energy device has been developed. Figure 3 shows the model of the compound energy system. The analysis method for operational planning using GA has been developed to minimize costs, and the estimated device capacities and an operational plan for a complex energy system are determined.

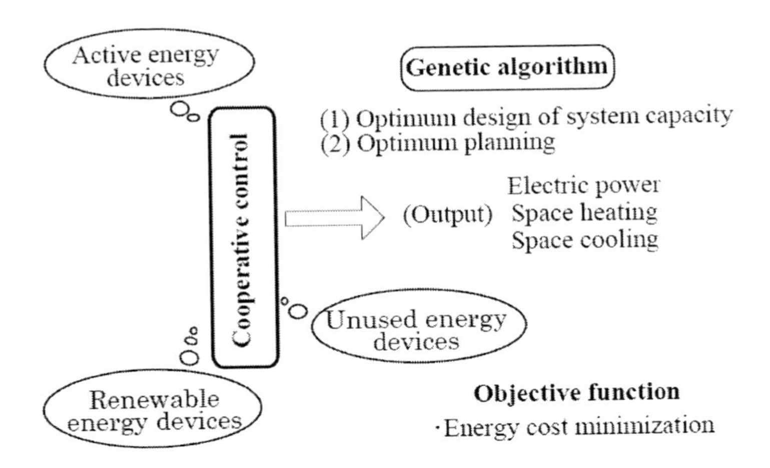

Figure 3. System optimal design.

2.3. Indication of Device Operation by the Chromosome Model

Figure 4 shows the chromosome model introduced in GA, and expresses information on electric energy output E_{D_i,t_k} for each time t_k of device D_i, heat output H_{D_i,t_k}, amount of electric energy storage S_{E,D_i,t_k}, amount of thermal storage S_{St,D_i,t_k}, and device selection switch S_{W,D_i,t_k} using the gene model with 0 and 1. When two or more devices do not yield a simultaneous energy supply, S_{W,D_i,t_k} is introduced in order to select the device that supplies energy. As the chromosome model determined above expresses the operational pattern of the device from t_k to t_{k+1}. as shown in figure 5, sets of the chromosome model of each sampling time of $k = 0,1,2,\cdots,R$ represent all the operational patterns for operational period R. Although a number of chromosome models N'_{dv} are created as an initial generation, either 0 or 1 is selected. If the value of the random number is less than 0.5, the gene model is set at 0, and 1 is selected if the random number is 0.5 or more.

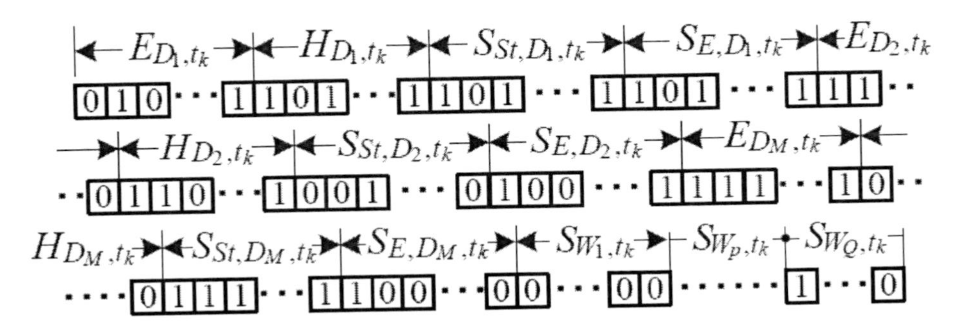

Figure 4. Chromosome code.

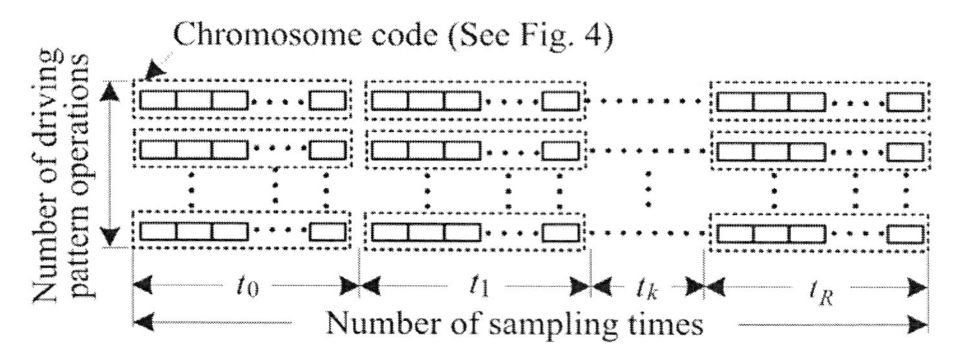

Figure 5. Driving pattern indicated by chromosome code.

2.4. Production, Selection and Reproduction

The fitness values of the number of N'_{dv} chromosome model groups (they indicate the patterns of the system operation) of an initial generation are calculated, and proliferation or selection is judged based on the values. The combination method is introduced in the calculation of the ranking selection [7] and roulette selection [2]. In the first reproduction calculation, the chromosome models of the initial generation are selected based on the number of N_{dv} (here, $N'_{dv} > N_{dv}$), and these chromosome models are used in subsequent calculations.

2.5. Crossover and Mutation of the Chromosome Model

The calculation process of crossover and mutation is given to the chromosome model group, and the diversity of genes is maintained. Using the calculation for the last generation, the chromosome model with the best fit is determined as the optimal operational pattern. However, the number of generations in the analysis is decided beforehand. When using the chromosome model group with the crossover process, only a specific chromosome model evolves beyond a certain generation, and a model with high fitness cannot be found beyond it. In the calculation of crossover, two parent chromosomes are chosen by probability P_{cros}, parent chromosomes are combined, and one child chromosome is generated in the intersection position decided at random. Subsequently, the calculation process of the mutation described below is added. In the mutation, parent chromosome models are chosen at random using probability P_{mut}, and the number and the position of the genes of the parent chromosomes are also decided at random. If the original value of a gene is 1, it has to change to 0, and if it is 0, it has to change to 1. In order to progress to the next generation, the fitness value is again evaluated with respect to all the operational patterns of number N_{dv} with added crossover and mutation. Proliferation and selection are performed using these results. The above analysis is repeated up to the number of the last generation, and the gene arrangement of the model that has the highest fitness value in the chromosome model group of the last generation is decoded, and the optimal operational pattern is decided.

2.6. Analysis Flow

The flow of calculation of the operational planning analysis of the complex energy system using GA is shown in figure 6. First, N'_{dv} chromosome model groups described in Section 2.4 are generated at random. The fitness values for each chromosome model are calculated, and the chromosome models of N_{dv} higher ranks are determined by the combination of ranking selection and roulette selection. Furthermore, the calculation of production and selection described in Section 2.5 is added to these N_{dv} chromosome models, and a chromosome model with a large fitness value is obtained, maintaining diversity by the calculation of crossover and mutation described in Section 2.6. This calculation is repeated, and the chromosome model with the highest fitness value when reaching the number of the last generation, decided beforehand, is determined as an optimal model. Operational planning of all energy devices for each sampling time is decided by decoding the optimal model.

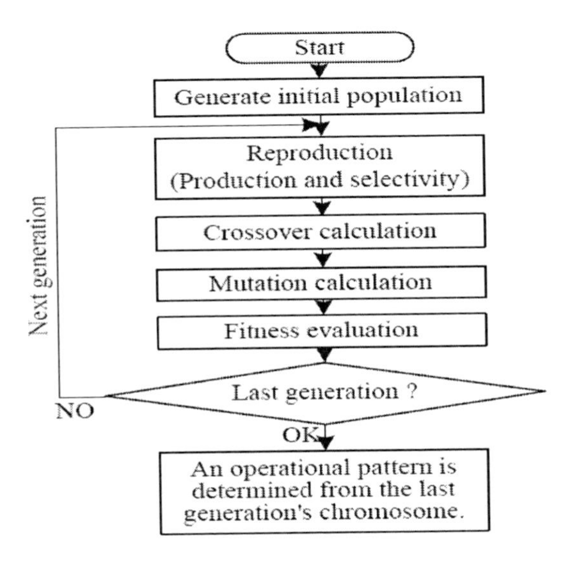

Figure 6. Genetic algorithm flow.

3. Independent Energy System for Cold-Region Houses

3.1. Characteristics of Weather in Sapporo in Japan

Sapporo is a cold, snowy region, and the annual average temperature for the past five years is 282 K [8]. The average temperature in February is 270 K, and the highest and the lowest temperatures on a representative day are 273 K and 266 K, respectively. Moreover, there is an average 25 days of snowfall in February. On the other hand, the highest and the lowest temperature on a representative July day for the past five years are 298 K and 290 K, respectively, and the average temperature is 293 K. Since air heat-source heat pumps cannot be used in winter, the use of a geothermal heat pump is assumed in this study. Air conditioning is not needed during summer.

3.2. Characteristics of Individual Houses in Sapporo

The average individual house in Sapporo is a 2-story wooden house with a 140-m^2 living area. The model of the average electric power and thermal energy demand of the representative February day and July day for individual houses in Sapporo is shown in figure 7. The thermoelectric ratio of representative days is 0.90 : 0.1 in February and 0.5 : 0.5 in July.

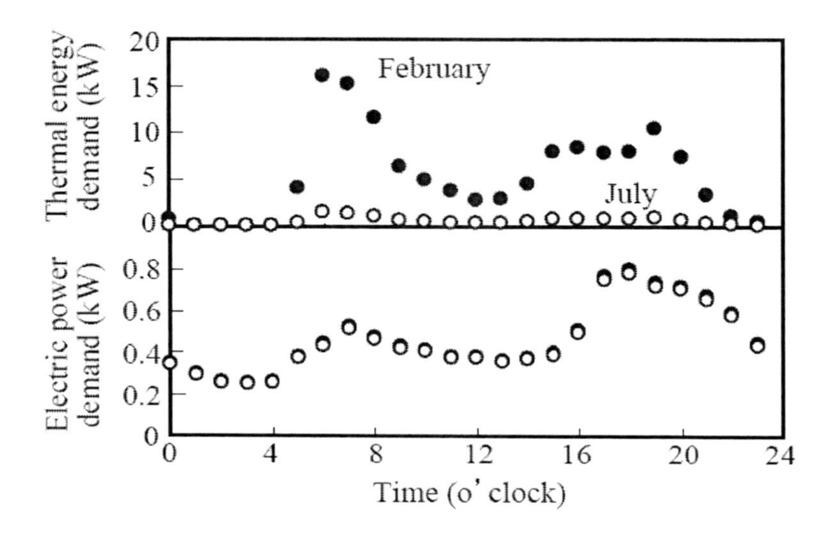

Figure 7. Energy demand of a house in Sapporo city [7].

3.3. The Combined Energy System to Be Assumed

A block diagram of the energy system for houses adopted in this study is shown in figure 1. In this system, methanol fuel is stored, and its distributed power supply is also possible in the residential areas of local cities where the city gas piping networks are not well developed. In individual houses and apartments, load changes are sharp and abrupt, and partial load operation of the energy system increases [9]. Therefore, to improve energy efficiency, a dynamic operational plan for the energy system is required, wherein thermal storage and electric energy storage are introduced. A water electrolytic bath and gas tanks are added to the electric power storage device, and electric power generated with a fuel cell and a solar cell is supplied to the water electrolysis bath. Since the output of a solar cell changes according to the weather, the operational plan is defined by considering the amount of power generation as a variable. Furthermore, when installing the fuel cell cogeneration system in houses in cold regions, since the supply of heat energy is insufficient, there is additional combined use of the geothermal heat pump system.

3.4. Operation Method of Combined Energy System

Methanol fuel (mole ratio of methanol/water = 1.0/1.4), which is contained in the methanol tank (3), is supplied to the reformer (2), and hydrogen and carbon dioxide are formed. The heat source of the reformer drives the catalytic combustion oam reformer is sent to the anodes, air is supplied to the cathodes by a blower, and electricity is generated by the proton exchange membrane fuel cell (1). The energy supply path for this system is shown in figure 8. The electric power generated by the fuel cell is supplied using one of the following methods: (a) Alternating current electric power is generated by the DC/AC converter (8), and demand is fulfilled; (b) Hydrogen and oxygen are generated in the water electrolyzer (5) and stored in a hydrogen tank (6) and an oxygen tank (7), respectively; (c) Electric power is changed into heat by an electric heater (9) in a thermal storage tank (10). It is possible to drive a fuel cell at any time using the stored hydrogen and oxygen. Selecting the appropriate energy supply path among (a) to (c) above is also possible for electric power generated by the solar cell.

However, electric power from the system to the demand side is supplied only via one of the following systems, without multiple supply sources: (a) Methanol fuel is reformed to generate hydrogen, which is supplied to a fuel cell, and electric power is generated; (b) Electric power is generated by the solar cell; (c) Stored hydrogen and oxygen, formed by water electrolysis, are used in the fuel cell to generate electric power. In order to reduce the discharge of carbon dioxide, methanol fuel is not used to the extent possible. Therefore, as many renewable energy supply sources as possible are used with priority set in decreasing order to be (b), (c) and (a).

A thermal storage tank has the following three heat input sources: (a) Exhaust heat from a fuel cell and the reformer; (b) Heat conversion of the electric power generated by the fuel cell and the solar cell; (c) Heat generated by the geothermal heat pump (11). However, when the heat input exceeds the thermal storage capacity, some of the surplus is released. After the heat from the thermal storage tank heats city water via the heat medium inside the thermal storage tank, it is supplied to the demand side.

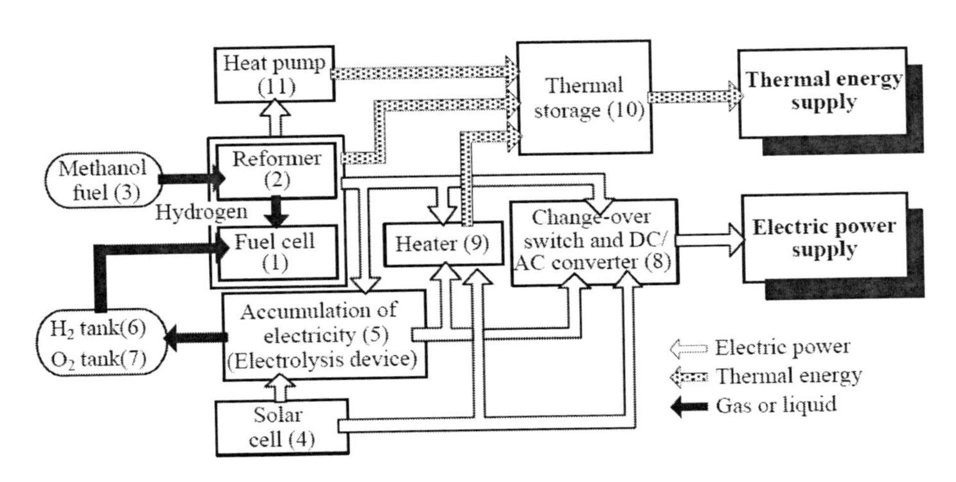

Figure 8. Energy supply path.

For the specifications of other system component devices, we used the values shown in table 1, which are typical for houses in cold regions such as Sapporo. Since a geothermal heat pump was used, the capacity of a fuel cell was set at 4.2 kW. With respect to device costs, installing a complex energy system such as shown for individual houses in figures 8 and 9 is difficult. In this study, the operational planning of such a complicated system is conducted as a prior study for the energy network of the compound energy system constituted from an active energy device, a renewable energy device, and an unutilized energy device.

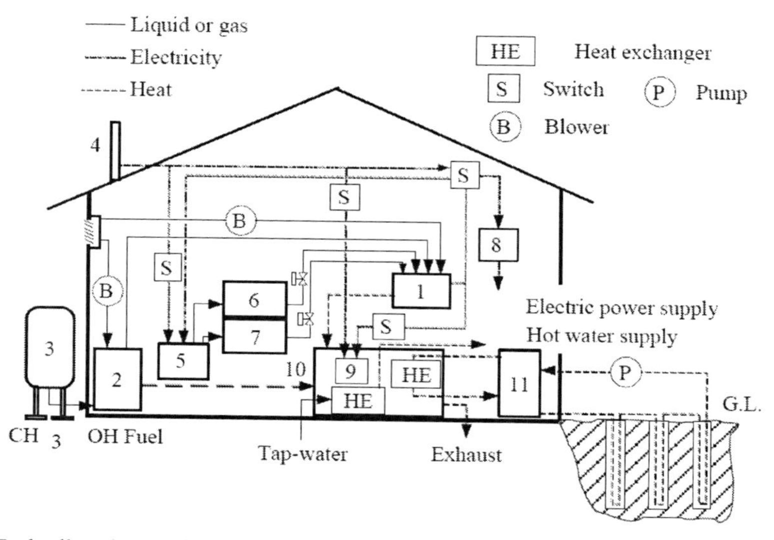

Figure 9. Combined system of fuel cell with reformer, geo-thermal heat pump, solar cell, water electrolysis bath and thermal storage tank, for individual houses.

4. Characteristics of Energy Devices

4.1. Fuel Cell Cogeneration

Figure 10 is the result of examining the relationship between electric power and thermal energy output, and the fuel amount of supply. The characteristic curve is divided into two or more regions, and each region is approximated by the least-squares method with an equation of secondary order. The characteristics of electric power shown in figure 10 are for a model that includes the power consumption of blowers and electric power loss of the DC/AC converter. Moreover, the values of the joule heat of the fuel cells, the battery reaction heat, and the exhaust heat of the reformer have been included in the heat characteristics in the figure. Methanol fuel using a burner for the heat sources of the reformer is also included on the horizontal axis of figure 10. In addition, to start the fuel cell system, consumption of methanol fuel equivalent to 900 kJ (250 Wh) is considered. In order to collect the hydrogen and oxygen generated by water electrolysis, tanks are installed in the electric energy storage device. For fuel cell systems using not the gas obtained by steam reforming of methanol but the hydrogen and oxygen in each tank, the power generation efficiency is 0.75 and the heat output is set at 0.05.

Table 1. Energy device initial specifications

Solar cell	
Area Electric power output	18.0 m2 3kW(Maximum)
Fuel cell with methanol reformer	
Type Electric power output Thermal energy output Fuel Efficiency (Pure H2,O2)	Proton-exchange membrane fuel cell Water/Methanol=1.4/1.0 (mole ratio) 4.2kW (Maximum) 15kW(Maximum) 0.75
Heat pump	
Type COP Thermal energy output Energy source	Geothermal heat source Electricity 12kW(Maximum) 3.0
Electrolysis device	
Electrolysis efficient Accumulation of electricity	Un-setting up 0.85
Thermal storage tank	
Thermal storage capacity Heat medium temperature Heat loss	Un-setting up 353K(Maximum) Per hour 1.0% (July) or 2.0% (February) of thermal storage value

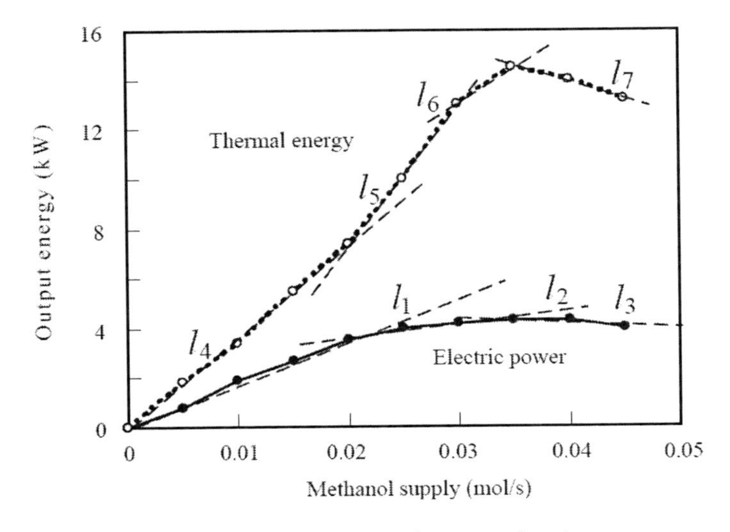

Figure 10. Characteristics of fuel cell stack with methanol steam reforming.

4.2. Geothermal Heat Pump

Based on the examination results of hydrocarbon binary vapor [10], we simplify the analysis by setting the temperature T_L $(=277K)$ of the low-temperature heat source and the condensation temperature T_H $(=347K)$ to be constant, and the coefficient of pump COP_{t_k} at 3.0.

4.3. Water Electrolyzer

E_{EL,t_k} indicates the amount of electric energy supplied to the water electrolysis bath, and ϕ_{EL} expresses the efficiency of the charge. The flow rate of hydrogen Q_{H_2,t_k} generated from sampling time t_k to Δt is calculated by the equation below. The oxygen flow rate is also determined by the same calculation. In a report on the water electrolysis bath for hydrogen generation, efficiency ϕ_{EL} of charge is given as 0.85 [11].

$$Q_{H_2,t_k} = \frac{E_{EL,t_k} \cdot E_c}{F_d \cdot E_V} \cdot \phi_{EL} \tag{1}$$

4.4. Thermal Storage Tank

$S_{St,\max}$ is the maximum thermal energy storage, and $T_{St,\max}$ is the maximum temperature of the heat medium. Equations (2) and (3) are restrictions for thermal storage. V is the capacity of a thermal storage medium volume (calcium chloride is assumed), C_p is the specific heat and T_∞ is the air temperature outside the thermal storage tank. The thermal storage temperature during sampling time t_k is calculated by $T_{St,t_k} = S_{St,t_k} / (\rho \cdot C_p \cdot V)$.

$$0 \le S_{St,t_k} \le S_{St,\max} \tag{2}$$

$$T_{\infty,t_k} \le T_{St,t_k} \le T_{St,\max} \tag{3}$$

The following equation is an expression of the thermal energy storage between time t_k and Δt.

$$S_{St,t_k} - S_{St,t_{k-1}} = \{H_{St,in,t_k} - H_{St,out,t_k} - \phi_{St} \cdot \rho \cdot C_p \cdot V \cdot (T_{St,t_k} - T_{\infty,t_k})\} \cdot \Delta t \tag{4}$$

H_{St,in,t_k} and H_{St,out,t_k} show the input and output heat energies of the thermal storage tank, respectively, and the loss of thermal storage is the 3rd term within { } on the right-hand side of Eq. (4) when it depends on open-air temperature T_{∞,t_k}. In this study, thermal storage loss at time t_k of the representative day will be considered as 1% of the value on the left-hand side of Eq. (4) in July and 2% in February.

4.5. Solar Modules

Figure 11 shows the results for a solar cell in Sapporo in winter (representative days in February) [12]. The solar cell is a roof installation-type device installed perpendicularly so that it does not become covered with snow. The characteristics for the representative days in July shown in the figure are the predicted results. Each characteristic curve is the amount of power generated during fair weather, and power generation falls during cloudy or rainy weather. Using the output characteristic performance of the solar cell as 0% in snowfall, 50% under cloudy conditions and 100% in fair weather in figure 11, the operational planning for the representative day for each month is analyzed.

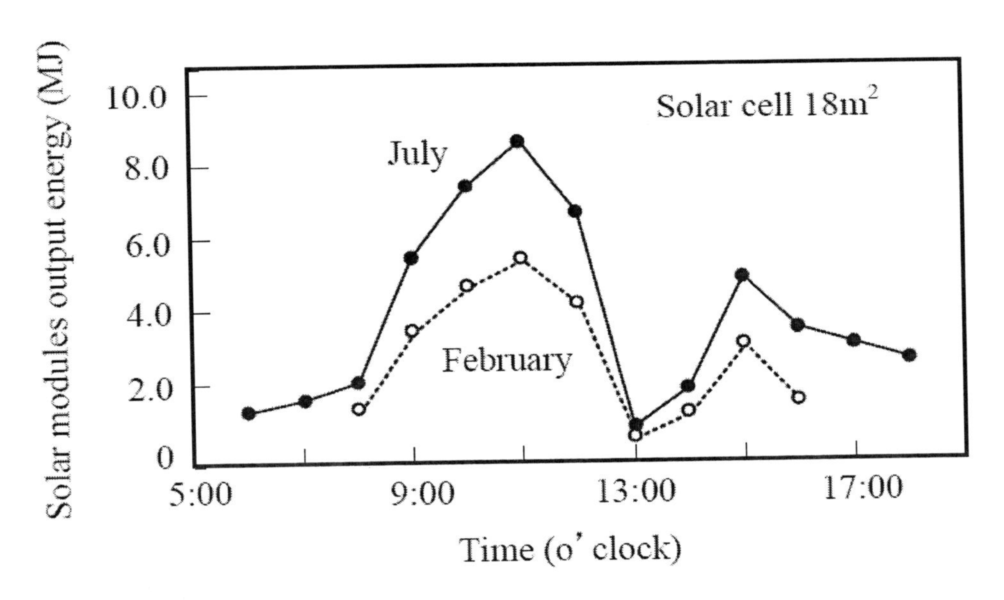

Figure 11. Time change of solar cell output [10].

5. Objective Function and Energy Equation

5.1. Objective Function

Only methanol is used as the fuel supplied to the system as shown in figure 9 . Therefore, operational planning to minimize costs requires an operational pattern where the methanol fuel consumption is minimum for each sampling time t_k . D_i represents the energy device and the subscript i ($i = 1,2,3,...,M$ and M is the total number of devices) corresponds to the device number used here. The operational costs of device D_i during sampling time t_k to Δt are equal to fuel input flow F_{D_i,t_k} of the device multiplied by unit fuel price C_{fuel} . The operational costs of the whole system are estimated using Eq. (5). Therefore, the total operational costs of all working periods of a system are calculated using Eq. (6) and called the "best-fit solution," so that the value of Eq. (6) should be small. In the example of the application of operational planning in the following section, C_{fuel} of methanol fuel is calculated to be 0.463 $/kg.

$$C_{System,t_k} = \sum_{i=1}^{M} \left(C_{fuel} \cdot F_{D_i,t_k} \cdot \Delta t \right) \quad (5)$$

$$C_{System,day} = \sum_{t_k=0}^{23} \sum_{i=1}^{M} C_{System,t_k} \quad (6)$$

5.2. Energy Balance

Equations (7) and (8) are the electric power and thermal energy balance equations of this system, respectively.

$$E_{FS,t_k} + E_{SL,t_k} = E_{System,t_k} + \Delta E_{EL,t_k} + \Delta E_{HP,t_k} + \Delta E_{H,t_k} \quad (7)$$

$$\alpha_{FS} \cdot F_{FS,t_k} \cdot \phi_{FS} + H_{HP,t_k} + H_{H,t_k} + H_{St,t_k} = H_{System,t_k} + \Delta H_{St,t_k} + \Delta H_{Rad,t_k} \quad (8)$$

The left-hand sides of Eqs. (7) and (8) correspond to the output energy from the system, and the right-hand sides correspond to the amount of consumption energy of the system. E_{System,t_k} and H_{System,t_k} are decided on the basis of energy demand patterns. The left-hand side of Eq. (7) shows electric power output from the fuel cell (E_{FS,t_k}) and solar cell (E_{SL,t_k}). The right-hand side of Eq. (7) shows the electric power consumption of the water electrolyzer ($\Delta E_{EL,t_k}$), electric power consumption of the heat pump ($\Delta E_{HP,t_k}$), and electric power converted into heat by the electric heater ($\Delta E_{H,t_k}$), respectively. The left-hand side of Eq. (8) shows thermal energy output from the fuel cell heat exhaust, heat pump (H_{HP,t_k}), electric heater (H_{H,t_k}) and thermal storage tank (H_{St,t_k}), respectively, The right-hand side of Eq. (8) shows heat loss from the thermal storage tank ($\Delta H_{St,t_k}$) and heat release from the radiator ($\Delta H_{Rad,t_k}$), respectively.

6. Analysis Conditions

6.1. Analysis Method of System Operational Planning

The "mixed-integer plan-making" method [1] has been studied to analyze the operational planning of an energy system. In this method, the nonlinear input-and-output characteristics of energy devices are expressed as a linear model and analyzed. An example of the test results of electric power and thermal energy output characteristics of a fuel cell with a reformer is shown in figure 10. If the nonlinear characteristics of an energy device can be made to fit a linear approximation problem, an increase in analysis error is predicted. In the mixed-integer plan-making method, the characteristics of the electric power output are approximated by three straight lines l_1 to l_3 in figure 10, and heat output is approximated by four straight lines l_4 to l_7. Generally, since the output of small energy equipment is nonlinear, we should use the nonlinear model for analysis. In the analysis of the operational planning of the system with a number of energy devices, many variables associated with each device operation are used. Therefore, if many variables can be calculated simultaneously, the efficiency of the analysis will increase. A genetic algorithm, where simultaneous calculations of many variables and the calculation of a nonlinear problem are possible, is introduced in the software developed in this study. However, neither the application of a GA to a small-scale energy system nor a design method that optimizes the operational pattern and device capacity has been studied previously. In particular, no research reports on the optimization of the operational plan for a compound system of an active energy device, a renewable energy device and unutilized energy device or their optimal capacity can be found in the literature.

6.2. Operation of a Chromosome Model

The chromosome model operated by the GA calculation needs to satisfy the energy balance in Eqs. (7) and (8). However, the chromosome model must also satisfy conditions (a) and (b) described below:

a. A quantity that excludes electric energy consumption (sum of all ΔE) from the amount of electric energy output of the fuel cell and the solar cell satisfies the electric energy demand.
b. A quantity excluding heat loss ($H_{Rad,t_k} + \Delta H_{St,t_k}$) from the sum total of the exhaust heat of the reformer and fuel cell, heat pump, electric heater, and the heat energy output of the thermal storage tank should satisfy the heat energy demand.

When an operational pattern (chromosome model) that does not fulfill one of these conditions arises, it is forced to a very low value of fit so that it cannot proceed to the next generation. Similarly, a low fitness value is given for an operational pattern that does not satisfy the energy balance of Eqs. (7) and (8).

For the chromosome models of the initial generation for the power generation of a fuel cell, the total power generated by the fuel cell is decided at random within the electric power capacity. In addition, the total power generated is distributed to the amounts of electric power output (E_{FS,t_k}), the quantity supplied to a water electrolyzer and stored as electricity ($\Delta E_{EL,t_k}$), and the quantity conducting heat conversion (E_{H,t_k}) in an electric heater at random. The total amount of power generated in fine weather in sampling time t_k in a solar cell is decided as shown in figure 12. The total power generated from the solar cell is distributed to the electric power supplied to the electric power output (E_{SL,t_k}), the amount of accumulated electricity ($\Delta E_{EL,t_k}$) and amount of electric power supplied to a heater (E_{H,t_k}) for every chromosome model at random. Moreover, the power consumption ($\Delta E_{HP,t_k}$) of the geothermal heat pump for every chromosome model is estimated from the amount of heat output that was decided at random within the limits of the device capacity and coefficient of performance (COP) .

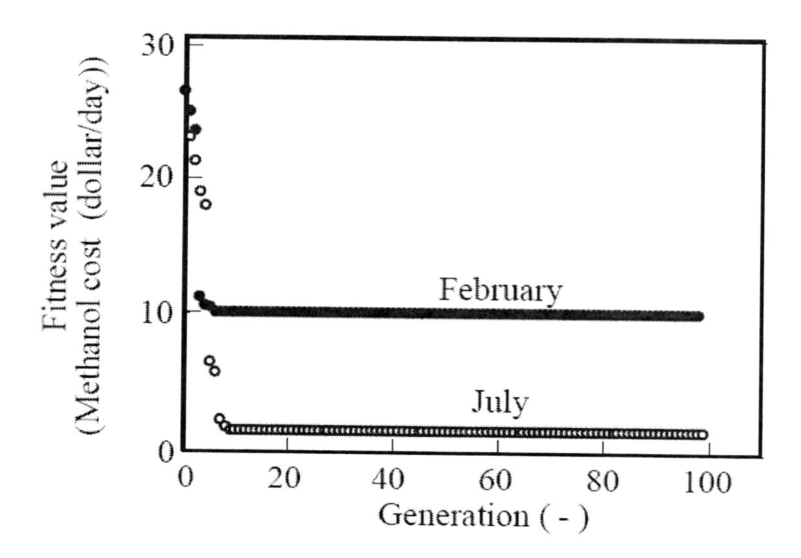

Figure 12. Convergence of GA calculations.

6.3. Analysis Conditions

Operational period R of a system is split into 23 parts, and t_k $(k = 0, 1, 2,....., 23)$ defines the sampling time in the case analysis of Chapter 7. Moreover, the number of devices M is set at five including the fuel cell with the reformer, solar cell, geothermal heat pump, water electrolyzer, and thermal storage tank. In the analysis described below, the number of initial-generation chromosome model groups N'_{dv} is set at 3000, and the number of chromosomes operational after reproduction N_{dv} is set at 2500. The last generation is analyzed as 100 generations. Moreover, considering the maintenance of the diversity of gene models, the number of intersections is selected randomly. That is, with one intersection, 0.2% of the total number of chromosomes is extracted as parent chromosomes, at maximum. The frequency of mutations governed is about 4% of the genes in a mutated chromosome model, at maximum. These parameters of the SGA confirm that the value of the optimal solution is in agreement within several percent, as a result of trial calculations with two or more parameters.

The minimum operational costs for every generation when performing an operational plan by GA to minimize the operational costs with the application of the energy demand pattern of representative days in July and February are shown in figure 12. In this calculation, we assumed fair weather and the electric power output of a solar module to be 100% (same as in figure 11). Although the fitness value of the representative day for both months decreases rapidly, the operational costs, to almost ten generations, shows a gradual change in successive generations. For the best-fitness solution after 10 generations, the mutation calculation is important. However, if analysis conditions are changed, the number of generations for convergence changes.

7. Case Study

7.1. Objective of Analysis Calculation

The operational planning of a system is analyzed using the energy demand pattern of the representative days in February and July in Sapporo. We assume that the active energy device (fuel cell cogeneration with a methanol reformer) is already installed in individual houses in Sapporo in this analysis. The renewable energy device (solar cell), unutilized energy device (geothermal heat pump), electricity accumulation device (water electrolyzer), and thermal storage tank are connected to an active energy device, and the complex independent energy system shown in figure 9 is built. It is difficult to introduce such a complex system into individual houses because of the device costs. This study investigates the energy network system of the distributed energy device. This study also examines, the operational planning method of the active energy device with the maximum use of renewable energy using the information obtained from the analysis. Furthermore, the capacity of devices to accumulate electricity and thermal storage is estimated.

7.2. Results of Operational Planning

Figure 13 shows the analytical results of operational planning assuming fair weather with the energy demand pattern for the representative day in each month, and it shows the device energy output for every time period. However, the output results of the fuel cell include the output values of both the electric energy and the heat energy. The breakdown of the energy output of the fuel cell is shown in figure 14. Hydrogen has been formed by the steam reformation of methanol in the time period with high thermal output (0:00 and 1:00 on the representative day for July and 0:00～9:00 and 19:00 and 21:00 on the representative day for February). The fuel cell is generated using hydrogen and oxygen, which were created by water electrolysis in the time period in which thermal output was small, except for that mentioned above. As described in Section 3.4, the reason for this is that the use of renewable energy (solar cell) is a top priority. Therefore, if the amount of power generated by a solar cell exceeds the electricity demand, the fuel cell will not operate by driving the reformer. If, from night to early morning, power generation from a solar cell is not conducted, the fuel cell operates by the driving reformer. On a representative day, power from a fuel cell by the operation of the reformer is generated from 0:00 at 9:00 in February, and the heat pump is operated using this electric power. From the analysis results of the operational plan, thermal storage of the heat generated by the heat pump occurs, and the stored thermal energy is used to conduct the time shift and to fulfill heat demand during the daytime.

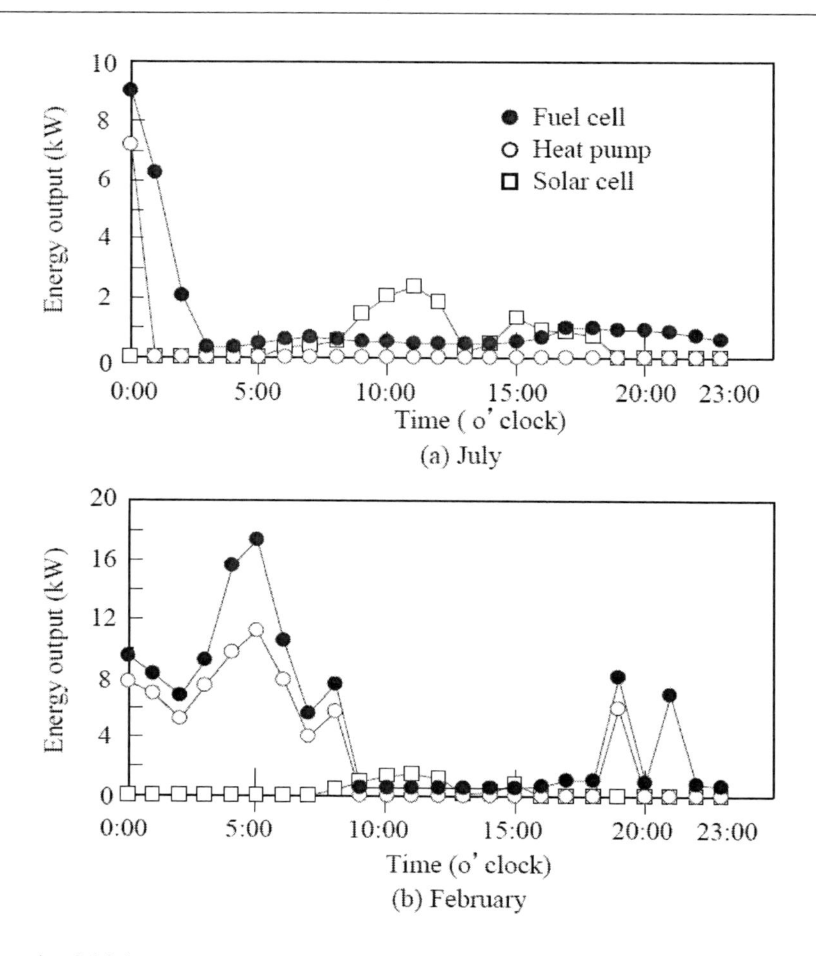

Figure 13. Result of driving pattern with 100% solar power generation output.

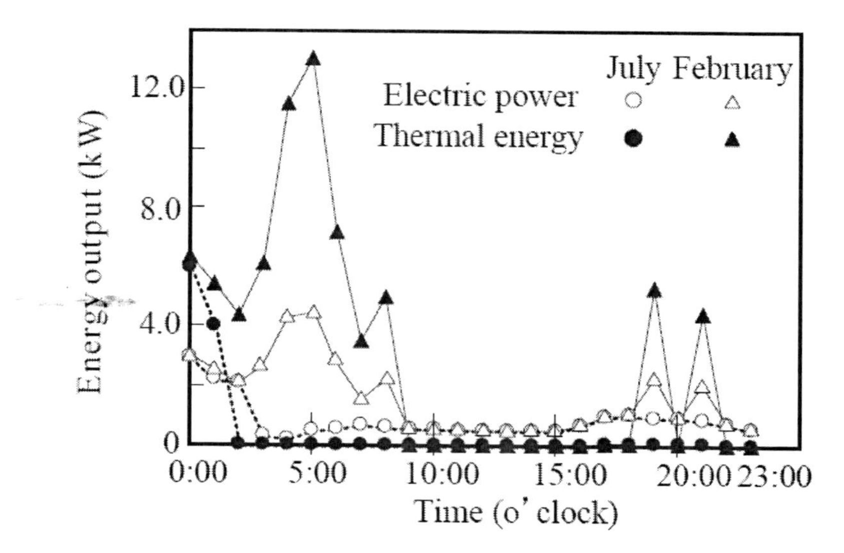

Figure 14. Output of Fuel Cell.

Figure 15 shows the analytical results of the system operational costs for conducting operational planning in which the output proportion of the solar cell is a variable, for the representative day in each month. Using the characteristic output performance of the solar cell of 0% in snowfall, 50% under cloudy conditions and 100% in fair weather in figure 11, operational planning for the representative day in each month is analyzed. The total value of one day of electric power and thermal energy demand is set at 100. The representative days in February and July of the total electric power obtained in the solar cell for fine weather are 95 and 28, respectively. On the other hand, from figure 15, the system operation costs during fine weather when setting the operation costs in case of rainy weather and snowfall at 100 for Februa ry and July representative days are 88 and 29, respectively.

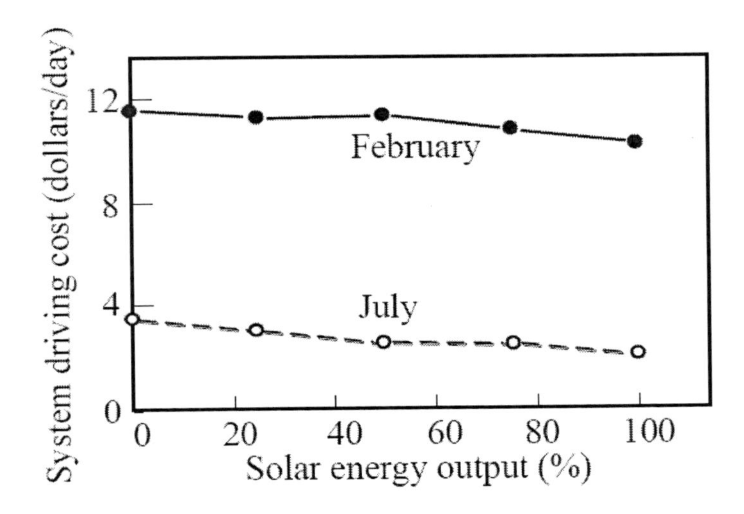

Figure 15. System driving cost.

The difference in the operational costs of the system according to the difference in weather for the representative days in February (100-88 = 12) is larger than the difference in solar cell output (100-95 = 5) due to the weather. The main reasons for this are that fuel cell efficiency improves by reducing operation because of low loads in figure 10, and because the driving period to operate a fuel cell by generating hydrogen and oxygen using a solar cell is long. On the other hand, for the representative July day, the generation of hydrogen and oxygen is conducted by a solar cell, and operation of a fuel cell in most periods is performed with high energy demand. Therefore, the difference in the output of a solar cell due to the weather directly affects the operational costs of the system.

7.3. Design of the Capacity of the Water Electrolyzer and Thermal Storage

Figure 16 shows the results of the operational planning of the amount of thermal storage, and the accumulation of electricity for a representative February day. The largest quantity of thermal storage and electric energy storage in the analytical results approximates the design capacity of each device in fair weather because it increases as the output proportion of the solar cell increases. Figure 17 shows the analytical results of the thermal storage quantity and electric energy quantity where the output proportion of the solar cell was 100% in the energy demand pattern for the representative day in each month. From these results, the maximum value of the thermal storage quantity is 308 MJ (representative February day), and the maximum value of electric energy storage quantity is 23 MJ (representative July day). Thermal storage capacity is reduced by lengthening the operational period of a fuel cell with the reformer and the heat pump.

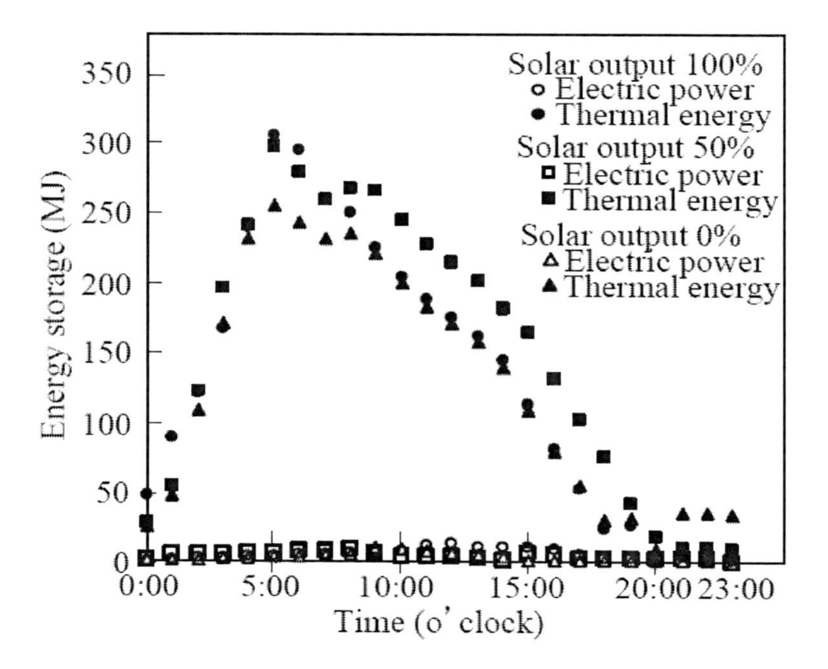

Figure 16. Quantity of energy storage in February.

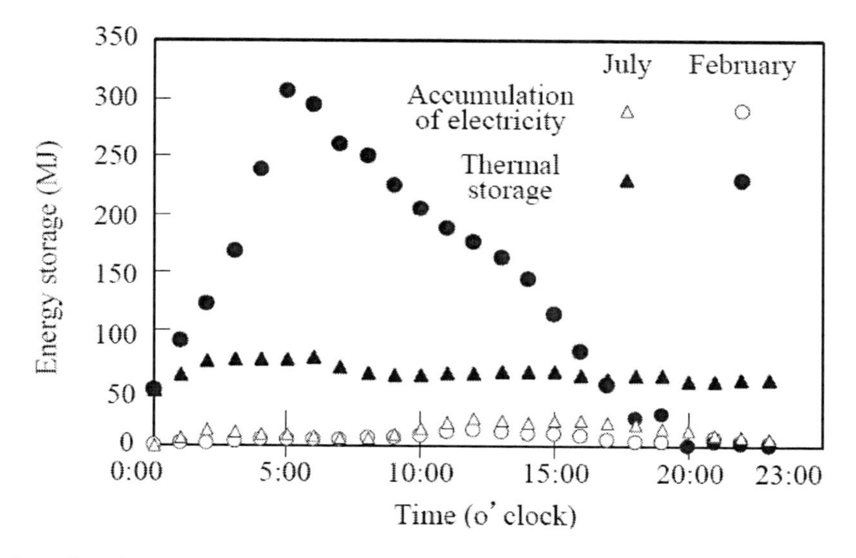

Figure 17. Quantity of energy storage with 100% solar generation.

8. Conclusions

An active energy device, a renewable energy device, and an unutilized energy device were connected by an energy network, and a preliminary survey of the energy network design required for cooperative operation was conducted. Analysis software for operational planning, when two or more energy devices were introduced in an individual house, was developed. A genetic algorithm (GA), which can analyze nonlinear problems and many variables at a time, was used for the analysis software for the operational planning of the complex energy system developed in this study.

The operational planning to introduce an active energy device (fuel cell cogeneration), a renewable energy device (solar cell), and an unutilized energy device (geothermal heat pump) in an individual house in Sapporo, Japan, which is a cold, snowy region, as an analysis example was conducted using the developed software. Furthermore, the device capacity of accumulation for an electricity storage device and a thermal storage tank were estimated. From the results of the analysis, a fuel cell with a reformer operates from night to early morning when a solar module is not operational. A heat pump also operates from night to early morning. The thermal storage of heat generated by the heat pump is conducted, and this heat is supplied to the heat load during the daytime.

References

[1] Ito K, Shibata T, Yokoyama R., 2002, "Optimal Operation of a Cogeneration Plant in Combination With Electric Heat Pumps," Trans. *ASME J. Energy Resource Technol.*, 116, pp. 56-64.

[2] Goldberg, D. E., 1989, Genetic Algorithms in Search, Optimization and Machine Learning, Addison Wesley.

[3] Hongmei, Y., Haipeng, F., Pingjing, Y., Yi, Y., 2000, "A Combined Genetic Algorithm/Simulated Annealing Algorithm for Large Scale System Energy Integration," *Computers and Chemical Engineering*, 24, pp.2023-2035.

[4] Srinivas M. and Patnaik L. M., 1994, "Genetic Algorithms: A Survey," *IEEE Computer*, Vol. 27, No. 6, pp. 17-26.

[5] Yu H, Fang H, Yao P, Yuan Y, 2000, "A combined genetic algorithm/simulated annealing algorithm for large scale system energy integration," *Comput. Chem. Eng.*, Vol. 24, No. 8, pp. 2023-2035.

[6] Fujiki K., Akagi S., Hirokawa T. and Yoshida K., 1997, "Optimal Planning Method of Energy Plant Configurations based on a Genetic Algorithm," Trans. *Jpn. Soc. Mech. Eng.*, C, 64-617, pp. 354-361 (In Japanese).

[7] Baker, J.E., 1985, "Adaptive Selection Methods for Genetic Algorithms," *Proc. 1st Int. Joint Conf. on Genetic Algorithms, ICGA85*, pp. 101-111.

[8] National Astronomical Observatory of Japan, 2003, *Chronological Scientific Tables*, Maruzen K.K.

[9] Obara, S., Kudo, K., Kuroda, A., 1999, "Study on Small-Scale Co-generation System for Domestic House Considering Partial Load and Load Fluctuation," Trans. *Jpn. Soc. Mech. Eng.*, B, 65-630, pp. 741-748 (In Japanese).

[10] HC-TECH Inc, 1997, HC12a and HC22a Properties and Performance Tests Data sheets.
[11] Kosaka, K., Tani, T., and Yoshida, S., 2000, "Thermal Analysis of Solid Polymer Water Electrolysis System," Trans. *Jpn. Soc. Mech. Eng.*, 66 (642), B, pp. 547-554 (In Japanese).
[12] Nagano, K., Mochida, T., Shimakura, K., Murashita, K., and Takeda, S., 2002, "Development of Thermal-photovoltaic Hybrid Exterior Wallboards incorporating PV cells in and Their Winter Performances," *Sol. Energy Mater. Sol. Cells* 77, pp. 265-282.

In: Air Conditioning Systems
Editors: T. Hästesko, O. Kiljunen, pp. 315-336
ISBN: 978-1-60741-555-8

Chapter 8

IMPROVING DESIGN IN REVERSIBLE HEAT PUMPS

***C. J. Renedo**[*1]**, A. Ortiz**[2] **and J. Carcedo**[2]*

Department of Electrical and Energy Engineering, University of Cantabria
[1] ETS Náutica, Gamazo 1, 39004, Santander Cantabria
[2] ETSI Industriales y Telecomunicación, Av Los Castros s/n,
39005 Santander, Cantabria

ABSTRACT

In temperate climates, the thermal supply for air conditioning systems can be carried out efficiently using reversible heat pumps. In this chapter, once the potential efficiency improvement of these machines is analyzed, several alternatives to the traditional designs, proposed by ASHRAE, are undertaken. Results for performance simulations are presented. The comparison shows that new designs do, indeed, lead to a considerable improvement in energy efficiency. Finally, two new systems are shown, the heat pump with total energy recovery and the variant refrigerant volume system, VRV. These are heat pumps capable of exploiting both the heat emitted by the condenser and the one absorbed by the evaporator, achieving high energy efficiency levels.

1. INTRODUCTION

Air conditioning aims to achieve thermal comfort and indoor air quality inside buildings, whatever the weather outside may be. This requires systems for heating, venting and air conditioning (HVAC). The continual improvement in the standard of living has caused the market for these systems in Spain and Europe to grow [1, 2].

The energy consumption of HVAC facilities corresponds to: (1) thermal generation of heat, (2) thermal generation of cold, and (3) fluid distribution throughout the building. It is the thermal generation to provide heat and cold that consumes the largest amount of energy. Thus, any energy efficiency improvement that is to be implemented in the generation system has a dramatic effect on installation efficiency.

[*] renedoc@unican.es; Tlfn 0034 942 20 13 82 Fax 0034 942 20 13 85

The cold supply is usually carried out via a machine that runs under a vapor compression cycle. These machines transfer heat from a low temperature source to a sink at high temperature. Thus, the cooling effect is achieved with the heat being absorbed in the evaporator of the machine [3].

The heat supply can be based on boilers or on heat pumps. This machine is similar to a chiller, and also works with the vapor compression cycle. However, rather than taking advantage of the thermal effect of the evaporator, they make use of the heat that the condenser gives off.

A heat pump is reversible when you can tap at will either the heat given off by the condenser or the cold produced in the evaporator.

In this way, both the definition of thermal capacity of a reversible heat pump (HP_{Th}) and the coefficient of performance (COP) vary from winter to summer. In winter HP_{TH} is the heat emitted by the condenser, while in summer it is the heat absorbed by the evaporator (Eqs 1). The COP in winter is the ratio between the heat emitted in the condenser and the work done by the compressor. In summer the COP is defined as the ratio of the heat absorbed by the evaporator and the work provided by the compressor (Eqs 2).

$$HP_{Th}\big|_W = \dot{Q}_{con}\big|_W \ ; \ HP_{Th}\big|_S = \dot{Q}_{eva}\big|_S \qquad \text{(Eqs. 1)}$$

$$COP\big|_W = \left.\frac{\dot{Q}_{con}}{W_{comp}}\right|_W \ ; \ COP\big|_S = \left.\frac{\dot{Q}_{eva}}{W_{comp}}\right|_S \qquad \text{(Eqs. 2)}$$

Thus, the closer the evaporation and condensation temperatures are, the higher the COP is. This is because the compressor has to carry out less work in order to complete the cycle [3].

With regard to the standard conditions of use, the greater the difference between evaporation and condensation temperature is in winter, the higher HP_{TH} and COP are in summer. However, this is counterbalanced by the heat emitted by the condenser being greater than that absorbed in the evaporator. In this way, both the HP_{TH} and the COP of a heat pump remain relatively constant throughout the year.

Currently, reversible heat pumps have great potential, especially in climates where winter is not very cold, as is the case of most Mediterranean countries [4]. This is due to the fact that a heat pump is able to carry out the thermal supply in summer and winter, its cost being about the same as that corresponding solely to the chiller, which only carries out the air conditioning in summer. In this way, the heat pump makes it unnecessary to install heat generators.

The design of reversible heat pumps is very similar to that of chillers. It is only necessary to include a reversing valve and a few more ancillary elements such as check valves, refrigerant receiver, ... [5, 6]. With a reversing valve the direction of refrigerant fluid is reversed, causing the condenser to become an evaporator, and the evaporator to become a condenser, figure 1, thereby providing heating in winter and cooling in summer.

This chapter studies several types of heat pumps from the energy point of view, introducing designs that improve the efficiency of the traditional ones proposed by ASHRAE [6].

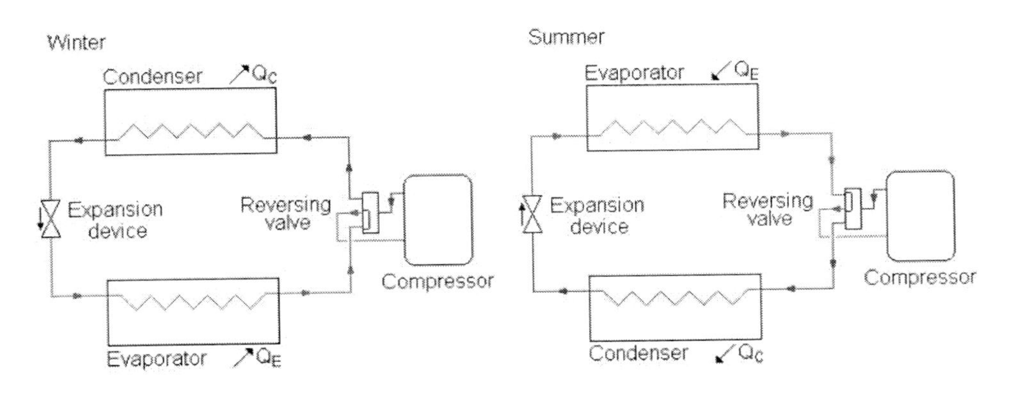

Figure 1. Reversible heat pump.

2. Possibilities of Improving Heat Pumps

The possibilities of improving the efficiency of heat pumps lie mainly in three aspects:

2.1. The Heat Exchanger Operation

Reversible heat pumps have two heat exchangers: in the summer cycle the main one works as an evaporator and the auxiliary as a condenser; in the winter operation cycle the opposite occurs (the main heat exchanger works as a condenser, and the auxiliary as an evaporator). It is known that heat exchanger operation is better in counterflow than in parallel flow [7].

Depending on the medium that exchanges heat with the refrigerant, the heat exchangers can be refrigerant-water or refrigerant-air.

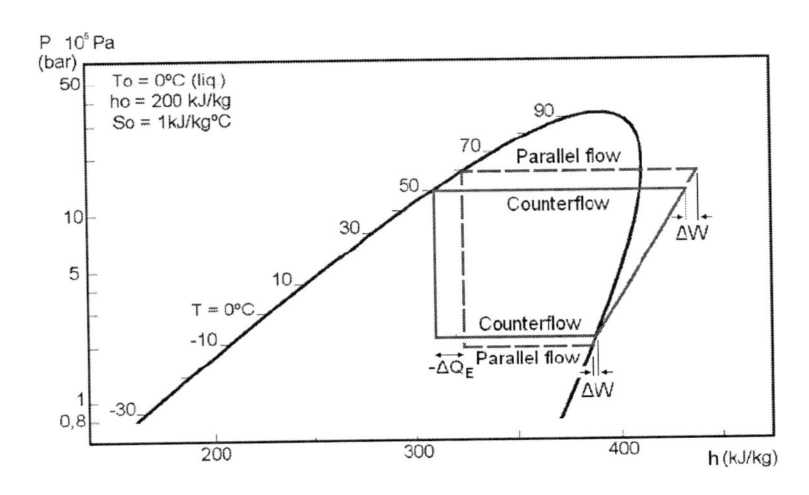

Figure 2. Effects of changing the refrigerant flow direction.

On inverting the operation cycle of a heat pump, the operation of the refrigerant-water heat exchangers are also inverted since the refrigerant flow direction is changed, maintaining the water flow direction. Therefore, there is an operation period for each heat exchanger, either summer or winter, in which they operate in parallel flow at a lower rate of efficiency than the maximum they could reach in counterflow.

On inverting the operation cycle of a heat pump, the operation of the refrigerant-air heat exchangers are not inverted since these heat exchangers have cross flow.

When the condenser and/or the evaporator are water heat exchangers, the effects of parallel flow operation on heat exchangers are: (1) it increases the high pressure, (2) it reduces the low pressure, (3) it increases the work required in compressor, (4) and it reduces the heat absorbed in the evaporator; these are shown in figure 2. All these effects contribute towards reducing the COP.

2.2. The Expansion Device

The expansion device of the reversible heat pump can operate in the two directions, but with one of them offering greater efficiency than the other.

By inverting the heat pump operation, the refrigerant flow direction is changed in the expansor, figure 1. Therefore, the efficiency of the heat pump, in one of the two operation modes, has been reduced from the maximum, since the expansion device does not operate in the optimum design conditions. In figure 1, if the preferential mode of the expansor is downward, the machines would offer good efficiency in winter and worse in summer.

In figure 3 the effect caused on the refrigerating cycle by changing the refrigerant flow direction in the expansion device is represented. The heat absorbed in the evaporator is reduced, keeping the compressor working, so the COP is gradually reduced.

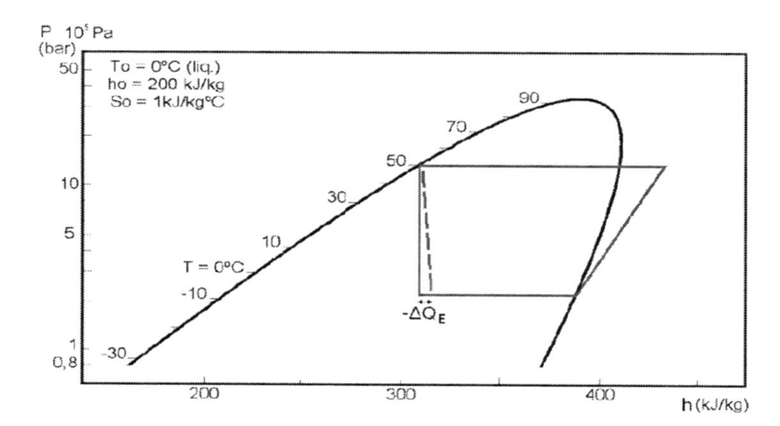

Figure 3. Effect of changing refrigerant flow direction in the expansion device.

2.3. Compressor

The energy absorbed by the compressor, according to the theoretical cycle of vapor compression, is dissipated as heat in the condenser. But the compressor, like any other real machine, presents inefficiencies. These may be caused, among other reasons, by: (1) heat generation by friction in the compressor, (2) pressure loss in the valves, (3) backflow through the valves, (4) vapor leakage inside the compressor, (5) volumetric efficiency, and so on [8].

Inefficiencies in the compressor involve increasing absorbed energy, which finally ends up in heat form. One part is absorbed by the refrigerant and is eliminated to the outside in the condenser, and the other is dissipated by the compressor housing, figure 4. When a heat pump works in summer the heat dissipated by the compressor housing is useless, but in winter this heat can be tapped.

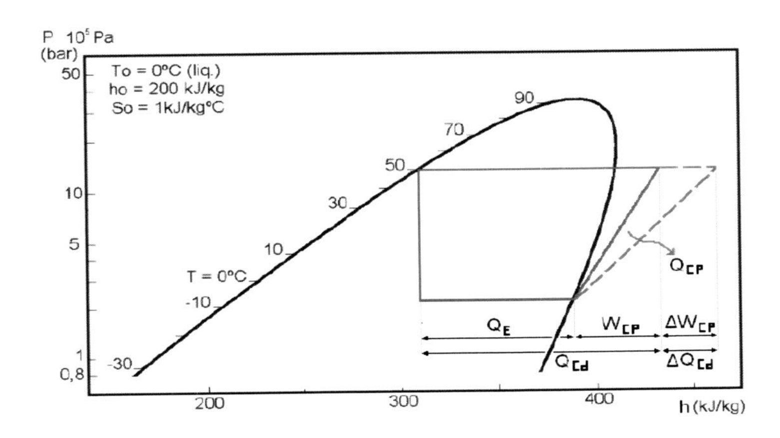

Figure 4. Effect of the compressor inefficiencies.

3. Air-Air Reversible Heat Pumps

Air-air heat pumps are machines that extract heat from an air mass flowing through the outside of its evaporator (air cooled), and transfer the heat to another air mass that circulates through the outside of its condenser (heating up this the second air mass). In the winter cycle the mass warmed by the condenser is useful (for the heating of a building, an industrial process, etc,...), whereas the one cooled by the evaporator is residual. In the summer cycle the mass cooled in the evaporator is useful for cooling purposes, whereas the mass heated in the condenser is residual.

3.1. ASHRAE Design for Air-Air Heat Pumps

ASHRAE design presents a cycle inversion by changing the flow direction of the refrigerant with a reversing valve. If the machine has a unique expansion valve, it could be adjusted to only one of the two cycles. Using this design in the two seasons, winter and summer, the yield moves away from the maximum possible. This disadvantage can be overcome with the installation of two expansion valves, and a check valve for each one in bypass configuration, figure 5. This design requires each of the expansion valves to work in only one of two periods of the year (summer or winter, it being in bypass in the other). Moreover, it will be able to adjust the operation of the heat pump in the two seasons, getting the optimum performance for each of them.

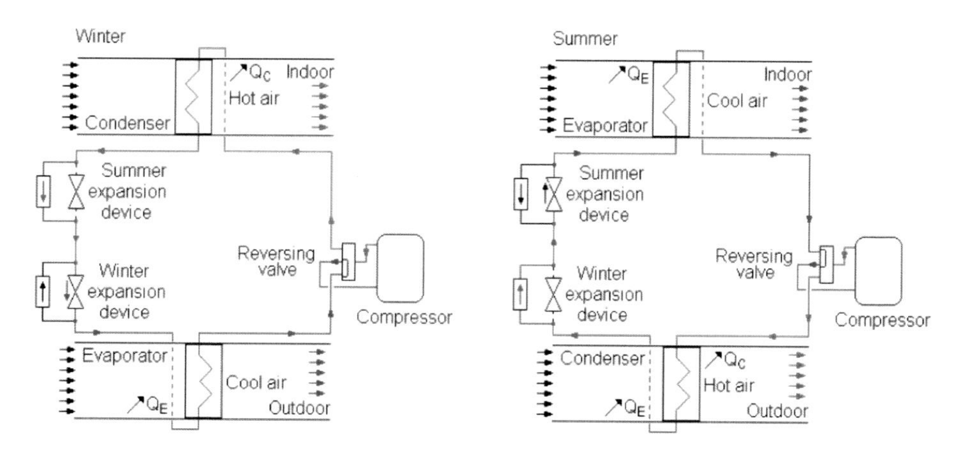

Figure 5. Air-air ASHRAE Design.

3.2. New Design for Air-Air Heat Pumps

In the New Design proposed, the cycle inversion is carried out in the duct system, figure 6, and not via the change of the refrigerant direction. This design has two improvements with respect to the ASHRAE design: (1) it permits, both in summer and in winter, the use of the expulsion air as a heat source, the temperature of which is more favourable than the outdoor air, and (2) it manages to use the heat emitted by the compressor housing in winter. This is achieved by increasing the machine efficiency, and therefore, reducing annual energy consumption. More details on this design can be found in [9].

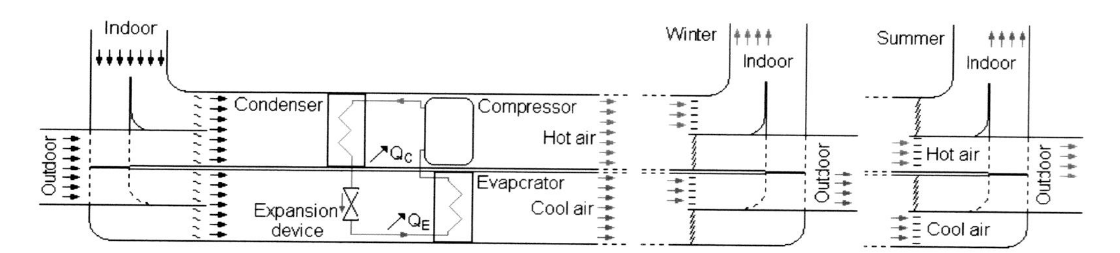

Figure 6. Air-air New Design.

3.3. Energy Study for Air-Air Heat Pumps

An energy study has been carried out simulating the operation of a heat pump according to the ASHRAE Design and the New Design proposed. Five different percentages of outdoor air have been checked: 0, 25, 50, 75 and 100%. The considered operation parameters are indicated in table 1.

Independently of the design of the heat pump, the input temperatures and the thermal difference in the heat exchangers determine the condensation temperature in winter and the evaporation temperature in summer. These are 55 and 2°C respectively.

Table 1. Parameters for comparison for air-air heat pumps

	Air temperature (°C)			ΔT air-refrigerant (°C)		Refrigerant
	Indoor	Outdoor	To indoor	Condenser	Evaporator	
Winter	22	0	40	15		R134a
Summer	24	40	17			

For the ASHRAE design, the evaporation temperatures in winter and condensation in summer are fixed by the outdoor temperature and the thermal difference in the heat exchangers. They are -15 and 55°C respectively.

For the New Design, the evaporation temperatures in winter and condensation in summer are not defined solely by the outdoor temperature and the thermal difference in the heat exchangers, since they are also influenced by the amount of recirculation air.

In winter, as the amount of outside air that enters the building increases, the indoor air that is sent over the evaporator also increases. This elevates the evaporation temperature, and thus the COP. With all outdoor air, the evaporation temperature becomes 7°C (equal to the indoor one 22°C minus the ΔT 15°C).

In summer, as the amount of outside air that enters the building increases, the indoor air that is sent over the condenser also increases. This reduces the condensation temperature, and thus improving the COP. With all outdoor air, the condensation temperature becomes 39°C (equal to the indoor one 24°C plus the ΔT 15°C).

In figure 7, the results of the simulation of the COP for both designs of heat pump are shown. The operation parameters of table 1 and five percentages of outdoor air are considered. A 70% isoentropic performance for the compressor has been taken into account. The heat emitted by the engine-compressor housing that can be used in winter should be 20% of the work absorbed by the compressor.

Figure 7 shows that the COP of the ASHRAE design does not depend on the percentage of outside air, while the New Design is always better. In the event that in summer only air recirculation is used in the building (0% outdoor air), both designs work at the same COP.

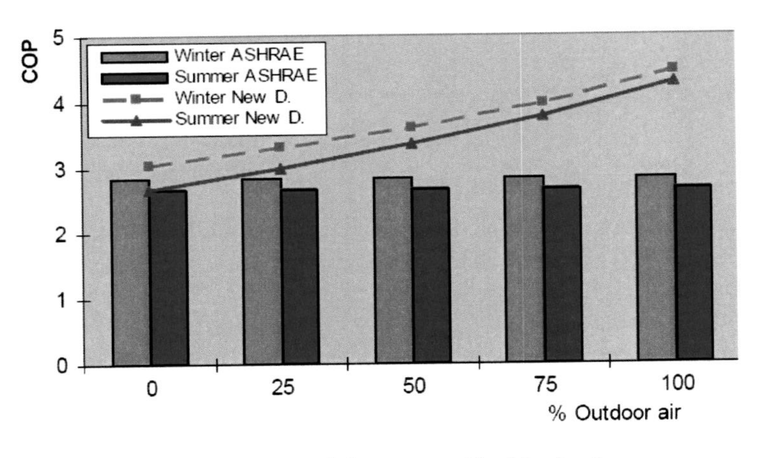

Figure 7. COP for the designs of ASHRAE and the proposed in this chapter.

4. Water-Air Reversible Heat Pumps

Water-air heat pumps are machines with a heat exchanger water-refrigerant and another air-refrigerant. In winter, they extract heat from the water (the water exchanger works as an evaporator), and warm an air mass, which is used in the building air conditioning system (the air exchanger works as a condenser). In summer, they warm the water (the water exchanger works as condenser), and cool the air mass for air conditioning (the air exchanger works as an evaporator).

4.1. ASHRAE Design for Water-Air Heat Pumps

The ASHRAE design, as in the air-air case, presents cycle inversion by changing the flow direction of the refrigerant with a reversing valve. The machine may have one or two valves for expansion, figure 8. Even so, in the summer cycle this design introduces the condenser operation in parallel flow (which is less effective than in counterflow) [7].

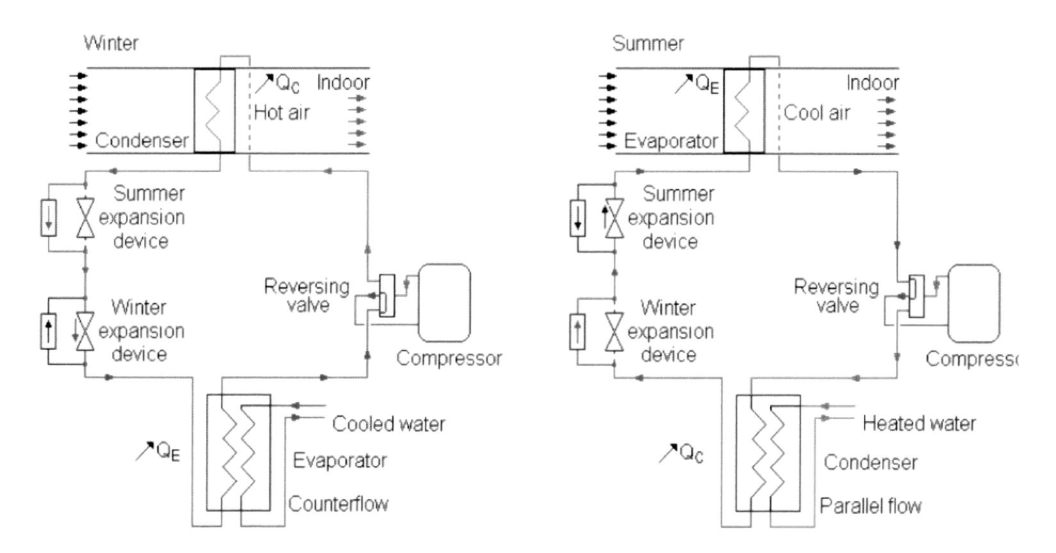

Figure 8. Water-air ASHRAE Design.

4.2. New Design for Water-Air Heat Pumps

In the New Design for such machines two main characteristics are proposed: (1) to include a reversing valve at the water exchanger outlet, which makes operation in counterflow possible, both in winter and in summer; (2) to use a system of ducts and dumpers that makes it possible to introduce a compressor into the duct system. This allows the winter cycle to take advantage of the heat emitted by the compressor housing, figure 9. This New Design includes the possibility of using outdoor air or recirculation air, which is not possible in the ASHRAE design.

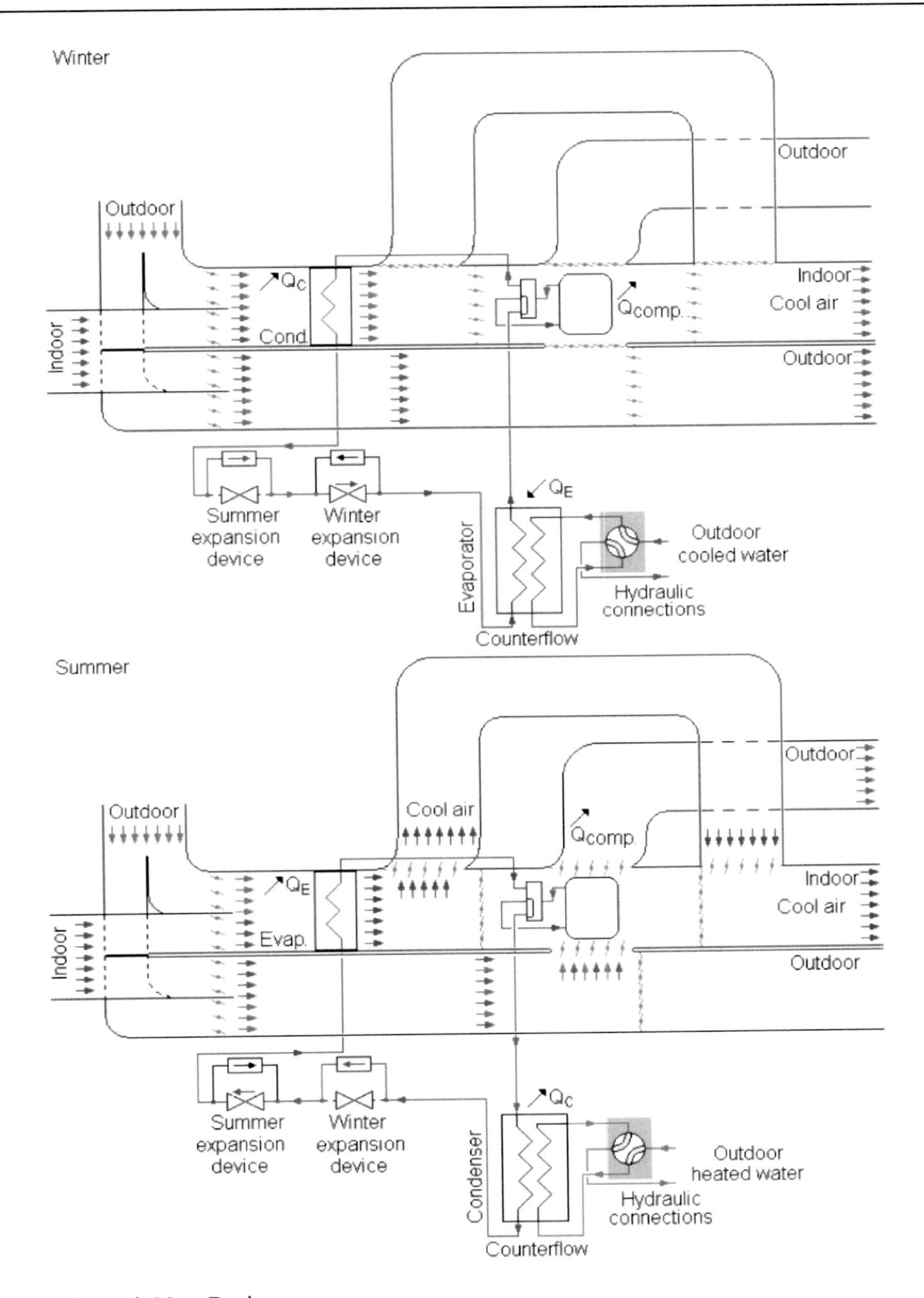

Figure 9. Water-air New Design.

4.3. Energy Study for Water-Air Heat Pumps

Simulations have been carried out on the COP of the ASHRAE design and of the New Design, considering the operating parameters of table 2. The heat emitted by the motor-compressor housing in winter is expected to be 20% of the absorbed energy. The results are shown in figure 10.

Table 2. Parameters for comparison of water-air heat pumps

<table>
<tr><th rowspan="2"></th><th colspan="3">Air temperature (°C)</th><th>Income water temperature (°C)</th><th colspan="2">ΔT heat exchanger (°C)</th><th rowspan="2">Refrigerant</th></tr>
<tr><th>Indoor</th><th>Outdoor</th><th>To indoor</th><th></th><th>Air</th><th>Water</th></tr>
<tr><td>Winter</td><td>22</td><td>0</td><td>40</td><td>5</td><td rowspan="2">15</td><td rowspan="2">7</td><td rowspan="2">R134a</td></tr>
<tr><td>Summer</td><td>24</td><td>40</td><td>17</td><td>30</td></tr>
</table>

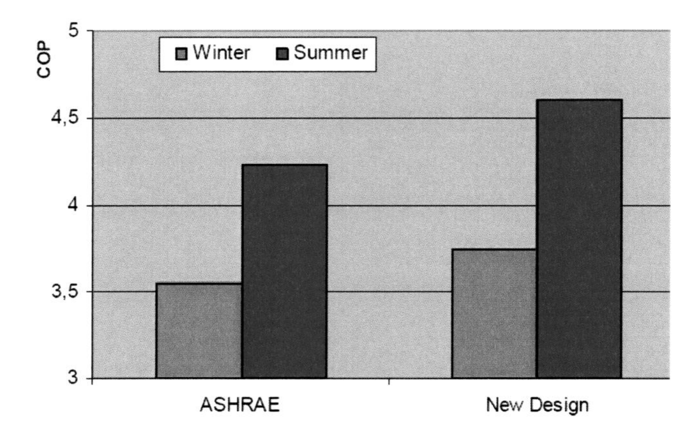

Figure 10. Water-air COP for the ASHRAE and the New design.

This increase of the heat pump COP implies that for the same thermal production, the energy consumption is reduced by around 5% in winter and 8% in summer.

5. WATER-WATER REVERSIBLE HEAT PUMPS

A water-water heat pump is a device that cools a water mass in the evaporator, and transfers heat to another water mass in the condenser. Similar to what happens in the air-air heat pump, in the winter cycle the mass warmed in the condenser is the useful one, whereas the one cooled in the evaporator is residual. In the summer cycle the opposite happens, the mass cooled in the evaporator is useful, whereas the one heated in the condenser is residual.

5.1. ASHRAE Design for Water-Water Heat Pumps

The ASHRAE design presents cycle inversion by changing the hydraulic connections, and thus the direction of both the water you want to use and the residual water. For this, four 3-way valves are necessary.

In this design the evaporator is always in counterflow. However, it has the disadvantage of having the condenser permanently in parallel flow, resulting in an efficiency reduction, figure 11.

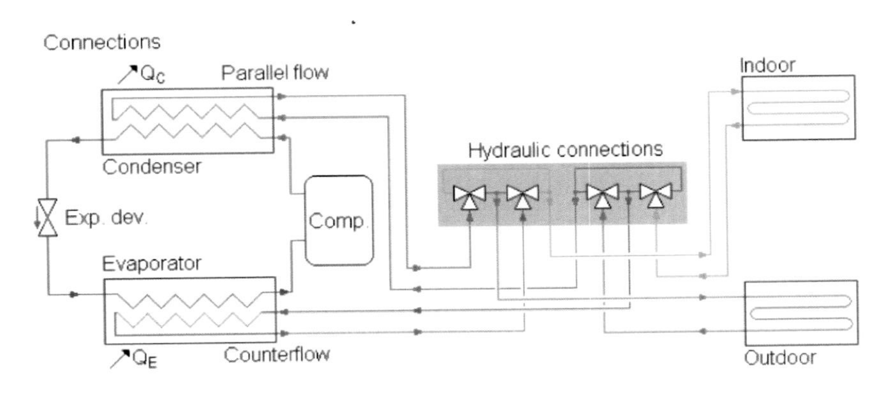

Figure 11. Water-water ASHRAE design improved.

5.2. New Design for Water-Water Heat Pumps

The New Design replaces the four 3-way valves with two reversing valves, keeping the condenser and evaporator in counterflow both in the winter cycle and in the summer cycle, figure 12. More details can be found in [10].

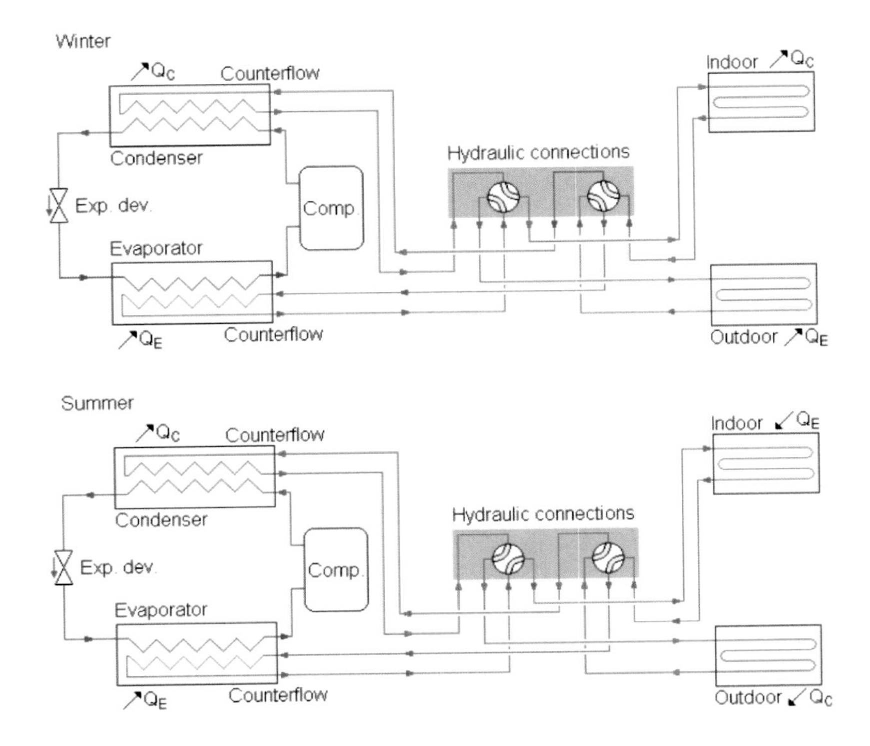

Figure 12. Water-water New Design.

5.3. Energy Study for Water-Water Heat Pumps

Simulations have been conducted on the capacity and COP from both ASHRAE and the New Design. The parameters are compared in table 3. The results are shown in figure 13.

Table 3. Parameters for comparison of water-water heat pumps

	Water temperature (°C)		ΔT water-refrigerant (°C)		Refrigerant
	Income from outdoor	To indoor	Condenser	Evaporator	
Winter	5	40	7		R134a
Summer	30	7			

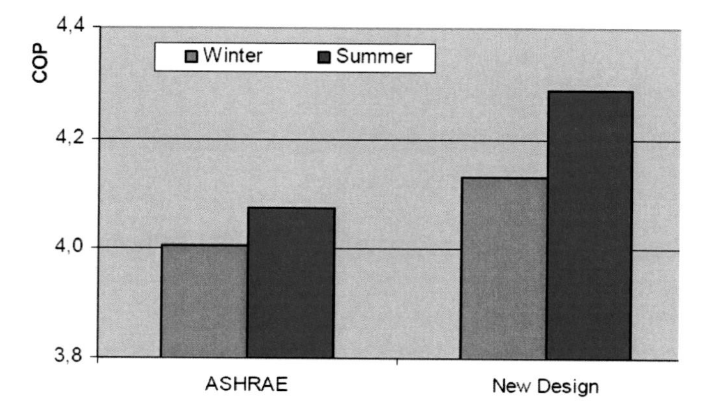

Figure 13. COP for the ASHRAE and New design.

This increase in the COP of the heat pump means that for the same heat output, energy consumption is reduced by approximately 3% in winter and 5% in summer.

6. Air-Water Reversible Heat Pumps

An air-water heat pump is a device with two heat exchangers, one an air-refrigerant and the other a water-refrigerant. In both the winter and the summer cycles, the water mass heated/cooled is useful, the air mass being just a heat source.

In winter it takes heat from an air mass (the air exchanger works as an evaporator), which is transferred to a water mass (the water exchanger works as a condenser). In summer it takes heat from the water mass (the water exchanger works as an evaporator) and it supplies it to an air mass (air exchanger works as a condenser).

6.1. ASHRAE Design for Air-Water Heat Pumps

The ASHRAE design presents the cycle inversion by changing the flow direction of the refrigerant with a reversing valve. Similar to the air-air and water-air cases, the machine can have one or two expansion valves, figure 14. Even so, in the summer cycle the COP in this design is limited by the operation of the condenser in parallel flow.

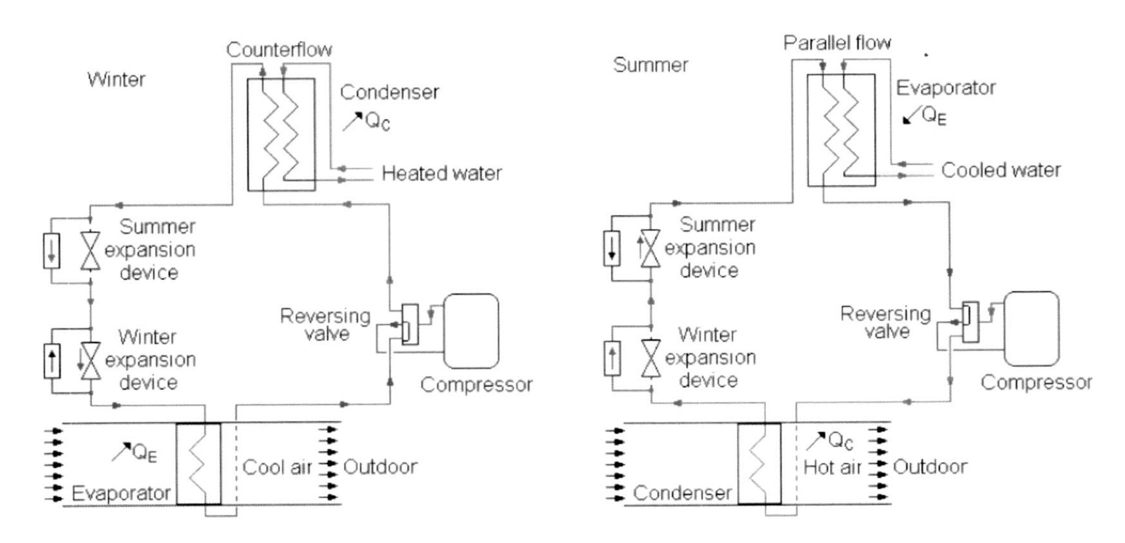

Figure 14. Air-water ASHRAE design.

6.2. New Design for Air-Water Heat Pumps

The New Design includes a reversing valve in the hydraulic connections, which keeps the water exchanger in counterflow, both in the winter and in the summer cycles, figure 15.

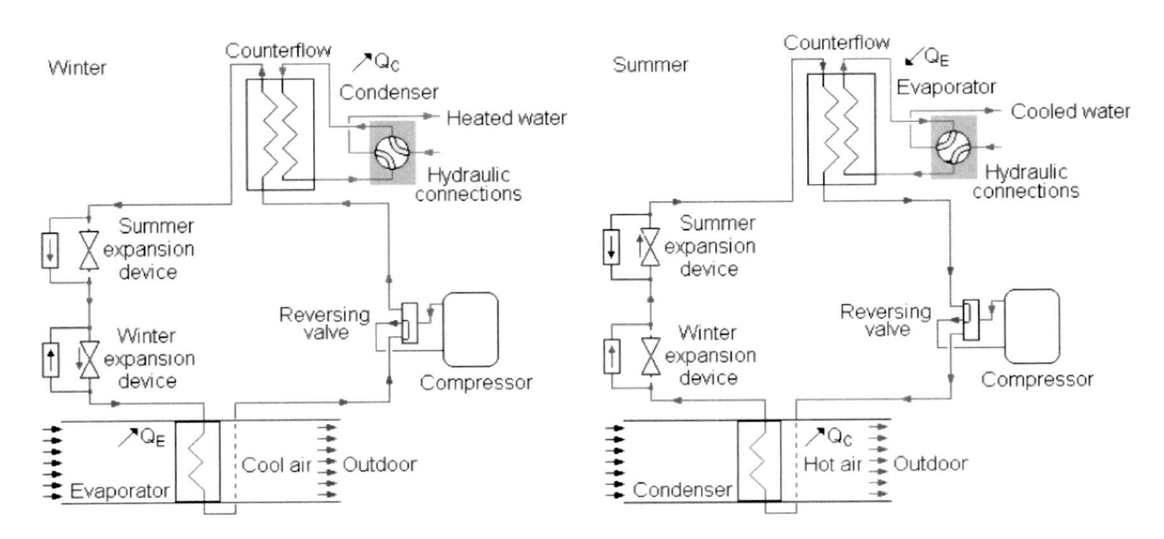

Figure 15. Air-water New Design.

It has not been considered in the design, but in the event that the building had air extraction, it could be used as a thermal focus in the air exchanger (like the air-air heat pump), which would further increase the heat pump performance.

6.3. Energy Study for Air-Water Heat Pumps

Simulations have been carried out on the COP of ASHRAE and New Design. The comparison parameters are given in table 4, and the results are shown in figure 16.

Table 4. Parameters for comparison of air-water heat pumps

	Outdoor air temperature (°C)	Water temperature to indoor (°C)	ΔT heat exchanger (°C)		Refrigerant
			Air	Water	
Winter	0	40	15	7	R134a
Summer	40	7			

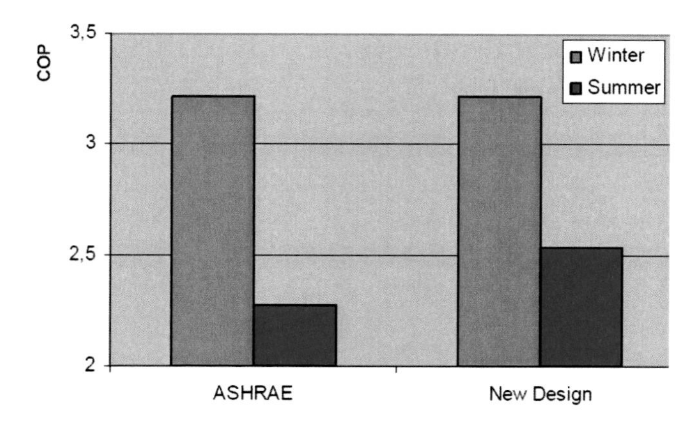

Figure 16. Capacity and COP for the ASHRAE and New Design.

Figure 16 shows that the winter operation in the two designs offers the same COP, the New Design being more efficient in summer.

7. Heat Pump with Total Heat Recovery

The heat pump with total energy recovery is a new design for water-water heat pumps. It seeks to exploit both the heat emitted by the condenser and that absorbed by the evaporator. In order to balance the two thermal demands, a new heat exchanger is included, refrigerant-air, which works "to rest", figure 17.

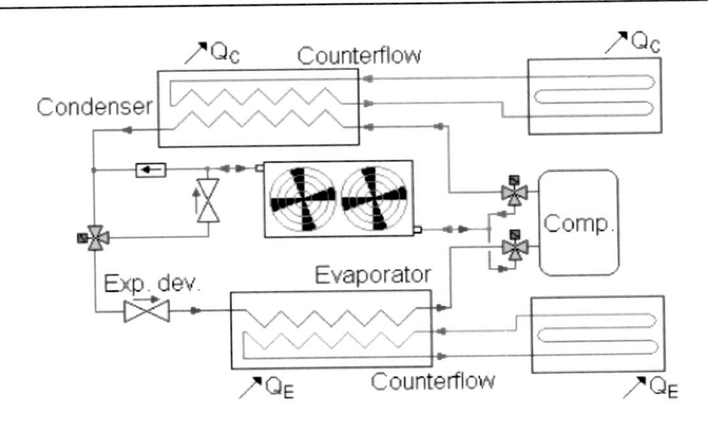

Figure 17. Water-water heat pump total energy recovery.

These heat pumps show five different operation modes:

- Balanced heat and cold demands; the air heat exchanger is not working, figure 18a.
- Heat demand without cooling demand; the air heat exchanger is the evaporator of the heat pump, figure 18b.
- Cold demand without heat demand; the air heat exchanger is the condenser of the heat pump, figure 18c.
- Demands of heat and cold unbalanced, with a greater heat demand, the air heat exchanger works as an evaporator parallel with the heat pump water evaporator, figure 18d.
- Demands of heat and cold unbalanced, with a greater cold demand, the air heat exchanger works as a condenser parallel with the heat pump water condenser, figure 18e.

As can be seen in figures 18, this heat pump is able to deliver the demands of heat and cold, whether the demand is balanced or not, so it offers great versatility. This type of heat pump is very energy-efficient with any kind of building demand. In these devices the thermal power (Eq 3) and the COP (Eq 4) are defined as:

$$HP_{Th} = \dot{Q}_{con}\Big|_{useful} + \dot{Q}_{eva}\Big|_{useful} \qquad \text{(Eq. 3)}$$

$$COP = \frac{\dot{Q}_{con}\Big|_{useful} + \dot{Q}_{eva}\Big|_{useful}}{W_{comp}} \qquad \text{(Eq. 4)}$$

Obviously the efficiency of this machine depends on the annual percentage of the five types of potential demands. Several simulations have been carried out on the energy consumption by the four different demand patterns and the operating parameters given in table 5. Demand for heating in summer corresponds to hot tap water. Figure 19 provides the results.

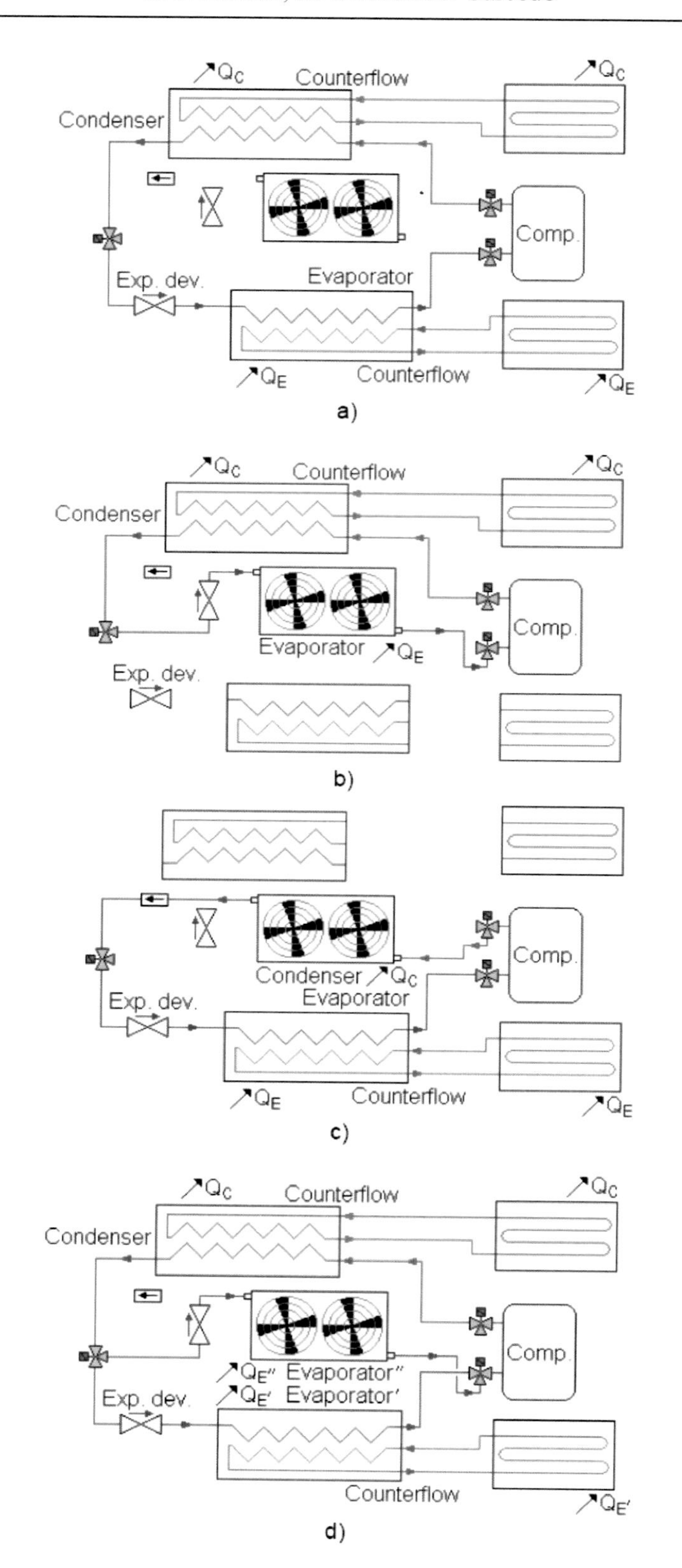

Figure 18. Continued on next page.

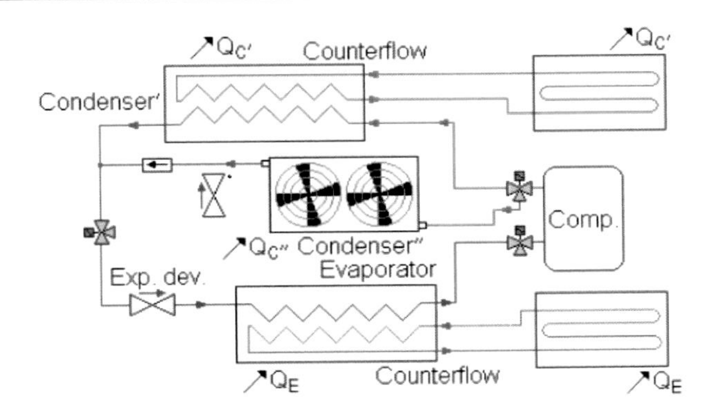

Figure 18. Different operation modes of the water-water heat pump with total energy recovery.

Table 5. Demand patterns and parameters for comparison of a water-water heat pump with total energy recovery

		% demand			
Type of demand		Spring	Summer	Autumn	Winter
a	Heating	20	0	20	60
	Cooling	80	100	80	40
b	Heating	35	5	35	70
	Cooling	65	95	65	30
c	Heating	50	10	50	80
	Cooling	50	90	50	20
d	Heating	65	15	65	90
	Cooling	35	85	35	10

	T (°C)				ΔT (°C) heat exchanger		Refrigerant
Outdoor air	20	40	20	0	Water	7	R134a
Heated water to indoor	40				Air	15	
Chilled water to indoor	7						

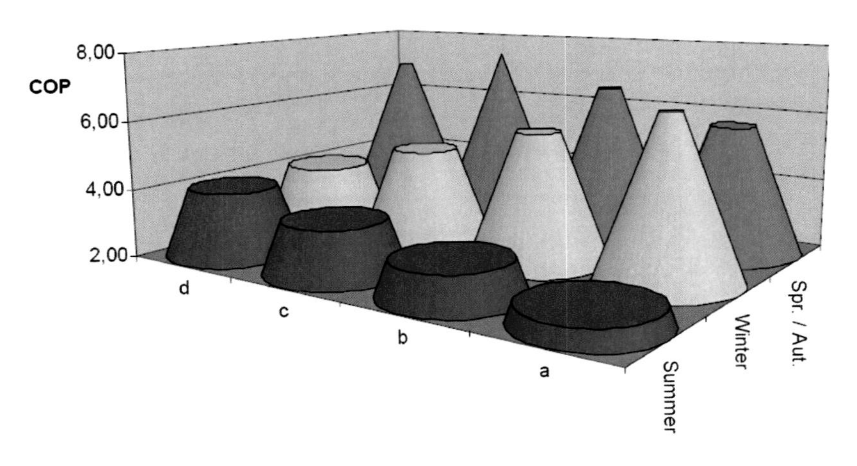

Figure 19. COP of a water-water heat pump with total energy recovery for different demand types.

As can be seen in figure 19, great efficiencies can be obtained with this kind of heat pumps in periods at which the thermal demand is balanced.

8. VRV Systems

The variant refrigerant volume systems, VRV, consist of an outdoor unit and multiple indoor devices. If these units are fed by three pipes, they can work individually as condensers or evaporators, allowing the device to be considered as a specific type of heat pump.

The main goal of the VRV system is to deliver to each indoor unit the refrigerant (thermal power) that the air-conditioned area demands. This way the instantaneous consumption is a function of the energy that is provided to compressor, and not of the thermal power supplied.

The original system has different variants, which have been appearing on the market as it has developed and incorporated the electronic control equipment for air conditioning. This has helped enormously to adjust thermal production to the variability of demand.

VRV devices allow a single outdoor unit (which can consist of several modules) to be connected to numerous indoor units (depending on the manufacturer and model this can be over 30) through a pipe circuit.

The outdoor unit includes an electronic expansion valve, which operates only in winter (a period in which this unit must be an evaporator), it remaining inactive when the unit operates as a condenser. The indoor units include an electronic expansion valve, which operates in summer (when these units must be evaporators), it remaining inactive when the units operate as condensers.

Before each indoor unit, or group of indoor units, is reached, there is a flow selector box. This box is joined by three pipes to the outdoor unit (liquid line, suction line and discharge line), and with two pipes to the indoor unit, or group of indoor units, figure 20.

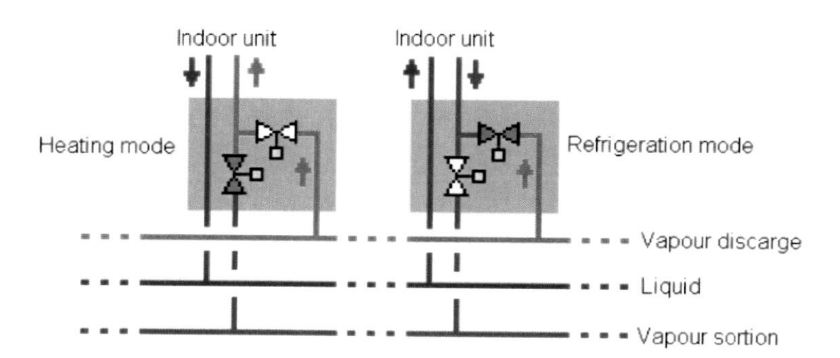

Figure 20. Selector box in a VRV system for cooling and heating modes.

Regarding the operation of the VRV system there may be three scenarios:

- Great refrigeration demand, figure 21, under this demand the solenoid valve S1 is open, while the S2 remains closed. The outdoor unit operates as a condenser, and its expansion valve is inactive.

- The selector units of the areas that require refrigeration supply the indoor units (evaporators) of the liquid line, the expansion valves of these units being active. The output is vapor that goes to the suction line.
- The selector units of the areas that require heating (if any) supply the indoor units (condensers) of the discharge line, the expansion valves of these units being inactive. The output is liquid, which obviously goes to the liquid line.

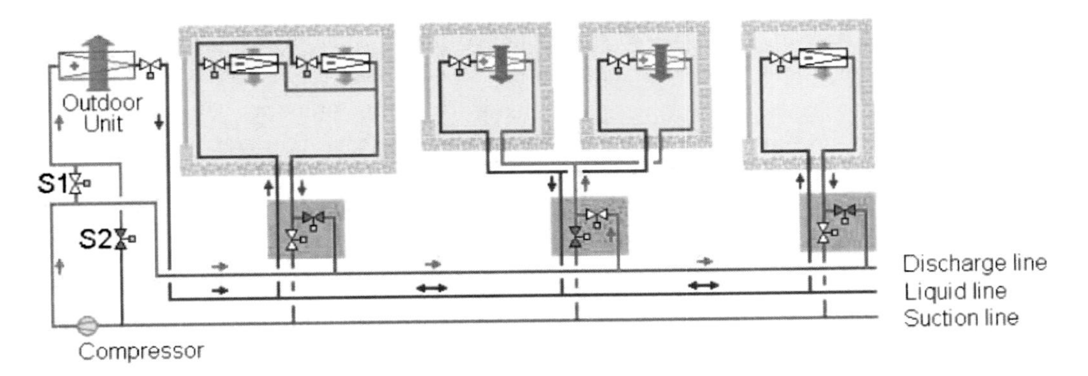

Figure 21. VRV system operation with cooling demand majority.

- Great heating demand, figure 22, under this demand the solenoid valve S1 is closed, while the S2 remains open. The outdoor unit operates as an evaporator, and its expansion valve is active.
- The selector units of the areas that require heating supply the indoor units (condensers) of the discharge line, the expansion valves of these units being inactive. The output is liquid which goes to the liquid line.
- The selector units of the areas that require refrigeration (if any) supply the indoor units (evaporators) of the liquid line, the expansion valves of these units being active. The output is vapor that goes to the suction line.

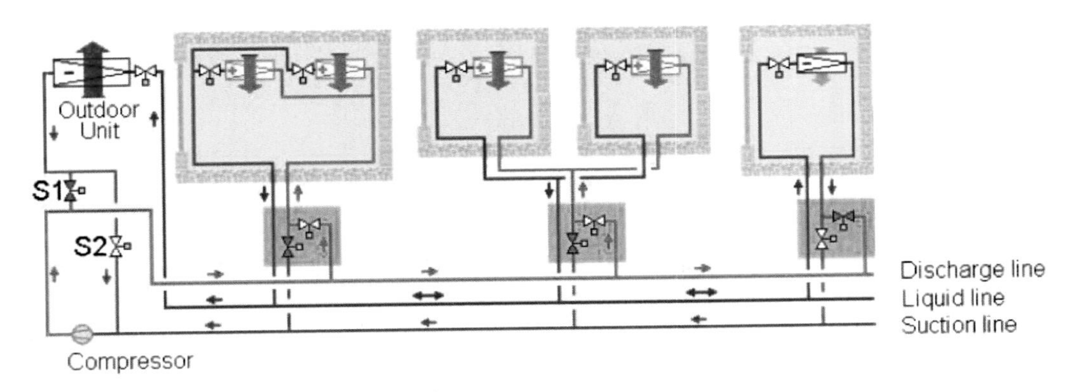

Figure 22. VRV system operation with heating demand majority.

- Balanced demand between heating and cooling, figure 23: under this demand the exchanger of the outdoor unit is inactive, and the solenoid valves S1 and S2 remain closed.
- The selector units of the areas that require heating supply the indoor units of the discharge line, making them work as condensers (the expansion valves of these units are inactive). The output is liquid that goes to the liquid line.
- The selector units of the areas that require refrigeration supply the indoor units of the liquid line, making them work as evaporators (the expansion valves of these units are active). The output is vapor that goes to the suction line.
- In short, with this kind of thermal demand, the indoor units operating inside rooms that require heating work as condensers, while those that require refrigeration operate as evaporators.

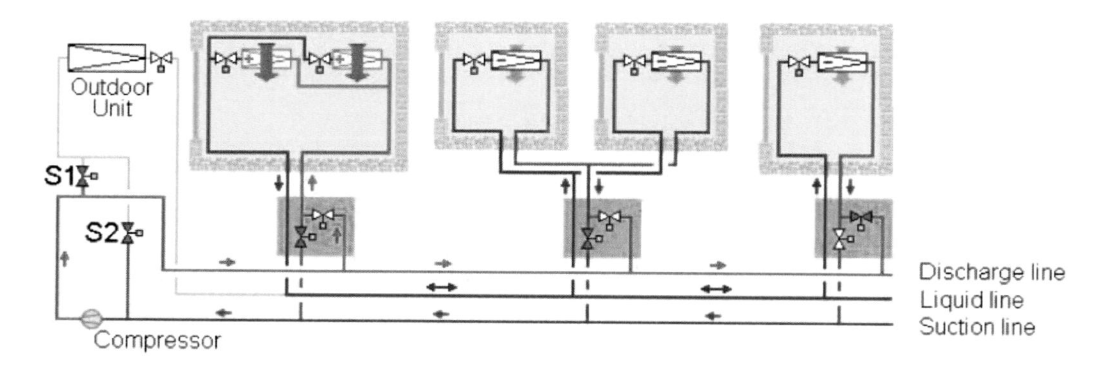

Figure 23. Operation of a VRV system with balanced demand between heating and cooling.

Table 6. Demand patterns and parameters for comparison of a VRV system

Type of demand		% demand Spring	Summer	Autumn	Winter
a	Heating	20	0	20	60
	Cooling	80	100	80	40
b	Heating	35	0	35	70
	Cooling	65	100	65	30
c	Heating	50	0	50	80
	Cooling	50	100	50	20
d	Heating	65	0	65	90
	Cooling	35	100	35	10

	T (°C)				ΔT (°C) heat exchanger	Refrigerant
Indoor air	23	24	23	22	15	R134a
Outdoor air	20	40	20	0		
Heated air to indoor	40					
Cooled air to indoor	17					

VRV systems are shown to be very energy efficient with any demand arising in the building. In these devices the thermal power and the COP are defined in a similar way to the heat pump with integral energy recovery (Eqs 3-4).

It is in climatic periods in which the buildings have a balanced thermal demand that the COP of the machine reaches its maximum value, since it takes advantage of the heat from the condenser and of the cold from the evaporator.

Obviously, the machine efficiency depends on the demand pattern for the building. Simulations carried out on the energy consumption corresponding to four different patterns and the operating parameters given in table 6, provide the results in figure 24.

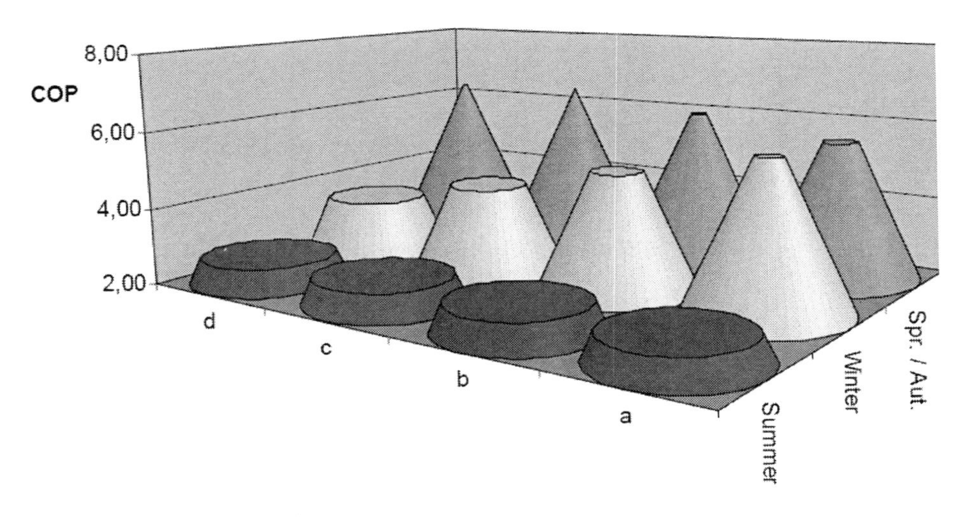

Figure 24. COP of a VRV system at different types of demand.

As can be seen in figure 24, and similar to what happens in the heat pump with total energy recovery, large efficiencies can be achieved with this type of heat pumps in the periods at which the thermal demand is balanced.

9. CONCLUSION

In this chapter an analysis of heat pump design has been carried out. Air-air, water-air, water-water and air-water configurations have been studied.

The designs suggested by ASHRAE were analyzed, and New Designs proposed. Energy studies have been conducted demonstrating that the New Designs are more efficient.

Moreover, the water-water heat pump designs with total energy recovery and the VRV system have been introduced. Both systems offer great potential in those buildings which demand heat and cold simultaneously.

Improvement in heat pump energy efficiency will be supported by the use of these machines in climates where they have so far not been energetically profitable.

REFERENCES

[1] AFEC, http://www.afec.es/.
[2] EUROVENT-CECOMAF, http://www.eurovent-cecomaf.org/web/eurovent/web/.
[3] Moran, M.J.; Shapiro, H.N. *Fundamentals of Engineering Thermodynamics*, Ed John Wiley and Sons, Inc., 2004; Ch 10.
[4] IDAE, http://www.idae.es/.
[5] ENEBEC, http://www.enebc.org/.
[6] ASHRAE, 2000 ASHRAE Handbook—HVAC Systems and Equipment; AHSRAE: Atlanta GA, 2000; Ch 8.
[7] Incropera, F.P.; DeWitt, D.P. *Fundamentals of Heat and Mass Transfer*, 4th Ed, John Wiley and Sons, Inc., 1996; Ch 11.
[8] Jakobsen; A., *Energy Optimisation of Refrigeration Systems*, PhD Dissertation, Technical University of Denmark, Kongens Lyngby, 1995; Ch 2.
[9] Renedo, C.J.; Ortiz, A.; Mañana, M.; Delgado, F. *Energy Build.* 2007, *vol* 39, 1244–1249.
[10] Renedo, C.J.; Ortiz, A.; Mañana, M.; Peredo, J. *Energy Build.* 2006, *vol* 38, 1240–1247.

In: Air Conditioning Systems
Editors: T. Hästesko, O. Kiljunen, pp. 337-351
ISBN: 978-1-60741-555-8

Chapter 9

LOAD PREDICTOR FOR AIR CONDITIONING SYSTEMS USING ARTIFICIAL NEURAL NETWORKS MODELS FOR INTELLIGENT BUILDINGS

M. Trejo-Perea, G. Herrera-Ruiz, D. Vargas-Vázquez, P. Talamantes-Contreras, E. A. Rivas-Araiza, R. Luna-Rubio and G. J. Ríos-Moreno*

School of Engineering, University of Queretaro,
Cerro de las Campanas s/n, 76010, Querétaro, México

ABSTRACT

The use of artificial neural networ (ANN) in several applications related to energy management in buildings has been increasing significantly in recent years. This work proposes and analyzes a load predictor for air conditioning systems in intelligent buildings. A linear autoregressive model with external input (ARX) is used to predict the indoor air temperature in a building. The accuracy of the load predictor can be improved by using a cascaded predictor architecture where the input variables for the ANN are: a) the output of a predicted temperature ARX model; b) historical hourly load data; c) day; and d) time. The load predictor uses a multi-layer perceptron (MLP) neural network, which is trained by Levenberg-Marquardt backpropagation (BP) algorithm. The performance of the load predictor was evaluated using real data obtained from a building located at the University of Queretaro, Mexico. The performance of the ANN model was evaluated by means of the analysis of variance (ANOVA). In addition, load values estimated by the ANN were compared, with regression-estimated and current values, using ANOVA and mean comparison procedures. The obtained results shows that the average of electric load estimated values for the selected ANN and the real data are very close, with a 95% of confidence level. Therefore, results from the ANN are significantly better than those obtained by conventional regression.

---

* E-mail: riosg@uaq.mx, Tel. 52 - (442) 192 1200 Ext. 6049

Keywords: Load prediction; Intelligent buildings; Neural networks; Air conditioning systems.

1. Introduction

An intelligent building is defined as the one that maximizes the efficiency of the service with a minimum cost [1]. This author enumerates a list of intelligent building components, placing the energy management system (EMS) as the top priority. This system controls the building energy consumption. For efficient operation of EMS, accurate information of consumption is needed in order to know how behaves in short-term and medium-term. This "short-term load forecasting" (STLF), can predict electric load in regions, countries and even in buildings or industries during a period of minutes, hours, days or weeks. The energy consumption by heating, ventilating and air-conditioning (HVAC) equipment in industrial and commercial buildings represents 50% of the world's energy consumption [2]. High energy consumption in the HVAC systems is due to the use of inefficient control operation sequences and system failures.

Experimental research and theoretical studies have demonstrated that a potential energy saving of about 20-30% can be achieved by improving energy management and by operating HVAC systems in an efficient way while guarantee a comfortable environment for the occupants [3]. The increase in electrical energy consumption in these equipments has caused that their use, administration and estimation to be essential issues.

Recent research regarding artificial neural networks (ANN) applications in the prediction of electric load shows that this technique is one of the most successful in different application areas (i.e countries or buildings). Their use in applications related to energy management began in the early 1990's. Kalogirou [4] provides a comprehensive overview of ANN applications in renewable energy systems and in buildings. Park [5] reported the use of a simple neural network using temperature information, was capable of predicting hourly, peak, and total energy consumption, more accurately than conventional techniques based on regression. Bacha [6] discussed why neural networks are appropriate for load prediction and proposed a cascaded sub-network system. Srinivasan [7] used a four layer multilayer perceptron (MLP) to predict hourly load in a power system.

A large amount of information on optimal temperature modeling in buildings development can be found in current literature. Some of these strategies and techniques are also applied in other areas, including: linear autoregressive models (ARX) [2], physical models [8, 9] and neural networks [10]. Generally, these temperature models are used for planning strategies for optimal control; however, it is important to consider the cost of electrical consumption in order to achieve the best energy saving. The climate variables have to be kept within an appropriate range while user comfort is increased, operating cost is minimized, and thus bigger energy savings are achieved. Estimating indoor air temperature in buildings is a complicated task which can be tackled with two different classes of models: statistical and not-statistical models. Both can be viewed as non-linear dynamical systems, and both use inputs and outputs [11-19]. Such studies demonstrate the importance of indoor temperature in buildings.

The advantage of ANN with respect to the statistical models is their ability to model multivariable problems with respect to complex relationships between variables. Also, ANN can extract the implicit non-linear relationships among these variables by means of "learning" with training data. Many excellent results in real applications have been accomplished with ANN in STLF's using a wide variety of ANN architectures. The works presented in [20-22] are good examples.

The purpose of this work is to develop a load forecaster for air conditioning systems in intelligent buildings. The predictor uses a cascaded architecture where the output of an ARX temperature model is used to predict the indoor air temperature in the buildings. This predictor is used as one of the inputs for the energy consumption prediction in addition to historical hourly load data, day and time. The energy consumption predictor uses a MLP ANN trained by the Levenbergh-Marquardt backpropagation (BP) algorithm. A validation model was performed by comparing the results with a nonlinear regression model and actual data, using analysis of variance (ANOVA) procedures and Duncan's multiple range test (DMRT).

2. THEORETICAL CONSIDERATIONS

Recently ANN's have emerged as a technology for load modeling and forecasting because of their ability to learn complex, nonlinear functions. They allow for the estimation of possible nonlinear models without the need to specify a precise functional form. ANN's can be viewed as parallel and distributed processing systems that consist of a huge number of simple and massively connected processors called neurons. Each individual neuron consists of a set of synaptic inputs, through which the input signals are received. Then, the incoming activations are multiplied by the synaptic weights and summed up. The outgoing activation is determined by applying a threshold function to the summation. The threshold function can be a linear, or a non-linear function that decides the output of the neuron. The structure of the neuron is shown in figure 1.

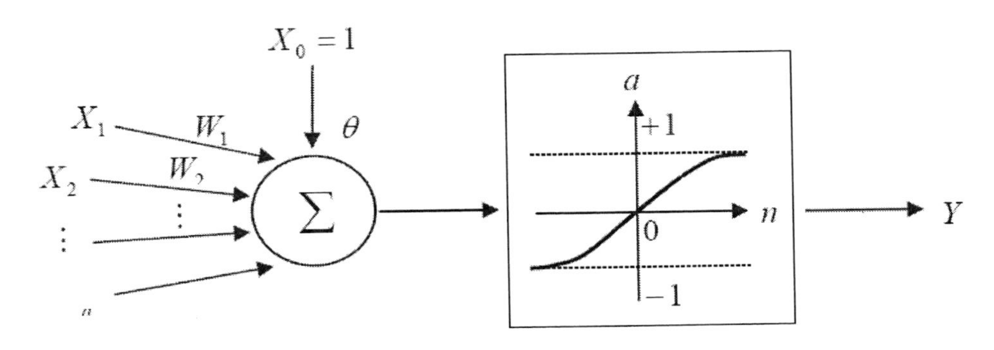

Figure 1. Neural Model.

In the above figure, X_1, X_2, ..., X_n present the input of the neuron. W_1, W_2, ..., W_n are their weights, θ is the threshold value and Y represents the output. The input to output relationship is characterized by

$$Y(X) = f(\sum_{i=1}^{n} w_i x_i + \theta) = f(W^T X + \theta) \quad (1)$$

where ***W*** is the vector of synaptic weights, ***X*** is the input vector, θ is a constant called offset or bias, and *f* is the activation function. The superscript T denotes the transpose operator and Y(X) is the output of the neuron. In this research, the activation function used is a sigmoid that has the following form:

$$f(x) = \frac{1}{1+e^{-x}} \quad (2)$$

The training of an ANN is undertaken for the most part using the BP based learning algorithm which is a supervised algorithm. This method requires a set of training patterns, and their corresponding desired outputs, and autonomously adjusts the connection weights among neurons. Correction of the weights is made according to imposed learning rules and thereby, obtains unique knowledge from the data.

Although it has been successfully used in many real-world applications, the standard backpropagation algorithm (SBP) suffers from a number of shortcomings, one of these being the rate at which the algorithm converges. Several iterations are required to train a small network, even for a simple problem. Reducing the number of iterations and speeding up learning times of ANN are recent subjects of research. Some improvements of the SBP algorithm are the gradient descent [23] and the Levenberg-Marquardt algorithms [24,25].

The model of neural network is determined by the following three factors: 1) topological structure of the network; 2) neuron characteristic; and 3) training algorithm. The ANN implemented in this study is a multilayer perceptron (MLP) with an input layer of 4 nodes, an hidden layer with a variable number of hidden nodes and an output layer with only one node. Several networks with variable number of hidden nodes were implemented and tested. The variables in the input layer are temperature, hour, day and load consumption. A representative schematic of the ANN used is depicted in figure 2.

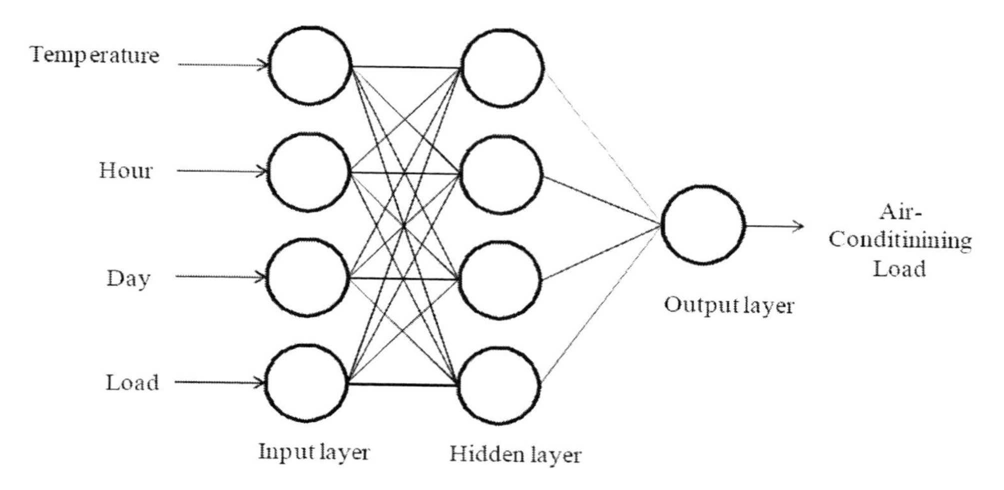

Figure 2. A typical feed forward neural network (MLP).

In buildings, the power consumption is highly dependent on climatic outdoor because this determines the operation sequences of air conditioning systems required to maintain the optimal indoor temperature. A model ARX for forecasting indoor air temperature of intelligent building was developed [2]. The predictor output is used as inputs for the ANN predictor for energy consumption. The inputs variables of forecast temperature model ARX were outside air temperature (T_o), global solar radiation flux (R_a), wind speed (V_w), and outside air relative humidity (R_{ho}). Information about these variables was gathered and recorded using the TUNA™ system SCII v5.0. This model provided the values of the environmental inside temperature which are inputs to the ANN-MLP model.

The time related information is considered as input to the ANN: the hour of the day, and the day of the week. The hour is coded as reported in current literature [26] by means of its sine and cosine values. Time is required because there are some tasks that must be accomplished according to a schedule. Consequently power consumption could vary from hour to hour and day to day depending on the time of the week. The last input variable is the power consumption in kWh. Since the predictor is designed to predict the consumption at time k+1, the value used for this input is the consumption value of time k, supplied by an energy consumption instrument. This instrument is called "New monitoring system and electric power saving for intelligent buildings" [27]. This information has great relevance since it reflects how the energy consumption of the installation behaves. A general view of the cascaded predictor is presented in figure 3.

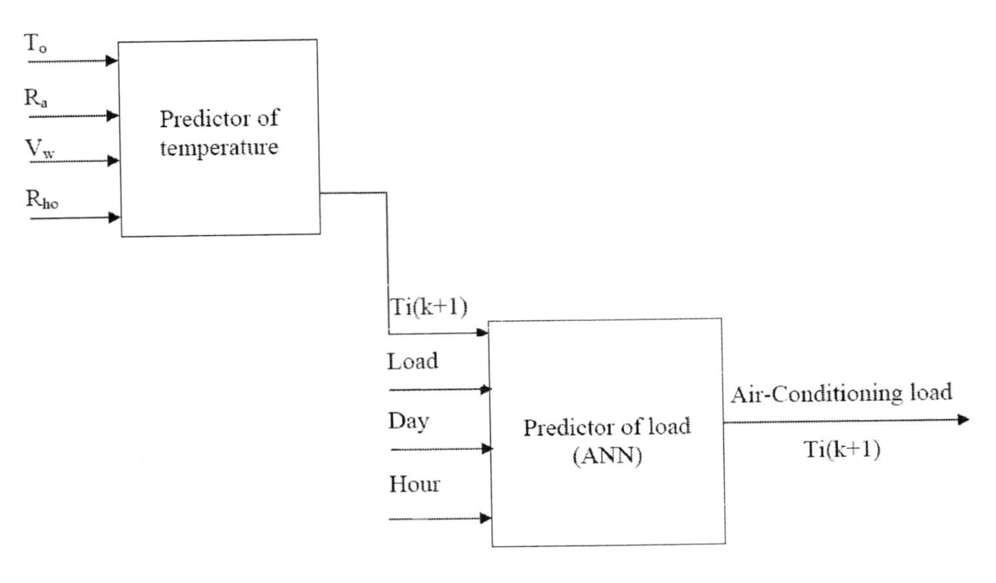

Figure 3. Proposed cascading predicting model structure.

In order to avoid the saturation in neurons condition, data normalization is required. If neurons get saturated, changes in the input value will produce a negligible change, and not just one change of the output value. For this reason, data must be normalized before being presented to the artificial neural network. Data normalization compresses the range of the training data between 0 and 1. The normalization in this study was carried out by means of the following expression:

$$X_n = \frac{(x - x_{\min}) * range}{x_{\max} - x_{\min}} + startingvalue \tag{3}$$

where X_n is the value of the normalized data and X_{min} y X_{max} are the minimum and maximum of the entire data set respectively.

3. Materials and Methods

3.1. Building Description

The building has a 12 cm thick concrete layer on top. The concrete layer was armed and has a resistance of 300 kg/cm^2. The lateral walls were built with 14 cm wide × 30 cm long cored bricks. With the purpose of providing the best illumination, the design include two 4 m long × 2 m high windows on the north side and one on the south side. The walls dividing the rooms are made of concrete blocks lined up with mortar with finished interiors and exteriors. It has a 0.5 mm and 0.7 cm thick coat of vinyl paint. The roof is covered with water proof Protexa layer and red sand. The description aforementioned is a general description of the architectonical characteristics for the standard buildings commonly used for classrooms in Mexico. This building has five classrooms, a meeting room, and a control room; the total covering an area of 304 m^2. The height of the entire structure is 3 m.

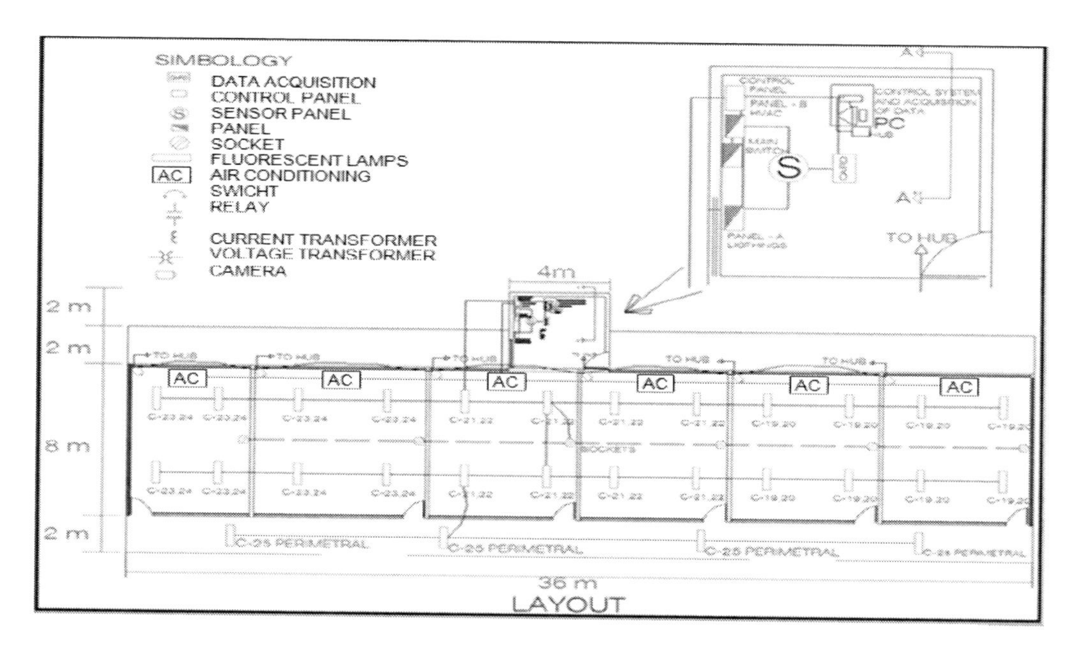

Figure 4. Location and the distribution of the electric loads of the building (dimensions in meters).

The location and distribution of the electrical loads in the building are shown in figure 4. Each one of the classrooms has two fluorescent lighting fixtures of 2x75 Watts and two fluorescent slim line lighting fixtures of 2 x 39 Watts equipped with electromagnetic ballast.

The air conditioning system is type mini split that is supplied by a three phase, 208 to 230 V, 60HZ power source. The lock rotor amps (LRA) of the compressor is 78 and the running load amps (RLA) is 10.9, while the fan motor is rated to ¼ HP. In the other side the building provide one socket for single phase of 127 volts, 10 Amp supply. The external lighting includes four fluorescent lighting fixtures of 2 x 39 Watts. Figure 2 shows the location of the loads, the instrumentation and monitoring and the energy saving system in the control room. In addition, the building has an indoor temperature prediction system that operates by means ARX models [7].

3.2. The Power Monitoring Description

The power distribution wires are linked with an external data acquisition (DAQ) board [27] through voltage and current sensors. The main features of the DAQ board are: 12 channels simultaneous sampling, 16 bit digital to analog converter (ADC), 250 k samples per second and USB interface. Power consumption measurements were gathered by using this DAQ board, and then resulting data are stored in a personal computer for further processing.

3.3. Algorithm's Description

A neural network algorithm for the energy consumption model in buildings was proposed. First, the mean absolute percent error (MAPE) was used to select the best network architecture ANN-MLP. Then, the best network was compared with actual and regression data using MAPE. Finally, analysis of variance (ANOVA) and Duncan's multiple range test (DMRT) procedures were used to compare, verify and validate the models. The description of the algorithm follows:

a. The input and output model variables are determined.
b. A group of data, namely B, of the input and output variables for past times describing the input/output relationship is collected.
c. B is divided into two subsets: training (B1) and testing (B2). This procedure requires independent validation data to be used in order to test the neuronal network capability for generalizing non-predicted data. Representative testing (validating) data are taken from the training data. It is necessary to find a balance between the size of training and validation data. The validation data are chosen near the actual period.
d. Once the (B1) and (B2) data subsets are correctly defined, a conventional regression model of the data training was performed. Then, the energy consumption for the testing periods was predicted.
e. Finally, the ANN method is used to estimate the inputs/outputs relationship. In order to find an appropriate number of hidden nodes, the aforementioned steps are repeated using different architecture and training parameters for the network, with one to (q) nodes in its hidden layer. If the (q) value is optional, it can be modified. If after applying the above steps the minimum relative error is not obtained, the following procedures have to be performed:

- Choose the architecture and the training parameters.
- Train the model using the learning data (B1).
- Evaluate the model using the testing data (B2).
- Select the best network architecture (ANN) for the testing data with the desired error.
- Apply ANOVA procedures to the formal testing data for verification and validation of the ANN results.

3.4. Accuracy Measurements

Multiple determination of coefficient (R^2) is a measurement of the correlation between the observed and predicted values [28]. Some measurements of variance are the Standard Error Prediction (SEP) [29], the Mean Square Error (MSE), and the Mean Absolute Percent Error (MAPE) [30]. These methods are used to determine the model's capability to explain total data variance. The MAPE was calculated for each model by using the following equation:

$$MAPE = \frac{1}{N}\sum_{J=1}^{N}\frac{|\,Ue_j - Ua_j\,|}{Ua_j} \tag{4}$$

where Ue_j is the estimated electrical power consumption, Ua_j is the actual energy consumption value, and N is the total number of generalized samples. The MSE, R^2 and SEP are determined by equations (5), (6) and (7), respectively.

$$MSE = \frac{\sum (Ua_j - Ue_j)^2}{N-2} \tag{5}$$

where N-2 correspond to the degrees of freedom,

$$R^2 = 1 - \frac{SSE}{SSTO}\ ;\ SSE = \sum (Ua_j - Ue_j)^2\ ;\ SSTO = \sum (Ua_j - \overline{U}a_j)^2 \tag{6}$$

where *SSTO* a measurement of the variability of mean observed values, *SSE* is a measurement of the relationship between the predicted and observed values. The Standard Error Prediction is defined by:

$$SEP = \frac{1}{\overline{U}a_j}\sqrt{\frac{\sum_{j=1}^{N}(Ua_j - Ue_j)^2}{N}} \tag{7}$$

In the following section, the application of the ANN algorithm in a greenhouse will be applied. Its advantages and superiority will be shown by using ANOVA y Duncan's Multiple Range Test (DMRT) procedures.

4. Results

The recorded data of the Air Conditioning electric load were captured from November 3, 2008 starting at 12:00 h and ended on November 8, 2008 at 7:00 h. Data were divided into two groups: first 70 hours were selected for network training (B1); the last 45 hours were chosen for testing the ANN (B2). Different MLP networks were generated and tested.

The Levenberg-Marquardt back propagation algorithm was used in order to adjust the learning procedure; in addition, the captured data from November 6 starting at 10:00 h. and ended on November 8, at 7:00 h. were used for testing the network.

Once the network was trained, it could be used to forecast associated data with validation set in order to obtain the corresponding prediction errors; in this way, the forecasting process performance could be studied. According with [7,26], the Mean Absolute Percentage Error (MAPE) has been widely used to examine the quality of the prediction models as a performance measure. In order to determine the best ANN model, MAPE values were computed for each one of them; obtained results are summarized in table 1.

Table 1. Different MLP results

Model MLP			
	R^2	*MAPE*	*SEP*
4-2-1	0.9355	0.1250	0.2659
4-3-1	0.9356	0.1327	0.2647
4-4-1	0.9344	0.1550	0.2673
4-5-1	0.9349	0.1626	0.2662
4-6-1	0.9310	0.1233	0.2740
4-7-1	0.9353	0.1386	0.2654
4-8-1	0.9471	0.1236	0.2321
4-9-1	0.9370	0.1391	0.2619

Table 1 shows the best models and their reliability and performance estimators. The MLP model with the best results was the (4-8-1) model, in comparison with the other seven (0.1236 MAPE, 0.9471 R2, and 0.2321 SEP). The results are shown in the ANN model graph (4-8-1) versus actual data figure 5.

Results of the MLP network were compared with the non-linear regression model (table 2). It can be observed that the MLP output has a smaller error than the non-linear regression model. The error comparison of MAPE, R^2 and SEP for the MLP and regression is shown at figure 6.

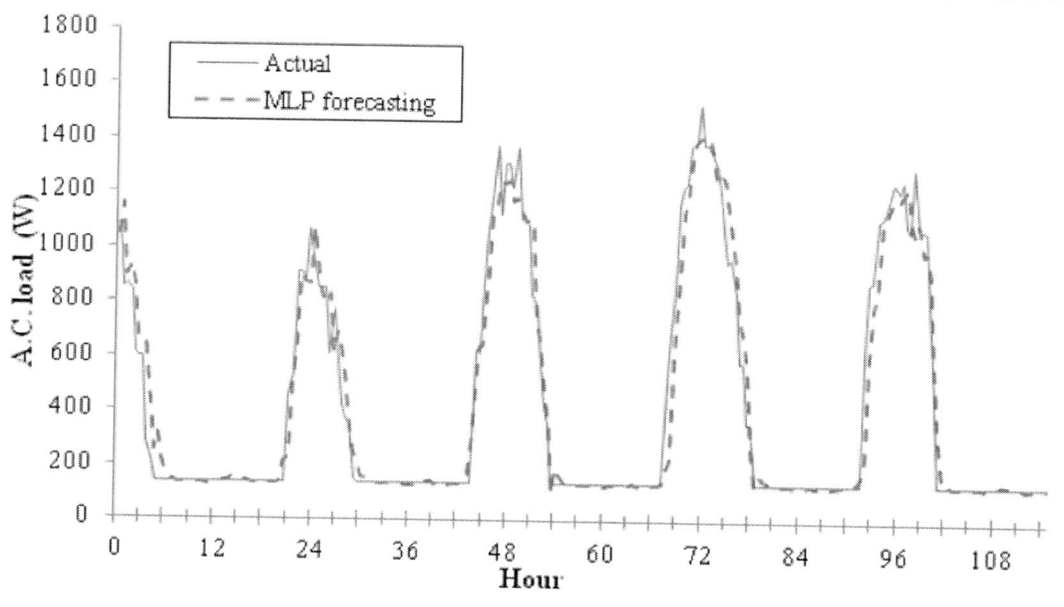

Figure 5. Measured and predicted values of load prediction of air conditioning systems.

Table 2. Comparing the two models (the first MLP and regression)

Time (Hr)	Actual	Neural Network	Regression
70.5	1238.67	1189.81	1105.06
71.0	1383.33	1320.28	1235.90
71.5	1396.67	1396.26	1308.37
72.0	1531.67	1405.33	1366.35
72.5	1385.00	1414.57	1465.96
73.0	1385.00	1420.05	1466.54
73.5	1361.67	1352.28	1462.57
74.0	1304.83	1254.21	1404.76
74.5	1215.33	1281.36	1377.51
75.0	1080.33	1249.89	1319.82
75.5	954.33	1135.47	1224.17
76.0	966.33	1044.83	1105.00
76.5	842.50	949.00	1020.34
77.0	583.83	760.98	939.48
77.5	583.33	658.07	813.51
.			
115.0	135.00	132.89	137.70
	MAPE error	0.089	0.2482

A forecasting neural network model that uses weather data was proposed in [32]. The model was capable to predict electrical energy monthly demand, where a MAPE of 2.03%

was accomplished; however, the prediction errors are more sensitive to temperature. The aforementioned work uses data sets obtained in different countries under different social and economical behaviors. For this reason, it is not easy to carry out a quantitative comparison because the data sets used in other works are different from the one considered here. The results presented in this work allow energy consumption prediction with a good level of reliability.

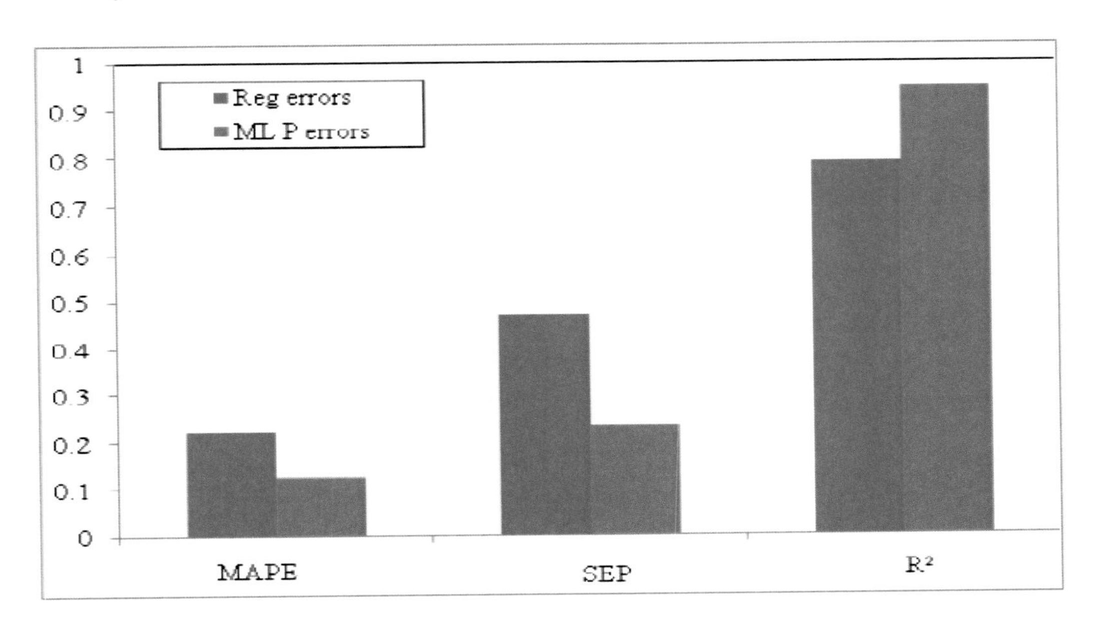

Figure 6. Comparing regression and MLP errors.

Analysis of Variance: Validation and Verification. The selected ANN's results were estimated and the regression method and the real data were compared by using ANOVA procedures. The experiment was designed in such way that variability was derived from external systematically-controlled sources. Time is the common external source of variability in the experiment and can be systematically controlled by blocking [31]. Therefore, one-way ANOVA was used. The results are shown in table 3. The hypothesis test was defined as follows:

$$H_0 : \mu_1 = \mu_2 = \mu_3, \qquad H_1 : \mu_i \neq \mu_j \quad i, j = 1,2,3, \quad i \neq j, \tag{8}$$

where μ_1, μ_2 and μ_3 are average values obtained from the regression model, actual data and the ANN model. From table 3, it is possible to concluded that the null-hypothesis is rejected with a significant level α = 0.025, since $f_{0.05,2,178} = 1.39$, $f_{0.025,2,178} = 3.69$ and $4.54 > 1.39, 3.69$.

Table 3. ANOVA table for comparison of actual, regression and neural network

Summary						
Groups	Count	Sum		Mean		
Measured	90	44116		490.18		
Neural network	90	43774		486.38		
Regression	90	47060		522.89		
Source of variation	Sum square	Degree of freedom	Mean square	F_0	$F_{0.25\,(2,178)}$	$F_{0.025(2,178)}$
Between groups (treatments)	72507	2	36253	4.54	1.39	3.69
Blocks (sampling)	61604893	89	692189			
Within groups	1421240	178	7984			
Total	63098641					

Duncan's Multiple Range Test (DMRT). Before the DMRT is performed, the standard deviation for each treatment has to be calculated as:

$$B_{\bar{y}i} = \sqrt{\frac{MS(error)}{a}} \tag{9}$$

where a is the number of replicates or observations for the three treatments (actual, ANN and regresión). Then, state values for R_p are calculated as:

$$R_p = r\alpha(p, f)B_{\bar{y}i} \tag{10}$$

$r\alpha(p, f)$ is obtained from the DMRT table. After means treatment classification, each treatment can be compared as follows:

$$B_{yi} = 9.41895$$

$$r_{0.05}(2{,}178) = 2.786$$

$$r_{0.05}(89{,}178) = 3.466$$

$$R_2 = r_{0.05}(2{,}178)B_{yi} = (2.178)(9.418) = 26.241$$

$$R_{89} = r_{0.05}(89{,}178)B_{yi} = (3.466)(9.418) = 32.646$$

Comparing treatments 2 and 3 = (522.89)-(486.38) = 36.506 > 32.646

$\rightarrow \mu 2 \neq \mu 3$

Comparing treatments 1 y 3 = (522.89)-(490.18) = 32.707 > 26.241

$\rightarrow \mu 1 \neq \mu 3$

Comparing treatments 1 y 2 = (490.18)-(486.38) = 3.799 < 32.646

$\rightarrow \mu 1 = \mu 2$

From the above results, it can be observed that only one third of the mean (actual data) and the second treatment (the selected ANN) equals to $\alpha = 0.05$. This indicates that the average of electric load estimated values for the selected ANN and the real data were approximately similar with a 95% confidence level. Therefore, results from the ANN are significantly better than those obtained by conventional regression.

5. Conclusion

A method for load prediction from a multilayer perceptron neural network was proven. The predictor uses a cascading architecture. Using ANOVA procedures, this study shows the advantages of the ANN compared with real data and conventional regression model data. This is one of the first studies presenting an algorithm based on ANOVA and ANN models to predict the electric load in air conditioning systems. The (4-8-1) MLP built model produced the best results, with an error estimation of 0.089 in the testing data. The ANOVA statistical method was used in order to compare the neuronal network and the regression model results versus real data. In addition, it was observed that using $\alpha = 0.05$, there was significant difference among treatments. Therefore, the DMRT was used to find the closest model to the real data, considering a significance level of 95%. The load prediction model presented in this work will be the base for the design of new intelligent climate controllers.

References

[1] Wong, J.K.W.; Li, H.; Wang, S.W. Intelligent building research: a review. *Automation in Construction,* 2005, 14, 143-159.

[2] Ríos, G.J.; Trejo, P.M.; Castañeda, C.R.; Herrera, R.G. Modelling temperature in intelligent buildings by means of autoregressive models. *Automation in Construction.* 2007, 16, 713-722.

[3] Yoshida, H.; Kumar, S.; Morita, Y. Online fault detection and diagnosis in VAV air handling unit by RARX modeling. *Energy and Buildings*, 2001, 33, 391-401.

[4] Kalogirou, S. A. Artificial neural networks in renewable energy systems applications: a review. *Renewable and Sustainable Energy Reviews,* 2001, 5, 373-401.

[5] Park, J.; Sandberg, I.W. Universal approximation using radial basis function networks. *Neural Computation*, 1991, 3, 246-257.

[6] Bacha, H.; Meyer, W. A neural network architecture for load forecasting. In: *Proc. IEEE Int. Joint Conf. Neural Networks,* 1992, 2, 442-447.
[7] Srinivasan, D.; Liew, A.C.; Chang, C.S. A neural network learning short-term load forecaster. *Electric Power Systems Research,* 1991, 28, 227-234.
[8] Castañeda, M.R.; Ventura, R.E.; Peniche, V.R.; Herrera, R.G. Fuzzy greenhouse climate control system based on a field programmable gate array. *Biosystems Engineering,* 2006, 2, 165-177.
[9] Lafont, F.; Balmat, J. F. Optimized fuzzy control of a greenhouse. *Fuzzy Sets Systems,* 2002, 128, 47-59.
[10] Ferreira, P.M.; Faria, E.A.; Ruano, A.E. Neural network models in greenhouse air temperature prediction. *Neurocomputing,* 2002, 43, 51-57.
[11] Kalogirou, S. Artificial neural networks in renewable energy systems applications: a review. *Renewable&Sustainable Energy Reviews,* 2001, 5, 373-401.
[12] Cetiner, C.; Halici, F.; Cacur, H. Generating hot water by solar energy and application of neural network. *Applied Thermal Engineering,* 2005, 25, 1337-1348.
[13] Argiriou, A.; Bellas-Velidis, I.; Balaras, C. Development of a neural network heating controller for solar buildings. *Neural Networks*, 2000, 13, 811–820.
[14] Ben-Nakhi, A.; Mahmoud, M. Energy conservation in buildings through efficient A/C control using neural networks. *Applied Energy,* 2002, 73, 5-23.
[15] Wang, Y.C. Fault-tolerant control for outdoor ventilation air flow rate in buildings based on neural network. *Building and Environment*, 2002, 37, 691–794.
[16] Mahmoud, M.; Ben-Nakhi, A. Architecture and performance of neural networks for efficient A/C control in buildings. *Energy conversion and Management,* 2003, 44, 3207-3226.
[17] Argiriou, A.; Bellas-Velidis I.; Kummert, M. A neural network controller for hydronic heating systems of solar buildings. *Neural Networks*, 2004, 17, 427-440.
[18] Yang, I.; Kim, K. Prediction of the time of room air temperature descending for heating systems in buildings, *Building and Environment,* 2004, 39, 19-29.
[19] Ben-Nakhi, A.; Mahmoud, M. Cooling load prediction for buildings using general regression neural networks. *Energy Conversion & Management,* 2004, 45, 2127–2141.
[20] Khotanzad, A.; Hwang, R.; Abaye, A.; Maratukulam, D. An adaptive modular artificial neural network hourly load forecaster and its implementation at electric utilities. *IEEE Transactions on Power System*, 1995, 3, 1716-1722.
[21] Khotanzad, A.; Afhkhami-Rohani, R.; Lu, T.L.; Abaye, A.; Davis, M.; Maratukulam, D. ANNSTLF a neural network-based electric load forecasting system. *IEEE Transactions on Neural Networks,* 1997, 4, 835-845.
[22] Khotanzad, A.; Afhkhami, R.; Maratukulam, D. ANNSTLF–artificial neural network short-term load forecaster generation three. *IEEE Transactions on Neural Networks,* 1998, 13, 1413-1422.
[23] Zhou, G.; Si, J. Advanced neural network training algorithm with reduced complexity based on Jacobian deficiency. *IEEE Transactions. Neural Networks*, 1998, 9, 448-453.
[24] Parisi, R.; Di, E.D.; Orlandi, G.; Rao, B.D. A generalized learning paradigm exploiting the structure of feedforward neural networks. *IEEE Transactions. Neural Networks,* 1996, 7, 1450-1459.
[25] Hagan, M.T.; Menhaj, M.B. Training feed-forward neural networks with the Marquardt algorithm. *IEEE Transactions. Neural Networks,* 1994, 5, 989-993.

[26] Dodier, R.; Henze, G. Statistical analysis of neural network as applied to building energy prediction. *Energy Systems Laboratory*. Technical Report ESL-PA-96/07. 1996.

[27] Trejo, P.M.; Ríos, G.J.; Rivas, A.E.; Vladimir, R.S. Savings and analysis of the consumption and quality of the energy, 1er International Congress of Engineering, Universidad Autónoma de Querétaro, México, 2005, 1, 253-261.

[28] Neter, J.; Kutner, M.H.; Nachtsheim, J.; Wasserman, W. Applied Linear Statistical Models, 4th edition, Irwin. 1996 266-279.

[29] Ventura, S.; Silvia. M.; Pérez P.B.; Hervás, C. Artificial neural networks for estimation of kinetic analytical parameters. *Anal. Chem.* 1995, 67, 1521-1525.

[30] Griño, R. Neural networks for univariate time series forecasting and their application to water demand prediction. *Neural Netw. World,* 1992, 437-450.

[31] Montgomery, D.C. Design and Analysis of Experiments, 3rd edition. New York: John Wiley & Sons. 1999, 177-199.

[32] Islam, S.M.; Alawi, S.M.; Ellity, K.A. Forecasting monthly electric load and energy for a fast growing utility using an artificial neural network. *Electric Power Systems*, 1995, 34, 1-9

In: Air Conditioning Systems
Editors: T. Hästesko, O. Kiljunen, pp. 353-371

ISBN: 978-1-60741-555-8

Chapter 10

PERFORMANCE OF A PASSIVE AQUARIUM'S CONDITIONING SYSTEM

Majdi Hazami*[*1], Sami Kooli[1], Mariem Lazaar[1], Abdelhamid Farahat[1] and Ali Belguith[2]

[1] Laboratoire de Maîtrise des Technologies de l'Energie;
Technopôle de Borj. Cedria Hammam-Lif BP 95
[2] Faculté des Sciences de Tunis ; Campus, El belvédère Tunis-1060

ABSTRACT

The objective of this work is to study the opportunity of exploiting seawater at a temperature of 18 °C, extracted from a well bored close to the museum (10 m) and close to the seashore (20 m), as a frigorific source for the cooling of SALAMMBO museum aquariums. Therefore an experimental prototype composed of an aquarium and a control basin containing a network of the capillary heat exchanger and a basin of seawater filtration is conceived. Experimental measurements were carried out under climatic conditions similar to those required by maritime species elevated in SALAMMBO museum's aquariums. A numerical study is also worked out. This numerical study allows the sizing of the capillary exchanger to be used by the passive cooling system in order to obtain the required temperature inside each aquarium. Theoretical analysis based on heat balance equations were testified to agree well with experimental results.

Keywords: Passive system, capillaries exchanger, seawater, thermal efficiency, electric analogy.

1. INTRODUCTION

The museum of SALAMMBÔ (Tunis) (figure 1) assures the raising of several maritime species that lives not far from the Tunisian coast (table 1). Besides its educational purpose for

* E-mail: Hazamdi321@ yahoo.fr

the thousands of visitors, the raise of these maritime species permit us to carry out biological studies on the fish's life, reproduction and behaviors in captivity. These studies also permit us to have an opportunity to achieve the goal of acclimatizing the fish in their new environment.

Figure 1. Photo of some aquariums in the museum.

Table 1. Required temperature in museum aquariums

Aquarium	Maritime species	T_{aq}(°C)
1	Mérou, Anémone	20.5 ± 1
2	Rascasse, Concombre de mer, oursin, Gobie	19 ± 1
3	Dente	20 ± 1
4	Congre, Mulet, Murène, Bar tachette, Anguille	21.5 ± 3.5
5	Daurade, Bar	21 ± 3
6	Sorbe et marbre	20 ± 0.5
7, 8, 9	Saule, Eponge, Etoile de mer	15 ± 2
10	Cigale	19 ± 2
11	Ombrine	17 ± 1
12	Poulpe	18 ± 2
13	Langouste	22 ± 2
14	Raie, Liche	16 ± 1.5
15	Baliste, Sar à museau pointu	21 ± 2

These maritime species require specific climatic conditions according to their nature, their origins and the depths in which these species lives. In fact, the water temperature inside the aquariums is one of the most important factors, which must be controlled carefully, that affect fish's survival. To assure all of the aquarium's climatic needs, SALAMMBÔ museum directors have installed a classic conditioning system functioning with electric energy. This system consumes a lot of electricity.

In this chapter we present a work which leads to a reduction in the cost of the conventional system. In this work we opted for the changeover of the classic system to a low cost passive system which exploits mainly cold seawater (at 18 °C). The seawater is drawn from a 20 m-depth well bored next to the museum (10 m) and close to the seashore (20 m). Cold water extracted from the well is used via a heat exchanger for cooling the water inside aquariums. The heated water is then rejected into the nature [1, 2]. The continuous working of the museum obliged us to elaborate an experimental prototype to not disrupt the normal functioning of the museum aquariums (figure 2).

This work is the result of a cooperation convention program between the CLINA-society, the Energies and Thermal Processes Laboratory (LEPT) and the Sciences and Sea Technologies National Institute (INSTM). This convention aims to study the feasibility of integrating the capillary polypropylene heat exchanger in a passive system used for the conditioning of SALAMMBÔ museum aquariums by using the seawater at 18 °C.

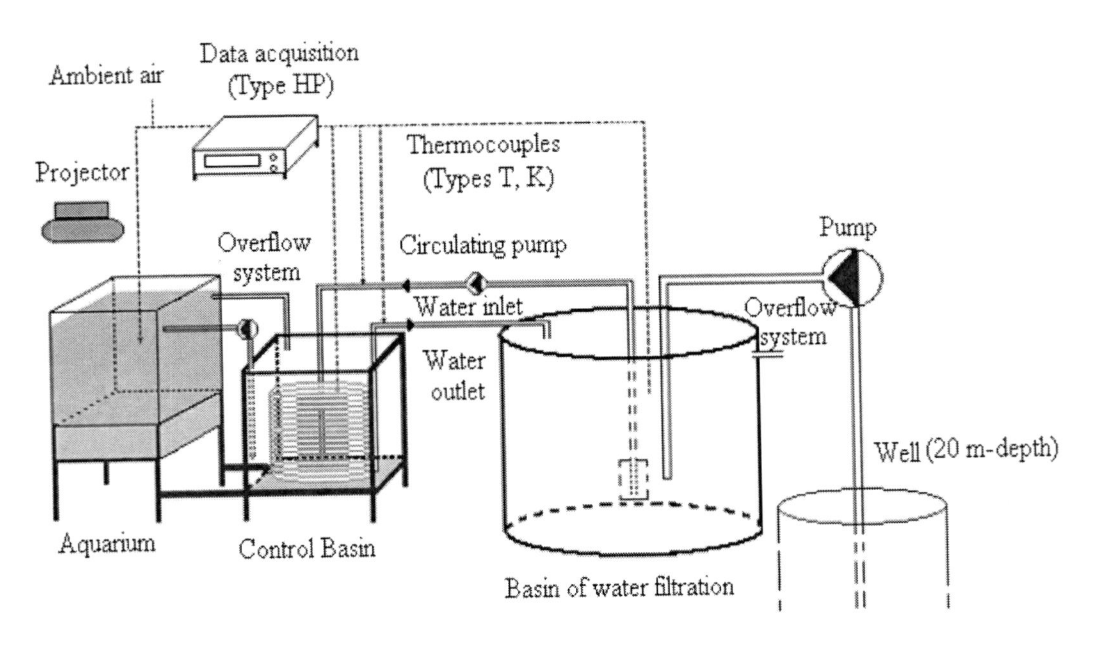

Figure 2. General view of the experimental set up.

2. Description of the Experimental Set-Up and Test Procedure

2.1. Description of the Experimental Set-Up

Figure 2 illustrates a schematic diagram of a passive aquarium's conditioning system designed and conceived in our laboratory (Laboratoire de Maîtrie des Technologies de l'Energies, LMTE, Tunisia). This passive aquarium's conditioning system consists of:

- *A water filtration basin*—a cylindrical tank with a capacity of 2 m^3 used for the filtration of the seawater extracted from a well bored close to the seashore. The water filtration basin allows us to avoid all risks of bacteriology contamination in the aquarium.
- *Control basin*—a 3 cm-thin Plexiglas cubic shape reservoir (1m x 1 m x 1m) with a capacity of 1m^3 filled with seawater. To reduce the thermal heat losses through the lateral sides of the control basin, this was covered by a 30 mm-thick polyurethane foam with a thermal conductivity and specific heat respectively equal to 0.32 $W.m^{-1}.K^{-1}$ and 2090 $J.kg^{-1}.K^{-1}$. Preliminary measurements carried out in the basin control before and after thermal insulation allow us to evaluate its thermal loss coefficient (K_B). This coefficient is determined by the standards that exist to heat water in the control basin and then the control basin is left during a period of 24 hours. The loss coefficient (K_B) is given by the following:

$$K_B = \frac{4180}{\Delta t.\ V_v} Ln\left[\frac{T_i - T_m}{T_f - T_m}\right]$$

with:

T_i: Initial temperature, °C.
T_f: Final temperature, °C.
T_m : Ambient air average temperature, °C.
Δt: Test time, s.
V_v: Basin capacity, m^3.

The results (table 2) show that by using the thermal insulation we have reduced the heat loss of the control basin. Indeed, the heat loss coefficient K_B decreases from 97 $W.°C^{-1}$ (without insulation) to 10 $W.°C^{-1}$ (after insulation).

Heat exchanger—In the side of the control basin a network of polypropylene capillary heat exchanger is placed (figure 3). This new type of capillary heat exchanger, which procures a certain number of substantial economic advantages, is classified according to its shapes, its size and its types (G types, S and U) (figure 4, table 3). This capillary heat exchanger, characterized by a very significant heat-transferring surface, can support temperatures and pressures respectively higher than 100°C and 10 bars. Contrary to a classical tubular metallic heat exchanger, the capillary heat exchangers don't become plugged by seawater. Seeing their very particular features, these capillary heat exchanger types can be integrated in several applications (local air-conditioning, heating of the sanitary water…).

- *Aquarium*—a 3 cm-thin Plexiglas cubic shape reservoir (0.5 m x 0.5 m x 1 m) with a capacity of 0.5 m^3 filled with sea water. The aquarium is covered on the three lateral sides by 2 cm-layer of polystyrene. The aquarium is placed on a 5 cm- thickness of wooden strip used to reduce the thermal losses by the bottom side of the aquarium. To assure a permanent renewal of water inside the aquarium an overflow system and an immersed pump are installed within the aquarium. The immersed pump allows homogenizing the aquarium's water to avoid the stratification phenomena. A projector (2000W) is used to illuminate the aquarium.
- *A data acquisition system*—A network of type T and K copper-constantan thermocouples permits the measures of temperatures at the inlet and the outlet of the capillary heat exchanger, the water inside the control basin, and the aquarium and filtration basin.

Another thermocouple is placed out of the water aquarium's cooling system to measure the ambient air temperature. These thermocouples were connected to a multi-channel digital data acquisition. All of these parameters were measured at an interval of 10 min with a total accuracy assumed to be about 3%.

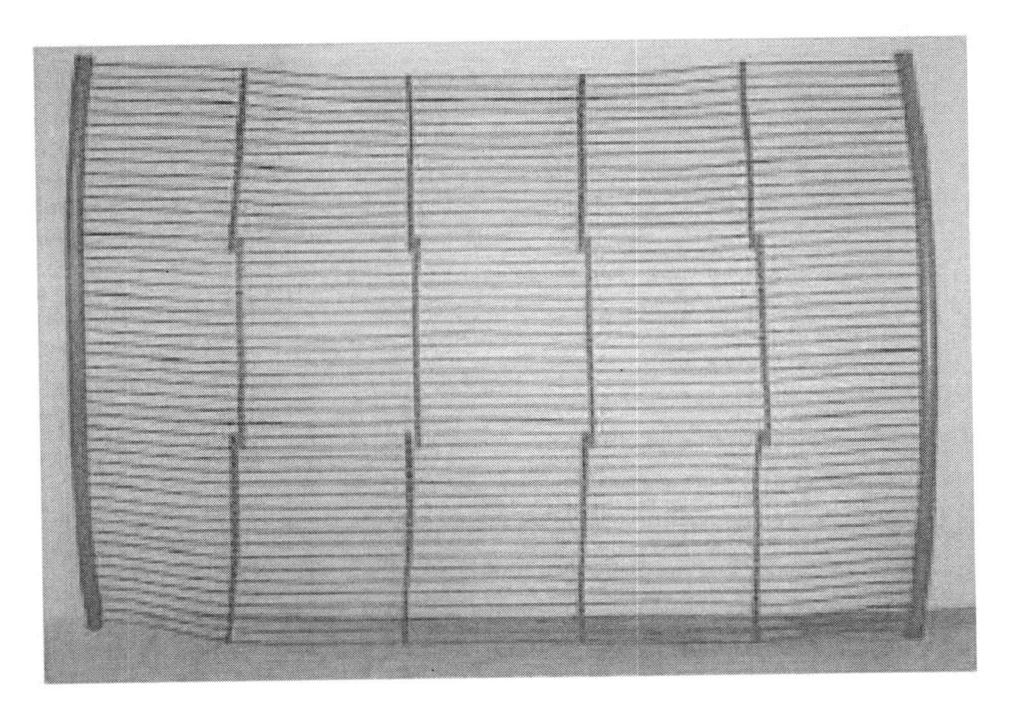

Figure 3. Capillary heat exchanger.

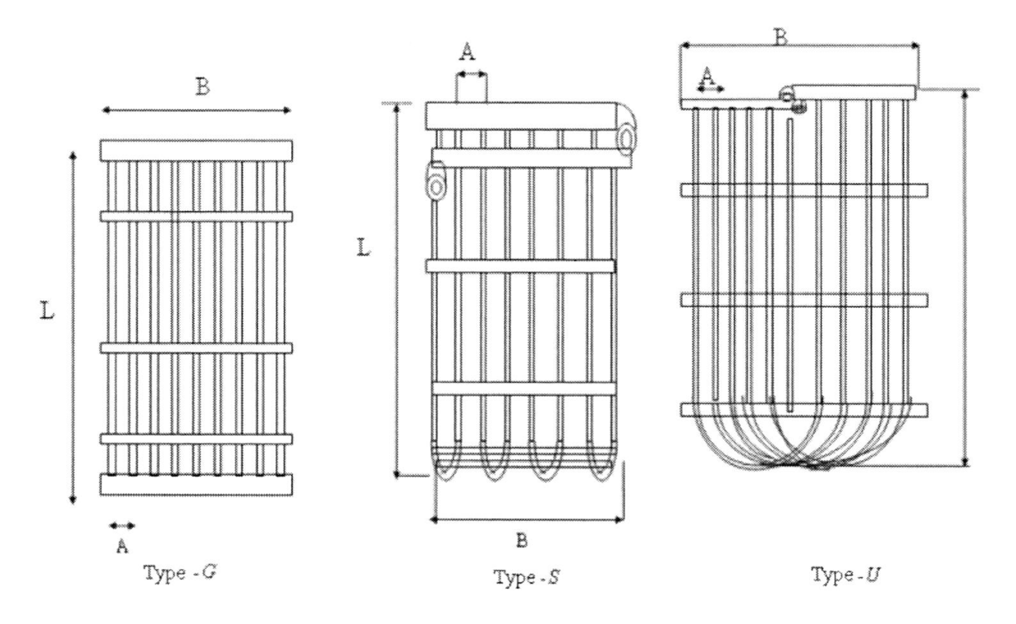

Figure 4. Different type of the capillary heat exchanger.

Table 2. Control basin heat loss coefficient (T_m = 18 °C)

	T_i (°C)	T_f (°C)	Δt (h)	K_b (W.°C^{-1})
Without insulation	32	27	24	97
After insulation	37	36	24	10

Table 3. Geometrical specification of the different types of capillary heat exchangers

Capillary tube spacing	10, 15, 20	mm
External diameter of the capillary tubes	3.4	mm
Thickness of the capillary tubes	0.55	mm
External diameter of the collector	20	mm
Length of the heat exchanger, L	600-15000	mm
Width of the heat exchanger , B	150	mm

2.1. Test Procedure

Indoor experiments were conducted in the SALAMMBO museum (North of Tunisia): Latitude 36° 50' N, Longitude 10 ° 44' E. During the experiments the water mass flow rates varies in the range of (50 l.h^{-1}–200 l.h^{-1}) by using a water circulating pump, a sliding valve and a flowmeter integrated in the experimental loop. Different heat exchanger surfaces with 1 m- width and different lengths are investigated during all the experimental testing. Water in the control basin is cooled by the cold water crossing the capillary exchanger and supplied by the water filtration basin.

3. THEORETICAL STUDY

In order to evaluate the thermal performances of the conceived system, a numerical study was elaborated. This numerical study allows the estimation of frigorific energy needs for each aquarium of the museum. It permits, also, the sizing of capillary heat exchanger surface and the optimal water cooling masse flow that permits the reaching of required temperature values inside evry aquariums.

3.1. Heat Exchanger Characterisation

The numerical study consist in:

- Considering that all capillary tubes have the same thermal behavior. Thus, it's possible to do a local study for one capillary tube and then a global study for all capillary tubes [3],
- Assuming that inside the capillary tube the cooling water temperature variation is linear and the local exchange coefficient remains constant,
- The water masse flow rates do not vary inside capillary tubes [4, 5],
- The capillary tube is divided into N equal thickness slices (figure 5) and an energy balance is written for each section.
-

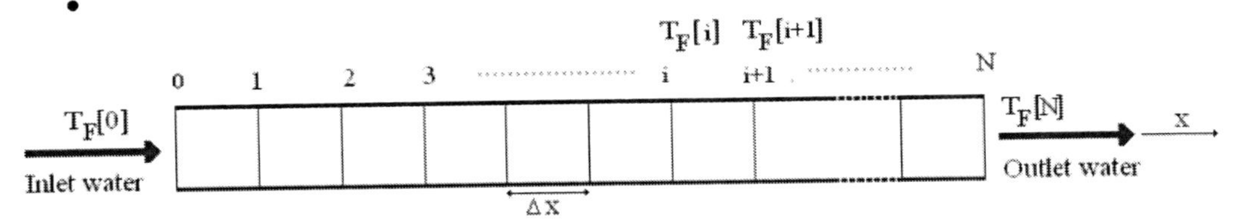

- Figure 5. Capillary tube diagram.
-

The discretization of the energy balance equation gives:

$$m\ C_p\ (T_{F,i+1} - T_{F,i}) = U_{i+1/2}\left[\left(\frac{T_{F,i+1} + T_{F,i}}{2}\right) - T_m\right] 2\pi\, r\, \Delta x \quad \text{avec} \quad 0 \leq i \leq N-1 \qquad (1)$$

The previous equation is written in the following form:

$$T_{F,i+1} = A\, T_{F,i} - \mathrm{B} \qquad \text{With} \quad 0 \leq i \leq N-1 \qquad (2)$$

with :

$$A = \frac{1 + \frac{U_{i+1/2}}{\dot{m}\,C_p}\pi r \Delta x}{1 - \frac{U_{i+1/2}}{\dot{m}\,C_p}\pi r \Delta x}$$

and

$$B = \frac{\frac{2\,U_{i+1/2}\,T_m\,\pi r \Delta x}{\dot{m}\,C_p}}{1 - \frac{U_{i+1/2}}{\dot{m}\,C_p}\pi r \Delta x}$$

where:

C_p : Water heat capacity, $J.K^{-1}$,
$\dot{m}$: Is the water masse flow inside the heat exchanger, $l.h^{-1}$,
T_F : Is the water temperature inside the heat exchanger, $°C$,
T_m : Is the surrounding average temperature, $°C$,
r :Capillary ray, m,

The overall heat exchange coefficient U is calculated by [6, 7, 8]:

$$U = \frac{1}{N}\sum_{i=0}^{N-1} U_{i+1/2} \tag{3}$$

The thermal efficiency, ε, is given by [9, 10, 11]:

$$\varepsilon = 1 - \exp\left\{-\left(\frac{US}{m\,C_p}\right)\right\} \tag{4}$$

3.2. Energy Need of the Aquarium

To evaluate the aquariums frigorific energy needs we have estimated the energy balance exchanged inside the aquarium. In figure 6 is represented a schematic diagram of the different heat energy exchanged within the aquarium.

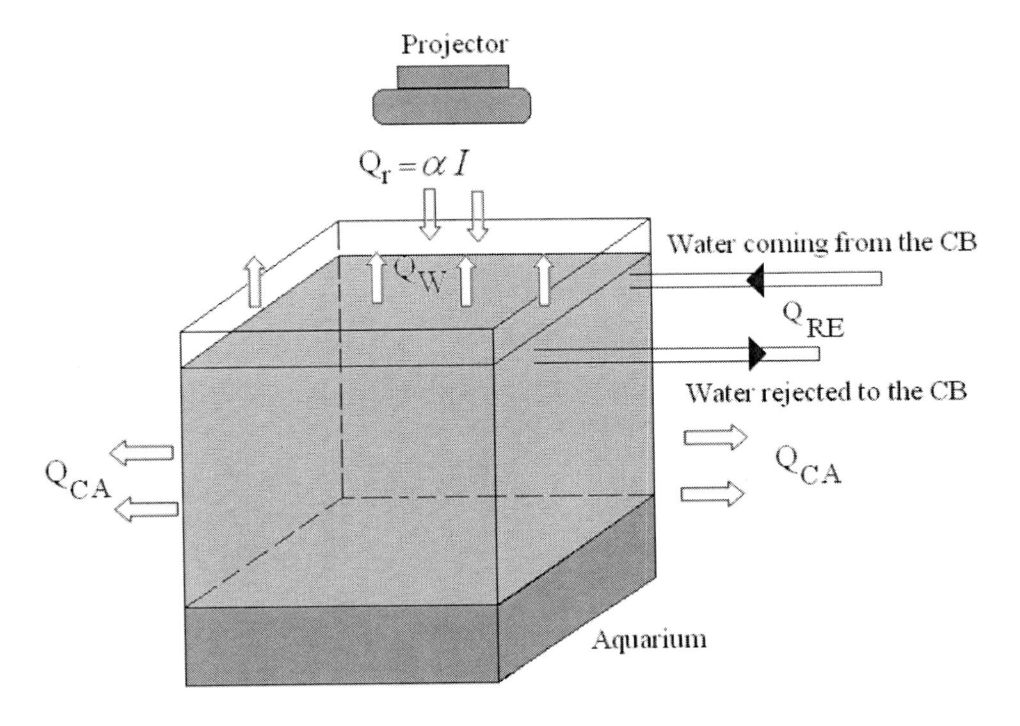

Figure 6. Schematic diagram of the heat exchanged in the aquarium.

The aquarium's energy balance in steady state is [12]:

$$Q_{RE} = Q_r + Q_{CA} + Q_W \tag{5}$$

Q_{RE} is the amount of heat supplied by the control basin, to maintain the aquarium temperature at the temperature for T_A, W:

$$Q_{RE} = C_p RE(T_A - T_B) \tag{6}$$

where

C_p :	Water heat capacity, $J.K^{-1}$,
RE :	The renewable water rate within the control basin, $kg.s^{-1}$.
T_B :	Basin control temperature, °C,
T_A :	Aquarium required temperature, °C.

Q_r refers to the energy radiation received by the projector:

$$Q_r = \alpha.\mathrm{I} \tag{7}$$

with

α :	The water absorption coefficient,
I :	Thermal power provided by the projector, $W.m^{-2}$.

Q_C refers to the energy exchanged through the aquarium's lateral sides with the external environment:

$$Q_{CA} = K_A.S_A\ (T_A - T_{ext}) \tag{8}$$

where

K_A :	Aquarium's heat losses coefficient, $W.m^{-2}.°C^{-1}$.
S_A :	Lateral sides of the aquarium, m^{-2}.
T_A :	Aquarium's required temperature, $°C$,
T_{ext} :	Ambient temperature, $°C$.

Q_W is the energy lost by water evaporation on the aquarium's upper surface, W [12] :

$$Q_W = \frac{L_v\ M_v\ P\ D_{v,a}}{R\ \Delta Z\ T_a} Ln\left[\frac{P-P_v\ (T_{air})}{P-P_{vs}(T_{eau})}\right] \tag{9}$$

where:

M_v :	Molar mass of water vapour, $kg.mol^{-1}$,
L_v :	Latent heat by water vaporisation, $J.kg^{-1}$,
$D_{v,a}$:	Coefficient of water vapour diffusivity in the air,
P :	Air pressure, Pa,
P_v :	Vapour pressure, Pa,
P_{vs} :	Pressure of vapour saturation in the air, Pa,
R :	Ideal gas constant, $J\ cal.mol^{-1}$,
ΔZ :	Distance between the water level and projector, m,
T_a :	Ambient air temperature, °C.

For $0 < \theta < 50°C$

$L_v = 597{,}3 - 0{,}566\theta$; $D_{v,a} = 2{,}3.10^{-5} \frac{P_r}{P}\left(\frac{T}{T_r}\right)^{1,81}$ and $P_r = 0.98$ bar and $T_r = 258$ K.

The vapour pressure is given by:

$P_{VS} = -16.037 + 1.8974\ T - 0.0639\ T^2 + 0.0012\ T^3 - 5.8511.\ 10^{-6}\ T^4$ (Pa)

where :

$20°C \leq T \leq 80\ °C$

The frigorific needs of the aquarium are calculated by:

$$Q_{RE} = \alpha I - K_A S_A (T_A - Text) - \frac{LvMvPDv,a}{R\Delta Z\ Ta} Ln\left[\frac{P - Pv\,(T\ air)}{P - Pvs(T\ \ eau)}\right] \quad (10)$$

Control basin water temperature is given by:

$$T_B = T_A\left(1 + \frac{K_A S_A}{Cp\,.RE}\right) - \frac{1}{Cp\,.RE}.\left(\alpha I + K_A S_A\ Text - \frac{LvMvPDv,a}{R\Delta\Delta ZT} Ln\left[\frac{P - Pv\,(T\ air)}{P - Pvs(T\ \ eau)}\right]\right) \quad (11)$$

3.3. Energy Needs of the Basin of Control

In figure 7 is represented a schematic diagram of the different types of heat exchanges that occur in the control (CB).

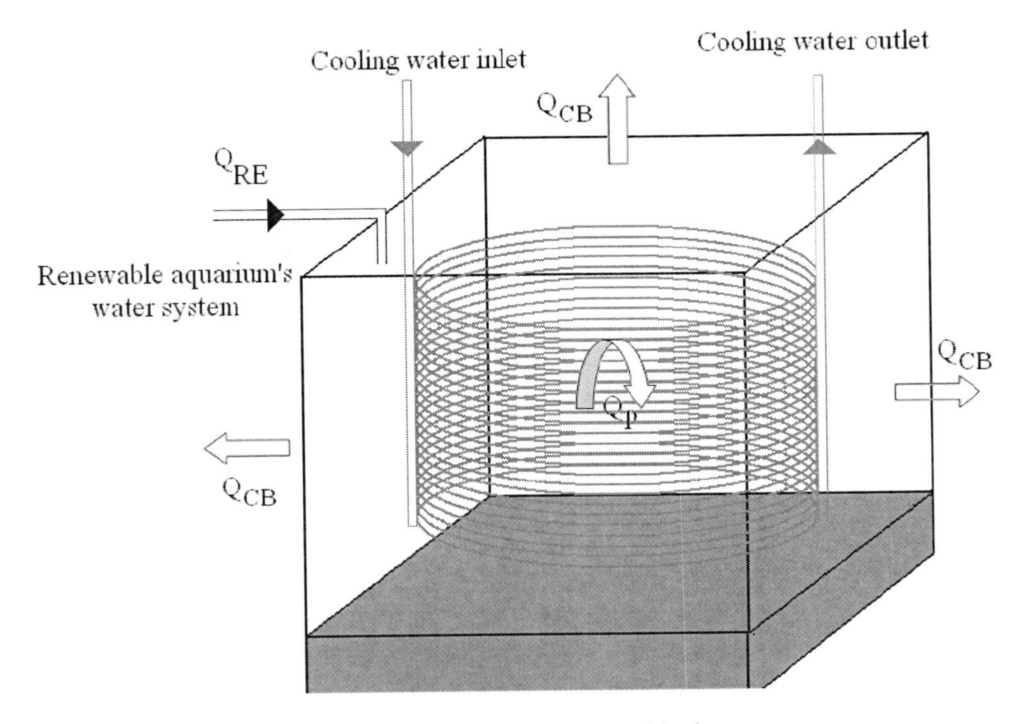

Figure 7. Schematic diagram of the heat exchanged in the control basin.

The energy balance in steady state within the basin of control is [12]:

$$Q_p = Q_{CB} + Q_{RE} \quad (12)$$

- Q_{CB} is the amount of heat lost through the basin of control sides, W:

$$Q_{CB} = K_B S_B (T_B - T_{ext}) \tag{13}$$

where

K_b :	Coefficient of heat lost by the control basin, $W.m^{-2}.C^{-1}$,
S_B :	Overall control basin sides surface, m^2,
T_B :	Water temperature inside the control basin, °C,
T_{ext} :	Ambient air temperature, °C.

- Q_p is the amount of heat exchanged through the capillary heat exchanger, placed inside the control basin, W:

$$Q_p = US\ \Delta T_{LM} \tag{14}$$

with

$$\Delta T_{LM} = \frac{T_e - T_s}{Ln\frac{T_e - T_B}{T_s - T_B}}$$

where

U :	Overall heat exchanger coefficient, $W.m^{-2}.C^{-1}$,
S:	Heat exchanger surface, m^{-2},
ΔT_{LM}	Average temperature logarithmic difference, °C,
T_e :	The inlet water temperature, °C,
T_s :	The outlet water temperature, °C,
T_B :	The control basin temperature, °C.

- The cooling power needed for the aquarium water conditioning is then written:

$$Q_p = K_B S_B (T_B - Text) + Q_{RE} \tag{15}$$

By replacing the energy exchanged by water renewal in the control basin, Q_{RE}, with its expression, we get the equation:

$$Q_p = K_B S_B (T_B - Text) + C_p RE (T_B - T_A) \tag{16}$$

- The temperature of the groundwater extracted from the well (T_e) needed to maintain the aquarium's temperature at the required temperature T_{aq} is then written:

$$T_e = T_s + (K_1 + K_2)\, T_B - K_1 . T_{aq} - K_2 . T_{ext} \tag{17}$$

with:

$$K_1 = \frac{C_p RE}{\dot{m}\, C_p};$$

and

$$K_2 = \frac{K_B S_B}{\dot{m}\, C_p}$$

$$T_B = T_{aq}\left(1 + \frac{K_A S_A}{Cp\,.RE}\right) - \frac{1}{Cp\,.RE}.\left(\alpha I + K_A S_A\, Text - \frac{LvMvPDv,a}{R\Delta\Delta ZT} Ln\left[\frac{P - Pv\,(T\,air)}{P - Pvs(T\;eau)}\right]\right)$$

4. Results and Discussions

4.1. Heat Exchanger Performances

To validate the numerical model concerning the total exchange coefficient, a measurement set was done in a temperature range spread out from 20°C–50°C. The results obtained have enabled us to calculate the heat exchanger capillary overall heat exchange coefficient, U, and the thermal power exchanged through the system. For a coolant masse flow, in side exchanger, equal to 100 $l.h^{-1}$, U decrease from 175 $W.m^{-2}.°C^{-1}$ to 110 $W.m^{-2}\,°C^{-1}$ when heat exchanger length passes from 1 m to 3 m (figure.8). The total exchange coefficient is inversely proportional to heat exchanger length.

The overall heat exchange coefficient variation according to the cooling water mass flow is shown in figure 9. We established that the overall heat exchange coefficient increases with mass flow increase. Although, this increase is slowed by the pressure drops in the capillary heat exchanger. The U coefficient tends asymptotically towards a value of about 175 $W.m^2.°C^{-1}$, for a length equal to 3 m. Any increase in the mass flow crossing the plaits beyond 180 $l.h^{-1}$ does not have an effect on the overall heat exchange coefficient and the maximum thermal power exchanged through the heat exchanger.

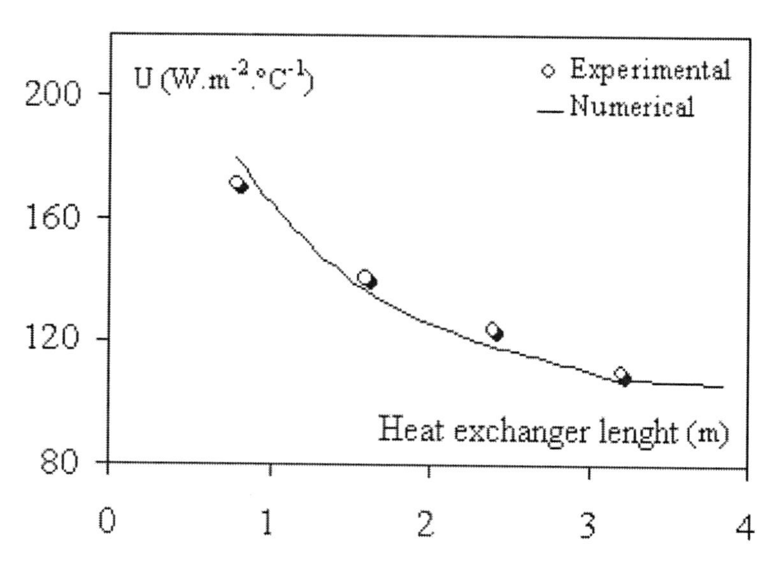

Figure 8. Variation of the heat exchange coefficient according to the braids length for a flow equal to 100 $l.h^{-1}$ and the temperature range [20-50°C].

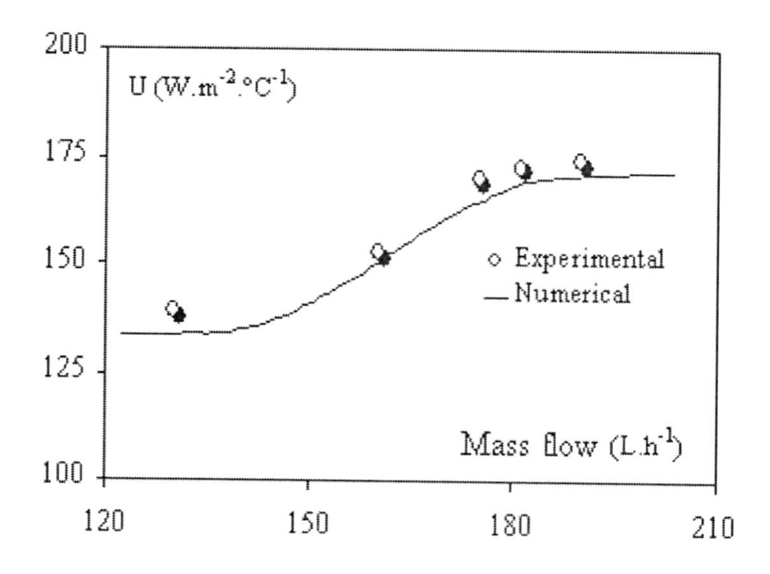

Figure 9. Variation of the overall heat exchange coefficient according to the water mass flow for a length equal to 3 m and the temperature [20-50°C].

For the sizing of the aquarium's system, we have evaluated the optimum conditions for a better function of the capillary exchanger (figure 10). Results show that there is an optimal heat exchanger length (3 m). Any increase of plait length beyond this value, does not improve capillary heat exchanger efficiency. The results also show that there is an optimal cooling water masse flow, about 180 $l.h^{-1}$.

The system performance is evaluated by the determination of the thermal efficiency E. In figure 11 is represented the efficiency of the passive conditioning system according to the cooling water mass flow. The results show that E increases when the flow crossing the exchanger increases. However beyond the optimal value of 180 $l.h^{-1}$ and a heat exchanger length equal to 4 m the effectiveness is constant, it is about 80 %. We established that in the temperature range, between 20°C and 70°C, the system has a good effectiveness compared with conventional systems based on tubular copper exchangers.

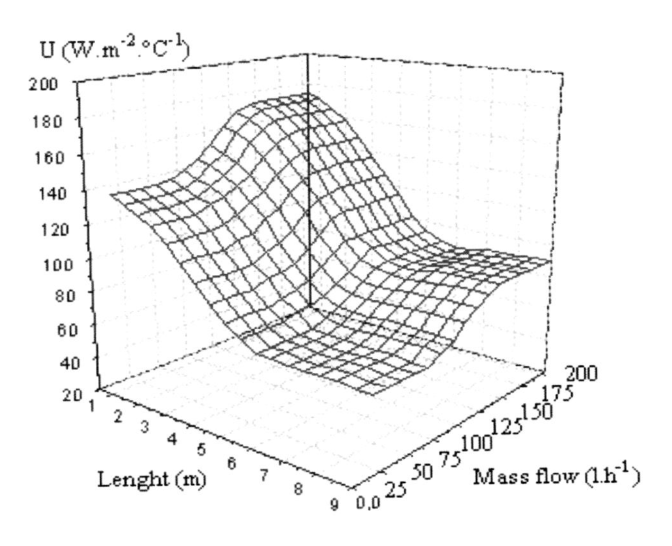

Figure 10. Numeric curve of the effect of the length of the heat exchanger and the masse flow on the overall heat exchange coefficient for the temperature range [20 °C - 70 °C].

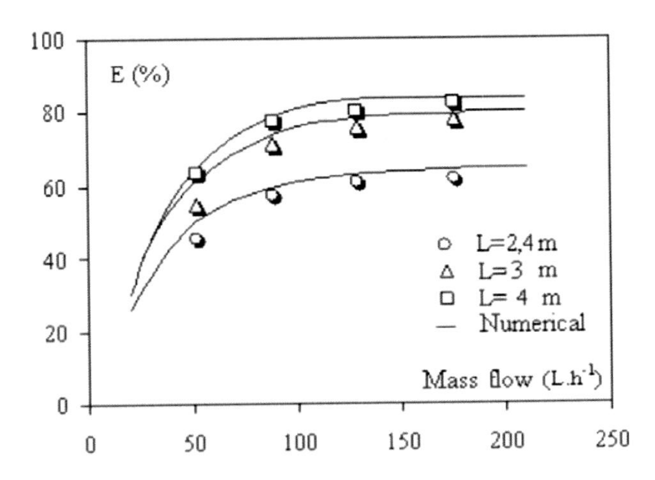

Figure 11. Efficiency of aquarium conditioning system.

4.2. Heat Exchanger Sizing

To study the feasibility of the passive aquariums' cooling system, we compared the required temperatures in every aquarium with the temperature of the frigorific source (18 °C).

As seen in table 1, the required temperatures inside aquariums (1, 2, 3, 4, 5, 6, 10, 12, 13 and 15) are higher than the frigorific source temperature. For these aquariums, the passive cooling system, which we have installed, is able to maintain the temperature of every aquarium at the required values. Indeed, there is a gap (2–4°C) between the required temperature of every aquarium and the cold source temperature.

To evaluate the rate of the thermal frigorific energy exchanged that can be supplied by the passive system as well as the frigorific aquariums' needs, we have used equation 5 of the theoretical study. We established that for the aquariums 2 and 10 having a required average temperature equal to 20 °C, the passive conditioning system must provide to the aquariums a frigorific energy about 2.2 kW (figure 12). We noted also that there is a difference between aquariums' frigorific energy needs and the supplied frigorific energy (which is about 2.7 kW). This gap corresponds to the heat losses inside the control basin.

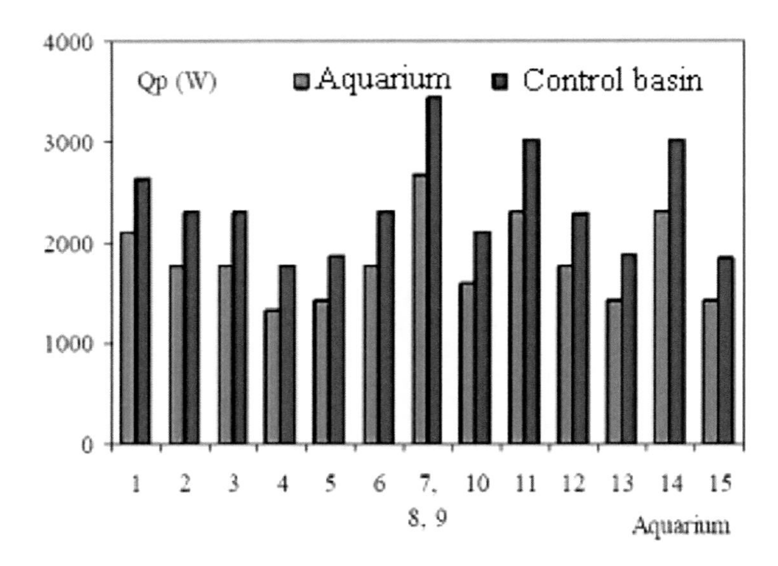

Figure 12. Aquariums frigorific power needs.

The theoretical study permits the estimation of the capillary heat exchanger surface to place in the control basin to maintain the temperature of every aquarium to the required values. The results show that the heat exchanger surface depends on the required temperature and every aquarium's capacity (figure 13). Thus, for aquarium 2 and 10 which possess the same required temperatures, it will be necessary to place a capillary heat exchanger with an exchange surface equal to 7 m^2 (2 batteries of capillary heat exchanger, 1m x 3.5 m).

For aquariums 7, 8, 9, 11 and 14, we note that the required temperatures are lower than the cold source temperature. In this case, the passive cooling system by using the sea water as a frigorific source is not able to reach the required temperatures in these aquariums.

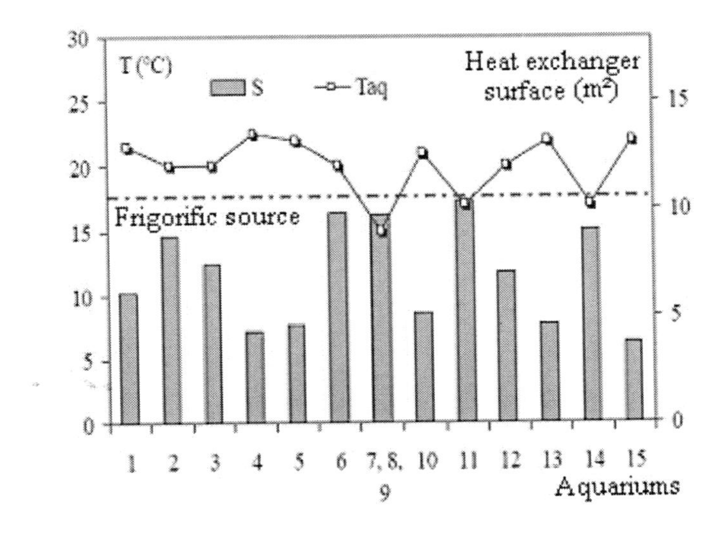

Figure 13. Estimated exchange surface for each aquarium.

4.3. Aquarium and Control Basin Behaviour

Other tests were conducted for an aquarium having the same capacity of aquarium number 10. Tests conducted on this aquarium have allowed us to follow the evolution of water temperature in the control basin and in the aquarium for an optimal exchange surface and cooling masse flow respectively equal to 7 m² and 180 lh^{-1}. In figure 14 is represented the evolution of water temperature in the control basin and in the aquarium. Results show that the passive aquarium's conditioning system decreases the water temperature of the aquarium to 20 ° C. It is seen that the steady state is established after two hours of cooling. We noted also that a temperature gap (1.5 ° C) persists between the control basin and the aquarium (figure 15).

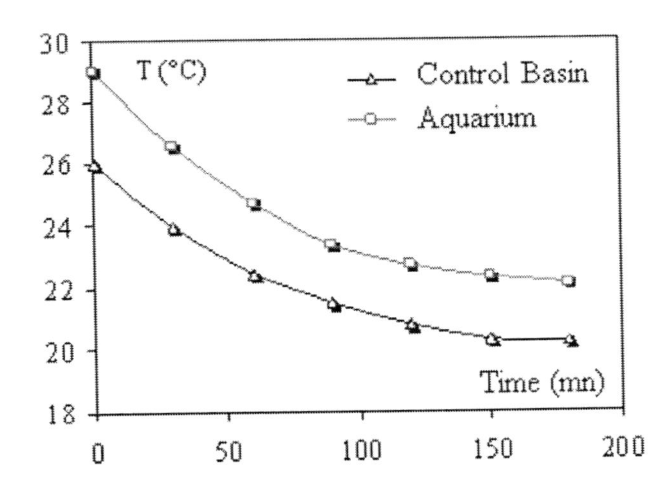

Figure 14. Evolution of water temperatures in the aquarium and control basin for a water mass flow equal to 180 lh^{-1} and an exchange surface equal to 7 m^2

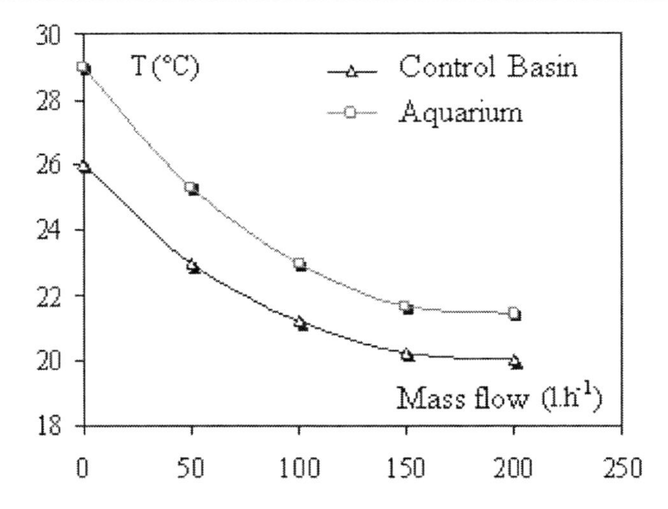

Figure 15. Evolution of water temperatures of the aquarium and control basin for an exchange surfaces equal to 7 m^2.

CONCLUSION

In this chapter we studied the possibility of integrating seawater, extracted from a well (18 ° C) bored next to the museum (10 m), in the cooling of SALAMMBÔ Museum aquariums. In this study we have identified the following points:

- The heat exchanger characterisation shows that there is an optimal length (4 m) and an optimal water masse flow (180 $l.h^{-1}$). Exceeding these optimal parameters does not improve the heat exchanger performances.
- The passive cooling aquariums Museum SALAMMBÔ system, using seawater as a cold source, has good thermal efficiency (about 80%) with a supplied frigorific energy ranging from 1.5 kW to 3.5 kW.
- Thus, 10 aquariums of SALAMMBÔ museum (whose temperature record is above 18 ° C) can be cooled by the passive system with a low cost.
- This work is considered to be a good support for a management of SALAMMBO museum aquariums with this passive cooling system which uses mainly the seawater as a frigorific source.

ACKNOWLEDGMENTS

The authors would like to express their gratitude for the financial support of the CLINA manufacture and ministry of research in this study. The input of our project partners, in particular, Institut National des Sciences et Technologies de la Mer, INSTM, is also gratefully acknowledged.

REFERENCES

[1] Mejdi HAZAMI, Refroidissement des aquariums du musée de Salammbô. DEA. 2001

[2] M. HAZAMI, Sami Kooli, Mariem Lazaar, Abdelhamid Farhat, Chekib Kerkani, Ali Belguith. (2004). Capillary polypropylene exchangers for conditioning of museum aquariums (Tunisia). *Desalination*. 166443-448

[3] R. L. Cornelissen, G. G. Hirs (1999). Thermodynamic optimisation of heat exchanger Universty of Twente. Departement of Mechanical Engineering, Chair of Energy Technology. *Intenational Journal of Heat and Mass Transfer*. 42 :951-959.

[4] M Mahfoud. 1987. Evolution du coefficient d'échange et des pertes de charges dans un faisceau de tubes en fonction de l'angle d'attaque. *International Journal of Heat and Mass Transfer,* 30.

[5] G. Fabbri effect of disuniformities in vapor satuartion pressure and coolant velocity on vapor back flow phenomenea in single pass air cooled condensers. *Heat and Mass Transfer,* Vol 49, (2000).

[6] S.A. Nada, H.H. El-Ghetany, H.M.S. Hussein(2004). Performance of a two-phase closed thermosyphon solar collector with a shell and tube heat exchanger. *Applied Thermal Engineering*. 24:1959–1968

[7] A. Nuntaphan, T. Kiatsiriroat C. C. (2000). Wang "Air side performance at low Reynolds number of cross-flow heat exchanger using crimped spiral fins" *International Communications in Heat and Mass Transfer*. 32:151-165.

[8] R.A. Seban, E.F. Mclaughlin. (1963). Heat transfer in tube coils with laminar and turbulent flow, Int. J. Heat Mass Transfer. 6:387–395.

[9] R. Romero-Méndez, M. Sen, K. T. Yang andt R. McClain. (2000). Effect of fin spacing on convection in plate fin and tube heat exchanger. *International Journal of Heat and Mass Transfer*. 43 :39-51.

[10] Th. Bes and W. Roetezel (1993). Thermal theory of the spiral heat exchanger. *Int. J. Heat Mass Transfer*. 36, (3):765-773.

[11] M. HAZAMI (2008). Etude expérimentale et numérique du phénomène de stockage et du déstockage de l'Energie thermique en utilisant un échangeur capillaire en polypropylène. These.

[12] M. Hazami, S. Kooli, M. Lazaar, A. Farhat, A. Belghith. (2004). Heat transfer characteristics of a capillary heat exchanger based air conditioning cupboard. *Desalination*. 166, 435-442.

In: Air Conditioning Systems
Editors: T. Hästesko, O. Kiljunen, pp. 373-383
ISBN: 978-1-60741-555-8

Chapter 11

New Progress in Liquid Desiccant Cooling Systems: Adsorption Dehumidifier and Membrane Regenerator

Xiu-Wei Li* ***and Xiao-Song Zhang***
School of energy and environment,
Southeast University, Nanjing 210096, China

Abstract

A liquid desiccant cooling system (LDCS) is a new type of air-conditioning system with good energy-saving potential. Its performance is dominated by dehumidification and regeneration processes. At present, few works have been done to propose a general principle for better dehumidifier design and most works about regeneration are only concentrated on the thermal regeneration method. For both aspects, new progress has been made and presented in this paper. On one hand, a new design method has been derived from the experiments: an adsorption dehumidifier, developed by integrating a solid desiccant with liquid dehumidifier, could greatly improve the dehumidification effects. On the other hand, a new regeneration style has been conceived: a membrane regenerator, which consists of many alternatively placed cation- and anion-exchange membranes, would regenerate the liquid desiccant in an electrodialysis way; while a solar photovoltaic generator provides electric power for fueling this process. This new regeneration method is immune from the adverse impact from outside high humidity, and it also has a pretty good performance, as well as the benefit that purified water can be obtained along with the regeneration process. These two developments can make LDCS more practical and competitive in the future market.

* good2007best@yahoo.com.cn, 0086-025-83792722

NOMENCLATURE

A	area (m^2)
$Con_{mol}^{\triangle}$	mole number change of the solute per kg solution (mol/kg)
d	diameter (m)
F	Faraday constant (s A/mol)
I	current (A)
m	mass flow rate (kg/s)
N	cell number (no units)
Num	number of silica gel bulbs (no units)
P_{ED}	electric power consumption (kW)
p_{ED}	electric power consumption for the solution of unit mass (kW)
r	radius (m)
U	voltage (V)
z	electrochemical valence (no units)

Greek Letters

ζ	current utilization (no units)

Subscripts

a	air
d	desiccant solution
deh	dehumidifier
ED	electrodialysis
PV	photovoltaic
TH	thermal energy
w	water

INTRODUCTION

Seeking comfortable living conditions in life is a popular trend today, which has led to the wide use of air-conditioners. However, that situation has a side effect, which upsets people by calling for a large amount of electric power to drive these "cooling machines". Many new types of air-conditioner have been developed to shoulder off the heavy dependence on electric power while still guaranteeing good comfort. Among these new faces, the liquid desiccant cooling system (LDCS), which typically consists of a liquid-dehumidification unit and an evaporation-cooling unit, has attracted ever increasing attention from both researchers and consumers. The reason for this is mostly from the fact that this system could be driven by heat sources with a relatively low temperature of around 70°C.

That means the work could be done by many renewable energies, such as solar power, geothermal power and waste heat [1-3].

The dehumidification process is the core part of this system. If the dehumidification effect is not good enough, the cooling capacity may be insufficient for the treated air to reach their expected states. The dehumidifier configuration has a great impact on this dehumidification effect, and many works have been done: Khan [4] made a theoretical analysis on an internally-cooled dehumidifier; Al-Farayedhi et al. [5] studied the geometric parameters of a gauze-type structured packing (dehumidifier); Abdul-Wahab et al. [6] took the influence of packing densities into consideration. Ani et al. [7] studied the influence of packing height on absorber performance, which was built from fibre-glass in pieces. Nevertheless, except for contacting area enlargement, it is far-fetched to say many substantial developments have been made. Recently in our experiments, an interesting attempt to improve the dehumidification effects has been made by embedding silica gel into the air-desiccant contacting surface of a liquid dehumidifier. The result shows the improved dehumidifier could achieve a more than 50% improvement in dehumidification effect than the original one, and this improvement was much greater than that could be reasonably predicted by just counting the area increase.

The energy consumption of LDCS mainly lies in the regeneration of desiccant solution. The conventional regeneration method is in a thermal energy (TH) style [8-13]: with the addition of thermal energy, the desiccant solution raises its temperature so that the water molecules can evaporate from the solution body to the surrounding atmosphere. Through this method, the solution concentration will increase (regenerated). Nevertheless, this regeneration pattern has some defects: For one thing, it heavily depends on the condition of the surrounding atmosphere and can be unreliable. It is well-known that the force of mass transfer is the positive vapour pressure difference between the desiccant solution and the surrounding atmosphere. When it is hot and humid, the vapour pressure difference will fall quickly or even become negative, which leads to the instability of regeneration. For another thing, the desiccant solution is "hot" after regeneration, but this high temperature is not good for the following dehumidification process. Excessive cooling measures have to be taken to remove this harmful heat, which inevitably causes a waste of energy. Besides, some droplets of desiccant solution may sneak into and pollute the air of the environment because this regeneration method always operates in open cycle. To overcome these problems, a new regeneration method has been conceived: the regenerator is designed as an electrodialysis (ED) stack composed by many ion-exchange membranes, and a solar photovoltaic (PV) generator provides energy for regeneration.

The ED method is a technique based on the transport of ions through selective membranes under the influence of an electrical field [14-16]. The conceived membrane regenerator is essentially an ED stack, in which cation- and anion-exchange membranes (CM and AM) are placed alternatively between the cathode and the anode. When a potential difference is applied between electrodes, the cations move toward the cathode, and anions move toward the anode. The cations go through the cation membrane, which have fixed groups with negative charges, and are retained by the anion-exchange membrane. On the other hand, the anions circulate through the anion-exchange membranes, which have fixed groups with positive charges, and are retained by the cation-exchange membranes. This movement produces a rise in the ions concentration in some compartments (concentrate compartments) and decrease in the adjacent ones (dilute compartments). The solar photovoltaic (PV) system is one of the most widespread and studied systems for electricity generation [17-18]. PV cells directly convert sunlight into electricity. The advantages of the use of this type of energy are that it is non-polluting—once the solar panels have been produced—it is also silent, abundant, decentralized, free and long lasting. The low maintenance cost of these systems is also another positive factor.

Inheriting the merits of the two major parts, this combined PV-ED regeneration system would have some advantages compared to the conventional TH method.

ADSORPTION DEHUMIDIFIER

The original dehumidifier, Dehumidifier No1, is a structured corrugated packing of inorganic material as shown in figure 1: the specific area of the packing is 396 (m^2/m^3). Figure 1 (a) and figure 1 (b) display the shapes viewed from the upside and the sidepiece of the packing. The definition of the packing dimension is shown in figure 1 (c). Its dimensions are height H=0.5m, width W=0.2m, and length L=0.5m. Dehumidifier No2 is an improved version of Dehumidifier No1. The transforming process is depicted in figure 2: silica gel self-indicators, the adsorbent, are filled in the wind tunnels of the packing material (original Dehumidifier No1) as shown in figure 2 (c). Described in figure 2 (a), the diameter of that silica gel self-indicator is of 3.2mm on average; the change of colour from blue to red symbolizes ample water molecules have been adsorbed and the silica gel self-indicator is saturated. Figure 2 (b) presents the surface of Dehumidifier No1 before being transformed. In the transforming process, silica gel are embedded and scattered on the surface in a random manner with uncertain distance from each other [shown in figure 2 (d)]. On average, around 50 silica gel bulbs are added in one single wind tunnel and totally 3/4 tunnels of the packing materials are treated in this way.

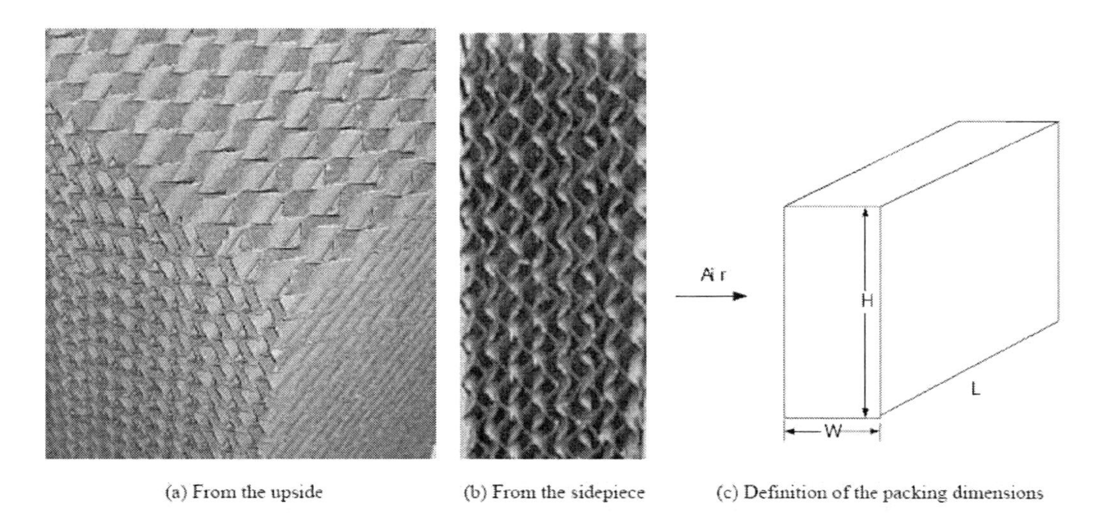

(a) From the upside (b) From the sidepiece (c) Definition of the packing dimensions

Figure 1. The shape and structure of the packing in the dehumidifier.

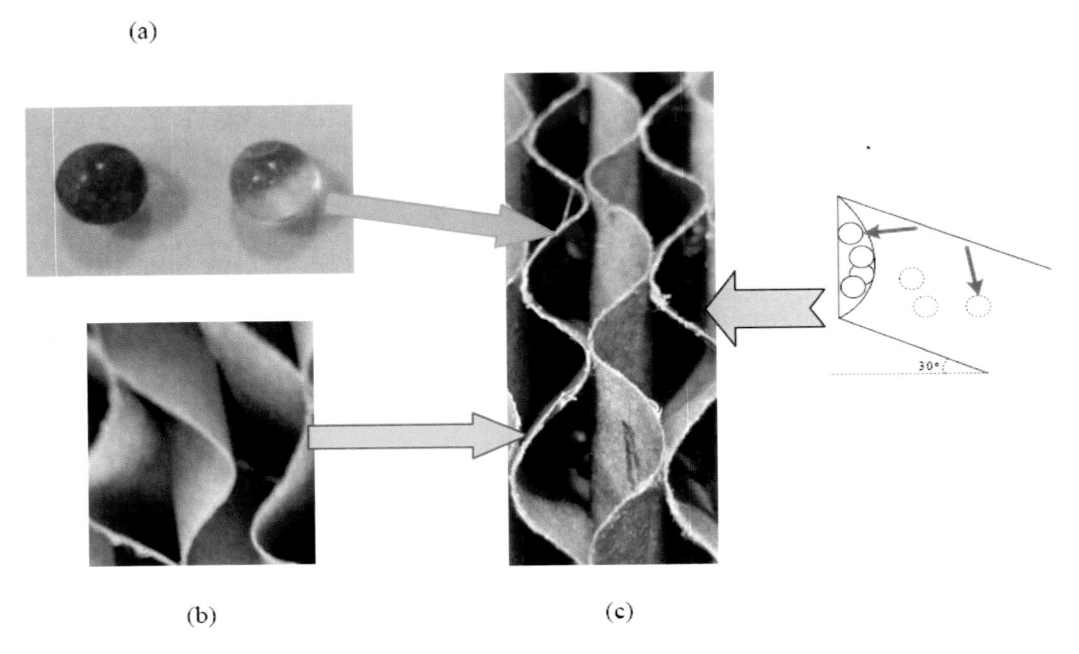

Figure 2. Improvement from Dehumidifier No2 to Dehumidifier No3.

To compare the dehumidification effects between the two dehumidifiers, experiments [2] were conducted: the air temperature was maintained 27±0.5°C; the initial air humidity was set as 14.5±0.3 g/ (kg air) and 15.5±0.3 g/ (kg air), respectively. Pure $CaCl_2$ solution with the mass concentration of 27% was taken as the liquid desiccant. The temperature of the desiccant solution was regulated as 19±0.5°C.

The experimental results are exhibited in figure 3 and figure 4, where the X axis represents the time and the Y axis represents the humidity reduction rate Δd. The results show that the moisture removal rate had been respectively enhanced by 66% and 70% by using Dehumidifier No2 under two initial humidity conditions. Since the liquid desiccant was same, this enhancement could only be attributed to the difference between the dehumidifiers. In the experiment, the pores of the silica gel self-indicators had been filled with desiccant solution and could not trap water molecules. Therefore, the reason for the improvement of dehumidification effects seems to be the contacting area enlargement by adding the adsorbent. The contribution of area enlargement could be evaluated by simple mathematical calculations: assume the silica gel self-indicators as bulbs, the total area enlargement ΔA could be calculated:

$$\Delta A = 4\pi r^2 Num = \pi d^2 Num \tag{1}$$

Num was approximate 50 for one single tunnel, and the enlarged area in one single tunnel was equal to 13.4% of the tunnel's original surface area. Because only 3/4 tunnels were stuffed, the total increased area was no more than 11% of the original total contacting area. It means the dehumidification effect improvement would not exceed 11%. However, that prediction result was much lower than the true improvement. Thus, there must be some other influential factors uplifting the dehumidification effect beyond the area enlargement.

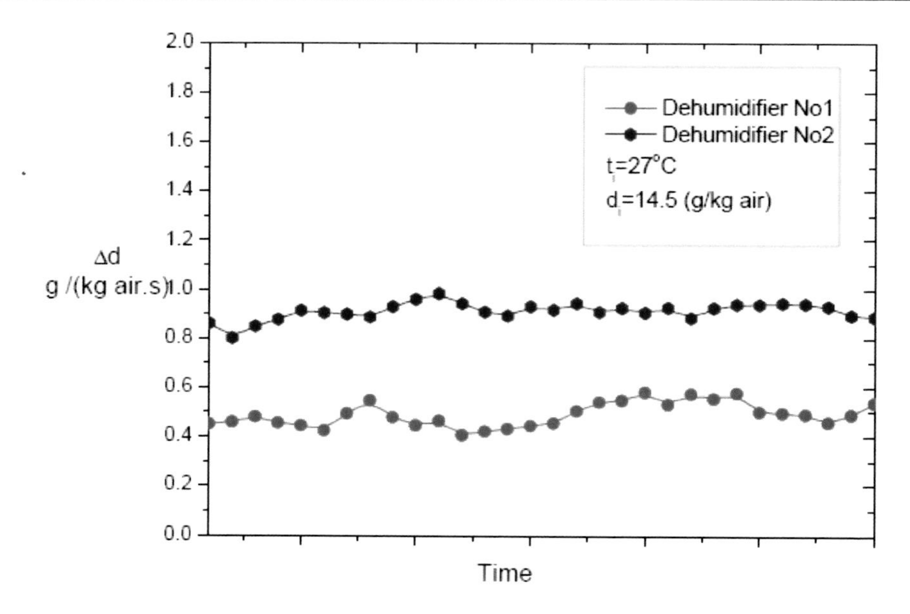

Figure 3. Dehumidification experiment with the initial humidity of 14.5 g /(kg air).

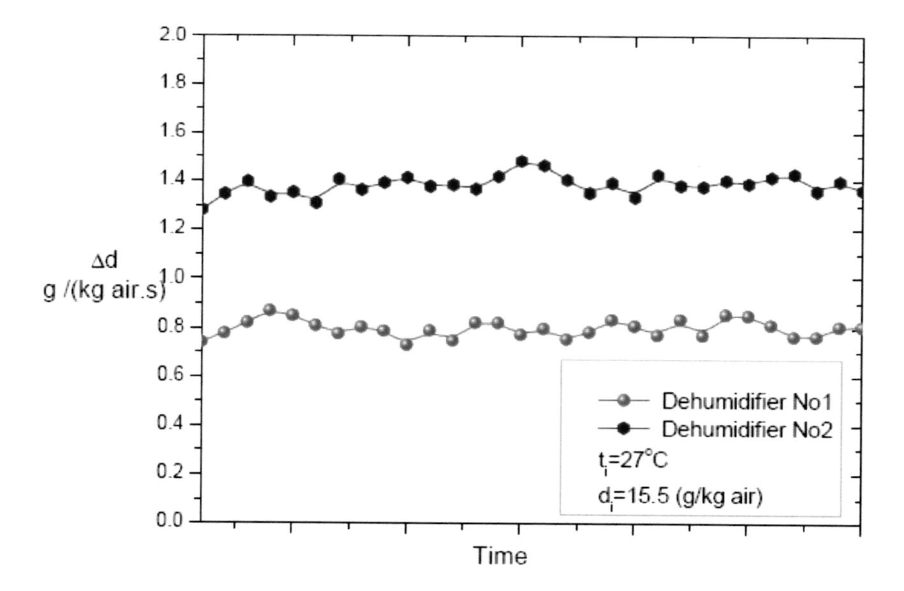

Figure 4. Dehumidification experiment with the initial humidity of 15.5 g /(kg air).

Literally, adsorption and absorption seems to be two different processes, but actually they came from the same word known as "Sorption" [19, 20]. In the meaning of "Sorption", the absorption process could be considered as the combination of two stages: in the first stage, the adsorption process happens, during which the molecules of the adsorbate (absorbate) are attracted by and approach the surface of the adsorbent; in the second stage, the diffusion process happens, during which some molecules penetrate the interface and proceed into the absorbent. With the "Sorption" conception, an assumption of adsorption potential superposition is taken to give a possible explanation for the experimental results. It is known that solid desiccant (like silica gel) could trap water molecules in its pores after it attracting these molecules with its adsorption potential. Definitely, the pores of the silica gel self-indicators were filled with desiccant solution in the experiments as the flowing film of liquid desiccant coated them. So the solid desiccant could not trap any more water molecules. Nevertheless, its adsorption potential still works for its intrinsic features, and this attracting power was hand in hand with the adsorption potential from the liquid desiccant. That came to be an overlapping of attracting power, which helped in forcing more water molecules close to the film of the desiccant. In terms of the "Sorption" definition, the first stage of absorption had been strengthened in this way. Naturally, more approaching molecules would provide more chances for successful penetration and diffusion process, so the absorption ability had been reasonably enhanced.

MEMBRANE REGENERATOR

The PV-ED regeneration process with the membrane regenerator is described in figure 5. The strong desiccant solution, whose mass flow rate is $m_{d,deh}$, absorbs m_w kg moisture from the to be treated air in the dehumidifier. The solution concentration decreases, and this weak solution is sent from the dehumidifier to the membrane regenerator. The regenerator (ED stack) consists of a multitude of cells placed in parallel between two electrodes [14]. In alternating cells the solution is concentrated and desalinated, respectively. The hydraulic circuits in two adjacent cells are referred to as dilute and concentrate flow streams and the adjacent cells are referred to as a cell pair (figure 6). At the beginning, all the dilute cells are feeding with the desiccant solution stream from Solution Storage Tank 1 (Valve 2 is open while Valve 4 is closed; Valve 1 and 3 are closed so that the water path is cut off); all the concentrate cells are feeding with the weak desiccant solution stream from the dehumidifier. The total mass flow rates of both these two streams are equal to ($m_{d,deh}+ m_w$). It should be noticed that the total amount of desiccant solution that has been regenerated is ($m_{d,deh}+ m_w$) in the ED method, but the needed amount of desiccant solution for dehumidification is $m_{d,deh}$. Therefore, for every regeneration process, excessive strong solution has been produced with the mass flow rate m_w. This excessive part will accumulate in the Solution Storage Tank 2 (Valve 7 is open while Valve 5 is closed). The diluted solution will be recycled from the regenerator to the Solution Storage Tank 1 (Valve 6 is open while Valve 8 is closed) and back to the dilute cells for regeneration until its concentration falls to zero, which implies it has finally become purified water. Then, Solution Storage Tank 2, which is full of strong desiccant solution takes the place of Tank 1 by feeding desiccant solution into the dilute cells of the regenerator; the purified water in Tank 1 will be removed out (Valve 1 is open while Valve 3 stays closed) to Water Storage Pool for later other use, and Tank 1 comes to play the role that Tank 2 played before (by manipulating the valves). Like this, Tank 1 and 2 keep on alternating their roles and both the strong desiccant solution and the purified water are acquired. Neglect the energy loss, the ideal electric power P_{ED} is given by [14]:

$$P_{ED} = UI = \frac{zF\left(m_{d,deh} + m_w\right)Con_{mol}^{\Delta}}{\zeta N}U \quad (2)$$

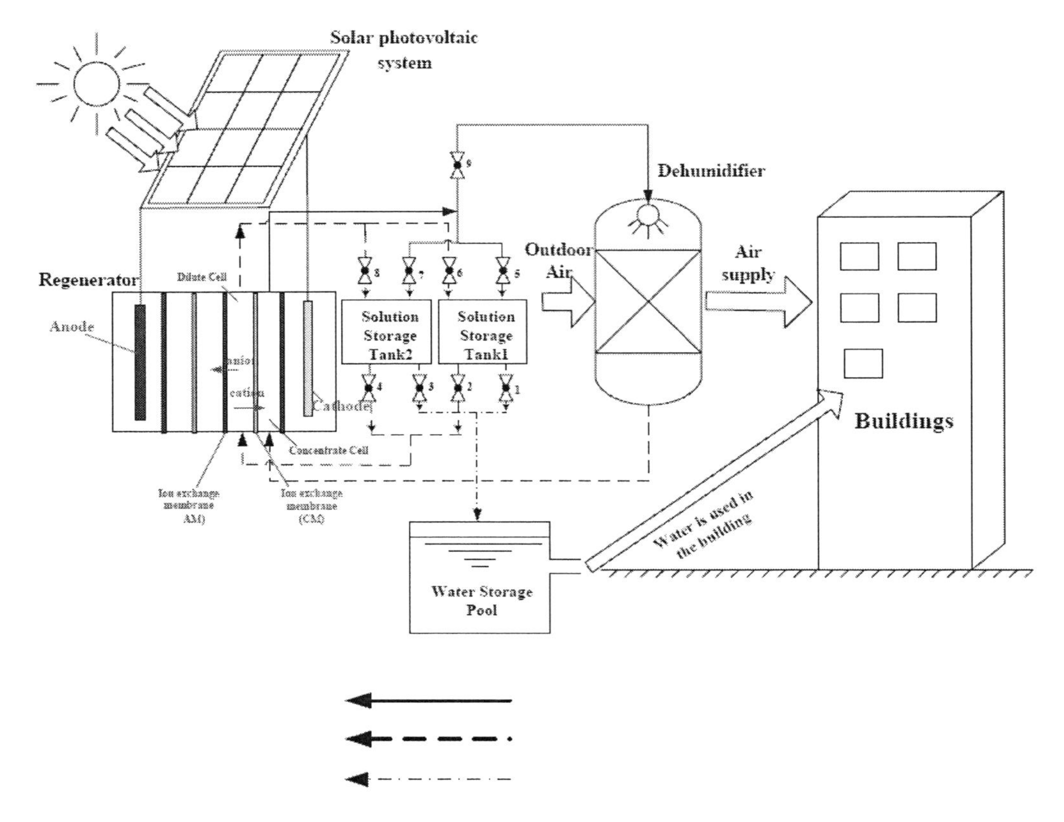

Figure 5. PV-ED-LDCS system.

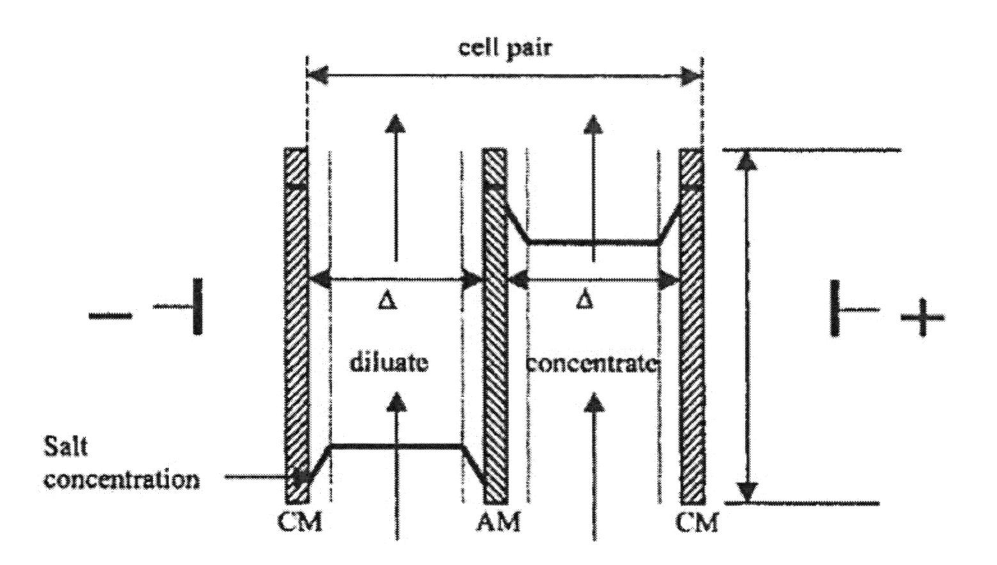

Figure 6. Configuration of an ED cell pair .

The energy consumption for regenerating the desiccant solution of one single kilogram is:

$$p_{ED} = \frac{zFCon_{mol}^{\Delta}}{\zeta N} U \qquad (3)$$

Under the same typical working conditions [9-18], the energy consumption for regenerating the desiccant solution of one kilogram is calculated for both the PV-ED system and the TH system. The results are presented in figure 7, which shows that the PV-ED system is immune against the change of relative humidity and has much lower energy consumption than that of the TH system. The TH system has to consume more energy for regeneration when the relative humidity rises. Nevertheless, it should be noticed the present PV generator has very poor solar energy conversion efficiency within the range of 5% ~20%. That will discount the performance of the whole PV-ED system, but even then, the performance of this new regeneration system will still be competitive with the conventional TH system in hot and humid days.

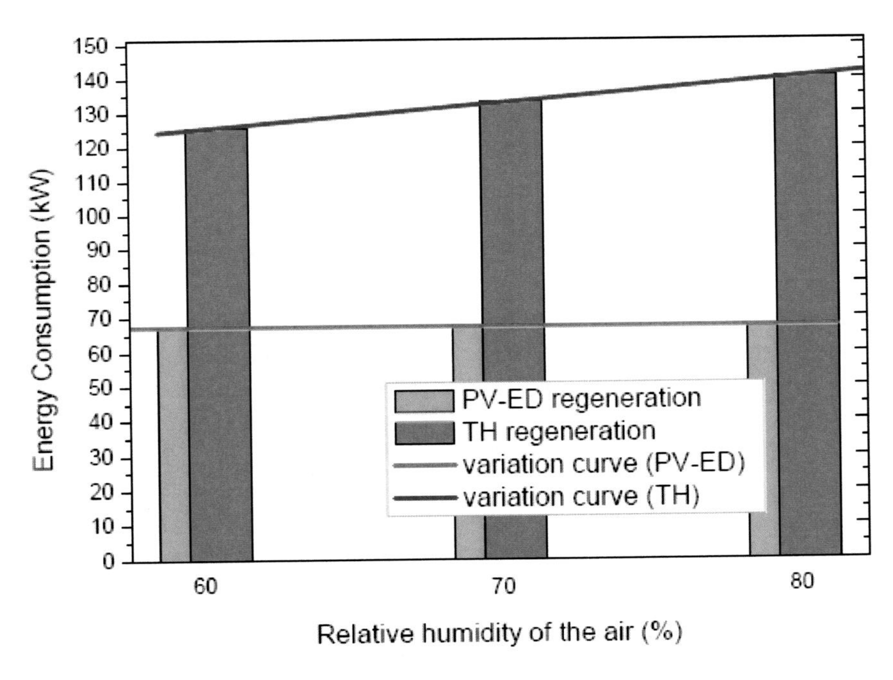

Figure 7. Energy consumption with different relative humidity of the air.

CONCLUSION

A new improvement in dehumidifier design for LDCS has been achieved by combining a normal packing material with solid adsorbent. The experimental result shows this improvement did help with the enhancement of the liquid dehumidification ability. The assumption of the "sorption" principle is proposed to explain the dehumidification effects enhancement by integrating liquid and solid desiccant. This assumption, buttressed by the experimental results, is reasonable and practical.

Compared to the conventional TH regeneration method, the PV-ED regeneration method has several advantages: it is immune against the instability caused by the high humidity of the atmosphere; it gets rid of the adding heat that is harmful to the dehumidification process, and thus saves the energy for producing a cooling effect; it is environmentally friendly by preventing the droplets of desiccant solution from sneaking into the surrounding atmosphere and introducing pollution; it can also produce purified water during the regeneration process. Moreover, exposed by theoretical analysis, the new regeneration system has a better performance while putting aside the low efficiency of the PV regenerator. The ED method has been proved to be a practical and useful technology for desalination; the energy-conversion efficiency for PV system has been improved every day with new emerging materials. Those will facilitate the development of a more competitive PV-ED system in the future.

REFERENCES

[1] Khalid Ahmed, C. S., Gandhidasan, P., & Al-Farayedhi, A. A.(1997). Simulation of a Hybrid Liquid Desiccant Based Air-conditioning System. *Applied Thermal Engineering*, 17(2), 125-134.

[2] Li, X. W., Zhang, X. S., Wang, G., Cao, & R. Q. (2008). Research on ratio selection of a mixed liquid desiccant: Mixed LiCl-$CaCl_2$ solution. *Solar Energy*, 82 (12), 1161-1171.

[3] Gommed, K., & Grossman, G. (2004). A liquid desiccant system for solar cooling and dehumidification. *Journal of Solar Energy Engineering*, 126, 879-885.

[4] Khan, A. Y. (1998). Cooling and dehumidification performance analysis of internally-cooled liquid desiccant absorbers. *Applied Thermal Engineering*, 18(5), 265-281.

[5] Al-Farayedhi, A. A., Gandhidasan, P., & Al-Mutairi, M.A. (2002). Evaluation of heat and mass transfer coefficients in a gauze-type structured packing air dehumidifier operating with liquid desiccant. *International Journal of Refrigeration*, 25, 330-339.

[6] Abdul-Wahab, S.A., Abu-Arabi, M.K., & Zurigat, Y.H. (2004). Effect of structured packing density on performance of air dehumidifier. *Energy Conversion and Management*, 45, 2539-2552.

[7] Ani, F.N., Badawi, E.M., & Kannan, K.S. (2005). The effect of absorber packing height on the performance of a hybrid liquid desiccant system. *Renewable Energy*, 30, 2247-2256.

[8] Nelson, F., & Goswami, D.Y. (2002). Study of an aqueous lithium chloride desiccant system: air dehumidification and desiccant Regeneration. *Solar Energy*, 72(4), 351-361.

[9] Gandhidasan, P. (2005). Quick performance prediction of liquid desiccant regeneration in a packed bed. *Solar Energy*, 79, 47-55.

[10] Ren, C. Q., Jiang, Y., Tang, G. F., & Zhang, Y. P. (2005). A characteristic study of liquid desiccant dehumidification/regeneration process. *Solar Energy*, 79, 483-494.

[11] Elsarrag, E. (2006). Performance study on a structured packed liquid desiccant regenerator. *Solar Energy*, 80, 1624-1631.

[12] Elsarrag, E. (2008). Evaporation rate of a novel tilted solar liquid desiccant regeneration system. *Solar Energy*, 82, 663-668.
[13] Liu, X. H., Jiang, Y., & Yi, X. Q. (2009). Effect of regeneration mode on the performance of liquid desiccant packed bed regenerator. *Renewable Energy*, 34, 209-216.
[14] Lee, H. J., Sarfert, F., Strathmann, H., & Moon, S. H. (2002). Designing of an electrodialysis desalination plant. *Desalination*, 142, 267-286.
[15] Tsiakis, P., & Papageorgiou, L. G. (2005). Optimal design of an electrodialysis brackish water desalination plant. *Desalination*, 173, 173-186.
[16] Ortiz, J. M., Expósito. E., Gallud. F., García-García, V., Montiel, V., & Aldaz. A. (2006). Photovoltaic electrodialysis system for brackish water desalination: Modeling of global process. *Journal of Membrane Science*, 274, 138-149.
[17] Zondag, H. A., de Vries, D. W., van Helden, W. G. J., van Zolingen, R. J. C., & van Steenhoven, A. A. (2003). The yield of different combined PV-thermal collector designs. *Solar Energy*, 74, 253-269.
[18] Charalambous, P. G., Maidment, G. G., Kalogirou, S. A., & Yiakoumetti, K. (2007). Photovoltaic thermal (PV/T) collectors: A review. *Applied Thermal Engineering*, 27, 275-286.
[19] Clark, A. (1970). The Theory of Adsorption and Catalysis. Academic Press.
[20] Aristov, Y.I., Tokarev, M. M., & Freni, A. (2006). Kinetics of water adsorption on silica Fuji Davison RD. *Microporous and Mesoporous Materials*, 96, 65-71.

INDEX

A

absorbents, 6, 8, 21, 27, 28, 29, 33, 37, 66, 76, 77, 99, 104, 135, 136
absorption, 4, 5, 6, 8, 9, 10, 21, 22, 23, 48, 66, 76, 100, 101, 103, 107, 117, 119, 130, 131, 136, 142, 143, 144, 145, 146, 147, 168, 193, 197, 198, 199, 273, 278, 378
acceleration, 137
accommodation, 157
accountability, 192
accounting, 53, 77, 152
accuracy, x, 20, 40, 46, 56, 85, 97, 98, 275, 295, 337, 357
achievement, 161
acid, 175, 232
acoustic, 171
activation, 339, 340
adaptation, 197
ADC, 343
additives, 21, 27, 76, 143
adjustment, viii, 195
administration, 338
adsorption, xi, 142, 373, 378, 383
aesthetics, 153, 171
afternoon, 185, 280, 288
Ag, 268
age, 161, 229, 230, 231
agent, 234
agents, 232
agriculture, ix, 2, 173, 175, 243, 244, 263
aid, 158, 181, 199
air pollutant, 168, 227
air pollutants, 168, 227
air quality, vii, viii, 153, 154, 155, 156, 157, 164, 225, 226, 227, 229, 230, 235, 239, 240, 241, 315
alcohol, 240
algorithm, x, 95, 207, 233, 234, 235, 293, 295, 299, 307, 313, 337, 339, 340, 343, 345, 349, 350
allergic reaction, 157
alpha, 168
alternative, 2, 7, 8, 9, 11, 13, 15, 31, 32, 65, 92, 95, 98, 103, 105, 107, 111, 113, 116, 117, 119, 124, 128, 130, 131, 133, 135, 136, 137, 139, 145, 193, 210, 232, 244, 270
alternative energy, 244
alternatives, vii, x, 149, 164, 233, 315
aluminium, 13, 21, 49, 72, 162
ambient air, 16, 125, 126, 156, 166, 245, 249, 253, 277, 357
ambient air temperature, 156, 166, 253, 357
amelioration, 151
ammonia, 140, 142
amplitude, 48
analog, 206, 343
analysis of variance, x, 337, 339, 343
animals, 168, 232
ankles, 231
ANN, vii, x, 337, 338, 339, 340, 341, 343, 344, 345, 347, 348, 349
annealing, 295, 313
anode, 375
anodes, 301
ANOVA, x, 337, 339, 343, 344, 345, 347, 348, 349
appraisals, 181
aqueous solution, 23, 27, 34, 76
aqueous solutions, 23, 27, 34, 76
Archimedes, 228, 231, 240
arid, 166, 185, 193, 240, 264
ASEAN, 145, 147
Asia, viii, 225
ASIC, 197
assessment, 162, 173, 174, 176, 193, 239
assimilation, 122
asymptotically, 365
Athens, 140

atmosphere, 1, 2, 11, 126, 128, 130, 152, 168, 170, 184, 191, 270, 375, 382
atmospheric pressure, 5
attitudes, 170
Australia, 3, 4
automation, 226
autoregressive model, x, 337, 338, 349
availability, 6, 42, 50, 159, 168
averaging, 40, 50, 81, 85, 154
awareness, 151, 179, 182, 235

B

back, 152, 184, 233, 249, 273, 345, 371, 379
bacteria, 158
ballast, 342
bananas, 264
banking, 183
barriers, 270
batteries, 3, 368
battery, 302
behavior, 85, 104, 200, 203, 230, 244, 263, 291, 359
Beijing, 232
benefits, ix, 170, 175, 178, 182, 191, 245, 267, 270
benzene, 227
bias, 340
biogas, 169
biomass, 3, 232, 245, 264
biosphere, 185
black hole, 168
blocks, 170, 189, 273, 342
body temperature, 184, 281
boilers, 7, 316
boiling, 291
Boltzman constant, 151, 188
boundary conditions, 41, 46, 69, 76, 77, 79, 80, 81, 84, 96, 160, 227
braids, 366
breakdown, 309
breathing, ix, 156, 225, 229, 230, 231
Britain, 175
broadband, 235
Brussels, 221
bubbles, 144
building blocks, 170
bulbs, 374, 376, 377
Bulgaria, 4, 141
burning, 152, 175, 183, 184, 232, 270
bypass, 319
by-products, 228

C

calcium, 21, 22, 143, 183, 304
calibration, 50, 275
campaigns, 180
Canada, 2, 5, 183
capacitance, 48
capacity building, 182
capillary, x, 50, 51, 56, 271, 273, 274, 275, 276, 280, 282, 283, 285, 286, 289, 290, 353, 355, 356, 357, 358, 359, 364, 365, 366, 368, 371
carbohydrates, 169
carbon, vii, viii, 1, 149, 152, 156, 157, 158, 160, 167, 168, 174, 175, 180, 182, 183, 184, 225, 226, 227, 232, 235, 238, 270, 295, 301
carbon dioxide, vii, viii, 1, 149, 152, 156, 158, 160, 167, 168, 174, 175, 184, 225, 226, 227, 232, 235, 238, 270, 295, 301
carbon emissions, 182
Carnot, 18, 179
carrier, 24, 33, 34, 77, 80, 85, 87, 91, 98, 130, 131
case study, x, 293
cathode, 375
cation, xi, 373, 375
cavities, 52, 59, 70, 138, 188
CEC, 10, 11, 16, 17, 65, 72, 99, 103, 104, 106, 107, 126, 127, 128, 129, 139
cell, x, 183, 236, 241, 275, 293, 294, 295, 296, 300, 301, 302, 303, 305, 306, 307, 308, 309, 311, 313, 374, 379, 380
cereals, 168
CFCs, 151
CFD, 160, 240
CH4, 168
channels, 9, 10, 31, 32, 33, 34, 37, 44, 48, 49, 52, 57, 59, 60, 71, 72, 73, 74, 75, 89, 90, 95, 105, 106, 110, 144, 207, 216, 238, 246, 343
chemical properties, 147
chemicals, 151, 157
chimneys, 166
China, 169, 227, 232, 241, 242, 373
chloride, 21, 22, 143, 304, 382
CHP, 176
chromosome, 294, 297, 298, 299, 307, 308
chromosomes, 298, 308
circulation, 77, 79, 82, 83, 84, 85, 87, 91, 98, 185, 198, 236, 279
Citrus aurantium, 263
classes, 338
classical, 5, 156, 162, 220, 356
classification, 244, 348
classrooms, 229, 230, 240, 342
clean air, 153, 156, 160

cleaning, 33, 156, 181, 249
clients, 171
climate change, viii, 1, 143, 149, 175, 190, 191, 192
climate extremes, 183
closed-loop, 207
clusters, 235
CMOS, 222
Co, 143, 144, 146, 186, 313
CO2, 2, 151, 168, 173, 174, 175, 182, 184, 228
coal, 2, 175, 183, 228, 232
codes, 226, 233, 239
coefficient of performance (COP), 36, 104, 138, 231, 308, 316
coil, 139, 229, 233
combined effect, 185
combustion, 227, 301
comfort zone, 99, 103
commodities, 245
communication, vii, 2, 149, 296
community, 1, 152, 197
compensation, 7
complexity, 165, 244, 350
complications, 186
components, 5, 8, 9, 10, 11, 21, 24, 36, 38, 71, 76, 77, 82, 85, 106, 135, 146, 171, 176, 187, 197, 201, 211, 220, 244, 338
composites, 142
composition, 103, 104, 209
computation, 211, 220
computational fluid dynamics, 160
computing, 213, 251
concentration, 1, 21, 23, 24, 25, 26, 34, 76, 95, 96, 97, 99, 104, 105, 112, 113, 123, 125, 136, 137, 155, 156, 170, 175, 226, 227, 235, 375, 377, 379
conception, 378
concrete, 171, 174, 273, 274, 275, 279, 280, 281, 282, 288, 342
concreteness, 171
condensation, 48, 65, 66, 69, 71, 227, 304, 316, 320, 321
conductance, 48, 49
conduction, 24, 32, 42, 50, 52, 53, 77, 78
conductivity, 72, 137, 150, 188, 246, 271, 273, 356
confidence, x, 337, 349
configuration, 15, 38, 104, 197, 200, 228, 246, 319, 375
conformity, 4, 6, 9, 11, 32, 53, 65, 76, 99
Congestion, 170
Congress, 4, 192, 193, 221, 351
conjugation, 77
consensus, 190
conservation, vii, 80, 147, 149, 170, 175, 192, 235, 239, 242, 280, 350
constant load, 216
constraints, 159
construction, 6, 7, 8, 11, 14, 32, 33, 53, 57, 76, 79, 85, 86, 95, 110, 136, 152, 158, 171, 172, 173, 174, 186, 239, 273, 274
construction materials, 6, 7, 76
consumers, 170, 182, 235, 374
contaminant, 155, 157, 158, 160
contaminants, 231
contamination, 157, 192, 231, 356
continuity, 38, 40, 44, 77, 82
contractors, 173
convection, 66, 68, 78, 79, 80, 81, 82, 83, 188, 231, 244, 263, 264, 277, 371
convective, 65, 166, 174, 188, 189, 228, 245, 263, 264
convergence, 309
conversion, 152, 183, 197, 199, 207, 218, 301, 307, 350, 381, 382
convex, 60, 234
cooking, 158, 181
cooling process, 21, 71, 73, 96, 100, 106
COP, 18, 19, 20, 21, 27, 85, 86, 87, 98, 104, 125, 126, 138, 234, 303, 316, 318, 321, 323, 324, 326, 327, 328, 329, 331, 335
copper, 357, 367
correlation, 5, 8, 52, 53, 110, 234, 344
corrosion, 7, 13, 33
corrosive, 21, 22, 23, 27, 76, 104, 136
cosine, 341
cost saving, 169, 171
cost-effective, 160, 175, 176, 185
costs, x, 33, 140, 153, 175, 177, 182, 191, 194, 236, 245, 269, 270, 293, 295, 297, 302, 305, 308, 309, 311
coughing, 227
coupling, viii, 195, 200, 209, 211, 213, 214, 215, 216, 220
covering, 3, 162, 280, 342
Cp, 150, 268, 360, 361
CPC, 140
CPU, 213
Crete, 3
critical value, 60, 139
crop drying, 169, 264
crops, ix, 168, 183, 243, 244, 263
cross-sectional, 59
crystallization, 21, 22, 23, 26, 28
cultivation, 168, 169
culture, 184
customers, 229, 235
Cybernetics, 223
cycles, 3, 8, 18, 20, 141, 143, 319, 326, 327

cycling, 178

D

daily living, 153
danger, 6
data processing, 275
data set, 342, 347
database, 231
debates, 182
decay, 2
decision makers, 180
decision making, 202
decision-making process, 173
decisions, 185
decoding, 299
decomposition, 1
defects, 375
deficiency, 350
definition, 173, 187, 201, 204, 316, 376, 379
deformation, 51
degradation, 226, 270
degrees of freedom, 200, 344
dehydration, ix, 243, 263
democracy, 182
Denmark, 3, 141, 142, 336
density, 22, 57, 59, 77, 97, 102, 105, 137, 146, 164, 169, 170, 271, 294, 382
density values, 57
Department of Energy (DOE), 4, 240
deposits, 22
depreciation, 36
desalination, 382, 383
designers, 159, 165, 171, 172, 226
desorption, 22, 23, 76, 136, 142
destruction, 177
detection, 349
developed countries, 2, 181
developing countries, 174, 181, 264
deviation, 47, 48
dew, 5, 9, 10, 19, 72, 103, 106, 138
diffusion, 137, 181, 250, 378
diffusion process, 378
diffusivity, 188, 362
discomfort, 154, 159, 184, 188, 189, 227, 229, 231, 237
discretization, 359
diseases, 183
dispersion, 46
displacement, viii, 155, 225, 227, 228, 229, 230, 239, 240
disposition, 173
dissociation, 168
distillation, 169
distribution, vii, viii, 32, 49, 53, 56, 59, 60, 63, 79, 103, 120, 121, 154, 160, 180, 186, 225, 227, 228, 229, 230, 237, 238, 270, 295, 315, 342, 343
diversity, 298, 299, 308
division, 106
dollar costs, 175
doors, 164, 172, 268
draft, 227, 229, 231
drought, 183
droughts, 175, 191
dry matter, 252
drying, ix, 76, 169, 243, 244, 245, 248, 249, 250, 251, 252, 253, 255, 257, 258, 259, 261, 262, 263, 264, 265
drying time, ix, 243, 244, 257, 258, 259, 261
DSM, 235
durability, 155, 173
duration, 79, 169, 200, 216
dust, 86, 156
dynamic viscosity, 137
dynamical system, 338
dynamical systems, 338

E

early warning, 154
earth, 152, 168, 175, 184, 270
East Asia, viii, 225
Eastern Europe, 141, 146
eating, ix, 168, 267, 270, 286, 288
ecology, 4, 8
economic crisis, 163
economic development, 173
economic growth, 191, 228
economics, 158
ecosystem, 151
ecosystems, 3
efficiency level, x, 315
electric energy, 269, 297, 300, 302, 304, 307, 309, 311, 355
electric power, xi, 2, 3, 4, 15, 35, 36, 196, 293, 294, 295, 300, 301, 302, 306, 307, 308, 309, 311, 341, 373, 374, 379
electric utilities, 350
electrical power, 7, 344
electrical resistance, 48
electricity, vii, 15, 149, 170, 175, 183, 189, 191, 194, 221, 225, 226, 228, 232, 234, 235, 237, 238, 239, 241, 295, 301, 303, 307, 309, 311, 313, 355, 376
electrodes, 49, 50, 53, 54, 375, 379
electrolysis, 300, 301, 302, 304, 309

emission, viii, 155, 160, 168, 174, 225, 226, 228, 232, 238, 270, 271
employment, 2, 182
encouragement, 167
energy audit, 176
energy efficiency, vii, viii, x, 151, 152, 155, 167, 169, 170, 171, 174, 175, 177, 178, 181, 182, 225, 226, 229, 230, 238, 242, 283, 300, 315, 335
energy management system, 338
energy recovery, 335
energy supply, 191, 228, 297, 301
energy transfer, 198
engagement, 182
enlargement, 375, 377
environment, vii, 3, 149, 150, 151, 153, 154, 157, 159, 161, 163, 165, 167, 168, 169, 171, 172, 173, 175, 178, 179, 183, 184, 185, 187, 198, 199, 226, 228, 229, 270, 288, 338, 354, 373, 375
environment control, vii, 149
environmental conditions, ix, 151, 186, 196, 267
environmental context, 270
environmental control, viii, 159, 174, 195, 196
environmental degradation, 226, 270
environmental factors, 173
environmental impact, vii, 149, 150, 151, 173, 178, 179, 191, 228, 239
environmental issues, vii, 149, 170, 177, 179
environmental protection, 1, 175
environmentalists, 175
epoxy, 50
equality, 39, 50
equilibrium, 168
error estimation, 349
ESL, 351
estates, 236
estimating, 295
estimators, 345
ethanol, 301
Europe, 2, 3, 4, 140, 145, 164, 172, 194, 196, 221, 226, 315
European Commission, 194, 221
European Community, 3
European Union (EU), 140, 143, 144, 145, 192, 221
evaporation, ix, 5, 62, 65, 68, 69, 72, 106, 137, 159, 197, 199, 227, 238, 243, 244, 250, 316, 320, 321, 374
evening, 157
evolution, 234, 239, 286, 289, 369
exaggeration, 161
exchange rate, 164
exchange rates, 164
exercise, 152
exhaust heat, 295, 302, 307
expenditures, 8, 10, 25, 30, 35, 36, 103, 117, 137
experimental condition, 255, 260
exploitation, viii, 149, 151, 153, 171
explosions, 168
exposure, 154, 170
expulsion, 320
external environment, 177, 188, 362
extraction, 91, 328
eye, 155, 227

F

fabric, 165, 175, 184
failure, 172
family, 153, 157, 164
farmers, 105
farming, vii, 149, 152
fatigue, 227
fault detection, 349
fax, 195
February, 274, 299, 300, 303, 305, 308, 309, 311, 312
feedback, 200, 216
feeding, 9, 13, 62, 64, 379
feelings, 227
field programmable gate array, 350
Field Programmable Gate Arrays, 197
film, 6, 7, 9, 10, 11, 31, 32, 37, 39, 41, 42, 43, 44, 45, 48, 49, 50, 51, 52, 53, 54, 55, 56, 59, 60, 61, 62, 63, 64, 65, 66, 67, 68, 72, 75, 76, 105, 106, 118, 139, 144, 379
film thickness, 32, 37, 41, 42, 48, 52, 138
filters, 157
filtration, x, 56, 153, 157, 353, 356, 357, 359
finance, 175
financial resources, 165, 182
financial support, 370
financing, 165, 182
finite differences, 70, 76, 85
fire, 160, 204, 205
fires, 204
firewood, 177
First World, 193
fish, 354
fitness, 298, 299, 307, 308
float, 13
flooding, 42, 44, 53, 59
fluctuations, 99
fluid, 140, 157, 160, 197, 227, 232, 234, 315, 316
fluid transport, 197
fluidized bed, 144
fluorescent lamps, 226
fluorescent light, 342

flushing, 181
focusing, 153, 181
food, 143, 146, 152, 168, 174, 175, 244, 253, 263, 264
food industry, 143
food production, 168
forecasting, 338, 339, 341, 345, 346, 350, 351
forestry, 2
forests, 183
formaldehyde, 227
fossil, vii, 1, 2, 149, 151, 152, 153, 159, 161, 170, 175, 177, 184, 194, 227, 232, 269, 270
fossil fuel, vii, 1, 2, 149, 151, 152, 153, 161, 170, 175, 177, 184, 227, 232, 269, 270
fossil fuels, vii, 149, 151, 152, 153, 161, 170, 175, 177, 184, 227, 232, 270
Fourier, 38
FPGA, 197, 222
France, 2, 4, 62
freedom, viii, 195, 200, 209, 210, 211, 213, 215, 220, 344, 348
freezing, 144
fresh water, 9, 11, 12, 16, 106
friction, 319
fruits, ix, 168, 243, 263, 264
fuel, x, 1, 2, 4, 169, 170, 175, 176, 177, 183, 232, 235, 239, 269, 270, 293, 294, 295, 300, 301, 302, 303, 305, 306, 307, 308, 309, 311, 312, 313
fuel cell, x, 183, 293, 294, 295, 300, 301, 302, 303, 306, 307, 308, 309, 311, 312, 313
fuel efficiency, 170
funding, 177
fungal spores, 157
fungi, 183, 227
fungicides, 183
fuzzy logic, viii, 195, 197, 200, 201, 209, 210, 211, 215, 216, 218, 220, 221
fuzzy sets, 201, 202, 203, 204, 205, 206, 209, 216

G

gamma radiation, 168
gases, vii, 1, 149, 152, 168, 170, 182, 184, 190, 191
GDP, 228
gel, 6, 233, 374, 375, 376, 377, 379
gene, 297, 298, 308
generalization, 47
generation, vii, 4, 9, 88, 146, 149, 175, 189, 228, 232, 297, 298, 299, 300, 303, 304, 305, 307, 308, 309, 310, 311, 312, 313, 315, 319, 350, 376
generators, 316
genes, 298, 308
Germany, 2, 5, 192
GHG, 182, 189
glass, 14, 78, 88, 93, 94, 98, 136, 152, 162, 176, 184, 236, 246, 251, 252, 255, 259, 260, 268, 273, 375
glasses, 251, 252, 255, 259, 260, 268
global climate change, 192
global demand, 168
global warming, 1, 2, 3, 4, 7, 152, 183, 184, 192, 270
goal setting, 235
government, 151, 174, 182, 191, 233, 235
government intervention, 235
grain, 265
grants, 177
grapes, 263, 265
graph, 345
gravity, 38, 42, 150, 205
Great Britain, 2
Greece, 195, 221
green buildings, 173
Green Revolution, 174
greenhouse, vii, 1, 139, 149, 151, 152, 162, 168, 169, 170, 175, 176, 182, 183, 184, 186, 189, 190, 191, 194, 226, 227, 232, 233, 264, 345, 350
greenhouse gas, vii, 1, 139, 149, 151, 170, 175, 182, 189, 190, 191, 226, 232, 233
greenhouse gases, vii, 1, 139, 149, 151, 170, 182, 190, 191, 226
groundwater, ix, 267, 270, 274, 285, 288, 289, 365
groups, 48, 176, 183, 298, 299, 308, 345, 348, 375
growth, 10, 51, 59, 60, 103, 104, 105, 152, 168, 172, 173, 183, 191, 227, 228, 235
guidance, 235
guidelines, 172, 173

H

H_2, 303
HA1, 195
habitation, 168
handling, 157, 349
harm, 169
harmony, 169
harvest, 250
hazards, 7
headache, 227
health, 153, 154, 155, 157, 158, 161, 175, 182, 191, 194, 227
health problems, 153
heart, 291
heat capacity, 20, 24, 28, 77, 85, 87, 360, 361
heat conductivity, 72
Heat Exchangers, 32, 130, 291
heat loss, 78, 87, 88, 98, 138, 170, 227, 273, 276, 279, 306, 307, 356, 358, 362, 368

heat pumps, x, 5, 140, 141, 145, 156, 299, 315, 316, 317, 319, 321, 322, 324, 326, 328, 329, 332, 335
heat release, 306
heat removal, 18, 230
heat transfer, 17, 25, 26, 33, 35, 36, 62, 71, 76, 97, 138, 142, 150, 160, 187, 188, 189, 232, 233, 234, 235, 245, 263, 273, 276, 277, 278, 282, 291
height, 32, 37, 48, 50, 53, 57, 60, 65, 68, 95, 105, 113, 138, 169, 174, 185, 230, 231, 271, 287, 288, 342, 375, 376, 382
helium, 168
high power density, 146
high pressure, 199, 318
high tech, 161
high temperature, 23, 27, 158, 184, 316, 375
Hong Kong, 225, 232, 233, 234, 235, 236, 239, 240, 242
horizon, 86, 197
hot water, viii, ix, 3, 9, 13, 88, 91, 92, 149, 151, 153, 180, 181, 226, 240, 267, 270, 274, 276, 281, 285, 286, 289, 295, 350
House, v, 194, 267, 313
household, 180, 270
households, 175, 177
housing, 14, 93, 94, 158, 184, 189, 192, 193, 236, 319, 320, 321, 322, 323
human, vii, 149, 158, 161, 168, 171, 172, 173, 175, 179, 184, 190, 191, 202, 227, 241
humanity, 1
humidity, ix, xi, 8, 19, 21, 57, 73, 95, 99, 122, 137, 146, 151, 158, 183, 185, 227, 241, 244, 245, 250, 251, 261, 262, 288, 341, 373, 377, 378, 381, 382
Hungary, 141
hurricanes, 175
hybrid, vii, 146, 147, 149, 151, 153, 181, 196, 245, 264, 382
hydrides, 6
hydrocarbon, 304
hydrodynamics, 144
hydrogen, 146, 295, 301, 302, 304, 309, 311
hydrogen gas, 295
hygienic, 99
hypothesis, 347
hypothesis test, 347

I

ice, 193
id, 139
ideology, 2
IEA, 4, 146, 226, 241
illumination, 186, 342
implementation, viii, 165, 179, 181, 192, 195, 197, 198, 201, 207, 213, 216, 222, 226, 350
imports, 2
inactive, 332, 333, 334
incentives, 173, 180, 235
incidence, 88
income, 2, 174, 177, 181
India, 169
indicators, 226, 376, 377, 379
indices, 226
industrial application, 5, 200
industry, vii, 2, 143, 149, 152, 156, 175, 183, 197, 244
inequality, 79
inertia, 189
infections, 227
Infiltration, 153, 164, 290
information exchange, 173
infrastructure, 3, 165, 169, 170, 182
inhibitors, 21, 76
injury, iv
instability, 46, 375, 382
insulation, 5, 8, 78, 87, 88, 93, 94, 155, 174, 176, 181, 186, 226, 228, 234, 238, 239, 246, 248, 280, 356, 358
insurance, 175
integration, 189, 237, 313
Intel, 222
intentions, 2
interaction, vii, 7, 32, 38, 44, 48, 50, 59, 60, 103, 106, 120, 121, 149, 171, 209, 251
interactions, 171, 220
interface, 65, 139, 216, 343, 378
Intergovernmental Panel on Climate Change, 175
internalised, 191
International Energy Agency, 4, 226
International Trade, 4
internet, 235
interval, 23, 24, 26, 27, 212, 213, 294, 357
intervention, 166, 235
intrinsic, 379
inversion, 319, 320, 322, 324, 326
investment, 176, 177, 228, 235, 269
ionic, 168
ionosphere, 168, 193
ions, 168, 375
irradiation, 171
irritation, 154, 155, 227
Islam, 351
island, 3, 169, 170, 196
isolation, 169, 172
isotherms, 23, 99, 103
Italy, 145, 183, 192

iteration, 77

J

Japan, 1, 4, 169, 293, 295, 299, 313
Japanese, 4, 51, 290, 313, 314
Jerusalem, 4, 146
Jiangxi, 142
jobs, 175
joints, 176
JOR, 144
Jordan, 291
Jordanian, 194
judgment, 153
Jun, 254

K

Kenya, 192
kindergarten, 236
kinematics, 137
kinetics, 244, 264
Korea, 189, 194
Kyoto Protocol, 1, 2, 181, 182

L

Lagrangian, 234
lamellar, 56
lamina, 25, 32, 43, 50, 51, 64, 66, 79, 83, 371
laminar, 25, 32, 43, 50, 51, 64, 66, 79, 83, 371
LAN, 296
land, 168, 173, 194
language, 171
Latin America, 140
law, 49, 184, 246, 257
leakage, 153, 157, 164, 174, 186, 250, 319
leaks, 164
learning, 200, 201, 340, 344, 345, 350
legislation, 226
leisure, 235
lethargy, 155
liberalisation, 177
life cycle, 161, 173, 179
limitation, 187
limitations, 229, 230
linear, x, 49, 76, 188, 200, 201, 210, 211, 216, 295, 307, 337, 338, 339, 345, 359
linear function, 339
linear model, 307
linear regression, 345
linguistic rule, 214, 215
linkage, 238
links, 9, 182
liquid film, 11, 32, 37, 38, 44, 48, 49, 50, 51, 53, 65, 66, 67, 138, 140, 141
liquid phase, 47
liquids, 33, 77
lithium, 21, 22, 27, 140, 142, 143, 145, 146, 382
livestock, vii, 149
living conditions, 374
living standard, 181
living standards, 181
loading, ix, 234, 239, 243, 249, 251, 256, 258, 259, 262
lobby, 175
localised, 158
location, 160, 184, 185, 193, 229, 235, 253, 270, 342
London, 140, 142, 143, 166, 192, 193, 194, 195, 221, 239
longevity, 180
losses, 21, 25, 26, 33, 61, 82, 83, 87, 106, 165, 178, 189, 248, 357
low power, 9, 196
low-income, 177, 181
low-temperature, 21, 145, 304
Luxembourg, 140, 290

M

machinery, 2, 158
machines, x, 143, 171, 315, 316, 318, 319, 322, 335, 374
Maine, 290
maintenance, 31, 155, 160, 174, 176, 192, 226, 296, 308, 376
management, ix, x, 173, 176, 177, 178, 192, 194, 197, 226, 235, 239, 240, 263, 264, 265, 267, 270, 337, 338, 370
manufacturer, 332
manufacturing, 172, 173
maritime, xi, 353, 354
market, xi, 2, 3, 5, 8, 151, 175, 177, 180, 182, 191, 196, 315, 332, 373
market access, 191
marketing, 173, 180
masonry, 271
mass loss, 250
mass transfer, 8, 9, 12, 15, 19, 21, 22, 23, 25, 27, 31, 32, 37, 44, 52, 56, 61, 62, 63, 64, 65, 66, 68, 70, 76, 95, 105, 117, 118, 136, 138, 139, 141, 144, 244, 375, 382
mass transfer process, 22, 65
material resources, 172

Mathematical Methods, 144
matrix, 201, 202, 203, 204, 206, 209, 273, 274, 275
meanings, 172, 179
mean-square deviation, 47
measurement, 171, 201, 206, 226, 237, 250, 253, 344, 365
measures, viii, 1, 2, 4, 149, 151, 152, 153, 174, 175, 176, 177, 181, 228, 357, 375
mechanical ventilation, 153, 155, 191
media, 4, 6, 7, 8, 12, 44, 99
Mediterranean, vii, 316
Mediterranean climate, vii
Mediterranean countries, 316
membership, 201, 202, 203, 204, 205, 211, 214, 216
membranes, xi, 227, 373, 375
memory, 197, 207, 210, 218, 220
messages, 171
metabolism, 153, 158
metals, 21
methane, 1, 2
methanol, 300, 301, 302, 303, 305, 309
Mexico, x, 140, 337, 342
MFC, viii, 196, 211, 213, 218, 219
microclimate, 6, 99, 169, 185
microcontrollers, 197
microorganisms, 21, 227
microprocessors, 197, 235
middle income, 174
migration, 191
milk, 168
mining, 175
missions, 146, 151, 155, 158, 175, 182, 184, 189, 191, 233
mixing, viii, 80, 156, 225, 227, 228, 229, 230, 240
mobility, 249
modeling, 9, 85, 93, 141, 144, 220, 241, 264, 265, 338, 339, 349
models, 73, 135, 144, 169, 214, 227, 233, 234, 242, 244, 294, 297, 298, 299, 307, 308, 338, 339, 343, 345, 346, 349, 350
modules, 332
mole, 24, 301, 303, 374
molecules, 375, 376, 377, 378
momentum, 160, 230, 231
money, 170
Moon, 383
morning, 157, 185, 189, 288, 309, 313
morphology, 174
Moscow, 86, 95, 99, 109, 110, 115, 140, 143, 144, 145, 146, 147
motion, 25, 81, 193
motivation, 235
motor control, 197, 200
motors, viii, 195, 200, 207, 216, 220
movement, 6, 7, 8, 10, 20, 31, 32, 34, 38, 53, 56, 57, 60, 66, 67, 70, 74, 77, 83, 85, 95, 98, 105, 106, 107, 155, 159, 160, 161, 184, 193, 228, 249, 375
MSW, 163
mucous membrane, 227
mucous membranes, 227
multiphase flow, 144
multiplicity, 95, 110
mutation, 233, 298, 299, 309
mutations, 308

N

nation, 183
natural, vii, 2, 7, 9, 14, 49, 77, 79, 81, 82, 83, 84, 85, 87, 91, 98, 149, 151, 153, 159, 160, 161, 163, 164, 166, 169, 170, 171, 173, 176, 177, 185, 186, 188, 191, 228, 232, 237, 244, 264
natural environment, 228
natural gas, 2, 9
natural resource management, 176
natural resources, vii, 149, 173, 177
nausea, 227
Navier-Stokes equation, 81
Netherlands, 3, 169
network, x, 271, 283, 293, 294, 295, 296, 302, 309, 313, 337, 338, 340, 341, 343, 344, 345, 346, 348, 349, 350, 351, 353, 356, 357
neural network, vii, x, 337, 338, 340, 341, 343, 346, 348, 349, 350, 351
neural networks, vii, 338, 349, 350, 351
neurons, 339, 340, 341
New York, 193, 290, 351
New Zealand, 2
next generation, 298, 307
NGOs, 182
nitrate, 140, 143
nitric oxide, 167
nitrogen, 167, 226
nitrogen dioxide, 226
nodes, 31, 340, 343
noise, 3, 151
nonlinear dynamics, 144
nonlinearities, 220
non-linearity, 205
non-renewable, 161, 164
non-renewable resources, 164
non-uniform, 221
normal, 34, 39, 44, 154, 215, 355, 381
normalization, 341
norms, 86, 95, 99, 226, 241
North America, 164

Norway, 2, 144
nose, 155, 227
nuclear, 167, 176, 183, 194, 232
nuclear energy, 194, 232
nuclear power, 176, 183, 232
numerical analysis, 69
Nusselt, 34, 138, 150

O

objective criteria, 177
obligation, 2, 177
observations, 59, 348
obstruction, 174
oil, 175, 183, 198, 207, 216, 218, 220, 232
oil spill, 175
oils, 168
old age, 161
olfactory, 171
Oman, ix, 243, 244, 245, 250, 253, 263
operator, 96, 340
optimization, 141, 226, 233, 234, 240, 295, 307
optimization method, 226
oral, 184
organic compounds, 227
orientation, 87, 103, 159, 172, 173, 174, 185, 186, 189
originality, 270
oscillation, 280, 288
oscillations, 206, 207, 220
oxide, 226
oxygen, 301, 302, 304, 309, 311
ozone, 3, 7, 147, 151, 168, 192

P

pacing, 32, 50, 67
packaging, 102
parameter, 57, 85, 106, 171, 187, 199, 202, 203, 205, 206, 207, 220, 226, 251, 252, 259, 291
parameter estimates, 291
Paris, 146
particles, 144, 157, 168
partition, 229
passive, vii, ix, xi, 3, 155, 158, 162, 163, 164, 166, 171, 172, 185, 189, 192, 193, 194, 225, 228, 232, 234, 235, 238, 239, 244, 267, 270, 289, 291, 353, 355, 356, 367, 368, 369, 370
passive techniques, 189, 192
pathogens, 227
payback period, 232, 233
PCM, 237, 241
per capita, 168, 169
perceptions, 171
permit, 25, 186, 283, 354
perturbation, 45
perturbations, 46, 50
pesticide, 183
PFC, 222
philosophy, 184
photosynthesis, 169, 238
physical environment, 171
physical factors, 158
physics, 146
physiology, 184
pilot studies, 164, 181
pipelines, 77, 80, 82, 83, 139
pitch, 37, 105, 138
planning, x, 2, 57, 152, 180, 293, 295, 296, 299, 302, 305, 306, 309, 311, 313, 338
plants, vii, 4, 11, 76, 107, 113, 137, 143, 149, 151, 152, 168, 169, 174, 180, 184, 200, 222, 232, 238, 242
plastic, 56
play, 2, 62, 170, 182, 183, 190, 191, 270, 379
PNG, 264
polarization, 49
policy makers, 191
politicians, 191
politics, 191
pollen, 157
pollutant, 153, 155, 156, 157, 158, 227, 240
pollutants, 2, 153, 154, 155, 156, 157, 158, 160, 168, 178, 193, 227
pollution, vii, 8, 149, 151, 153, 157, 158, 160, 167, 171, 184, 189, 191, 196, 382
polymer, 295
polypropylene, 271, 273, 275, 276, 277, 286, 355, 356, 371
polystyrene, 246, 275, 357
polyurethane, 273, 279, 356
polyurethane foam, 273, 279, 356
pond, 186
poor, 154, 157, 161, 164, 169, 172, 185, 229, 381
poor health, 157
poor performance, 172
population, 168, 169, 171, 172, 173, 175, 179, 180, 234
population density, 169
population growth, 173
pores, 377, 379
porous, 50, 56
Portugal, 143, 192, 193
power plant, 197, 221, 235
power plants, 197, 235